WHITE NOISE ANALYSIS

MATHEMATICS and APPLICATIONS

WHITE NOISE ANALYSIS

MATHEMATICS and APPLICATIONS

9–15 July 1989
Bielefeld

Editors

T. Hida
Nagoya University

H. H. Kuo, J. Potthoff
Louisiana State University

L. Streit
Bielefeld University

World Scientific
Singapore • New Jersey • London • Hong Kong

Published by

World Scientific Publishing Co. Pte. Ltd.
P O Box 128, Farrer Road, Singapore 9128
USA office: 687 Hartwell Street, Teaneck, NJ 07666
UK office: 73 Lynton Mead, Totteridge, London N20 8DH

WHITE NOISE ANALYSIS – MATHEMATICS AND APPLICATIONS

ISBN 981-02-0242-3

Printed in Singapore by Loi Printing Pte. Ltd.

Preface

For more than a decade, the Center of Interdisciplinary Research ("ZiF") at Bielefeld has been the host of meetings, workshops and conferences in which "White Noise Analysis" has been one of the major themes. In 1989 we organized an international conference at the ZiF which was essentially devoted to White Noise. The mathematical theory of functionals of the time derivative of Brownian motion – "white noise" – and its applications to various domains had been developed to a degree which called for and justified such a meeting.

In accordance with the traditional spirit of the ZiF – interdisciplinary discussion, dialogue and comparison of different approaches –, a number of talks on related subjects were also invited. We had excellent lectures on so diverse but connected topics as abstract Wiener spaces, large deviation, machine learning, Malliavin calculus, molecular biology, non–standard analysis, quantum theory, and stochastic analysis. We believe that it is hard to overestimate the stimulus for the field which has emerged out of this exciting and fruitful exchange of ideas.

By now, we may consider white noise analysis as a well–established mathematical framework for problems and phenomena which require calculus on infinite dimensional spaces ("infinitely many degrees of freedom"). The theory of generalized functionals of white noise in its various formulations, together with the calculus of differential operators forms a flexible and quite intuitive mathematical apparatus for such problems. It allows for efficient applications in domains such as quantum field theory, quantum mechanics, infinite dimensional Dirichlet forms and (anticipating) stochastic analysis.

We would like to thank the participants of the conference for their contribution to its success and the authors of the present volume for their beautiful articles. Also, we would like to express our deep gratitude to the staff at the ZiF. It is their efficiency and competence that made the meeting a memorable experience. Finally, we acknowledge gratefully the generous financial support for the meeting from the Center of Interdisciplinary Research and the Japan Society for the Promotion of Science.

T. Hida, H.-H. Kuo, J. Potthoff, L. Streit

Nagoya, Baton Rouge, Bielefeld, February 1990

Table of Contents

WHITE NOISE ANALYSIS

MATHEMATICS and APPLICATIONS

Infinite dimensional diffusions connected with positive generalized white noise functionals

by

SERGIO ALBEVERIO,
Institut für Mathematik, Ruhr-Universität Bochum
4630 Bochum 1, West Germany

and

MICHAEL RÖCKNER,
Department of Mathematics, University of Edinburgh,
The King's Buildings, Mayfield Road, Edinburgh EH9 3JZ, Scotland.

0. Introduction

In a previous paper [AR 89d] we used the theory of Dirichlet forms to (weakly) solve stochastic differential equations of type

$$dX_t = dW_t + \beta(X_t)\,dt \qquad X_o = z. \tag{0.1}$$

Here $(X_t)_{t\geq 0}$ takes values in a locally convex topological vector space E, $z \in E$, $\beta : E \to E$ and $(W_t)_{t\geq 0}$ is an E-valued Brownian motion over some Hilbert space H (i.e. its covariance is given by the inner product in H) with $H \subset E$ continuously and densely. Of course, we have to assume that E is "large enough" such that $(W_t)_{t\geq 0}$ exists. The proper class of Dirichlet forms to consider are so-called *classical Dirichlet forms* $(\mathcal{E},D(\mathcal{E}))$ of type (closure of)

$$\mathcal{E}(u,v) = \int_E \langle\nabla u, \nabla v\rangle_H \, d\mu \ , \quad u,\ v \text{ bounded, smooth cylinder functions} \tag{0.1}$$

on $L^2(E;\mu)$ (cf. section 1 below). Here μ is a finite positive measure on E which has to be chosen in a particular way depending on β. If $\dim E < \infty$, $E = \mathbb{R}^d$ say, the relation between β and μ is given as follows:

$$\mu = \rho\, dx \quad \text{and} \quad \beta = \nabla \log \rho \tag{0.3}$$

where dx denotes Lebesgue measure on $\mathbb{R}^d$. In the case $\dim E = \infty$, there is no direct analogue of (0.3) because of the lack of an infinite dimensional Lebesgue measure. However, as shown in [AR 89d], (0.3) can be reformulated in terms of conditional densities of μ (cf. section 4 below; in particular, Definition 4.1, (A.1) and Theorem 4.3) so that it carries over to the infinite dimensional situation. In [AR 89d] it was then proved that a diffusion process $\mathcal{X} = (\Omega,\mathcal{F},(\mathcal{F}_t)_{t\geq0}, (X_t)_{t\geq0}, (P_z)_{z\in E})$ associated with $(\mathcal{E},D(\mathcal{E}))$ (cf.Definition 2.2 below) solves (0.1) provided $\mathcal{X}$ exists and the capacity associated with $(\mathcal{E},D(\mathcal{E}))$ is tight. Both the existence of $\mathcal{X}$ and the tightness of the capacity was proved in [AR 89b] if E is a Hilbert space or E is conuclear. If E is a Banach space this was proved in [K 82] in case μ is quasi-invariant and under some additional technical assumptions. In recent articles by Schmuland [S 89] resp. Lyons/Röckner [LR 90], however, it was shown that if E is a Banach space, there always exists a diffusion process $\mathcal{X}$ associated with $(\mathcal{E},D(\mathcal{E}))$ resp. the associated capacity is always tight. Therefore, in the Banach space case the results of [AR 89d] simplify considerably. To illustrate this in some detail is one main purpose of this paper, another is to explain the role of positive generalized white noise functionals within this scheme. Let us briefly summarize the contents of this work.

In section 1 we recall the framework of classical Dirichlet forms on $L^2(E;\mu)$. We also look at the particular case where $E = \mathcal{S}^*$ and μ comes from a positive generalized white noise functional and explain the special and useful properties of the corresponding Dirichlet form. In section 2 we consider the associated diffusion processes recalling Schmuland's existence result mentioned above. We stress that though we confine ourselves here to the situation where E is a Banach space the case wher μ comes from a positive generalized white noise functional is still covered since all these are supported by some Hilbert subspace of $\mathcal{S}^*$ (cf. Remark 2.1 below). In section 3 we give a detailed account on capacities and quasi-continuity. In particular, we reprove the tightness result in [LR 90] but using a slightly different method. We would also like to draw attention to the new proof of Proposition 3.4. Finally, in section 4 we present one of the main results in [AR 89d] on solving stochastic differential equations of type (0.4) using classical Dirichlet forms and their associated diffusion processes in the now simplified Banach space case (cf. Theorem 4.3 below). There are only two (resp. three, cf. the discussion following Remark 4.4) conditions left that have to be checked in applications (cf. (A.1), (A.2) resp. (A.3) below).

Finally, we want to emphasize that condition (A.2) always holds if E is some Banach (or Hilbert) subspace of $\mathcal{S}^*$ and μ is given by a positive generalized white noise functional. We believe that in case (A.1) also holds for such a μ, i.e. Theorem 4.3 is applicable, the further analysis of the corresponding solution $(X_t)_{t\geq0}$ of (0.1) provides a potentially rich field to apply methods and results from white noise analysis.

1. Classical Dirichlet forms and generalized white noise functionals

Let E be a locally convex Hausdorff topological vector space (over $\mathbb{R}$) which is Souslinean (i.e. the continuous image of a Polish space). Let E' denote its (topological) dual and $\mathscr{B}(E)$ the σ-algebra of its Borel subsets. Let $\mathscr{F}C^\infty$ denote the set of all functions $u : E \to \mathbb{R}$ of the form

$$u(z) = f(l_1(z),\dots,l_m(z)) \ , \quad z \in E, \tag{1.1}$$

for some $m \in \mathbb{N}$, $l_1,\dots,l_m \in E'$, $f \in C^\infty(\mathbb{R}^m)$. Below we shall also consider the following subsets of $\mathscr{F}C^\infty$, namely $\mathscr{F}C_b^\infty$ and $\mathscr{P}_b$ being defined as those functions in $\mathscr{F}C^\infty$ for which f in (1.1) is together with all its partial derivatives bounded resp. polynomially bounded. Furthermore, let $\mathscr{P}$ and $\mathscr{A}$ be the algebras of functions on E generated by E' resp. $\{\sin\circ l, \cos\circ l : l \in E'\}$. For $u \in \mathscr{F}C^\infty$ and $z, k \in E$ we define

$$\frac{\partial u}{\partial k}(z) := \frac{d}{ds}\, u(z+sk)\Big|_{s=0} \ . \tag{1.2}$$

Fix a finite positive measure μ on $(E,\mathscr{B}(E))$ and let $L^2(E;\mu)$ denote the corresponding space of (classes of) μ-square integrable functions on E with usual inner product $(\,,)_\mu$. Consider the following condition on $k \in E$

$$\frac{\partial u}{\partial k} = \frac{\partial v}{\partial k} \ \mu\text{- a.e.} \quad \text{if} \quad u,\ v \in \mathscr{F}C^\infty \quad \text{with} \quad u = v \quad \mu\text{- a.e.} \tag{1.3}$$

Note that (1.3) is always fulfilled if $\operatorname{supp}\mu = E$, i.e. $\mu(U) > 0$ for all $U \subset E$, U open, but fails to hold if e.g. $\mu = \delta_z$ i.e. the Dirac measure with mass in $z \in E$. For a subset $\mathscr{L}$ of $\mathscr{B}(E)$-measurable functions we denote the corresponding set of μ-classes by $\tilde{\mathscr{L}}$.

<u>Definition 1.1</u>. Let K_o be a finite or countable subset of E such that (1.3) holds for each $k \in K_o$ and

$$\sum_{k\in K_o} l(k)^2 < \infty \quad \text{for each} \quad l \in E'. \tag{1.4}$$

Then the quadratic form $(\mathscr{E},\tilde{\mathscr{F}C}_b^\infty)$ on $L^2(E;\mu)$ defined by

$$\mathscr{E}(u,v) := \sum_{k\in K_o} \int \frac{\partial u}{\partial k}\frac{\partial v}{\partial k}\, d\mu \ ; \quad u,\ v \in \tilde{\mathscr{F}C}_b^\infty \tag{1.5}$$

is called a *classical pre-Dirichlet form on* E. If $(\mathscr{E},\tilde{\mathscr{F}C}_b^\infty)$ is *closable* on $L^2(E;\mu)$ (i.e. if $\tilde{\mathcal{F}}C_b^\infty$ is equipped with $\mathcal{E}_1 := \mathcal{E} + (\,,\,)_\mu$ and $L^2(E;\mu)$ with $(\,,\,)_\mu$ and if $\overline{\tilde{\mathcal{F}}C_b^\infty}^{\mathcal{E}_1}$ denotes the abstract completion of $\tilde{\mathcal{F}}C_b^\infty$ w.r.t. $\mathcal{E}_1$, then the unique continuous extension $i : \overline{\tilde{\mathcal{F}}C_b^\infty}^{\mathcal{E}_1} \longrightarrow L^2(E;\mu)$ of the continuous embedding $\mathcal{F}C_b^\infty \subset L^2(E;\mu)$ is one to one), we call its closure (i.e. smallest closed extension) a *classical Dirichlet form on* E and denote it by $(\mathcal{E}, D(\mathcal{E}))$.

Note that because of (1.4) $\mathcal{E}$ in (1.5) is real valued. The notion Dirichlet form for $(\mathcal{E}, D(\mathcal{E}))$ in Definition 1.1 is justified since it has the *Dirichlet property* i.e. for any map $T : \mathbb{R} \to \mathbb{R}$ with $T(0) = 0$ and $|T(x) - T(y)| \leqslant |x-y|$, $x, y \in \mathbb{R}$, we have that $T \circ u \in D(\mathcal{E})$ if $u \in D(\mathcal{E})$ and $\mathcal{E}(T \circ u, T \circ u) \leqslant \mathcal{E}(u,u)$ (cf. [F 80], [AR 88]).

Remark 1.2. (i) In [AR 88] a sufficient condition for $(\mathcal{E}, \tilde{\mathcal{F}}C_b^\infty)$ to be closable was proved which is also necessary if all summand-forms in (1.5) are closable.
(ii) $(\mathcal{E}, D(\mathcal{E}))$ in Definition 1.1 is always densely defined since $\tilde{\mathcal{F}}C_b^\infty$ is dense in $L^2(E;\mu)$ because E is Souslinean (cf. [AR 88, Remark 3.1]).

A particularly nice classical pre-Dirichlet form is obtained as follows. Assume there exists a separable real Hilbert space $(H, \langle\,,\,\rangle_H)$ such that $H \subset E$ continuously and densely. Identifying H with its dual H' we have that

$$E' \subset H \subset E \quad \text{continuously and densely} \tag{1.6}$$

where E' is equipped with the strong topology. H should be thought of as a "tangent space" to E at each point in E. Let for $u \in \mathcal{F}C^\infty$, $z \in E$, $\nabla u(z)$ be the unique element in H representing the continuous linear functional $h \mapsto \frac{\partial u}{\partial h}(z)$, $h \in H$, on H. If K_o is any orthonormal basis in H then (1.4) holds by (1.6) and if (1.3) holds for each $k \in K_o$ we have that

$$\mathcal{E}(u,v) = \int_E \langle\nabla u, \nabla v\rangle_H \, d\mu = \sum_{k \in K_o} \int \frac{\partial u}{\partial k}\frac{\partial v}{\partial k}\, d\mu; \quad u, v \in \tilde{\mathcal{F}}C_b^\infty \tag{1.7}$$

is a classical pre-Dirichlet form on $L^2(E;\mu)$.

Remark 1.3. For concrete examples of classical pre-Dirichlet forms (1.5), (1.7) in the case $\dim E = +\infty$ we refer to [AR 88,89a-d] and [AH-K 77a,b]. In these references classical (pre-) Dirichlet forms with both Gaussian measures μ as well as non-Gaussian measures μ constructed in quantum field theory have been discussed in detail.

Now we come to those Dirichlet forms studied within white noise analysis as introduced in [HPS 88]. It has been proved in [AHP 89b] (cf. Corollary 1.6 below) that these form a special class of classical Dirichlet forms, namely those for which $E = \mathcal{S}^*(\mathbb{R}^d) = : \mathcal{S}^*$, $d \in \mathbb{N}$ (i.e the space of Schwarz (tempered) distributions equipped with the strong topology) and μ is given by a positive generalized white noise functional. In this case $\tilde{\mathcal{F}C}_b^\infty$ as a core can be replaced by $(\mathcal{S})$ i.e. the space of white noise test functionals. Let us recall the necessary notions and describe these forms more precisely within the above framework of classical Dirichlet forms.

Let $E = \mathcal{S}^*$ (cf. [Sch 73] for a proof that $\mathcal{S}^*$ is Souslinean) and let $\mathcal{S} := \mathcal{S}(\mathbb{R}^d)$ be the corresponding space of test functions. Let $\langle\, ,\rangle$ denote the dualisation between $\mathcal{S}^*$ and $\mathcal{S}$. For the "tangent space" H take $H := L^2(\mathbb{R}^d;\, dx)$ where dx denotes Lebesgue measure on $\mathbb{R}^d$, and H is embedded in $\mathcal{S}^*$ in the usual way. Let μ_o denote the white noise measure on $(\mathcal{S}^*, \mathcal{B}(\mathcal{S}^*))$ (which will serve as a reference measure for the rest of this section) i.e. μ_o is the unique mean zero Gaussian measure on $(\mathcal{S}^*, \mathcal{B}(\mathcal{S}^*))$ such that for all $l_1, l_2 \in \mathcal{S}$

$$\int_{\mathcal{S}^*} \langle l_1, z\rangle \langle l_2, z\rangle \, \mu_o(dz) = \int_{\mathbb{R}^d} l_1(x)\, l_2(x)\, dx.$$

We have the well-known decomposition $L^2(\mathcal{S}^*;\mu_o) = \bigoplus_{n=0}^{\infty} \mathcal{H}^{(n)}$ into homogeneous chaos' of degree n. Consider the closure A of the operator on $L^2(\mathbb{R}^d;dx)$ given on $\mathcal{S}$ by $-\Delta + |x|^2 + 1$, and denote its second quantization by $\Gamma(A)$ (which is a self-adjoint operator on $L^2(\mathcal{S}^*;\mu_o)$). $(\mathcal{S})_p$ is the domain of $\Gamma(A)^p$ in $L^2(\mathcal{S}^*;\mu_o)$, which is a Hilbert space with Hilbertian norm $\|F\|_{2,p} := \|\Gamma(A)^p F\|_2$, $(\mathcal{S})$ is the projective limit of the chain $((\mathcal{S})_p : p \geq 0)$, $(\mathcal{S})^*$ its dual and $\langle\, ,\rangle$ denotes the corresponding dualisation. $(\mathcal{S})$ is a nuclear Frechet algebra. For more details we refer to [HPS 88] and [HKP].

Remark 1.4. By [Yo 87] we know that every $u \in (\mathcal{S})$ has a (μ_o-) version $\tilde{u}$ which is continuous on $\mathcal{S}^*$ (cf. also [AHP 89b, Appendix]).

By [AR 88, 5.5(ii)] and [AR 89c, Proposition 2.7] it follows that supp $\mu_o = \mathcal{S}^*$. In particular, $\tilde{u}$ defined above is unique. This fact is crucial for the following.

It follows by [AHP 89 b, Proposition 3.2] that the map $u \mapsto \|\nabla u\|^2_{L^2(\mathbb{R}^d;dx)}$, $u \in \mathcal{P}$, extends to a continuous map from $(\mathcal{S})$ to $(\mathcal{S})$ (equipped with the above nuclear topology) which coincides with $\|\nabla u\|^2_{L^2(\mathbb{R}^d;dx)}$ for each $u \in (\mathcal{S}) \cap \mathcal{FC}^\infty$. Therefore, we denote this extension also by $\|\nabla u\|^2_{L^2(\mathbb{R}^d;dx)}$, $u \in (\mathcal{S})$. Finally, we note that both $\mathcal{P}$ and $\mathcal{A}$ are dense subspaces of $(\mathcal{S})$ (cf. [AHP 89b]).

<u>Remark</u>. In [HPS 88], $\|\nabla u\|^2_{L^2(\mathbb{R}^d;dx)}$ (there denoted by $|Du|^2$) was defined for $u \in (\mathcal{S})$ in a different way, but it was proved in [AHP 89b] that the two definitions coincide.

Now we fix a positive finite measure μ on $(\mathcal{S}^*, \mathcal{B}(\mathcal{S}^*))$ which comes from a positive generalized white noise functional, i.e. there exists $\Phi \in (\mathcal{S})^*$, $\langle u,\Phi\rangle \geqslant 0$ for each $u \in (\mathcal{S})$ with $u \geqslant 0$ μ_o-a.e. on $\mathcal{S}^*$ such that

$$\int_{\mathcal{S}^*} \tilde{u}(z)\ \mu(dz) = \langle u,\Phi\rangle \quad \text{for each } u \in (\mathcal{S}). \tag{1.8}$$

Here $\tilde{u}$ is as in 1.4. We assume in addition that supp $\mu = \mathcal{S}^*$ so that (1.3) holds and that $(\mathcal{S})(\subset L^2(\mathcal{S}^*;\mu_o))$ naturally embeds into $L^2(\mathcal{S}^*;\mu)$ by associating to $u \in (\mathcal{S})$ the μ-class corresponding to $\tilde{u}$. Note that μ is in general singular w.r.t. to μ_o. Examples of such measures μ, in particular those constructed in quantum field theory, can be found in [AHP 89a]. For any $u \in (\mathcal{S}) \cup \mathcal{FC}^\infty$ we define

$$\mathcal{E}(u,u) = \int_{\mathcal{S}^*} \|\widetilde{\nabla u}\|^2_{L^2(\mathbb{R}^d;dx)}\, d\mu. \tag{1.9}$$

Now we can restate one of the main results in [AHP 89b], [PR 89].

<u>Theorem 1.5</u>. Let $\overline{}^{\mathcal{E}_1}$ denote abstract completion w.r.t. $\mathcal{E}_1 := \mathcal{E} + (\,,)_\mu$. Then $\overline{\mathcal{P}}^{\mathcal{E}_1} = \overline{\mathcal{P}_b}^{\mathcal{E}_1} = \overline{(\mathcal{S})}^{\mathcal{E}_1} = \overline{\mathcal{A}}^{\mathcal{E}_1} = \overline{(\mathcal{S})}^{\mathcal{E}_1}$.

As an immediate consequence we obtain:

Corollary 1.6. $(\mathcal{E},(\mathcal{S}))$ is closable on $L^2(\mathcal{S}^*;\mu)$ if and only if $(\mathcal{E},\mathcal{F}C_b^\infty)$ is. In this case their closures coincide and in particular, the closure $(\mathcal{E},D(\mathcal{E}))$ of $(\mathcal{E},(\mathcal{S}))$ is a classical Dirichlet form on $\mathcal{S}^*$.

Classical Dirichlet forms $(\mathcal{E},D(\mathcal{E}))$ associated with a white noise test functional (as in Corollary 1.6) are particularly nice since there are several nice function spaces that can serve as a core. Furthermore, one can apply methods and results from the white noise calculus to them.

Finally, we want to recall a well-known representation formula for $(\mathcal{E},\mathcal{F}C_b^\infty)$ ($\mathcal{E}$ as in (1.9)) that does not require the gradient ∇, but uses instead the "functional" or "Hida" derivative ∂_x, $x \in \mathbb{R}^d$. Since for each $x \in \mathbb{R}^d$ the Dirac measure δ_x on $\mathbb{R}^d$, belongs to $\mathcal{S}^*$ we can define (cf. (1.2)) for $u \in \mathcal{F}C_b^\infty$, $x \in \mathbb{R}^d$

$$\partial_x u := \frac{\partial u}{\partial \delta_x} \tag{1.10}$$

If $u = f(l_1,\dots,l_m)$, $l_1,\dots,l_m \in E' = \mathcal{S}$, $f \in C_b^\infty(\mathbb{R}^m)$ and $k \in \mathcal{S}^*$ then

$$\frac{\partial u}{\partial k} = \sum_{j=1}^{m} \frac{\partial f}{\partial x_j}(l_1,\dots,l_m) \quad \langle l,k_j\rangle .$$

Hence for $z \in \mathcal{S}^*$ fixed, $\partial_\cdot u(z) \in \mathcal{S}$ and for all $k \in \mathcal{S}^*$

$$\frac{\partial u}{\partial k}(z) = \langle \partial_\cdot u(z),k\rangle . \tag{1.11}$$

In particular, for any orthonormal basis $K_o \subset \mathcal{S}$ of $H = L^2(\mathbb{R}^d;dx)$ we have for $u, v \in \mathcal{F}C_b^\infty$ by Fubini's theorem

$$\begin{aligned}\mathcal{E}(u,v) &= \int_{\mathcal{S}^*} \langle \nabla u, \nabla v\rangle_H \, d\mu \\ &= \sum_{k\in K_o} \int_{\mathcal{S}^*} \frac{\partial u}{\partial k}\frac{\partial v}{\partial k}\, d\mu \\ &= \int_{\mathcal{S}^*}\int_{\mathbb{R}^d} \partial_x u(z) \quad \partial_x v(z)\, dx \quad \mu(dz).\end{aligned} \tag{1.12}$$

2. Construction of an associated diffusion

For the rest of this paper we consider the situation described at the beginning of section 1, but we confine ourselves to the case where E is a separable real Banach space. We fix a classical pre-Dirichlet form on $L^2(E;\mu)$ of type (1.7), i.e.

$$\mathscr{E}(u,v) = \frac{1}{2}\int \langle\nabla u, \nabla v\rangle_H \, d\mu, \quad u, v \in \tilde{\mathscr{F}}C_b^\infty \tag{2.1}$$

where as before H is a separable real Hilbert space continuously and densely embedded in E and we assume that (1.3) holds for all k in one (and hence any) orthonormal basis K_o of H. We assume furthermore that

$$(\mathscr{E},\tilde{\mathscr{F}}C_b^\infty) \text{ is closable on } L^2(E;\mu) \tag{2.2}$$

and denote its closure by $(\mathscr{E},D(\mathscr{E}))$. Let $(T_t)_{t\geqslant 0}$ be the associated Markovian semigroup.

<u>Remark 2.1</u>. We emphasize that though $\mathscr{S}^*$ is, of course, not a Banach space, the classical Dirichlet forms coming from a positive generalized white noise functional (cf. Corollary 1.6) are still special cases of the above form $(\mathscr{E},D(\mathscr{E}))$. Indeed, since in this case $\int |u| \, d\mu < \infty$ for all $u \in \mathscr{P}$ hence in particular

$$\int_{\mathscr{S}^*} |l(z)| \, \mu(dz) < \infty \quad \text{for all} \quad l \in \mathscr{S},$$

an easy application of the uniform boundedness principle implies that $\mu(\mathscr{S}_{-p}) = 1$ for some $p \geqslant 1$ (cf. [AR 89b, Proposition 3.7]). Then the Dirichlet form can be considered as a form on $L^2(\mathscr{S}_{-p};\mu)$ (cf. [AR 89b, Section 3(c)] for details.

<u>Definition 2.2</u>. A diffusion process $\mathscr{X} := (\Omega,\mathscr{F},(\mathscr{F}_t)_{t\geqslant 0}, (X_t)_{t\geqslant 0}, (P_z)_{z\in E})$ (cf. [D 65]) is called *associated with* $(\mathscr{E},D(\mathscr{E}))$ if its state space is E and for all $u : E \to \mathbb{R}$, $\mathscr{B}(E)$-measurable, bounded, and all $t \geqslant 0$

$$\int_\Omega u(X_t) \, dP_z = T_t u(z) \quad \mu\text{- a.e.} \quad z \in E.$$

Since $T_t 1 = 1$, $t \geqslant 0$, we obviously have that μ is the invariant measure of $\mathscr{X}$. Under some additional assumptions on E and H a diffusion process associated with $(\mathscr{E},D(\mathscr{E}))$ has been constructed in [K 82] and more generally in [AR 89b]. It turns out, however, that all these additional assumptions are superfluous. The following theorem which is a direct consequence of a recent result by B. Schmuland (cf. [S 89]) holds in our general situation as described preceding Remark 2.1.

<u>Theorem 2.3.</u> (B. Schmuland) There exists a diffusion process $\mathfrak{X} = (\Omega, \mathcal{F}, (\mathcal{F}_t)_{t\geq 0}, (X_t)_{t\geq 0}, (P_z)_{z\in E}$ associated with $(\mathcal{E}, D(\mathcal{E}))$.

<u>Proof.</u> The assertion follows directly from [S 89, Proposition 5.2 and its proof] and [F 80, Theorem 4.2.1 and Theorem 4.1.2]. □

In [K 82], [AR 89b] the construction of the diffusion $\mathfrak{X}$ was based on first proving that the capacity associated with $(\mathcal{E}, D(\mathcal{E}))$ is tight (cf. Section 3 below). This was a non-trivial task and only possible under some additional assumptions on E and H. Schmuland, however, constructs $\mathfrak{X}$ directly without using the tightness. Now an interesting and important question is whether this tightness follows from the existence of $\mathfrak{X}$. This problem will be solved in the next section (cf. Theorem 3.1).

3. Tightness of capacities

Now let us recall the definition of the *capacity* associated with $(\mathcal{E}, D(\mathcal{E}))$:

For $U \subset E$, U open, define

$$\mathrm{Cap}(U) := \inf\{\mathcal{E}_1(u,u) : u \in D(\mathcal{E}),\ u \geq 1 \text{ on } U\ \mu\text{- a.e.}\} \tag{3.1}$$

and for an arbitrary subset $A \subset E$ define

$$\mathrm{Cap}(A) := \inf\{\mathrm{Cap}(U) : A \subset U,\ U \text{ open}\}. \tag{3.2}$$

Observe that Cap takes values in $[0,1]$ since $1 \in D(\mathcal{E})$ and that $\mathrm{Cap}(A) = 0$ implies $\mu(A) = 0$ for any $A \in \mathcal{B}(E)$. As in [F 80, Section 3.1] it follows that Cap is a Choquet capacity, i.e. it has the following two properties:

If $A_n \subset E$, $n \in \mathbb{N}$, are increasing then

$$\mathrm{Cap}(\bigcup_{n\in\mathbb{N}} A_n) = \sup_{n\in\mathbb{N}} \mathrm{Cap}(A_n). \tag{3.3}$$

If $K_n \subset E$, $n \in \mathbb{N}$, are compact and decreasing then

$$\mathrm{Cap}(\bigcap_{n\in\mathbb{N}} K_n) = \inf_{n\in\mathbb{N}} \mathrm{Cap}(K_n). \tag{3.4}$$

Furthermore, we have that

$$\mathrm{Cap}(\bigcup_{n\in\mathbb{N}} A_n) \leq \sum_{n=1}^{\infty} \mathrm{Cap}(A_n) \quad \text{for } A_n \subset E,\ n \in \mathbb{N} \tag{3.5}$$

(cf. [F 80, Section 3.1]) and that

$$\mathrm{Cap}(A) = \sup\{\mathrm{Cap}(K) : K \subset A,\ K \text{ compact}\} \quad \text{for } A \in \mathcal{B}(E) \tag{3.6}$$

(since E is Souslinean, cf. [AR 89b, Proposition 2.2]).

In sufficiently regular cases sets of capacity zero are exactly the sets which are not hit by the diffusion process $\mathcal{X}$ associated with $(\mathcal{E},D(\mathcal{E}))$ (cf. [F 80, Theorem 4.3.1]).

Theorem 3.1. Cap is *tight*, i.e. there exist compact sets $K_n \subset E$, $n \in \mathbb{N}$, such that $\lim_{n\to\infty} \mathrm{Cap}(E\setminus K_n) = 0$.

This theorem is just a special case of a general result in [LR 90] whose proof uses some nice results of E. B. Dynkin on the potential theory of Dirichlet forms associated with symmetric fine Markov processes on standard Borel spaces contained in [D 82]. In this paper we will present a slightly different proof using a compactification of the state space E in order to be able to apply Fukushima's potential theory of Dirichlet forms on locally compact spaces. This compactification method has been used extensively in our earlier work (cf. [AR 89b] and in particular [AR 89d]). It serves as a general and easy machinery to carry over well-known theorems of potential theory on locally compact spaces to our situation where the state space E is not locally compact if it is infinite dimensional. The basic underlying result is Proposition 3.2 below. First we need some preparations.

Let $\hat{E}$ be a *compactification* of E, i.e. $\hat{E}$ is a compact Hausdorff space with E continuously and densely embedded into $\hat{E}$ by some map i. Define the image measure of μ under i by $\hat{\mu} := \mu \circ i^{-1}$. It can be seen (cf. [AR 89d, Section 1]) that the map $\hat{u} \mapsto u \circ i$, $\hat{u} \in L^2(\hat{E};\hat{\mu})$, defines an isometry between $L^2(\hat{E};\hat{\mu})$ and $L^2(E;\mu)$. Define the image form $(\hat{\mathcal{E}},D(\hat{\mathcal{E}}))$ of $(\mathcal{E},D(\mathcal{E}))$ on $L^2(\hat{E};\hat{\mu})$ as follows

$$D(\hat{\mathcal{E}}) := \{\hat{u} \in L^2(\hat{E};\hat{\mu}) \;:\; \hat{u} \circ i \in D(\mathcal{E})\} \tag{3.7}$$

$$\hat{\mathcal{E}}(\hat{u},\hat{v}) := \mathcal{E}(\hat{u} \circ i,\ \hat{v} \circ i),\quad \hat{u},\ \hat{v} \in D(\hat{\mathcal{E}}) \tag{3.8}$$

Clearly, $(\hat{\mathcal{E}},D(\hat{\mathcal{E}}))$ is a Dirichlet form on $L^2(\hat{E};\hat{\mu})$, i.e. a densely defined positive definite symmetric bilinear form on $L^2(\hat{E};\hat{\mu})$ having the Dirichlet property.

Proposition 3.2. There exists a compactification $\hat{E}$ of E such that

(i) $\hat{E}$ is a (complete separable) metric space.

(ii) The image $(\hat{\mathcal{E}},D(\hat{\mathcal{E}}))$ of $(\mathcal{E},D(\mathcal{E}))$ on $L^2(\hat{E};\hat{\mu})$ is a *regular* Dirichlet form on $L^2(\hat{E};\hat{\mu})$, i.e. $D(\hat{\mathcal{E}}) \cap C(\hat{E})$ is dense both in $D(\hat{\mathcal{E}})$ w.r.t. $\hat{\mathcal{E}}_1$ $(:= \hat{\mathcal{E}} + (\,,\,)_{\hat{\mu}})$ and in $C(\hat{E})$ (:= all continuous functions on $\hat{E}$) w.r.t. uniform norm.

<u>Proof</u>. (cf. [AR 89d, Proposition 3.4]) By [AR 89b, Proposition 2.9] there exists a countable $\mathbb{Q}$-algebra $D \subset \mathcal{F}C_b^\infty$ containing the (rational) constants and separating the points of E which is dense in $D(\mathcal{E})$ w.r.t. $\mathcal{E}_1$. By Gelfand's representation theorem applied to the uniform closure $\overline{D}$ of D there exists a compact Hausdorff space $\hat{E}$ such that E is densely and continuously embedded into $\hat{E}$ and the restriction to E gives rise to an isomorphism from $C(\hat{E})$ to $\overline{D}$ (cf. e.g. [G 69, Chap.1, Section 8]). Since $\overline{D}$ is separable, $C(\hat{E})$ is a separable metric space, hence second countable. Consequently, $\hat{E}$ is second countable hence completely metrizable as a compact space. Thus (i) is proved. Since every function in D extends to a continuous function on $\hat{E}$ whose correspondig $\hat{\mu}$-class is by definition in $D(\hat{\mathcal{E}})$, assertion (ii) follows. □

From now on we fix a compactification $\hat{E}$ of E satisfying (i), (ii) in Proposition 3.2 and identify E with $i(E)$ i.e. $E \subset \hat{E}$. Let $\hat{\text{Cap}}$ denote the capacity associated with $(\hat{\mathcal{E}}, D(\hat{\mathcal{E}}))$ defined analogously to (3.1), (3.2). The following Lemma is then immediate (cf. [AR 89d, Lemma 1.3]):

<u>Lemma 3.3</u>. (i) $\hat{\text{Cap}}(U) = \text{Cap}(U \cap E)$ for any $U \subset \hat{E}$, U open (in $\hat{E}$). In particular, $\hat{\text{Cap}}(\hat{E}\backslash K) = \text{Cap}(E\backslash K)$ for any $K \subset E$, K compact (in E).

(ii) $\text{Cap}(A) \leqslant \hat{\text{Cap}}(A)$ for any $A \subset E (\subset \hat{E})$.

We recall that a property of points in E is said to hold *quasi-everywhere* (abbreviated "q.e.") if it holds outside a set $N \subset E$ with $\text{Cap}(N) = 0$. A q.e. defined function $u : E \to \mathbb{R}$ is called *quasi-continuous* if for each $\varepsilon > 0$ there exists a closed set $F \subset E$ such that $\text{Cap}(E\backslash F) < \varepsilon$ and the restriction $u_{|F}$ of u to F is (everywhere defined and) continuous. Since both $(E,\mathcal{B}(E))$ and $(\hat{E},\mathcal{B}(\hat{E}))$ are standard Borel, we have that $E \in \mathcal{B}(\hat{E})$ (cf. [P 67, Chapter V, Definition 2.2 and Theorem 2.4]). Hence it follows by [F 80, Theorem 4.1.3] that the diffusion $\mathcal{X}$ associated with $(\mathcal{E}, D(\mathcal{E}))$ (cf. Theorem 2.3 above) is the *restriction* of a Hunt process $\hat{\mathcal{X}} := (\hat{\Omega}, \hat{\mathcal{F}}, (\mathcal{F}_t)_{t\geqslant 0}, (\hat{X}_t)_{t\geqslant 0}, (P_z)_{z\in\hat{E}})$ defined on $\hat{E}$ in such a way that each $z \in \hat{E}\backslash E$ is a *trap* for the process and E is $\hat{\mathcal{X}}$*-invariant* (cf. [F 80] for the exact definition of $\hat{\mathcal{X}}$). In addition, we know by assumption that $\mathcal{X}$ has continuous sample paths even w.r.t. the finer (original) topology on E. Clearly, by definition $\hat{\mathcal{X}}$ is associated with the image Dirichlet form $(\hat{\mathcal{E}},D(\hat{\mathcal{E}}))$ on $\hat{E}$.

Below we denote as usual, the expectation w.r.t. P_z by E_z. We denote quantities corresponding to $(\hat{\mathcal{E}},D(\hat{\mathcal{E}}))$ and $\hat{\mathcal{X}}$ by an additional " ^ " (e.g. $\hat{E}_z$, $\hat{\text{Cap}}$, $\hat{\text{q.e.}}$, $\hat{\text{quasi-continuous}}$ etc.). We consider any function u on E as a function on $\hat{E}$ by putting $u \equiv 0$ on $\hat{E}\backslash E$.

We recall that since $(\hat{\mathcal{E}},D(\hat{\mathcal{E}}))$ is regular each $\hat{u} \in D(\hat{\mathcal{E}})$ has a $\hat{\text{quasi-continuous}}$ (μ-) version $\tilde{u}$ on $\hat{E}$ (cf. [F 80, Theorem 3.1.3]). Now we are prepared to prove Theorem 3.1.

<u>Proof of Theorem 3.1</u>. Since $T_t 1 = 1$, $t \geq 0$, i.e $(\mathcal{E},D(\mathcal{E}))$ is *conservative*, we may assume that $\mathcal{X}$ is *canonical* in the following sense: $\Omega = C([0,\infty),E)$ equipped with the topology given by "locally univorm convergence", $\mathcal{F}$ = Borel σ-algebra on Ω, $X_t : \Omega \to E$ denotes evaluation at $t \geq 0$ and $\mathcal{F}_t = \sigma\{X_s : s \leq t\}$.

Let $P_\mu := \int P_z \, \mu(dz)$. Since Ω is Polish and $P_\mu(\Omega) < \infty$ there exist compact sets $\Omega_n \subset \Omega$, $n \in \mathbb{N}$, such that $\Omega_n \subset \Omega_{n+1}$, $n \in \mathbb{N}$, and $\lim_{n\to\infty} P_\mu(\Omega\backslash\Omega_n) = 0$. Define $\Lambda : [0,\infty) \times \Omega \to E$ by $\Lambda(t,\omega) = \omega(t)$, $t \geq 0$, $\omega \in \Omega$. Then Λ is continuous with respect to the product topology on $[0,\infty) \times \Omega$. Let

$$K_n := \Lambda([0,n] \times \Omega_n) , \quad n \in \mathbb{N}.$$

Then each K_n is compact in E. Let for $A \subset E$, $\sigma_A := \inf\{t>0 : X_t \in A\}$ be the *first hitting time* of A then we have for $U_n := E\backslash K_n$, $n \in \mathbb{N}$, that σ_{U_n} increases with n and for every $M > 0$

$$\begin{aligned} P_\mu[\sup_n \sigma_{U_n} < M] &\leq \inf_{n\geq M} P_\mu[\sigma_{U_n} < M] \\ &\leq \inf_{n\geq M} P_\mu[\Omega\backslash\Omega_n] = 0. \end{aligned} \tag{3.9}$$

But if $\hat{e}_{U_n} \in D(\hat{\mathcal{E}})$ is the 1-equilibrium potential of U_n (note that $U_n = E \cap(\hat{E}\backslash K_n) \in \mathcal{B}(\hat{E}))$ we know by [F 80, Theorem 3.3.1] that

$$\hat{\text{Cap}}(U_n) = \hat{\mathcal{E}}_1(\hat{e}_{U_n},1) = \int \hat{e}_{U_n} \, d\hat{\mu} \ , \tag{3.10}$$

since $T_t 1 = 1$, $t \geq 0$, hence $\hat{\mathcal{E}}(v,1) = 0$ for all $v \in D(\hat{\mathcal{E}})$. By [F 80, Theorem 4.3.5]

$$\tilde{e}_{U_n}(z) := \int_{\hat{\Omega}} \exp[-\hat{\sigma}_{U_n}]\, d\hat{P}_z, \quad z \in \hat{E},$$

is a quasi-continuous $(\hat{\mu}-)$ version of $\hat{e}_{U_n}$. By (3.10) and the definition of $\hat{\mathfrak{X}}$ it follows that

$$\hat{\mathrm{Cap}}(U_n) = \int_{\Omega} \exp[-\sigma_{U_n}]\, dP_\mu.$$

But by Lemma 3.3(ii), $\mathrm{Cap}(U_n) \leq \hat{\mathrm{Cap}}(U_n)$ and by (3.9) $\sigma_{U_n} \uparrow +\infty$ P_μ-a.s. as $n \to \infty$, hence the theorem is proved. □

As a consequence we shall obtain that Cap and $\hat{\mathrm{Cap}}$ are equal on E. We already know by Lemma 3.3 (ii) that $\mathrm{Cap}(A) \leq \hat{\mathrm{Cap}}(A)$ for all $A \subset E (\subset \hat{E})$. The dual inequality was proved previously using the diffusion process $\mathfrak{X}$ (cf. [K 82, Lemma 3.1] and also [AR 89d, Proposition 4.1]). The following purely analytic proof we learnt from W. Hansen whom we like to thank for it at this point.

<u>Proposition 3.4</u>. $\mathrm{Cap}(A) = \hat{\mathrm{Cap}}(A) = \hat{\mathrm{Cap}}(A \cup (\hat{E}\backslash E))$ for every $A \subset E$.

<u>Proof</u>. Let $K_n \subset E$, $n \in \mathbb{N}$, be as in Theorem 3.1. By (3.3) and Lemma 3.3 (i) we obtain that

$$\hat{\mathrm{Cap}}(\hat{E}\backslash E) \leq \hat{\mathrm{Cap}}(\hat{E}\backslash K_n) = \mathrm{Cap}(E\backslash K_n) \xrightarrow[n\to\infty]{} 0.$$

Hence the second equality of the assertion follows from (3.3) and (3.5). Because of Lemma 3.3 (ii) it remains to show that for $A \subset E$

$$\mathrm{Cap}(A) \geq \hat{\mathrm{Cap}}(A) \tag{3.11}$$

Let $\varepsilon > 0$. There exists $K \subset E$, K compact in E such that $\mathrm{Cap}(E\backslash K) < \varepsilon$ and there exists $U \subset E$, U open in E with $A \subset U$ such that $\mathrm{Cap}(U) < \mathrm{Cap}(A) + \varepsilon$. Observe that the trace topologies induced by E and $\hat{E}$ on K coincide, hence there exists $\hat{U} \subset \hat{E}$, $\hat{U}$ open in $\hat{E}$ such that

$$U \cap K = \hat{U} \cap K. \tag{3.12}$$

Consequently,

$$
\begin{aligned}
\widehat{\mathrm{Cap}}(A) &\leqslant \widehat{\mathrm{Cap}}(A \cap K) + \widehat{\mathrm{Cap}}(E\backslash K) && \text{(by (3.5) and (3.3)} \\
&= \widehat{\mathrm{Cap}}(A \cap K) + \mathrm{Cap}(E\backslash K) && \text{(by Lemma 3.3(i))} \\
&\leqslant \widehat{\mathrm{Cap}}(U \cap K) + \varepsilon && \text{(by (3.3) since } A \subset U) \\
&= \widehat{\mathrm{Cap}}(\hat{U} \cap K) + \varepsilon && \text{(by (3.12))} \\
&\leqslant \widehat{\mathrm{Cap}}(\hat{U}) + \varepsilon && \text{(by (3.3))} \\
&= \mathrm{Cap}(\hat{U} \cap E) + \varepsilon && \text{(by Lemma 3.3(i))} \\
&\leqslant \mathrm{Cap}(\hat{U} \cap K) + \mathrm{Cap}(E\backslash K) + \varepsilon && \text{(by (3.5) and (3.3))} \\
&\leqslant \mathrm{Cap}(U \cap K) + 2\varepsilon && \text{(by (3.12))} \\
&\leqslant \mathrm{Cap}(A) + 3\varepsilon && \text{(by (3.3) and the choice of } U).
\end{aligned}
$$

Since ε was arbitrary (3.11) is proven. □

Finally as an immediate consequence we obtain:

Corollary 3.5. Let $u : E \to \mathbb{R}$. Then u is quasi-continuous if and only if it is $\widehat{\text{quasi}}$-continuous.

4. Solution of stochastic differential equations in infinite dimensions via Dirichlet forms.

In this section we want to prove that (under some mild assumptions) the diffusion process $\mathcal{X}$ associated with $(\mathcal{E}, D(\mathcal{E}))$ weakly solves a stochastic differential equation of type

$$
\begin{aligned}
dX_t &= dW_t + \beta(X_t)\, dt \\
X_o &= z \ (\in E) \qquad \text{under } P_z
\end{aligned}
\tag{4.1}
$$

where $(W_t)_{t \geqslant 0}$ is an E-valued Brownian motion over H and $\beta : E \to E$. To obtain β we need a partial integration formula.

Definition 4.1. (cf. [AKR 88]) $k \in E$ is called *well-(μ-) admissible* if there exists $\beta_k \in L^2(E;\mu)$ such that

$$
\int \frac{\partial u}{\partial k}\, d\mu = - \int u\, \beta_k\, d\mu \quad \text{for all} \quad u \in \mathcal{F}C_b^\infty. \tag{4.2}
$$

Note that (4.2) is, of course, equivalent with

$$
\int \frac{\partial u}{\partial k}\, v\, d\mu = - \int u \frac{\partial v}{\partial k}\, d\mu - \int uv\, \beta_k\, d\mu \quad \text{for all} \quad u, v \in \mathcal{F}C_b^\infty \tag{4.3}
$$

(cf. also Theorem 4.6 below). From now on we assume that there exist "sufficiently many" well-(μ-) admissible elements in E, i.e. the following condition holds:

There exists a dense linear subspace K of $E'(\subset H \subset E)$ consisting of well-(μ-) admissible elements in E. (A.1)

Clearly, $k \mapsto \beta_k$ is a linear map from K to $L^2(E;\mu)$. It is also obvious that (1.3) automatically holds for each $k \in K$, hence (2.1) is well-defined as a quadratic form on $L^2(E;\mu)$. But (A.1) also implies (2.2):

<u>Proposition 4.2</u>. $(\mathcal{E},\tilde{\mathcal{F}}C_b^\infty)$ is closable on $L^2(E;\mu)$.

<u>Proof</u>. Let $k \in K$ and $u_n \in (\mathcal{E},\tilde{\mathcal{F}}C_b^\infty)$, $n \in \mathbb{N}$, such that $u_n \to 0$ in $L^2(E;\mu)$ as $n \to \infty$ and $\left[\frac{\partial u_n}{\partial k}\right]_{n \in \mathbb{N}}$ is a Cauchy sequence in $L^2(E;\mu)$. Let $f \in L^2(E;\mu)$ such that $\frac{\partial u_n}{\partial k} \to f$ in $L^2(E;\mu)$ as $n \to \infty$. Then by (4.3) for each $v \in \tilde{\mathcal{F}}C_b^\infty$

$$\int f\, v\, d\mu = - \lim_{n\to\infty} \int u_n\left[\frac{\partial v}{\partial k} + \beta_k\, v\right] d\mu = 0.$$

Hence $f = 0$, consequently the form $(\mathcal{E}_k,\tilde{\mathcal{F}}C_b^\infty)$ defined by

$$\mathcal{E}_k(u,v) := \frac{1}{2}\int \frac{\partial u}{\partial k}\,\frac{\partial v}{\partial k}\, d\mu\ , \quad u,\ v \in \tilde{\mathcal{F}}C_b^\infty \tag{4.4}$$

is closable on $L^2(E;\mu)$. Therefore, if $K_o \subset K$ is an orthonormal basis of H then for all $u,\ v \in \tilde{\mathcal{F}}C_b^\infty$

$$\mathcal{E}(u,v) = \frac{1}{2}\int \langle\nabla u,\ \nabla v\rangle_H\, d\mu = \sum_{k\in K_o} \mathcal{E}_k(u,v)$$

and hence $(\mathcal{E},\tilde{\mathcal{F}}C_b^\infty)$ is closable on $L^2(E;\mu)$ by [AR 88, Theorem 1.2] (or [AHP 89a, Theorem 4.1]). □

In order to prove that $\mathcal{X}$ solves a stochastic differential equation of type (1.1) we need one more assumption.

$$\int {}_{E'}\langle k,z\rangle_E^2\ \mu(dz) < \infty \quad \text{for all} \quad k \in K. \tag{A.2}$$

Here ${}_{E'}\langle\ ,\ \rangle_E$ denotes the dualisation between E' and E.

<u>Theorem 4.3</u>. Assume that for one (and hence all) $t > 0$ there exists a probability measure μ_t on $(E,\mathcal{B}(E))$ such that

$$\int \exp(i\ {}_{E'}\langle k,z\rangle_E)\ \mu_t(dz) = \exp(-\tfrac{1}{2} t\ \|k\|_H^2) \quad \text{for all} \quad k \in E'. \tag{4.5}$$

Then there exist maps $W : \Omega \to C([0,\infty),E)$ and $N : \Omega \to C([0,\infty),E)$ such that:

(i) $\omega \mapsto W_t(\omega) := W(\omega)(t)$ and $\omega \mapsto N_t(\omega) := N(\omega)(t)$, $\omega \in \Omega$, are both $\mathcal{F}_t/\mathcal{B}(E)$-measurable for all $t \geq 0$.

(ii) For q.e. $z \in E$ under P_z, $(W_t)_{t\geq0}$ is an $(\mathcal{F}_t)_{t\geq0}$- Brownian motion starting at $0(\in E)$ with covariance $\langle\ ,\ \rangle_H$.

(iii) ${}_{E'}\langle k, N_t\rangle_E = \frac{1}{2}\int_0^t \beta_k(X_s)\ ds$, $t \geq 0$, P_z- a.s. for q.e. $z \in E$.

(iv) $X_t = z + W_t + N_t$, $t \geq 0$, P_z - a.s. for q.e. $z \in E$.

<u>Proof</u>. By Theorem 3.1 resp. Theorem 2.3 above conditions (C.2) resp. (C.3) in [AR 89d] are fulfilled. Therefore, by (A.1), (A.2) all conditions (C.1)′, (C.2) - (C.4) for Theorems 6.2 and 6.5 in [AR 89d] are fulfilled which therefore apply and prove the assertion. □

<u>Remark 4.4</u>. (i) The assumption in Theorem 4.3 that the Gaussian measures μ_t, $t \geq 0$, with covariance given by $\langle\ ,\ \rangle_H$ in (4.5) exist on $(E,\mathcal{B}(E))$, is entirely natural since it just says that there exists a Brownian semigroup on E with covariance $\langle\ ,\ \rangle_H$, i.e. there exists a Brownian motion on E over H. This is, of course, necessary for (4.1) to make sense. Hence Theorem 4.3 is the best result one could hope for.

(ii) For necessary and sufficient conditions for μ_t, $t \geq 0$, as in (4.5) to exist we refer to [Gr 65] (see also [Ku 75], [Y 89] and [AR 89d]). Roughly speaking μ_t, $t \geq 0$ exist on $(E,\mathcal{B}(E))$ if E is "big enough".

Theorem 4.3 (iv) tells us that $\mathcal{X}$ solves in a sense the stochastic differential equation (4.1) except that we have not defined β. And in fact, in our general situation there is no guarantee that the map $k \mapsto \beta_k(z)$, $k \in K$, is represented by an element in E for μ - a.e. $z \in E$. We introduce the following condition:

There exists a $\mathcal{B}(E)/\mathcal{B}(E)$- measurable map $\beta : E \to E$ (A.3)
such that

(i) ${}_{E'}\langle k,\beta\rangle_E = \beta_k$ μ - a.e. for each $k \in K$ $(\subset E')$

(ii) $\int \|\beta\|_E^2\ d\mu < \infty$.

We have the following sufficient condition for (A.3).

Proposition 4.5. Assume that E is a Hilbert space and that there exists another Hilbert space $\langle H_o, \langle\ ,\ \rangle_{H_o}\rangle$ such that $E' \subset H_o$ densely by a Hilbert Schmidt map and there exists $c \in (0,\infty)$ such that

$$\int \beta_k^2 \, d\mu \leqslant c\|k\|_{H_o}^2 \quad \text{for all} \quad k \in K.$$

Then (A.3) holds.

Proof. [AR 89d, Proposition 6.9]. □

The assumptions in Proposition 4.5 are fulfilled in many examples in particular those from quantum field theory (cf. [AR 89d, Section 7]), and those where $\beta : E \to E$ comes from a linear operator on H (cf. [AR 90]). (A.3) implies that the process $N = (N_t)_{t\geqslant 0}$ in Theorem 4.3 is given as a Bochner integral. Indeed, by [AR 89d, Lemma 4.2] we know that for any $u \in L^2(E;\mu)$, $t \geqslant 0$,

$$E_z[\int_0^t |u|(X_s)\, ds] < \infty \quad \text{for} \quad \text{q.e.} \quad z \in E.$$

Hence, defining

$$\Omega_o := \{\int_0^t \|\beta\|_E(X_s)\, ds < \infty \quad \text{for all} \quad t > 0\}$$

we have that $P_z(\Omega_o) = 1$ for q.e. $z \in E$. Consequently,

$$N_t = 1_{\Omega_o} \frac{1}{2} \int_0^t \beta(X_s)\, ds, \quad t \geqslant 0,$$

(where the integral is in the sense of Bochner) and Theorem 4.3(iv) is then just the integrated version of (4.1).

So far we have proved that given a classical Dirichlet form $(\mathcal{E},D(\mathcal{E}))$ of type (closure of) (2.1) on $L^2(E;\mu)$ such that (A.1) - (A.3) is fulfilled and if E is big enough to support the measures μ_t, $t \geqslant 0$, in (4.5) then the diffusion process $\mathfrak{X}$ associated with $(\mathcal{E},D(\mathcal{E}))$ solves (4.1). In applications, however, (4.1) and in particular $\beta : E \to E$ is given a priori and we have to find the appropriate classical Dirichlet form $(\mathcal{E},D(\mathcal{E}))$, i.e. we have to find the measure μ such that the β_k's in Definition 4.1 (corresponding to the well-(μ-) admissible elements) are the components of the given β. Below we will state our final theorem which characterizes the well-(μ-) admissible elements in E and provides a technique to construct μ corresponding to a given β. First we need some preparations.

For $k \in E\backslash\{0\}$ fixed, let E_k be a "complementing subspace", i.e. a closed subspace of E such that $E = E_k \oplus k\mathbb{R}$. For each $z \in E$, $z = x + sk$, where $x \in E_k$, $s \in \mathbb{R}$ are uniquely determined. Let $\pi_k : E \to E_k$ be the canonical projection. Since E and E_k are Souslinean we can disintegrate μ w.r.t. $\pi_k : E \to E_k$, i.e. there exists a kernel $\rho_k : E_k \times \mathcal{B}(\mathbb{R}) \to [0,1]$ such that for all $u : E \to [0,\infty[$, $\mathcal{B}(E)$-measurable

$$\int_E u(z)\, \mu(dz) = \int_{E_k} \int_{\mathbb{R}} u(x+sk)\, \rho_k(x,ds)\, \nu_k(dx) \tag{4.6}$$

where $\nu_k := \mu \circ \pi_k^{-1}$ (i.e. the image measure of μ under π_k). $\rho_k(\cdot,ds)$ is ν_k - a.e. uniquely determined. Correspondingly, $L^2(E;\mu)$ can be written as a direct integral of L^2-spaces over $\mathbb{R}$:

$$L^2(E;\mu) = \int^{\oplus} L^2(\mathbb{R};\rho_k(x,ds))\, \nu_k(dx), \tag{4.7}$$

in the sense that each $u \in L^2(E;\mu)$ corresponds to a "field of vectors" $(u_x)_{x \in E_k}$ where $u_x := u(x + \cdot\, k)$, $x \in E_k$ (cf. [AR 88] for details and references).

<u>Theorem 4.6</u>. $k \in E\backslash\{0\}$ is well-(μ-) admissible if and only if for one (hence all) complementary subspace(s) E_k the following holds: for ν_k - a.e. $x \in E_k$, $\rho_k(x,ds) = \rho_k(x,s)\, ds$ for some $\mathcal{B}(\mathbb{R})$-measurable function $\rho_k(x,\cdot) : \mathbb{R} \to [0,\infty)$ such that $\frac{\partial}{\partial s} \rho_k(x,\cdot) \in L^1_{loc}(\mathbb{R};ds)$ and $\left[\frac{\partial}{\partial s} \rho_k(x,\cdot)/\rho_k(x,\cdot)\right]_{x \in E_k} \in \int^{\oplus} L^2(\mathbb{R};\rho_k(x,ds))\, \nu_k(dx) = L^2(E;\mu)$. (Here $\rho_k(x,ds)$ is as in (4.6) and ds denotes one dimensional Lebesgue measure. The derivative $\frac{\partial}{\partial s}$ is in the sense of Schwartz distributions and we use the convention $\frac{a}{o} := (\text{sign } a)\cdot(+\infty)$ for $a \in \mathbb{R}$ and $(\pm\infty)\cdot 0 = 0$.)

<u>Proof</u>. [AKR 88, Theorem 2.5]. □

Acknowledgement

We would like to thank the organizers for this interesting and very stimulating conference on White Noise Analysis. We would also like to thank W. Hansen for his comments on Section 3 of this paper, in particular for the proof of Proposition 3.4.

References

[AHP 89a] Albeverio, S., Hida, T., Potthoff, J., Röckner, M., Streit, L.: Dirichlet forms in terms of white noise analysis I: Construction and QFT examples. Preprint (1989). To appear in Reviews in Math. Phys.

[AHP 89b] Albeverio, S., Hida, T., Potthoff, J., Röckner, M., Streit, L.: Dirichlet forms in terms of white noise analysis II: Closability and diffusion processes. Preprint (1989). To appear in Reviews in Math. Phys.

[AH-K 77a] Albeverio, S., Høegh-Krohn, R.: Dirichlet forms and diffusion processes on rigged Hilbert spaces. Z. Wahrscheinlichkeitstheorie verw. Gebiete 40, 1-57 (1977).

[AH-K 77b] Albeverio, S., Høegh-Krohn, R.: Hunt processes and analytic potential theory on rigged Hilbert spaces. Ann. Inst. Henri Poincaré Sect. B 13, 269-291 (1977).

[AKR 88] Albeverio, S., Kusuoka, S., Röckner, M.: On partial integration in infinite dimensional space and applications to Dirichlet forms. Preprint (1988). To appear in J. London Math. Soc.

[AR 88] Albeverio, S., Röckner, M.: Classical Dirichlet forms on topological vector spaces - closability and a Cameron-Martin formula. BiBoS - preprint 312/88. To appear in J. Funct. Anal.

[AR 89a] Albeverio, S., Röckner, M.: Dirichlet forms, quantum fields and stochastic quantization. Stochastic analysis, path integration and dynamics. Research Notes in Math. Vol. 200, 1-21. Editors: K.D. Elworthy, J.C. Zambrini. Harlow:Longman 1989.

[AR 89b] Albeverio, S., Röckner, M.: Classical Dirichlet forms on topological vector spaces - construction of an associated diffusion process. BiBoS - preprint 342/88. Probab. Th. Rel. Fields 83, 405-434 (1989).

[AR 89c] Albeverio, S., Röckner, M.: New developments in theory and applications of Dirichlet forms. To appear in Proc. "Ascona July 1988".

[AR 89d] Albeverio, S., Röckner, M.: Stochastic differential equations in infinite dimensions: solutions via Dirichlet forms. Preprint Edinburgh (1989). Publication in preparation.

[AR 90] Albeverio, S., Röckner, M.: Stochastic differential equations in infinite dimensions: the linear case. In preparation.

[D 65] Dynkin, E. B.: Markov processes Vols. I and II. Berlin-Heidelberg-New York: Springer 1965.

[D 82] Dynkin, E. B.: Green's and Dirichlet spaces associated with fine Markov processes. J. Funct. Anal. 47, 381-418 (1982).

[F 80] Fukushima, M.: Dirichlet forms and Markov processes. Amsterdam-Oxford-New York: North Holland 1980.

[G 69] Gamelin, T.W.: Uniform algebras. Englewood Cliffs: Prentice Hall 1969.

[Gr 65] Gross, L.: Abstract Wiener spaces. Proc. 5th Berkeley Symp. Math. Stat. Prob. 2, 31-42 (1965).

[HKP] Hida, T., Kuo, H.-H., Potthoff, J., Streit, L.: White noise: an infinite dimensional calculus. Monograph in preparation.

[HPS 88] Hida, T., Potthoff, J., Streit, L.: Dirichlet forms and white noise analysis. Commun. Math. Phys. 116, 235-245 (1988).

[Ku 75] Kuo, H.: Gaussián measures in Banach spaces. Lecture Notes in Math. 463, 1-224. Berlin-Heidelberg-New York: Springer 1975.

[K 82] Kusuoka, S.: Dirichlet forms and diffusion processes on Banach spaces. J. Fac. Sci. Univ. Tokyo Sect. IA 29, 79-95 (1982).

[LR 90] Lyons, T.J., Röckner, M.: A note on tightness of capacities associated with Dirichlet forms. Preprint Edinburgh (1990). Publication in preparation.

[P 67] Parthasarathy, K.R.: Probability measures on metric spaces. New York-London: Academic Press 1967.

[Sch 73] Schwartz, L.: Radon measures on arbitrary topological spaces and cylindrical measures. London: Oxford University Press 1973.

[S 89] Schmuland, B.: An alternative compactification for classical Dirichlet forms on topological vector spaces. Preprint (1989).

[Y 89] Yan, J.A.: Generalizations of Gross' and Minlos' theorems. To appear in Séminaire de Probabilités.

[Yo 87] Yokoi, Y.: Positive generalized functionals. Preprint (1987).

PROBABILISTIC ASPECTS of MACHINE LEARNING

Bruno Apolloni
Diego de Falco

Dipartimento di Scienze dell' Informazione
Università di Milano
I-20133 Milano, Italy

ABSTRACT

"Making the right noise", namely training a machine to reproduce at run time significant statistical features of the environmental signal to which it has been exposed in the learning stage, is an important aspect of the problem of machine learning. In this perspective, we revisit the time honoured Little-Shaw-Vasudevan model as a parallel asymmetric Boltzmann machine, providing a variational characterization of the corresponding Hebbian learning rule.

1. Learning

A widely accepted operational definition of what is to be meant by "learning" was provided by Hebb's neurophysiological postulate [1]:
"When an axon of cell A is near enough to excite a cell B and repeatedly and persistently takes part in firing it, some growth process or metabolic change takes place in one or both cells such that A's efficiency as one of the cells firing B is enhanced".

Mathematical aspects of models implementing the above idea of learning as adiabatic updating of the coupling constants of highly non linear deterministic dynamical systems have been studied by Caianiello since 1961 [2].

An important addition to Hebb's postulate came from the observation [3] and the implementation in a model [4,5] of the neurophysiological evidence [6] that the process of mutual excitation among neurons shows intrinsic statistical fluctuations due to the fact that the number of quanta of chemical transmitter released when an action potential reaches the synapse follows a Poisson law and each quantum has a Gaussian distribution of transmitter molecules.

Together with the accepted modellization of a neuron as a two level system, these developments have, since an early stage, brought this hardware oriented, neuromimetic line of research on machine learning in contact with the statistical mechanics of spin glasses (for extensive reviews see References 7 an 8).

The emerging problem in mathematical physics is that of studying strongly interconnected arrays of spins capable of modifying, through the interaction with (the collection of statistics from) a noisy environment, their coupling constants in such a way as to reproduce, in the distribution of their configurations, given features (say, support properties or suitable conditional or marginal properties) of the distribution of the environmental random signal to which they have been exposed in the learning stage.

It is to be stressed that probabilistic notions appear also in the software oriented, symbolic approach to machine learning advocated by artificial intelligence.

By the way of example, consider the following problem posed by Valiant [9] in his exploration of the notion of learnability:

let $f: \mathfrak{x} \in \{0,1\}^m \rightarrow f(\mathfrak{x}) \in \{0,1\}$ be a Boolean function of m Boolean variables.
Call $\Phi = f^{-1}(\{1\})$ the support of f, and consider a probability measure D on $\{0,1\}^m$ with $D(\Phi)=1$.
D is to be thought of as the probability distribution with which the environment exhibits "positive examples" to the machine.
The issue is to construct, from a random sample $\mathfrak{x}^1,..., \mathfrak{x}^\ell$ drawn from the population D, an approximation $\Gamma_\ell(\mathfrak{x}^1,..., \mathfrak{x}^\ell)$ to Φ in such a way that $\Gamma_\ell \subseteq \Phi$ and for given ε,δ in $(0,1)$ it is, with respect to the sample distribution D_ℓ:

$$D_\ell(\{\mathfrak{x}^1,\ldots,\mathfrak{x}^\ell\} \in \Phi^\ell : D(\Phi - \Gamma_\ell) > \varepsilon) < \delta$$

The function f will be said to be learnable if the construction of Γ_ℓ runs in a time bounded above by a polynomial in the number m of variables and in the accuracy ε,δ required.

The qualitative idea behind this approach is that the classification that f imposes of patterns $\mathfrak{x} \in \{0,1\}^m$ into "yes" or "no" patterns can be learned if a reasonable amount ℓ of examples leads to a negligible D_ℓ-probability of misclassifying those examples that the environment exhibits with significant D-probability. Most proofs in the literature of the existence of classes of learnable functions focus on the construction of approximations f_ℓ to f: suppose for instance one knows that f is the product of some (Boolean) sums of up to a fixed number k of variables; starting from the zero order approximation f_0=product of all sums of up to k variables, the i-th order approximation f_i is obtained by pruning out of f_{i-1} those factors which vanish on the i-th example $\mathfrak{x}^i$: one recognizes such factors should not be in f as $\mathfrak{x}^i$ is a <u>positive</u> example. Γ_ℓ is consequently taken to be $f_\ell^{-1}(\{1\})$.

It is an interesting open research problem to explore whether a proposal for Γ_ℓ can be reached at the subsymbolic level of letting a spin glass follow a fast Glauber dynamics [10] in contact with the environment, by a slow adiabatic process adjusting its parameters in such a way as to exhibit at the interface nodes configurations following a distribution which approximates the support properties of the environmental one.

The theoretical prejudice, and the partial evidence provided by models, that this should be possible is at the basis of the treatment we give below of the Boltzmann machine [11] not as an input-output machine but rather as a tool to build subsymbolic models of noisy environments.

2. The Little-Shaw-Vasudevan model

A set $\Lambda_\nu = \{1,...,\nu\}$ of neurons is coupled by a real n x n matrix α of synaptic weights.

If the machine is in the configuration $\underline{\tau}=(\tau_1,\ldots,\tau_\nu) \in \{0,1\}^\nu \equiv Q_\nu$ the signal X_i present on the i-th neuron, and on which it will base its future action is determined by α and by the bias vector $\underline{\gamma}=(\gamma_1,\ldots,\gamma_\nu) \in \mathbb{R}^\nu$ by:

$$X_i(\underline{\tau}) = \sum_{j=1}^{\nu} \alpha_{ij}\tau_j + \gamma_i$$

The configuration $\underline{T}(t)$ of the machine performs a random walk on Q_ν, with discrete time parameter $t = 0,1,\ldots$

The analysis of Shaw and Vasudevan requires that the conditional expectation of $T_i(t+1)$ given $\underline{T}(t)$ be of the form

$$E(T_i(t+1)|\underline{T}(t)) = \frac{1}{1+e^{-\beta X_i(\underline{T}(t))}}$$

where the inverse-temperature-like parameter $\beta>0$ measures the scale of fluctuations in the firing mechanism.

For the synchronous parallel activation mechanism considered here, in which, given the current configuration, each neuron is activated simultaneously to and independently of all the others, the random walk $\underline{T}(t)$ is ruled by the Markov transition matrix

$$p_{\underline{\tau},\underline{\tau}'} = P(\underline{T}(t+1)=\underline{\tau}'|\underline{T}(t)=\underline{\tau}) = \prod_{i\in\Lambda_\nu}[1+e^{\beta(1-2\tau'_i)X_i(\underline{\tau})}]^{-1}$$

As in the deterministic counterpart ($\beta\to+\infty$ limit) of the above dynamics, considerable understanding is gained by introducing the characteristic bilinear form of the network [12]:

$$\mathcal{V}: (\underline{x}',\underline{x})\in \mathbb{R}^\nu\times\mathbb{R}^\nu \to \mathcal{V}(\underline{x}',\underline{x}) = -\frac{1}{2}\underline{x}'\cdot\alpha\underline{x} - \frac{1}{2}\gamma\cdot(\underline{x}'+\underline{x})$$

in terms of which p can be written as:

$$p_{\underline{\tau},\underline{\tau}'} = \frac{e^{-2\beta\mathcal{V}(\underline{\tau}',\underline{\tau})}}{\sum_{\underline{\sigma}} e^{-2\beta\mathcal{V}(\underline{\sigma},\underline{\tau})}}$$

The corresponding representation of the n-step transition matrix

$$p^{(n)}_{\underline{\tau}(0),\underline{\tau}(n)} = \sum_{\underline{\tau}(1)\ldots\underline{\tau}(n-1)} \frac{1}{\displaystyle\sum_{\underline{\tau}'(1)\ldots\underline{\tau}'(n)} \exp\,[2\beta \sum_{k=1}^{n} \mathcal{V}(\underline{\tau}(k)-\underline{\tau}'(k),\underline{\tau}(k))]}$$

gives a clear insight into the dynamical phenomena the model can present.

The contribution of a given trajectory

$$\Gamma = (\underline{\tau}(0),\ \underline{\tau}(1),\ldots,\ \underline{\tau}(n))$$

is obtained by comparison with all virtual alternatives

$$\Gamma' = (\underline{\tau}(0),\ \underline{\tau}'(1),\ldots,\ \underline{\tau}'(n)).$$

Those trajectories are , in the $\beta\to+\infty$ limit, followed with overwhelming probability for which, at each step k, $\underline{\tau}(k)$ is taken in such a way as to minimize $\mathcal{V}(\cdot,\underline{\tau}(k-1))$; this by direct inspection of $\mathcal{V}$ requires that

$$\tau_i(k) = \begin{cases} 1 \text{ if } X_i\,(\underline{\tau}(k-1)) > 0 \\ 0 \text{ if } X_i\,(\underline{\tau}(k-1)) < 0 \end{cases}$$

so that the LSV model provides, in fact, the biological motivation of the deterministic threshold dynamics of Ref. 2.

These considerations bring into play the function

$$L(\underline{\tau}) = \min_{\underline{\tau}'} \mathcal{V}(\underline{\tau}',\underline{\tau})$$

as the most likely value of $\mathcal{V}$ on a given $\underline{\tau}$ and its successor in the evolution.

If α is symmetric (a condition against which all neurophysiological evidence points), and therefore $\mathcal{V}$ is, these considerations very simply imply that the deterministic evolution takes place along trajectories of decreasing L. The counterpart of this phenomenon for the Markovian evolution follows from the observation that

$$\frac{p_{\underline{\tau},\underline{\tau}'}}{p_{\underline{\tau}',\underline{\tau}}} = \frac{\sum_{\underline{\rho}} e^{-2\beta\mathcal{V}(\underline{\rho},\underline{\tau}')}}{\sum_{\underline{\sigma}} e^{-2\beta\mathcal{V}(\underline{\sigma},\underline{\tau})}}\, e^{\beta[\underline{\tau}'\cdot\alpha\underline{\tau}\ -\underline{\tau}\cdot\alpha\underline{\tau}']}$$

so that if α is symmetric the stationary distribution for p, call it π_β, satisfies the detailed balance condition

$$\pi_\beta(\underline{\tau})p_{\underline{\tau},\underline{\tau}'} = \pi_\beta(\underline{\tau}')p_{\underline{\tau}',\underline{\tau}}$$

and can be directly read from the above ratio to be [13,14]:

$$\pi_\beta(\underline{\tau}) = \frac{\sum_{\underline{\sigma}} e^{-2\beta\mathcal{V}(\underline{\sigma},\underline{\tau})}}{\sum_{\underline{\sigma},\underline{\rho}} e^{-2\beta\mathcal{V}(\underline{\sigma},\underline{\rho})}}$$

or, more expressively [15]:

$$\pi_\beta(\underline{\tau}) = \frac{1}{Z(\beta)} \prod_{i \in \Lambda_v} \frac{1+e^{-\beta|X_i(\underline{\tau})|}}{2} e^{-2\beta L(\underline{\tau})}$$

The simple, unrealistic, case of the symmetric LSV model just considered, therefore, already shows computational capabilities in its showing, at equilibrium at large β, with high probability configurations which minimize L [16].

With an obvious extension of the name adopted in Ref. 11 for the sequentially activated version of the SLSV model, whose stationary distribution is

$$s_\beta(\underline{\tau}) = \frac{e^{-\beta V(\underline{\tau})}}{Z(\beta)},$$

with $V(\underline{\tau}) = \mathcal{V}(\underline{\tau}, \underline{\tau})$, we will refer to the whole class of models considered here as to parallel Boltzmann machines.

3. Learning by parallel Boltzmann machines

For the parametric family $s_\beta(\underline{\tau}) = s_\beta(\underline{\tau}\,; \alpha, \underline{\gamma})$ of equilibrium distributions of the sequential SLSV model (in fact for suitable marginals of such family; but here we will not consider, for simplicity the role of "hidden nodes") the problem was posed, in Ref .11, of determining the maximum likelihood estimators A and $\underline{\Gamma}$ of the parameters, based on a random sample $\underline{\tau}^1, \ldots, \underline{\tau}^\ell$ drawn from a given environmental law $\phi(\underline{\tau})$.

As the derivatives of the log-likelihood function

$$\mathcal{L} = \log \left(\prod_{h=1}^{\ell} s_\beta(\underline{\tau}^h\,; \alpha, \underline{\gamma}) \right)$$

are

$$\frac{\partial \mathcal{L}}{\partial \alpha_{ij}} = \beta \left[\sum_{h=1}^{\ell} \tau_i^h \tau_j^h - \ell\, E_{s_\beta(\cdot\,;\alpha,\underline{\gamma})}(\tau_i \tau_j) \right]$$

$$\frac{\partial \mathcal{L}}{\partial \gamma_i} = \beta\,[\sum_{h=1}^{\ell} \tau_i^h - \ell\, E_{s_\beta(.;\alpha,\gamma)}(\tau_i)]$$

A and Γ are determined by the equations

$$E_{s_\beta(.;A,\Gamma)}(\tau_i\tau_j) = \frac{1}{\ell}\sum_{h=1}^{\ell} \tau_i^h \tau_j^h$$

$$E_{s_\beta(.;A,\Gamma)}(\tau_i) = \frac{1}{\ell}\sum_{h=1}^{\ell} \tau_i^h$$

Moreover, the direction in parameter space of increasing $\mathcal{L}$ turns out to be determined precisely by Hebb's rule: increase a given parameter if the corresponding sampled correlation between nodes is larger than the one predicted by the current value of the parameters.

The conceptual role of the Boltzmann machine is precisely in its characterization of Hebb's rule in terms of a variational principle, namely the one which underlies the maximum likelihood method [17]: for every pair ϕ and ϕ' of strictly positive stochastic vectors

$$E_\phi(\log \phi\,) \geq E_\phi\,(\log \phi'\,).$$

Here we are going to pose the same question for the full LSV model, releasing therefore the unrealistic hypothesis of symmetry of α and considering parallel instead of sequential activation (The thermodynamic equivalence discussed in Ref.18 between parallel and sequential machine is a mean field indication for a particular choice of parameters).

The technical difficulty to be overcome is that, as compared to the symmetric case discussed at the end of the previous section and in Ref. 15, in the asymmetric case the stationary distribution $\pi_\beta(\underline{\tau}\,;\,\alpha,\,\underline{\gamma})$ under $p(\alpha,\,\underline{\gamma})$ is not explicitly known.

We work, instead, with

$$\pi^{(n)}(\underline{\tau}) = \pi^{(n)}(\underline{\tau};\alpha,\underline{\gamma},\psi) = \sum_{\underline{\tau}(0)} \psi(\underline{\tau}(0))\; p^{(n)}_{\underline{\tau}(0),\underline{\tau}}(\alpha,\underline{\gamma}),$$

namely the distribution reached after n evolution steps from a given distribution ψ in which the network was initialized at time 0.

Coherently with the maximum likelihood principle, we look for the choice of parameters which minimizes the relative entropy [19]

$$I^{(n)} = \sum_{\underline{\tau}} \phi(\underline{\tau}) \lg \frac{\phi(\underline{\tau})}{\pi^{(n)}(\underline{\tau})}$$

between $\pi^{(n)}$ and the given law ϕ we want to simulate.

It is easy to compute that

$$\frac{\partial I^{(n)}}{\partial \alpha_{ij}} = \beta \sum_{\underline{\tau}(0),\dots,\underline{\tau}(n)} r(\underline{\tau}(0),\dots,\underline{\tau}(n)) \sum_{k=1}^{n} \left[\tau_j(k-1) \sum_{\underline{\tau}'} (\tau'_i - \tau_i(k)) p_{\underline{\tau}(k-1),\underline{\tau}'}\right]$$

where

$$r(\underline{\tau}(0),\dots,\underline{\tau}(n)) = \frac{\psi(\underline{\tau}(0)) p_{\underline{\tau}(0),\underline{\tau}(1)} \cdots p_{\underline{\tau}(n-1),\underline{\tau}(n)} \phi(\underline{\tau}(n))}{\pi^{(n)}(\underline{\tau}(n))}$$

This ratio can be easily expressed in terms of the backward transition probabilities:

$$p^{*}_{\underline{a},\underline{b}}(h) = p_{\psi,p}(\underline{T}(h) = \underline{b} \mid \underline{T}(h+1) = \underline{a}) = \frac{\pi^{(h)}(\underline{b}) p_{\underline{b},\underline{a}}}{\pi^{(h+1)}(\underline{a})}$$

and the above expression for $\frac{\partial I^{(n)}}{\partial \alpha_{ij}}$ can therefore be resummed as:

$$\frac{\partial I^{(n)}}{\partial \alpha_{ij}} = \beta \sum_{k=1}^{n} \sum_{\underline{\tau}(n),\underline{\tau}(n-1),\dots,\underline{\tau}(k-1)} \phi(\underline{\tau}(n)) \cdot p^{*}_{\underline{\tau}(n),\underline{\tau}(n-1)}(n-1) \dots\dots p^{*}_{\underline{\tau}(k),\underline{\tau}(k-1)}(k-1) \cdot$$

$$\cdot \tau_j(k-1) \sum_{\underline{\tau}'} (\tau'_i - \tau_i(k)) p_{\underline{\tau}(k-1),\underline{\tau}'} =$$

$$= \beta \sum_{\underline{\tau}} \phi(\underline{\tau}) \cdot \left[E_{\psi,p}\left(\sum_{k=1}^{n} T_j(k-1)\, E(T_i(k) \mid \underline{T}(k-1)) \,\middle|\, \underline{T}(n)=\underline{\tau}\right) - E_{\psi,p}\left(\sum_{k=1}^{n} T_j(k-1) \cdot T_i(k) \,\middle|\, \underline{T}(n)=\underline{\tau}\right)\right]$$

The resulting Hebbian rule is: for each sample path <u>ending</u> in a configuration $\underline{\tau}$ exhibited by the environment with significant probability $\phi(\underline{\tau})$, the error signal to be collected and on whose sign to base the updating of α_{ij} is the time average along the <u>past history</u> of the difference between the actual "consensus" $T_j(k-1)\ T_i(k)$ of the receiver neuron i with the transmitter neuron j and the best prediction of such consensus given the state of the transmitters.

As a final remark we wish to rederive from the above general observations the results known for the case of a symmetric α [11,15].

In this case, and with the additional assumption $\psi=\pi_\beta(\cdot;\alpha,\underline{\gamma})$ (so that one is in fact, coherently with the adiabatic hypothesis [2], studying the effect on I of letting the state which is already in dynamical equilibrium under $p(\alpha)$ evolve for n more steps according to $p(\alpha+d\alpha)$), because of the detailed balance condition it is

$$p^*_{\underline{a},\underline{b}} = p_{\underline{a},\underline{b}}$$

so that :

$$\frac{\partial I^{(n)}}{\partial \alpha_{ij}} = \beta \sum_{k=1}^{n} \sum_{\underline{\sigma}(0),\underline{\sigma}(1)\ldots,\underline{\sigma}(k),\underline{\sigma}(k+1)} \phi(\underline{\sigma}(0))\, p_{\underline{\sigma}(0),\underline{\sigma}(1)}\, p_{\underline{\sigma}(1),\underline{\sigma}(2)} \cdots p_{\underline{\sigma}(k-1),\underline{\sigma}(k)} p_{\underline{\sigma}(k),\underline{\sigma}(k+1)} \cdot$$

$$[\sigma_j(k)\sigma_i(k+1) - \sigma_j(k)\sigma_i(k-1)] +$$

$$+ \frac{\partial I^{(n)}}{\partial \alpha_{ji}} \frac{\partial \alpha_{ji}}{\partial \alpha_{ij}} =$$

$$= \beta \sum_{k=1}^{n} E_{\phi,p(\alpha)} \left[T_j(k)T_i(k+1)+T_i(k)T_j(k+1) - T_i(k-1)T_j(k) - T_j(k-1)T_i(k)\right] =$$

$$= \beta E_{\phi,p(\alpha)} \left[T_j(n)T_i(n+1)+T_i(n)T_j(n+1)\right] - \beta E_{\phi,p(\alpha)}\left[T_j(0)T_i(1)+T_i(0)T_j(1)\right] =$$

Symmetry plays two roles in deriving the above result:

a. detailed balance has the effect of substituting the operationally obscure conditional expectations with respect to the final configuration $\underline{T}(n)$ with expectations $E_{\phi,p(\alpha)}$ for the process which evolves according to $p(\alpha)$ from the <u>initial</u> state ϕ at t=0;

b. The $\dfrac{\partial I^{(n)}}{\partial \alpha_{ji}}$ term provides intermediate cancellations, so that only the final and initial steps of the trajectory become relevant.

REFERENCES

1. D.O. Hebb, "The organization of behaviour", Wiley, New York (1949).
2. E.R. Caianiello, "Outline of thought processes and thinking machines", J. Theor. Biol. 2, 204 (1961).
3. G.L. Shaw, R. Vasudevan, "Persistent states of neural networks and the random nature of synaptic transmission", Math. Biosciences 21, 207 (1974).
4. W. A. Little: "The existence of persistent states in the brain", Math. Biosciences 19, 101 (1974).
5. W. A. Little, G.L. Shaw: "Analytic study of the memory storage capacity of a neural network", Math Biosciences 39, 281 (1978).
6. B. Katz, "Nerve, muscle and synapse", McGraw-Hill, New York (1963).
7. J.W. Clark, "Statistical mechanics of neural networks", Physics Reports 158, 91 (1988).
8. M. Mezard, G. Parisi, M.A. Virasoro, "Spin glass theory and beyond", World Scientific, Singapore (1987).
9. L.G. Valiant, "A theory of the learnable", Communications of ACM 27, 1134 (1984).
10. R.J. Glauber, " Time-dependent statistics of the Ising model ", J. Math Phys 4, 294 (1963).
11. D. H. Ackley, G.E. Hinton, T.J. Sejnowski, "A learning algorithm for Boltzmann machines", Cognitive Science 9, 147 (1985).
12. J. Bruck, J.W. Goodmann, "A generalized convergence theorem for neural networks", Stanford preprint (1988), to appear in IEEE Transactions in Information Theory.
13. P. Peretto, "Collective properties of neural networks: a statistical physics approach", Biological Cybernetics 50, 51, (1984).
14. E. Aarts, J. Korst, "Simulated annealing and Boltzmann machines", Wiley (1989).
15. B. Apolloni, D. de Falco, "Learning by parallel Boltzmann machines", Milano preprint (1989).
16. A. Bertoni, P. Campadelli, F. Grassani, "Full parallelism in Boltzmann machines", Proceeding of Neuro-Nimes '89, 361 (1989).
17. M.G. Kendall, A. Stuart, "The advanced theory of statistics, Vol 2, Inference and relationship", Hafner Publishing Co, New York (1961).
18. D.J. Amit, H. Gutfreund, H. Sompolinsky, "Spin glass models of neural networks", Phys. Rev. A 32,1007 (1985).
19. S. Kullback, "Information theory and statistics", Wiley (1959).

This research was supported in part by Consiglio Nazionale delle Ricerche under Grant 88.03556.12.

DIFFERENTIAL FORMS WITH VALUES IN $(L^2)^-$

DAVID BETOUNES and MYLAN REDFERN

Mathematics Department
University of Southern Mississippi
Hattiesburg, MS 39406–5045 USA

ABSTRACT
A calculus is presented for generalized stochastic processes and generalized stochastic differential forms which allows one to express various well known stochastic integrals and formulas in terms of white noise analysis.

1. Introduction

In this paper we develop a differential and integral calculus for multiparameter generalized stochastic processes $\phi(t)$, that is, for vector–valued functions $\phi : \mathbb{R}^N \to (L^2)^-$ taking their values in the Hilbert space $(L^2)^-$ of generalized Wiener functionals. The white noise integral $\int \phi(t)dt$ of such processes is defined as an element of $(L^2)^-$ in a way which is similar to the Pettis integral for functions with values in a Banach space. We also find it convenient to use a similar Pettis–type definition for the partial derivatives $\partial\phi(t)/\partial t_j$ of such processes. These operations together with the symmetric tensor product: $\phi(t) \hat{\otimes} \zeta(t)$ prove to be instrumental in expressing the stochastic integrals of Ito, Wong and Zakai, and Imkeller as white noise integrals, and obtaining, in a natural fashion, a differential and integral version of Ito's formula for multiparameter generalized Wiener functionals $F(W(t))$. These results are presented in another paper.[1]

To obtain the complete calculus associated with the white noise space $(L^2)^-$, we formulate in an analogous (Pettis–type) fashion a theory for differential forms:

$$\theta = \theta_{j_1 \cdots j_p} dt_{j_1} \wedge \ldots \wedge dt_{j_p}$$

with values in $(L^2)^-$, that is, with component functions $\theta_{j_1 \cdots j_p}$ being vector–valued functions (generalized stochastic processes) $\theta_{j_1 \cdots j_p} : \mathbb{R}^N \to (L^2)^-$. We adapt the standard tools: exterior derivative $d\theta$, exterior product $\theta \wedge \tau$, integrals $\int_\Omega \theta$ of θ over submanifolds, and Stokes' theorem from the general theory of vector–valued differential forms to the white noise setting here, and show, among other things, how the stochastic path integral of Cairoli and Walsh can be quite simply expressed (and generalized) in the white noise setting by

$$\int_C \phi_j(t) \,\hat{\otimes}\, \frac{\partial W}{\partial t_j}(t)\, dt_j$$

2. Preliminaries

We work with the space $\mathscr{S}^*(\mathbb{R}^N)$ of tempered distributions on $\mathbb{R}^N$ endowed with the probability measure μ for which

$$\int \exp(i\langle x,\xi\rangle)\, d\mu(x) = \exp(-\tfrac{1}{2}\|\xi\|^2)$$

for every $\xi \in \mathscr{S}(\mathbb{R}^N)$. Here $\langle\cdot,\cdot\rangle$ is the pairing between $\mathscr{S}^*$ and $\mathscr{S}$. As is well–known the Hilbert space $L^2(\mathscr{S}^*(\mathbb{R}^N),\mu) = \Sigma_{n=0}^{\infty} K_n$ obtains its Fock space (particle) structure via the isometry

$$I_n : \sqrt{n!}\, \hat{L}^2(\mathbb{R}^{Nn}) \to K_n$$

which sends each symmetric L^2 function $f(u^1,\ldots,u^n)$ to its multiple Wiener integral $I_n(f)$. To obtain a space of generalized Wiener functionals with the necessary degree of differentiability, we consider the Sobolev spaces

$$H_{\pm(n,k)} \equiv \sqrt{n!}\, \hat{H}^{\pm(n(N+1)/2+k)}(\mathbb{R}^{Nn}),$$

and let $H_n^{\pm} \equiv \Sigma_{k=0}^{\infty} H_{\pm(n,k)}$.(For most applications we just take $H_n^{\pm} = H_{\pm(n,0)}$).

Then

$$\hat{C}_c^{\infty}(\mathbb{R}^{Nn}) \subseteq H_n^{+} \subseteq \sqrt{n!}\, \hat{L}^2(\mathbb{R}^{Nn}) \subseteq H_n^{-}$$

and by restricting/extending I_n one obtains some corresponding spaces

$$C_n \subseteq K_n^{+} \subseteq K_n \subseteq K_n^{-}.$$

Taking the appropriate (Hilbert space) direct sums one obtains the space $(L^2)^{-} \equiv \Sigma_{n=0}^{\infty} K_n^{-}$ of generalized Wiener functionals and its dual $(L^2)^{+} \equiv \Sigma_{n=0}^{\infty} K_n^{+}$. Rather than using the direct sum (product) $\Sigma_{n=0}^{\infty} C_n = \{ \psi = \Sigma_{n=0}^{\infty} \psi_n \mid \Sigma_{n=0}^{\infty} \|\psi_n\|^2 < \infty \}$ of the the C_n's, we use the subspace (coproduct) $C = \Sigma_{\text{finite}} C_n$ consisting of those ψ with only finitely many non–zero terms. One considers $C \subseteq (L^2)^{+} \subseteq (L^2) \subseteq (L^2)^{-}$ where $(L^2) \equiv L^2(\mathscr{S}^*(\mathbb{R}^N))$ and denotes the pairing between $\phi \in (L^2)^{-}$ and $\psi \in (L^2)^{+}$ by $\langle\phi,\psi\rangle$. Note that for $\phi = I_n(f) \in K_n^{-}$ and $\psi = I_n(\xi) \in C_n$ this pairing reduces to the distributional pairing $\langle\phi,\psi\rangle = \langle f,\xi\rangle_{\text{dist}}$. This is not true in general and motivates the choice of the C_n's. The S–transform extends to a transform on K_n^-, also denoted by S as well.

Finally we note that the product giving the graded algebra (Fock space) structure on $(L^2) = \Sigma K_n$ extends to $(L^2)^- = \Sigma K_n^-$ by using tensor products of distributions.[2] In particular for $f \in H^{-\alpha}(\mathbb{R}^n)$ and $g \in H^{-\beta}(\mathbb{R}^k)$, one has $f\otimes g \in H^{-(\alpha+\beta)}(\mathbb{R}^{n+k})$. The action of the symmetric groups on test functions extends to an action on distrbutions in the natural fashion, and letting S_n denote the symmetrization operator, one defines the symmetric tensor product by $f\hat{\otimes}g = S_n(f\otimes g)$. Thus one gets a product $\hat{\otimes} : K_n^- \times K_k^- \to K_{n+k}^-$ naturally by defining $\phi\hat{\otimes}\zeta = I_{n+k}(f\hat{\otimes}g)$, when $\phi = I_n(f)$ and $\zeta = I_k(g)$. This product is just the extension of the usual Wick product of random variables: if $\phi_j = I_1(\eta_j) = \langle .,\eta_j\rangle \in K_1$ for $j = 1,...,p$, then $\phi_1\hat{\otimes}...\hat{\otimes}\phi_p = :\phi_1...\phi_p:$.

3. Integrability

Various theories for the integrals $\int\phi(t)dt$ of vector–valued functions $\phi:R \to V$ on a measure space $(R, \mathscr{M}, dt)$ were developed by Pettis, Birkoff, Dunford, Bochner and others in the period 1930–1940 .[3] The motivating idea behind these theories lies in the special case when V is finite dimensional, say $V = \mathbb{R}^k$, so that $\phi(t) = (\phi_1(t),...,\phi_k(t))$, and a natural definition of integrability ϕ is that $\phi_1,...,\phi_k$ each be integrable, in which case one defines

$$\int\phi(t)dt = (\int\phi_1(t)dt,..., \int\phi_k(t)dt)$$

Move generally if V is a Banach spacc then ϕ is Pettis integrable if (1) $v^*\phi$ is integrable for each continuous linear functional $v^* \in V^*$, and (2) there exists a element in V, denoted by $\int\phi(t)dt$, such that $v^*(\int\phi(t)dt) = \int v^*(\phi(t))dt$ for every $v^* \in V^*$. (If V is reflexive then one can drop the condition (2) from the definition).

For certain technical reasons we find it convenient to use the following modification of Pettis integrability in our setting with $V = (L^2)^-$. Also for the most part $R = \mathbb{R}^N$ is a Euclidean space with Lebesque measure $dt = dt_1dt_2...dt_N$.

<u>Definition</u>: A function $\phi : R \to (L^2)^-$ is called integrable if

(1) $\langle\phi(t),\psi\rangle$ is intcgrable for every $\psi \in C$

(2) there exists an element of $(L^2)^-$, denoted by $\int\phi(t)dt$, such that
$$\langle\int\phi(t)dt , \psi\rangle = \int\langle\phi(t),\psi\rangle dt$$
for every $\psi \in C$.

Of course a Pettis integrable function is integrable in this sense, but not conversely (in general). We will refer to this notion of integrability as a Pettis type integrability, and one sees that the notion applies in any Banach space V if a subset $C \subseteq V^*$ of its dual is specified.

Integrability of $\phi : R \to (L^2)^-$ is essentially equivalent to the integrability of each of its components ϕ_n. Namely since $\phi(t) = \Sigma_{n=0}^{\infty} \phi_n(t)$, one gets vector–valued functions $\phi_n : R \to K_n^- \subseteq (L^2)^-$ and $f_n : R \to H_n^-$ with $\phi_n(t) = I_n(f_n(t))$. A direct application of the various definitions then gives the following proposition.

<u>Proposition 1</u>

(a) ϕ is integrable if and only if

(i) ϕ_n is integrable $\forall n$, and

(ii) $\Sigma_{n=0}^{\infty} [\int \phi_n(t)dt]$ is in $(L^2)^-$.

In this case $\int \phi(t)dt = \Sigma_{n=0}^{\infty} \int \phi_n(t)dt$

(b) ϕ_n is integrable if and only if f_n is integrable. In this case

$$\int \phi_n(t)dt = \int I_n(f_n(t))dt = I_n(\int f_n(t)dt)$$

(c) If ϕ is integrable then $\forall\, \xi \in C_c^{\infty}(\mathbb{R}^N)$ $S(\phi)(\xi)$ is integrable and

$$S(\int \phi(t)dt)(\xi) = \int S(\phi(t))(\xi)dt$$

4. Differentiability

For our development of the white noise analysis we need a notion of differentiability which is similar to the Pettis type integrability introduced in the last section, but weaker than Frechet differentiability.

<u>Definition</u>: A vector–valued function $\phi : \mathbb{R}^N \to (L^2)^-$ is called differentiable if

(i) $\langle \phi(t), \psi \rangle$ is continuously differentiable $\forall\, \psi \in C$

(ii) For each $j \in \{1,...N\}$ there exists an element in $(L^2)^-$, denoted by $\frac{\partial \phi}{\partial t_j}(t)$, such that for every $\psi \in C$:

$$\left\langle \frac{\partial \phi}{\partial t_j}(t), \psi \right\rangle = \frac{\partial}{\partial t_j} \langle \phi(t), \psi \rangle = \lim_{h \to 0} \left\langle \frac{\phi(t+h\epsilon_j)-\phi(t)}{h}, \psi \right\rangle$$

As a primary example consider the multi–time parameter Wiener process $W(t) = \langle \cdot, 1_{[0,t]} \rangle$, with $t = (t_1,...t_N) \in \mathbb{R}_+^N$. Then since $\langle W(t), \psi \rangle = \langle 1_{[0,t]}, \xi \rangle = \int_{[0,t]} \xi(u)du$, the fundamental theorem of calculus gives

$$\frac{\partial}{\partial t_j}\langle W(t),\psi\rangle = \int_{[0,t]_j}\xi(u_1,\ldots,t_j,\ldots,u_N)(du)_j \quad (*)$$

where $[0,t]_j$, $(du)_j$ arise from products $[0,t] = [0,t_1]\times\ldots\times[0,t_N]$ and $du = du_1\ldots du_N$ by deleting the jth factor. Letting $T(\xi)$ denote the expression at the right hand side of (*) one finds that T is a distribution in H_1^-, and so $T = \frac{\partial}{\partial t_j}1_{[0,t]}$.

Hence $\frac{\partial W}{\partial t_j}(t) \equiv I_1(\frac{\partial}{\partial t_j}1_{[0,t]})$. The following proposition shows that differentiability has properties analogous to those for integrability.

Proposition 2

Suppose $\phi:\mathbb{R}^N\to(L^2)^-$ is a vector–valued function with representation $\phi = \Sigma_{n=0}^{\infty}\phi_n$ with $\phi_n = I_n(f_n)$. Then

(a) ϕ is differentiable if and only if

(i) ϕ_n is differentiable $\forall$ n

(ii) For each j, $\Sigma_{n=0}^{\infty}\frac{\partial\phi}{\partial t_j}(t)$ is in $(L^2)^-$

In this case $\frac{\partial\phi}{\partial t_j}(t) = \Sigma_{n=0}^{\infty}\frac{\partial\phi}{\partial t_j}(t)$

(b) ϕ_n is differentiable if and only if f_n is differentiable. In this case

$$\frac{\partial}{\partial t_j}\phi_n(t) = \frac{\partial}{\partial t_j}I_n(f_n(t)) = I_n\frac{\partial}{\partial t_j}(f_n(t))$$

(c) If ϕ is differentiable then $\forall\ \xi\in C_c^{\infty}(\mathbb{R}^N)$, $S(\phi)(\xi)$ is differentiable and

$$S(\frac{\partial\phi}{\partial t_j}(t))(\xi) = \frac{\partial}{\partial t_j}[S(\phi(t))(\xi)] .$$

(d) The product rule holds:

$$\frac{\partial}{\partial t_j}[\phi(t)\hat{\otimes}\zeta(t)] = \frac{\partial\phi}{\partial t_j}(t)\hat{\otimes}\zeta(t) + \phi(t)\hat{\otimes}\frac{\partial\zeta}{\partial t_j}(t)$$

whenever the product $\phi(t)\hat{\otimes}\zeta(t)$ is everywhere defined and each of the factors is differentiable.

Examples. We give here some examples, applications, and formulas for this differential and integral calculus of $(L^2)^-$–valued functions. The basic theme is that all the usual rules and formulas from calculus apply (provided one uses $\hat{\otimes}$ for the product).

(1) For $W^{\hat{\otimes}n}(t) \equiv W(t)\hat{\otimes}\cdots\hat{\otimes}W(t)$ (n–times)

$$\frac{\partial}{\partial t_j}W^{\hat{\otimes}n}(t) = n\,W^{\hat{\otimes}n-1}(t)\hat{\otimes}\frac{\partial W}{\partial t_j}(t)$$

The formula holds more generally for any differentiable $\phi : \mathbb{R}^N \to (L^2)^-$ provided $\phi^{\hat{\otimes} n}$ is defined.

(2) Suppose $\phi(t) = I_1(f(t)) = \langle \cdot, f(t) \rangle$ with $f(t) \in L^2(\mathbb{R}^N) \subseteq H_1^-$ and assume that f is differentiable in the above sense. Since $\phi(t)$ is in $L^2(\mathscr{S}^*)$ one can consider its nth power $\phi(t)^n = \phi^n(t)$ (ordinary product of $\phi(t)$ with itself n–times). Then ϕ^n is differentiable and $\frac{\partial}{\partial t_j} \phi^n$ is easily computed by decomposing ϕ^n into sums of multiple Wiener integrals:

$$\phi^n(t) = \Sigma_{k=0}^{[n/2]} \frac{n!}{(n-2k)!k!2^k} a(t)^k \phi^{\hat{\otimes} n-2k}(t) ,$$

where $a(t) = \langle \phi(t), \phi(t) \rangle = \langle f(t), f(t) \rangle$ (and is assumed to be differentiable). For example

$$W^2(t) = (t_1..t_N) + W(t) \hat{\otimes} W(t)$$

so

$$\frac{\partial}{\partial t_j} W^2(t) = \frac{\partial}{\partial t_j}(t_1..t_N) + 2W(t) \hat{\otimes} \frac{\partial W}{\partial t_j}(t) .$$

(3) A Fubini–type theorem holds, namely the integral: $\int \phi(t)dt = \int \phi(t_1,...,t_N)dt_1...dt_N$ can be viewed as an iterated integral and the order of integration does not affect the result.

(4) $\int_{[a,b]} \frac{\partial \phi}{\partial t_j}(t)dt = \int_{[a,b]_j} [\phi(t_1,...,b_j,...,t_N) - \phi(t_1,...,a_j,...,t_N)](dt)_j$

(5) As an application of (4) one has

$$\int_{[a,b]} \frac{\partial^N \phi(t)}{\partial t_1 \ldots \partial t_N} dt = \Delta_{[a,b]}\phi = \phi(t) \Big|_{a_1}^{b_1} \ldots \Big|_{a_N}^{b_N} .$$

In particular for $N = 2$, consider the 2–time parameter Wiener process $W(t) = \langle \cdot, 1_{[0,t]} \rangle$ $t = (t_1, t_2)$. Then the Pettis calculus gives

$$\int_{[a,b]} \frac{\partial^2 W}{\partial t_1 \partial t_2}(t)dt = \Delta_{[a,b]}W$$
$$= W(b_1,b_2) - W(a_1,b_2) + W(a_1,a_2) - W(b_1,a_2)$$
$$= \langle \cdot, 1_{[a,b]} \rangle = \int_{[a,b]} dW(t) .$$

The latter integral is the Ito integral with respect to W. To put this in perspective recall that for an ordinary function W of two variables (with suitable properties) the Stieltjes measure of a rectangle [a,b] is defined to be the number $\Delta_{[a,b]}W$. (This is motivated by the special case when W is the product function: $W(t_1,t_2) = t_1 t_2$).When W is differentiable, the measure on $\mathbb{R}^2$ generated by W is $(\partial^2 W / \partial t_1 \partial t_2)dt$.

5. Vector–valued differential forms

There is a standard calculus [4,5,6] from differential geometry for differential forms with values in a vector space V, which we adopt here for our case $V = (L^2)^-$. However, since $(L^2)^-$ is an infinite dimensional vector space and our Pettis type notion of smoothness (differentiability) of vector–valued functions $\phi:\mathbb{R}^N \to (L^2)^-$ is not standard, we must take care to appropriately interpret the various constructions.

For a manifold R, let $F(R,V)$ denote the set of vector–valued functions $\phi : R \to V$. These are the 0 forms on R with values in V. A differential p–form θ on R with values in V is generally regarded as a function (section) $\theta : R \to A^p(R,V)$ of a certain vector bundle $A^p(R,V)$ over R. Thus $\theta(t) = (t,\theta_t)$ where θ_t is a V–valued, multilinear, antisymmetric function on the tangent space T_tR:

$$\theta_t : T_tR \times \ldots \times T_tR \to V$$

(with p factors in the cartesian product). For vector fields $X^1,\ldots,X^p$ on R one gets a V–valued function $\theta(X^1,\ldots,X^p) : R \to V$ defined by $\theta(X^1,\ldots,X^p)(t) \equiv \theta_t(X^1_t,\ldots,X^p_t)$. Thus θ can also be viewed as a map $\theta : \Gamma(TR) \times \ldots \times \Gamma(TR) \to F(R,V)$ with is antisymmetric and multilinear with respect to the $C^\infty(R)$–module structure of $\Gamma(TR)$. On a coordinate chart U with coordinate functions $t_j:U \to \mathbb{R}$, the local vector fields $\partial/\partial t_j$ give the components:

$$\theta_{j_1\cdots j_p} \equiv \theta(\frac{\partial}{\partial t_{j_1}},\ldots,\frac{\partial}{\partial t_{j_p}}) \in F(R,V)$$

of θ and locally θ has the representation:

$$\theta = \theta_{j_1\cdots j_p}\, dt_{j_1} \wedge\ldots\wedge dt_{j_p} ,$$

where there is implied summation over repeated indices $j_1,\ldots,j_p$.

Usually one wishes to have some degree of differentiability (or C^∞–smoothness) in the component functions $\theta_{j_1\cdots j_p}$, and for finite dimensional vector spaces V this can be defined intrinsically by either putting a smooth manifold structure on $A^p(R,V)$ and requiring that $\theta : R \to A^p(R,V)$ be a smooth map, or alternatively requiring that $\theta(X^1,\ldots,X^p)$ be a smooth vector–valued function for each choice of vector fields $X^1,\ldots,X^p$. For our case with $V = (L^2)^-$ we interpret the smoothness of $\theta(X^1,\ldots,X^p)$ to be in the Pettis sense (with the order of differentiability, C^k, $k = 0\ 1,\ldots,\infty$, being specified or inferred from the context). With these preliminaries out of the way, we now specialize to the case $R = \mathbb{R}^N$ and $V = (L^2)^-$ (and $t_j:\mathbb{R}^N \to \mathbb{R}$ is projection on the jth factor).

The definition of the exterior derivative $d\theta$ is the standard one once we specify the meaning of the Lie derivative $X(\phi)$ of a 0–form $\phi : \mathbb{R}^N \to (L^2)^-$ along a vector field $X = X_j \frac{\partial}{\partial t_j}$. For this we naturally take:

$$X(\phi) \equiv X_j \frac{\partial \phi}{\partial t_j}$$

Then for 0–forms, $d\phi(X) \equiv X(\phi)$ so that

$$d\phi = \frac{\partial \phi}{\partial t_j} dt_j .$$

For p–forms θ

$$d\theta(X^1,\dots,X^{p+1}) \equiv \sum_{i=1}^{p+1} (-1)^{i+1} X^i(\theta(X^1,\dots,\hat{X}^i,\dots,X^{p+1}))$$
$$+ \sum_{i<j} (-1)^{i+j} \theta([X^i,Y^j],X^1,\dots,\hat{X}^i,\dots,\hat{X}^j,\dots,X^{p+1})$$

In terms of coordinates

$$d\theta = \frac{\partial}{\partial t_k} (\theta_{j_1 \dots j_p})\, dt_k \wedge dt_{j_1} \dots \wedge dt_{j_p}$$

The definition of the exterior product $\theta \wedge \tau$ of θ with a V–valued k–form τ requires a product operation on V. Thus on $(L^2)^-$ we use the natural product $\hat{\otimes}$ (where defined) and define

$$(\theta \wedge \tau)(X^1,\dots,X^{p+k}) = \Sigma\, (-1)^\sigma\, \theta(X^{\sigma_1},\dots,X^{\sigma_p}) \,\hat{\otimes}\, \tau(X^{\sigma(p+1)},\dots,X^{\sigma(p+k)})$$

the summation being over all permutations σ of $\{1,\dots,p+k\}$. In terms of coordinates:

$$\theta \wedge \tau = \theta_{j_1 \dots j_p} \,\hat{\otimes}\, \tau_{i_1 \dots i_k}\, dt_{j_1} \wedge \dots \wedge dt_{j_p} \wedge dt_{i_1} \wedge \dots \wedge dt_{i_k} .$$

For a 0–form ϕ, $\phi \wedge \tau \equiv \phi \,\hat{\otimes}\, \tau$.

6. Integrals over submanifolds–Stokes' theorem

Stokes' theorem for differential p–forms θ with values in $(L^2)^-$ follows easily from the traditional Stokes' theorem, once we define the notion of the integral of θ over a p–chain Ω in $\mathbb{R}^N$. For this note that if $\psi \in C$, then one obtains an ordinary ($\mathbb{R}$–valued) p–form $\langle \theta,\psi \rangle(X^1,\dots,X^p)(t) \equiv \langle \theta(X^1,\dots,X^p)(t),\psi \rangle$. In terms of coordinates one has $\langle \theta,\psi \rangle = \langle \theta_{j_1 \dots j_p},\psi \rangle\, dt_{j_1} \wedge \dots \wedge dt_{j_p}$.

Definition: The p–form θ is said to be integrable over the p–chain Ω if

(i) $\langle \theta,\psi \rangle$ is integrable over Ω $\forall\, \psi \in C$

(ii) there exists an element of $(L^2)^-$, denoted by $\int_\Omega \theta$, such that

$$\langle \textstyle\int_\Omega \theta,\psi \rangle = \int_\Omega \langle \theta,\psi \rangle$$

for every $\psi \in C$.

Theorem 1. (Stokes' Theorem)

Suppose θ is a p–form and Γ is a $p+1$ chain then

$$\int_\Gamma d\theta = \int_{\partial\Gamma}\theta .$$

Proof: The identity $d\langle\theta,\psi\rangle = \langle d\theta,\psi\rangle$ follows easily from the definitions. Thus by the usual Stokes' theorem

$$\langle\int_\Gamma d\theta,\psi\rangle = \int_\Gamma\langle d\theta,\psi\rangle = \int_\Gamma d\langle\theta,\psi\rangle = \int_{\partial\Gamma}\langle\theta,\psi\rangle = \langle\int_{\partial\Gamma}\theta,\psi\rangle .$$

Hence since C is dense in $(L^2)^+$ we get that $\int_\Gamma d\theta = \int_{\partial\Gamma}\theta$.

To illustrate the computation of $\int_\Omega\theta$ and the application of Stokes' theorem in particular examples, let us assume that Ω is a submanifold parameterized by $\alpha{:}R \subseteq \mathbb{R}^p \to \mathbb{R}^N$, $\Omega = \alpha(R)$. The pullback $\alpha^*(\theta)$ of θ by α then gives a p–form on $\mathbb{R}^p$ with values in $(L^2)^-$:

$$\alpha^*(\theta)(Y^1,\dots,Y^p)(r) \equiv \theta_{\alpha(r)}(d\alpha_r\, Y^1_r,\dots,d\alpha_r\, Y^p_r) .$$

In terms of coordinates this is simply

$$\alpha^*(\theta) = \alpha^*(\theta_{j_1\cdots j_p}\, dt_{j_1}\wedge\dots\wedge dt_{j_p})$$

$$= (\theta_{j_1\cdots j_p}\circ\alpha)\,(\partial\alpha_{j_1}/\partial r_{i_1}\, dr_{i_1})\wedge\dots\wedge(\partial\alpha_{j_p}/\partial r_{i_p} dr_{i_p})$$

$$= (\theta_{j_1\cdots j_p}\circ\alpha)\det(D\alpha_{j_1\cdots j_p})\, dr_1\dots dr_p$$

where $D\alpha_{j_1\cdots j_p}$ denotes the Jacobian matrix of the map $\alpha_{j_1\cdots j_p}{:}R\to\mathbb{R}^p$ defined by $\alpha_{j_1\cdots j_p}(r) \equiv (\alpha_{j_1}(r),\dots,\alpha_{j_p}(r))$. Thus since $\alpha^*\langle\theta,\psi\rangle = \langle\alpha^*\theta,\psi\rangle$ one gets

$$\langle\int_\Omega \theta,\psi\rangle = \int_\Omega\langle\theta,\psi\rangle = \int_{\alpha(R)}\langle\theta,\psi\rangle = \int_R\langle\alpha^*\theta,\psi\rangle$$

$$= \int_R\langle\theta_{j_1\cdots j_p}\circ\alpha\,,\psi\rangle\det(D\alpha_{j_1\cdots j_p})\, dr_1\dots dr_p .$$

This shows that $\int_\Omega\theta$ coincides with the integral of a vector–valued function:

$$\int_\Omega\theta = \int_R\theta_{j_1\cdots j_p}(\alpha(r))\det(D\alpha_{j_1\cdots j_p}(r))\, dr$$

and this could be used as an alternative definition of $\int_\Omega\theta$.

Example The differential 1–form

$$\theta = \phi_j(t)\,\hat{\otimes}\,\frac{\partial W}{\partial t_j}(t)\, dt_j$$

with $\phi_j:\mathbb{R}^N \to (L^2)^-$, is a primary example of an $(L^2)^-$ valued 1–form (a white noise 1–form). One can show that the corresponding white noise path integral (when it exists):

$$\int_C \theta = \int_{r_0}^{r_1} \phi_j(\alpha(r)) \mathbin{\hat{\otimes}} \frac{\partial W}{\partial t_j}(\alpha(r))\, \alpha_j'(r)dr$$

generalizes the stochastic path integral of Cairoli and Walsh[6] and to a certain extent is a natural analog of the Ito integral for $N = 1$ time parameters. More specifically:

(1) For $N = 1$ and $\phi(t) \in (L^2)$ a non–anticipating process, a result of Kubo shows that $\int \phi(t) \otimes W'(t)dt$ coincides with the Ito integral.

(2) The exterior derivative of the Wiener process W is $dW = \frac{\partial W}{\partial t_j} dt_j$, and by Stokes' theorem

$$\int_{C_{a,b}} dW(t) = W(t)|_a^b$$

for any path $C_{a,b}$ between the points $a,b \in \mathbb{R}^N$. It is also instructive, rather than using Stokes' theorem, to verify this directly from the definitions.

(3) More generally

$$\int_{C_{a,b}} W^{\hat{\otimes} n}(t) \mathbin{\hat{\otimes}} dW(t) = \frac{1}{n+1} W^{\hat{\otimes} n+1}(t)\, |_a^b$$

(4) On the otherhand, using a formula from Ref.1:

$$\int_{C_{a,b}} W^n(t) \mathbin{\hat{\otimes}} dW(t) = \frac{1}{n+1} W^{n+1}(t)\, |_a^b - \frac{1}{2} \int_{C_{a,b}} nW^{n-1}(t) \frac{\partial}{\partial t_j}(t_1 \ldots t_N)dt_j$$

(5) Let $Y(t) = e^{W(t) - \frac{1}{2}(t_1 \ldots t_N)} = \sum_{n=0}^{\infty} \frac{W(t)^{\hat{\otimes} n}}{n!}$

Then $Y(t)$ satisfies the stochastic PDE

$$\frac{\partial Y}{\partial t_j}(t) = Y(t) \mathbin{\hat{\otimes}} \frac{\partial W}{\partial t_j}(t)$$

In terms of exterior derivatives $dY = Y \hat{\otimes} dW$ and so by Stokes' Theorem:

$$\int_{C_{a,b}} Y(t) \mathbin{\hat{\otimes}} dW(t) = Y(t)|_a^b$$

(6) To see that the white noise path integral

$$\int_C \theta = \int_C \phi_j(t) \mathbin{\hat{\otimes}} \frac{\partial W}{\partial t_j}(t)dt_j \in (L^2)^-$$

is a generalization of the Cairoli and Walsh stochastic path integral (which they only define for $N = 2$), we note that if each $\phi_j(t)$ is a stochastic process, *i.e.*, $\phi_j(t) \in (L^2)$, then $\int_C \theta \in (L^2)$ as well. To prove this it suffices to just consider the case when each $\phi_j = I_n(f_j) \in K_n$. Then for $\psi = I_n(\xi)$ one finds

$$\int_C\langle\theta,\psi\rangle = \int_{r_0}^{r_1} \langle\, \theta_j(\alpha(r)) \hat{\otimes} \frac{\partial W}{\partial t_j}(\alpha(r)),\, \psi\,\rangle \alpha_j'(r)dr$$

$$= \sum_{j=1}^{N} \int_{\mathbb{R}^{Nn}}\int_{r_0}^{r_1} f_j(\alpha(r),z)\int_{[0,\alpha(r)]_j}\xi(u_1,\ldots,\alpha_j(r),\ldots,u_N,z)(du)_j\alpha_j'(r)drdz$$

$$= \sum_{j=1}^{N} \int_{\mathbb{R}^{N(n+1)}} f_j^{\alpha}(u,z)\xi(u,z)dz\ .$$

Here f_j^{α} is defined as follows: Let

$$M_{j\,r}^{\alpha} = [0,\alpha_1(r)]\times\ldots\times\{\alpha_j(r)\}\times\ldots\times[0,\alpha_N(r)]$$

and let $\mathrm{sgn}(x) = x/|x|$ be the sign function. Then

$$f_j^{\alpha}(u,z) \equiv \begin{cases} \mathrm{sgn}(\alpha'(t))\, f_j(\alpha(r),z) & \text{if } u \in M_{j\,r}^{\alpha} \text{ for some } r \in [r_0\,,\, r_1] \text{ and } \alpha_j'(r) \neq 0 \\ 0 & \text{otherwise} \end{cases}$$

Thus one finds that

$$\int_C \theta = I_{n+1}(\Sigma_{j=1}^{N} f_j^{\alpha})\ .$$

Concluding Remarks: We have investigated white noise differential forms of other degrees as well, for example the surface integrals:

$$\int_S \phi_{ij}(t) \hat{\otimes} \frac{\partial^2 W}{\partial t_i\, \partial t_j}(t)\ dt_i \wedge dt_j$$

$$\int_S \phi_{ij}(t) \hat{\otimes} \frac{\partial W}{\partial t_i}(t) \hat{\otimes} \frac{\partial W}{\partial t_j}(t)\ dt_i \wedge dt_j$$

The former has many properties analogous to those for the line integral in the above example. The degree = N case is noteworthy since then

$$\int_{\mathbb{R}} \phi(t) \otimes \frac{\partial^N W}{\partial t_1 \cdots \partial t_N}(t)\ dt$$

(with $dt = dt_1\wedge\ldots\wedge dt_N$) coincides with the Ito integral of ϕ, when ϕ is non-anticipating. Various other choices for the integrand in the degree N case yield interesting stochastic integrals as well. This analysis is presented in another paper.

REFERENCES

[1] Mylan Redfern and David Betounes, "A generalized Ito formula for N–dimensional time." (Part of these proceedings).

[2] J.Diestel and J.J.Uhl, *Vector Measures*, AMS Mathematical Surveys,no.**15** (American Math. Soc.,Providence,1977).

[3] Francois Treves, *Topological Vector Spaces, Distributions, and Kernals*, (Academic, New York,1977).

[4] D. Bleecker, *Gauge Theory and Variational Principles*, (Addison–Wesley, 1981).

[5] M. Modugno, "Systems of vector–valued forms on a manifold and applic–ations to gauge theories," Lect. Notes. Math., vol.**1251** (1982).

[6] Gabor Toth, *Harmonic and Minimal Maps with Applications to Physics*, (Wiley,New York, 1985).

[7] R. Cairoli and J.B. Walsh, "Stochastic integrals in the plane," Acta Math. **134** (1975),111–183.

On Sharp L^p Estimates for Iterated Stochastic Integrals and Some Related Topics

by

Eric A. Carlen[1] and Paul Krée[2]

[1]Department of Mathematics, Princeton Univ., Princeton NJ 08544 U.S.A.

[2]Départemant de Mathématiques, Paris VI, 75252 Paris cedex 05 France

Abstract: We sketch our recent proof of certain L^p estimates for iterated stochastic integrals. We then briefly discuss applications of our result; and as an illustration of our methods, we give a new proof of Nelson's sharp hypercontractivity inequality in the framework of our investigation.

Let M_t be a continuous real valued martingale on some filtered probability space $(\Omega, \mathcal{B}_t, Pr)$ with quadratic variation process $\langle M, M\rangle_t$. Define the iterated stochastic integrals $I_n(t)$ by

$$I_n(t) = \int_0^t I_{n-1}(s)dM_s \quad I_0(t) = 1 \tag{1}$$

for all $n \geq 1$.

We begin by sketching a proof of the following inequalities which we have proved in [1]:

$$A_{n,p}\|\langle M, M\rangle_t^{1/2}\|_{np}^n \leq \|I_n(t)\|_p \leq B_{n,p}\|\langle M, M\rangle_t^{1/2}\|_{np}^n \tag{2}$$

where the left side holds for all $p > 1$ and the right side holds for all $p \geq 1$ and $A_{n,p}$ and $B_{n,p}$ are both asymptotic to $-\frac{n}{2}\ln n$ for large n. Clearly then these inequalities are qualitatively sharp.

Our result is an extension of the Burkholder-Davis-Gundy inequalities [2][3]:

$$a_p\|\langle M, M\rangle_t^{1/2}\|_p \leq \|M_t\|_p \leq b_p\|\langle M, M\rangle_t^{1/2}\|_p \tag{3}$$

where again the left side holds for all $p > 1$, and the right side holds for all $p \geq 1$.

The inequalities (3) are closely related to some moment inequalities of Zakai [4] for stochastic integrals with respect to a Brownian motion B_t; one part of these is

$$(E|\int_0^t \alpha_s dB_s|^q)^{1/q} \leq (q-1)^{1/2}(\int_0^t (E|\alpha_s|^q)^{2/q} ds)^{1/2} \tag{4}$$

for all adapted processes α_t and all $q \geq 2$.

Since $\langle \int_0^\cdot \alpha_s dB_s, \int_0^\cdot \alpha_s dB_s \rangle_t = \int_0^t \alpha_s^2 ds$, the qualatative form of Zakai's inequalities follows from (3) by an application of the Ingham-Minkowski inequality (see p.31 of [5]):

$$(\int_Y (\int_X |f(x,y)|^p dm(x))^{q/p} dn(y))^{1/q} \leq (\int_X (\int_Y |f(x,y)|^q dn(y))^{p/q} dm(x))^{1/p}$$

for all $0 < p \leq q$, all measure spaces (X, dm),(Y, dn), and all jointly measurable functions f on $(X \times Y, dm \otimes dn)$. However, such a derivation leads to a constant larger than $(q-1)^{1/2}$ in (4) even when one uses the sharp constant b_p in (3) due to Davis [2]. Suprisingly, Zakai does not note that his constant $(q-1)^{1/2}$ is sharp. We will show here that not only is it sharp, but that within the framework of our proof of (2), it implies Nelson's sharp hypercontractivity inequality for the Ornstein-Uhlenbeck semigroup $\{P_t | t \geq 0\}$:

$$\|P_t f\|_{L^q(\mathbf{R}, d\mu)} \leq \|f\|_{L^2(\mathbf{R}, d\mu)} \quad \forall q \geq 2$$

for all t with $e^{-t} \leq (q-1)^{1/2}$. Here $d\mu(x) = \frac{1}{\sqrt{2\pi}} e^{-x^2/2} dx$ and

$$P_t f(x) = \int_{\mathbf{R}} f(e^{-t}x + (1-e^{-2t})^{1/2} y) d\mu(y) \quad .$$

While the result is classic, the proof is new and illustrates our methods.

We now turn to our proof of (2). We will make essential use of the fact that the sharp constant b_p in (3) is asymptotic to $\sqrt{p}$ for large p. In fact, we have proved in the appendix to [1] that

$$sup\{b_p/\sqrt{p} \quad | \quad 1 \leq p < \infty\} = 2 \tag{5}$$

and so we have the simple explicit bound $b_p \leq 2\sqrt{p}$. The sharp constant a_p in (3) is much less important to us. While we must apply the right side of (3) n times to estimate $B_{n,p}$ in (2), we apply the left side of (3) only once, together with our estimate for $B_{n,p}$, to estimate $A_{n,p}$.

To prove the right side of (2), we start from the Kailath-Segall identity for continuous martingales [7]:

$$I_n(t) = \frac{1}{n}\left(I_{n-1}(t)M_t - I_{n-2}(t)\langle M,M\rangle_t\right) \quad . \tag{6}$$

By Hölder's inequality with $\frac{1}{p} = \frac{n-1}{np} + \frac{1}{np}$ and then (3) and (5)

$$\|I_{n-1}(t)M_t\|_p \le \|I_{n-1}(t)\|_{np/(n-1)}\|\langle M,M\rangle_t^{1/2}\|_{np}$$

$$\le 2\sqrt{np}\|I_{n-1}(t)\|_{np/(n-1)}\|\langle M,M\rangle_t^{1/2}\|_{np} \quad .$$

By Hölder's inequality with $\frac{1}{p} = \frac{n-2}{np} + \frac{2}{np}$,

$$\|I_{n-2}(t)\langle M,M\rangle_t\|_p \le \|I_{n-2}(t)\|_{np/(n-2)}\|\langle M,M\rangle_t\|_{np/2} \quad .$$

Thus

$$\|I_n(t)\|p \le \left(\frac{2p}{\sqrt{n}}\|\langle M,M\rangle_t^{1/2}\|_{np}\|I_{n-1}(t)\|_{np/(n-1)} + \left(\frac{1}{\sqrt{n}}\|\langle M,M\rangle_t^{1/2}\|_{np}\right)^2\|I_{n-2}(t)\|_{np/n-2}\right) \quad .$$

Notice that on account of (5) and (6), each factor of $\|\langle M,M\rangle_t^{1/2}\|_{np}$ on the right comes with a factor of $\frac{1}{\sqrt{n}}$ and a constant independent of n. It is now easy to see that an induction argument leads to the result we desire. Carefully carrying out this argument, one obtains

$$\|I_n\|_p \le \frac{1}{n!}(np)^{n/2}\left(1+(1+1/p)^{1/2}\right)^n\|\langle M,M\rangle_t^{1/2}\|_{np}^n \quad .$$

Many proofs of the Burkholder-Davis-Gundy inequalities give an estimate on b_p of order p instead of $\sqrt{p}$ for large p. Such an estimate would give us no decay at all in $B_{n,p}$ with increasing n. The $\sqrt{p}$ behavior was first explicitly obtained by Barlow and Yor [3], though of course it is implicit in Davis's identification of b_p as the last zero of a certain parabolic cylinder function of order p.

To prove the left side of (2), we first prove an identity

$$I_n(t)I_{n-2}(t) = I_{n-1}^2(t) - \sum_{m=1}^{n}\frac{(n-m)!}{n!}I_{n-m}^2(t)\langle M,M\rangle_t^{m-1} \quad \forall n \ge 2$$

for iterated stochastic integrals of continuous martingales. Like all stochastic integral identities, it follows from Ito's formula, though one can shorten the proof by exploiting its relation to (6).

The main virtue of this identity is that all of the terms in the sum on the right are positive so that there is no cancelation in the sum. Keeping only the first and last terms in the sum, and integrating with respect to $d\langle M,M\rangle_t$ leads to

$$\frac{1}{n!}\langle M,M\rangle_t^n \leq (n-1)\langle I_n,I_n\rangle_t - n\int_0^t I_n(s)I_{n-2}(s)d\langle M,M\rangle_s$$

so that

$$\frac{\langle M,M\rangle_t^{n/2}}{\sqrt{n!}} \leq \sqrt{n-1}\langle I_n,I_n\rangle_t^{1/2} + \sqrt{n}(I_n^*(t)I_{n-2}^*(t)\langle M,M\rangle_t^{1/2})^{1/2}$$

where as usual, the star denotes the maximal process.

By the Schwarz inequality, Doob's inequality, and the right side of (2) together with simplifying estimates on the constants we arrive at the quadratic inequality

$$\frac{1}{\sqrt{n!}}\|\langle M,M\rangle_t^{1/2}\|_{np}^n \leq \frac{\sqrt{n-1}}{a_p}\|I_n(t)\|_p + \sqrt{n}(\frac{p}{p-1})B_{n-2,p}^{1/2}\|I_n(t)\|_p^{1/2}\|\langle M,M\rangle_t^{1/2}\|_{np}^{n/2} \quad .$$

Since $B_{n-2,p}^{1/2}$ is asymptotic to $(n!)^{-1/4}$, we easily conclude that the left side of (2) holds with a constant asymptotic to $\frac{1}{\sqrt{n!}}$. The details can be found in [1].

The inequalities (2) can be used to obtain necessary and sufficient conditions for the convergence of

$$\sum_{n=0}^{\infty} I_n(t)$$

in L^p. This is of interest because the sum is the exponential martingale

$$e^{M_t - \frac{1}{2}\langle M,M\rangle_t}$$

which is the solution to the stochastic differential equation

$$dZ_t = Z_t dM_t \ , \qquad Z_0 = 1 \quad .$$

Similar equations where M_t takes values in a non trivial Lie algebra arise in mathematical physics. In these cases, an explicit solution is generaly absent, and estimates on the series $\sum_{n=0}^{\infty} I_n(t)$ are then very helpful. We have such applications in progress, and shall report on them elsewhere.

We conclude this note by proving Nelson's sharp hypercontractivity inequality within the framework of our proof of (2). The Hermite polynomials provide us with the connection between the two results.

First, it is well known that the Hermite polynomials $h_n(x)$ given by

$$h_n(x) = e^{x^2/2}(-\frac{d}{dx})^n e^{-x^2/2} \quad \forall n \geq 0$$

are an orthonormal basis of $L^2(\mathbf{R}, d\mu)$ and that $P_t h_n(x) = e^{-nt} h_n(x)$ for all n and t.

Second, it is well known that if we take a standard Brownian motion B_t for M_t in (1), we have

$$I_n(t) = \frac{1}{n!} H_n(B_t, t)$$

where $H_n(x,t) = (\sqrt{t})^n h_n(\frac{x}{\sqrt{t}})$. Thus

$$\|h_n\|_{L^q(\mathbf{R},d\mu)} = \|I_n(1)\|_q \leq B_{n,q}$$

since $\langle B, B\rangle_1 = 1$ a.s.

For $p > 2$ we can obtain a better estimate by using Zakai's inequality (4) applied inductively to $I_n(t) = \int_0^t I_{n-1}(s) dB_s$ which is possible since the quadratic variation process of B_t is deterministic. We clearly obtain

$$\|h_n\|_{L^q(\mathbf{R},d\mu)} = \|I_n(1)\|_q \leq (q-1)^{1/2} \quad .$$

Now suppose $f(x) = \sum_{n=0}^{\infty} c_n h_n(x)$ where $\sum_{n=0}^{\infty} c_n^2 = 1$. Then

$$\|P_t f\|_{L^q(\mathbf{R},d\mu)} \leq \sum_{n=0}^{\infty} e^{-nt} c_n \|h_n\|_{L^q(\mathbf{R},d\mu)}$$

$$\leq \sum_{n=0}^{\infty} e^{-nt} c_n (q-1)^{n/2} \leq \Big(\sum_{n=0}^{\infty} (e^{-2nt}(q-1)^n)\Big)^{1/2} \quad .$$

Of course this last series converges provided $e^{-2t}(q-1) < 1$. Thus, for such t, $P_t : L^2(\mathbf{R}, d\mu) \to L^q(\mathbf{R}, d\mu)$ is bounded. Let C denote the norm. Now since P_t is positivity preserving, Segal's lemma [8] implies that $P_t \otimes P_t : L^2(\mathbf{R}^2, d\mu \otimes d\mu) \to L^q(\mathbf{R}^2, d\mu \otimes d\mu)$ is bounded with norm C^2. Considering functions of the form $f \otimes 1$ we have

$$\|(P_t \otimes P_t) f \otimes 1\|_{L^q(\mathbf{R}^2, d\mu\otimes d\mu)} = \|P_t f\|_{L^q(\mathbf{R},d\mu)}$$

since $P_t 1 = 1$. We conclude that $C \leq C^2$.

However, any function F on $\mathbf{R}^2$ is $d\mu \otimes d\mu$ equimeasurable to a unique function F^o of the form $f \otimes 1$ with f monotone decreasing, and by a well known Gauss space rearrangement inequality

$$\|P_t \otimes P_t F\|_{L^q(\mathbf{R}^2, d\mu \otimes d\mu)} \leq \|P_t \otimes P_t F^o\|_{L^q(\mathbf{R}^2, d\mu \otimes d\mu)} = \|f\|_{L^q(\mathbf{R}, d\mu)} \quad .$$

A proof is given in the appendix to [9]. From this we easily conclude that $C^2 \leq C$.

Therefore we must have $C = 1$ so that $\|P_t f\|_{L^q(\mathbf{R}, d\mu)} \leq \|f\|_{L^2(\mathbf{R}, d\mu)}$ for all t such that $e^{-2t}(q-1) < 1$. Considering first bounded functions f, a simple dominated convergence argument shows that P_t is also contraction when $e^{-2t}(q-1) = 1$. Segal's lemma now extends the result to the mutidimensional case $P_t \otimes \cdots \otimes P_t$ as usual.

References

[1] Carlen, E. and Krée, P.: Some sharp L^p estimates for iterated stochastic integrals, Ann. of Prob., in press.

[2] Davis, B.: On the L^p norms of stochastic integrals and other martingales, Duke Math. Jour., **43** (1976) 697-704

[3] Barlow, M. and Yor, M.: Semimartingale inequalities via the Garsia-Rodemich-Rumsey lemma, Jour. Func. Anal., **49** (1982) 198-229

[4] Zakai, M.: Some moment inequalities for stochastic integrals and for solutions of stochastic differential equations, Israel Jour. Math., **3** (1967) 170-177

[5] Hardy, G., Littlewood, J.E. and Polya, G.: *Inequalities*, Cambridge Univ. Press, Cambridge, 1959

[6] Nelson, E.: The free Markov field, Jour. Func. Anal., **12** (1973) 221-227

[7] Kailath, T. and Segall, A.: Orthogonal functions of independent increment processes, IEEE Trans. on Info. Theory, **22** (1976) 287-298

[8] Segal, I.: Construction of non-linear local quantum processes: I, Ann. of Math., **92** (1970) 462-481

[9] Carlen, E. and Loss, M.: Extremals of functionals with competing symmetries, Jour. Func. Anal., **88** (1990) 335-355

INTRINSIC CLOCK AS A TOOL FOR PATH INTEGRATION: AN EXAMPLE

DOMENICO P. L. CASTRIGIANO

and

FLORIAN STÄRK

Mathematisches Institut
Technische Universität München, D-8000 München 2, F.R.G.

ABSTRACT

The intrinsic clock of a continuous local martingale is emphasized as the basic stochastic fact underlying the new time transformation introduced by physicists as a tool for path integration. It is used to derive equalities of measures on path spaces, which contain the functional integrational essence of the computations of the Coulomb propagator via the Kustaanheimo-Stiefel transformation and via the radial path integral expansion as well. In deriving the latter some new insight is gained into its relation to Bessel processes.

1. Introduction

Two ways are known how to solve the path integral in configuration space for the (Euclidean) Coulomb propagator. The first one uses the Kustaanheimo-Stiefel transformation (KS) [1-6]. The other one exploits the spherical symmetry directly by a partial wave expansion of the Coulomb path integral and applies the square map, which is the one-dimensional KS-transformation, to the radial path integrals [7-9]. In both cases, a subsequent "new time" transformation adapted to the KS-transformations converts the transformed path integrals to Gaussian ones, the evaluation of which is standard. We want to point out that the actual path integrational content of both approaches is concentrated in two equalities of measures on the path spaces. These equalities can be derived straightforwardly in the language of stochastic differential equations introducing the concept of the intrinsic clock. By this a continuous local martingale (M_t) can be considered as a Brownian motion which runs according to an altered time. This time is the intrinsic clock of (M_t), which turns out to be the quadratic variation process of (M_t). Putting it the other way, a continuous local martingale becomes a Brownian motion if run according to the inverse of the intrinsic clock. This fact is the probabilistic background of the so-called new time technique developed by physicists, see mainly Duru / Kleinert [1] and Pak / Sokmen [10]. It was applied successfully also to other potentials, see Refs. 11, 12, and the literature cited therein. The correspondence between time changes in stochastic differential equations and in path integrals was pointed out by Blanchard / Sirugue [3] and Young / DeWitt-Morette [5]. Our intention is to investigate the relations between the measures on path spaces which follow from the procedure of space and time transforming. Besides, the probabilistic content of the radial path integral treatment is elucidated considering the mutual densities of Bessel path measures.

2. Intrinsic clock of continuous local martingales

The key result, on which our computations rely, is the following theorem of Dambis, Dubins-Schwarz (DDS), see e.g. Ref. 13, Chap. 3.4.:
Let $(M_t, \mathcal{F}_t, P)_{t\geq 0}$ be a continuous local martingale in $\mathbb{R}$ with quadratic variation process $\langle M\rangle_t$, such that

$$\lim_{t\to\infty} \langle M\rangle_t = \infty \qquad \text{P.a.s.} \tag{1}$$

Define the family of stopping times for $u \geq 0$

$$\tau(u) := \inf\left\{t \geq 0 \mid \langle M\rangle_t > u\right\} \tag{2}$$

which is the right continuous inverse of $\langle M\rangle_t$. Then the time changed process

$$(B_u, \mathcal{G}_u, P)_{u\geq 0} := (M_{\tau(u)}, \mathcal{F}_{\tau(u)}, P)_{u\geq 0} \tag{3}$$

is a one-dimensional Brownian motion.
The DDS-Theorem is proved by verifying the premises of the Levy Criterion for (B_u). Since conversely

$$M_t = B_{\langle M\rangle_t}, \quad t \geq 0 \tag{4}$$

holds P.a.s., every continuous local martingale is a time changed Brownian motion run according to its ***intrinsic clock*** $\langle M\rangle_t$.
Now if the continuous local martingale is given by a stochastic integral

$$M_t = \int_0^t H_s \, dN_s \tag{5}$$

where the integrator (N_t) is some other continuous local martingale, its quadratic variation process is given by

$$\langle M \rangle_t = \int_0^t H_s^2 \, d\langle N \rangle_s. \tag{6}$$

In particular, for a Brownian integrator $(N_t) = (b_t)$ one obtains

$$\langle M \rangle_t = \int_0^t H_s^2 \, ds. \tag{7}$$

In case the local continuous martingale (M_t) is given as a solution (X_t) of a one-dimensional time-homogeneous stochastic differential equation (SDE)

$$dX_t = \sigma(X_t)\, db_t \tag{8}$$

with (b_t) Brownian and $\sigma : \mathbb{R} \to \mathbb{R}$ continuous, then

$$\langle X \rangle_t = \int_0^t \sigma(X_s)^2 \, ds \tag{9}$$

follows. Hence, according to the DDS-Theorem,

$$\begin{aligned} (B_u) &:= (X_{\tau(u)}) \text{ is Brownian for} \\ \tau(u) &:= \inf\Big\{t \ge 0 \mid \int_0^t \sigma(X_s)^2 \, ds > u\Big\}. \end{aligned} \tag{10}$$

It may be mentioned that the relation in Eq. (4) of the DDS-Theorem is used in the theory of one-dimensional diffusions to obtain a weak solution of the SDE (8), see Ref. 14.

The statement of Eq. (10) is a special case of the general result concerning time transformations for SDEs as treated for instance in Ref. 14: Let $(X_t)_{t \ge 0}$ be a solution of the time-homogeneous SDE

$$dX_t = \sigma(X_t)\, db_t + a(X_t)\, dt \tag{11}$$

where (b_t) is Brownian with respect to a filtration $(\mathcal{F}_t)$. If $c : \mathbb{R} \to \mathbb{R}$ is a nonnegative continuous function satisfying

$$\lim_{t \to \infty} \int_0^t c(X_s)\, ds = \infty \qquad \text{P.a.s.} \tag{12}$$

define a family of stopping times for $u \ge 0$ by

$$\tau(u) := \inf\left\{t \geq 0 \mid \int_0^t c(X_s)\,ds > u\right\}. \tag{13}$$

Then

$$B_u := \int_0^{\tau(u)} \sqrt{c(X_s)}\,db_s \tag{14}$$

yields a Brownian motion with respect to the filtration $(\mathcal{G}_u) := (\mathcal{F}_{\tau(u)})$, and $Y_u := X_{\tau(u)}$ is a solution of the SDE

$$dY_u = \frac{\sigma(Y_u)}{\sqrt{c(Y_u)}}\,dB_u + \frac{a(Y_u)}{c(Y_u)}\,du \tag{15}$$

if $c(Y_u) > 0$ for all $u \geq 0$ P.a.s.

Note that the time differential dt transforms according to ordinary calculus: Because of $u = \int_0^{\tau(u)} c(X_s)\,ds$ one obtains $1 = c(X_{\tau(u)})\,\frac{d\tau(u)}{du}$ and hence

$$dt = \frac{du}{c(Y_u)}\,. \tag{16}$$

The multidimensional generalization of the DDS-Theorem is due to F.B. Knight (1971), see Ref. 13, Chap. 3.4.:

Let $(M_t)_{t\geq 0} = (M_t^1, \ldots, M_t^d;\ \mathcal{F}_t, P)_{t\geq 0}$ be a continuous vector valued local martingale with

$$\lim_{t\to\infty} \langle M^k\rangle_t = \infty \quad \text{P.a.s.}, \tag{17}$$

and vanishing mutual variation processes

$$\langle M^\ell, M^k\rangle_t = 0, \qquad 1 \leq k \neq \ell \leq d. \tag{18}$$

Define for each coordinate a family of stopping times for $u \geq 0$

$$\tau^k(u) := \inf\left\{t \geq 0 \mid \langle M^k\rangle_t > u\right\}. \tag{19}$$

Then the time changed processes

$$(B_u^k, \mathcal{G}_u^k, P)_{u\geq 0} := (M^k_{\tau^k(u)}, \mathcal{F}_{\tau^k(u)}, P)_{u\geq 0} \tag{20}$$

define a d-dimensional Brownian motion $(B_u) := ((B_u^1, \ldots, B_u^d))$ i.e. the (B_u^k), $1 \leq k \leq d$, are independent. In case the local continuous martingale (M_t) is given as a d-dimensional stochastic integral

$$M_t^k = \sum_{i=1}^n \int_0^t H_s^{ki}\,db_s^i =: \int_0^t H_s^k\,db_s, \tag{21}$$

with respect to a n-dimensional Brownian motion $(b_t) = \left((b_t^1, \ldots, b_t^n)\right)$, the coordinate quadratic variation processes are given by

$$\langle M^k \rangle_t = \int_0^t \sum_{i=1}^n (H_s^{ki})^2 \, ds =: \int_0^t (H_s^k)^2 \, ds . \tag{22}$$

3. Equalities of measures

The properties of the intrinsic clock we sketched in the foregoing section will be used now to derive some equalities of measures on spaces of paths.

3.1. The Kustaanheimo-Stiefel transformation (KS) is a mapping $k : \mathbb{R}^4 \to \mathbb{R}^3$ given by $y = k(x)$ with

$$\begin{aligned} y_1 &= x_1^2 - x_2^2 - x_3^2 + x_4^2 \\ y_2 &= 2x_1 x_2 - 2x_3 x_4 \\ y_3 &= 2x_1 x_3 + 2x_2 x_4 . \end{aligned} \tag{23}$$

Particularly interesting in the sequel is its property

$$|y| = x^2. \tag{24}$$

Now let $(X_t)_{t \geq 0}$ be a 4-dimensional Brownian motion and consider the KS-transformed process $Y_t := k(X_t)$. By the Ito formula, its stochastic differential is

$$dY_t^j = (\nabla Y)_t^j \, dX_t , \qquad j = 1, 2, 3 . \tag{25}$$

There is no time differential since k is harmonic. Explicitly one has

$$\begin{aligned} (\nabla Y)^1 &= 2(X^1, -X^2, -X^3, X^4) \\ (\nabla Y)^2 &= 2(X^2, X^1, -X^4, -X^3) \\ (\nabla Y)^3 &= 2(X^3, X^4, X^1, X^2) . \end{aligned} \tag{26}$$

Thus the corresponding quadratic variation processes are according to Eqs. (22) and (24)

$$\langle Y^j\rangle_t = \int_0^t ((\nabla Y)_s^j)^2\, ds = 4\int_0^t X_s^2\, ds = 4\int_0^t |Y_s|\, ds . \tag{27}$$

From transience of 4-dimensional Brownian motion follows

$$\lim_{t\to\infty} \langle Y^j\rangle_t = \lim_{t\to\infty} 4\int_0^t X_s^2\, ds = \infty \qquad \text{P.a.s.} \tag{28}$$

The mutual variation processes vanish. Indeed, for $1 \le \ell, k \le 3$

$$\langle Y^\ell, Y^k\rangle_t = \left\langle \sum_{i=1}^4 \int_0^t (\nabla Y)_s^{\ell i}\, dX_s^i, \sum_{j=1}^4 \int_0^t (\nabla Y)_s^{kj}\, dX_s^j \right\rangle$$

$$\overset{(a)}{=} \sum_{i,j=1}^4 \int_0^t (\nabla Y)_s^{\ell i}\, (\nabla Y)_s^{kj}\, d\langle X^i, X^j\rangle_s$$

$$\overset{(b)}{=} \sum_{i=1}^4 \int_0^t (\nabla Y)_s^{\ell i}\, (\nabla Y)_s^{ki}\, ds \overset{(c)}{=} 0 \tag{29}$$

where (a) applies the general rule for the computation of mutual variations of stochastic integrals, see Ref. 13, Chap. 3.2., (b) follows from the fact that (X_t) is Brownian, and (c) holds true since the vectors $(\nabla Y)^j$, $j = 1, 2, 3$, are orthogonal, see Eq. (26). Thus Knight's theorem tells us that the time changed process

$$(Z_u) := (Y_{\tau(u)}) \text{ with } \tau(u) := \inf\left(t \ge 0 \,\middle|\, 4\int_0^t |Y_s|\, ds > u\right) \tag{30}$$

is a 3-dimensional Brownian motion.

This result will now be expressed equivalently as an equality of two measures on corresponding spaces of paths. To this end realize the 4-dimensional Brownian motion (X_t) as the canonical process $X_t\,\varphi := \varphi(t)$ on the probability space (C_x^4, W_x^4) of continuous paths starting in $x \in \mathbb{R}^4$ with respect to the Wiener measure W_x^4.

The Brownian motion (X_t) gets transformed to the process (Y_t) by means of the KS-transformation

$$\begin{aligned} &Y_t\,\varphi := k(X_t\,\varphi) = (K\varphi)(t), \quad \text{where} \\ &K : C_x^4 \to C_y^3, \quad K\varphi := k\circ\varphi, \quad y = k(x) \end{aligned} \tag{31}$$

denotes the transformation of paths corresponding to k. The process (Y_t)

on (C^4_x, W^4_x) is, of course, equal by law to the canonical process $(\psi, t) \mapsto \psi(t)$ on $(C^3_y, K(W^4_x))$, $y = k(x)$, where $K(W^4_x) = W^4_x \circ K^{-1}$ denotes the image of W^4_x under K.

The passage from the original time to the intrinsic clock time induces a transformation T of C^3_y. For $\psi \in C^3_y$ and $u \geq 0$ let

$$\tau_\psi(u) := \inf\left\{t \geq 0 \mid 4\int_0^t |\psi(s)|\, ds > u\right\}. \tag{32}$$

Then the time transformation is given by

$$T : C^3_y \to C^3_y\,, \quad T\psi := \psi \circ \tau_\psi\,. \tag{33}$$

For the computations which will follow in Sec. 4 note that τ_ψ is bijective for almost all paths ψ [15] and that

$$\tau_\psi^{-1}(t) = 4\int_0^t |\psi(s)|\, ds\,, \quad d\tau_\psi^{-1}(t)/dt = 4\,|\psi(t)|\,. \tag{34}$$

The fact that (Z_u) defined in Eq. (30) is a 3-dimensional Brownian motion, can now be expressed by the following equality of measures [16] on C^3_y

$$TK(W^4_x) = W^3_{k(x)} \tag{35}$$

where TK is the composition of the transformations T and K defined above, and the right side denotes the image of the Wiener measure W^4_x on C^4_x under TK.

3.2. A series of equalities of measures on spaces of paths we want to show now is based on a relation holding between Bessel processes. These processes occur naturally when studying rotationally invariant systems. The d-dimensional ($d \geq 2$) Bessel process $(X_t)_{t\geq0}$ (a misleading name since (X_t) is a realvalued stochastic process) can be defined as the radial part of a d-dimensional Brownian motion $(W_t)_{t\geq0}$, namely

$$X_t := |W_t| \equiv \left[(W^1_t)^2 + \ldots + (W^d_t)^2\right]^{1/2}. \tag{36}$$

Ito calculus yields the following SDE for (X_t) starting at a positive point, see Ref. 13, Chap. 3.3, and Ref. 17,

$$dX_t = db_t + \tfrac{1}{2}(d-1)\,\frac{1}{X_t}\,dt\,. \tag{37}$$

Here $b_t := \sum_{i=1}^{d} \int_0^t \frac{1}{X_s} W_s^i \, dW_s^i$ is a one-dimensional Brownian motion [18].

From Eq. (37) it follows [19] that

$$\begin{aligned} &\text{(i)} \quad P(X_t > 0, \forall t > 0) = 1 \qquad \text{if } d \geq 2 \\ &\text{(ii)} \quad P(X_t \to \infty \text{ for } t \to \infty) = 1 \qquad \text{if } d > 2 . \end{aligned} \tag{38}$$

Consider now the transformation

$$k : \mathbb{R} \to \mathbb{R}, \quad k(x) := x^2 \tag{39}$$

which is the one-dimensional analogue of the KS-transformation introduced in Sec. 3.1. The squared process $(Y_t) := (k(X_t))$ satisfies the SDE

$$dY_t = 2X_t \, db_t + d \; dt . \tag{40}$$

Its quadratic variation is $\langle Y \rangle_t = \int_0^t (2X_s)^2 \, ds = 4 \int_0^t Y_s \, ds$ defining the family of stopping times for $u \geq 0$

$$\tau(u) := \inf \left\{ t \geq 0 \;\middle|\; 4 \int_0^t Y_s \, ds > u \right\} . \tag{41}$$

According to Eq. (15) the time changed process $(Z_u) := (Y_{\tau(u)})$ satisfies the SDE

$$dZ_u = dB_u + \frac{d}{4} \frac{1}{Z_u} du \tag{42}$$

where $B_u := \int_0^{\tau(u)} 2 X_s \, db_s$ is a Brownian motion. This shows that (Z_u) is a $\tilde{d}$-dimensional Bessel process with $\frac{1}{2}(\tilde{d} - 1) = \frac{1}{4} d$, i.e. $\tilde{d} = \frac{1}{2} d + 1$ (provided $d > 2$ is even, see however the remark at the end of this section).

Proceeding as in Sec. 3.1 this result is expressed equivalently by an equality of measures on spaces of paths. Let C_x^+ denote the set of continuous paths in $]0, \infty[$ starting at $x > 0$. The transformation of paths induced by k is given by

$$K : C_x^+ \to C_y^+, \quad K\varphi := k \circ \varphi, \quad y := k(x) = x^2 \tag{43}$$

and the corresponding time transformation T reads

$$T : C_y^+ \to C_y^+, \quad T\psi := \psi \circ \tau_\psi, \quad \text{with}$$
$$\tau_\psi(u) := \inf\left\{t \geq 0 \,\Big|\, 4\int_0^t \psi(s)\,ds > u\right\}. \tag{44}$$

Note that, due to Eq. (38), for $d > 2$ the time change τ_ψ is bijective for almost all ψ and that

$$\tau_\psi^{-1}(t) = 4\int_0^t \psi(s)\,ds, \quad d\tau_\psi^{-1}(t)/dt = 4\,\psi(t). \tag{45}$$

Because of the uniqueness in law of the solutions of Bessel SDEs one thus obtains [16]

$$TK(\mathrm{Bes}_x^d) = \mathrm{Bes}_{x^2}^{1+d/2} \tag{46}$$

for $d > 2$ and $x > 0$, where Bes_x^d denotes the measure on C_x^+ induced by the canonical d-dimensional Bessel process starting at x.

It should be mentioned that Bessel processes can be defined by Eq. (37) for all real $d \geq 2$ [17] and that Eq. (46) still holds for non-integer values of $d > 2$ by quite the same reasoning.

4. Coulomb path integral

The propagator $\varrho(y, y', u)$ for the Coulomb potential is the fundamental solution of the diffusion equation

$$\frac{\partial}{\partial u}\varrho(y, y', u) = \left(\frac{1}{2m}\,\Delta + \frac{-e^2}{|y|}\right)\varrho(y, y', u) \tag{47}$$

with the initial distribution $\varrho(y, y', 0) = \delta^3(y - y')$. By the Feynman-Kac formula it can be expressed as the Wiener path integral

$$\varrho(y, y', u) = \int_{C_y^3} \delta(\psi(u) - y')\,\exp\left[\int_0^u \frac{e^2}{|\psi(s)|}\,ds\right] dW_y^3(\psi). \tag{48}$$

Its evaluation is not trivial. However, having the equalities of measures Eq. (35) or Eq. (46) at ones disposal, it is easy to convert the Coulomb path integral into a Gaussian one or a series of Gaussian ones respectively, the evaluation of which is standard.

4.1. Evaluation based on $TK(W_x^4) = W_{k(x)}^3$.
Consider the path integral

$$w(y,u) \equiv \int_{C_y^3} w(\psi(u)) \exp\left[-\int_0^u V(\psi(s))\, ds\right] dW_y^3(\psi) \tag{49}$$

for some potential V on $\mathbb{R}^3$ and some initial distribution $w(y) \equiv w(y,0)$, where $u > 0$ and $y \in \mathbb{R}^3$. Due to the equalities of measures Eq. (35) one obtains

$$w(y,u) = \int_{C_x^4} w(TK\varphi(u)) \exp\left[-\int_0^u V(TK\varphi(s))\, ds\right] dW_x^4(\varphi)$$

which yields by Eqs. (33), (34), and with $\chi := K\varphi$

$$\int_{C_x^4} w(\chi(\tau_\chi(u))) \exp\left[-\int_0^{\tau_\chi(u)} 4\,|\chi(s)|\, V(\chi(s))\, ds\right] dW_x^4(\varphi) =$$

$$= \int_0^\infty dt \int_{C_x^4} \delta(\tau_\chi(u)-t)\, w(\chi(t)) \exp\left[-\int_0^t 4\,|\chi(s)|\, V(\chi(s))\, ds\right] dW_x^4(\varphi)\,.$$

Representing $\delta(\tau_\chi(u)-t) = 4\,|\chi(t)|\,\delta(u-\tau_\chi^{-1}(t))$ by a Fourier integral it follows

$$w(y,u) = \frac{1}{2\pi}\int_0^\infty dE e^{iEu}\int_0^\infty dt \left\{\int_{C_x^4} \tilde{w}(\varphi(t)) \exp\left[-\int_0^t V_E(\varphi(s))\, ds\right] dW_x^4(\varphi)\right\} \tag{50}$$

where by Eq. (24)

$$V_E(x) := 4\,x^2(V(k(x)) + iE)\,, \quad \tilde{w}(x) := 4\,x^2\, w(k(x)) \tag{51}$$

are the transformed potential and the new initial distribution, respectively, for the path integral within the curly brackets. Now, in the Coulomb case, the potential

$$V_E(x) = -4\,e^2 + 4\,iE\,x^2 \tag{52}$$

becomes quadratic and, therefore, can be treated by path integration. The propagator for Eq. (52) is well known to be

$$\varrho_d(E;x,x',t) = e^{4e^2t}\left[2\,iEm\,\pi^{-2}\,\mathrm{csh}^2(t\sqrt{8\,iE/m})\right]^{d/4}\cdot$$
$$\exp\left\{-\sqrt{iEm/2}\left[(x-x')^2\,\mathrm{cth}(t\sqrt{2\,iE/m}) + (x+x')^2\,\mathrm{th}(t\sqrt{2\,iE/m})\right]\right\} \tag{53}$$

with the dimension d equal to 4. Hence, in the case $w(y) = \delta^3(y-y')$ the path integral in Eq. (50) within the curly brackets yields explicitly

$$w(E;y,y',t) \equiv \int 4\,x'^2\,\delta^3(k(x')-y')\,\varrho(E;x,x',t)\,d^4x'$$

$$= \frac{2iEm\exp(4e^2t)}{\pi\,\mathrm{sh}^2(t\sqrt{8iE/m})}\exp\left[-\sqrt{2iEm}\,\mathrm{cth}(t\sqrt{8iE/m})(|y|+|y'|)\right]\cdot \tag{54}$$

$$\cdot\, I_0\left(\frac{\sqrt{8iEm}}{\mathrm{sh}\,(t\sqrt{8iE/m})}\sqrt{\tfrac{1}{2}(|y|\,|y'|+yy')}\right).$$

As expected, the integral in Eq. (54) does not depend on x but only on its image $y = k(x)$ under the KS-transformation. It is not hard to do with the aid of the integration formula

$$\int_{\mathbb{R}^4} g(x)\,d^4x = \int_{\mathbb{R}^3}\int_{-\pi}^{\pi} g(x(y,\varphi))\,d\varphi\,\frac{d^3y}{8|y|} \tag{55}$$

where the KS-coordinates of $x = x(y,\varphi) \in \mathbb{R}^4$ are given by

$$x_1 = \frac{1}{2\sigma}\left(y_2\cos\varphi + y_3\sin\varphi\right),\quad x_2 = \sigma\cos\varphi$$

$$x_3 = \sigma\sin\varphi,\quad x_4 = \frac{1}{2\sigma}\left(y_3\cos\varphi - y_2\sin\varphi\right) \tag{56}$$

$$\sigma := \sqrt{\tfrac{1}{2}(|y|-y_1)}\,.$$

I_0 denotes the modified Bessel function of the first kind. Combining Eqs. (50) and (54) the Coulomb propagator comes out to be

$$\varrho(y,y',u) = \frac{1}{2\pi}\int_{-\infty}^{\infty} dE e^{iEu}\int_0^{\infty} dt\; w(E;y,y',t)\,. \tag{57}$$

It should be mentioned that because of the factor $\exp(4e^2t)$ the integrals in Eq. (57) do not exist in the Lebesgue sense. So there is the question about the meaning of Eq. (57). For instance, it is not hard to show by means of Fubini's theorem and the theorem of monotone convergence that

$$\varrho(y,y',u) = \lim_{\varepsilon\downarrow 0}\frac{1}{2\pi}\int_{-\infty}^{\infty} dE e^{iEu}\int_0^{\infty} dt\, w(E;y,y',t)\,e^{-\varepsilon t^2}\,.$$

Following Ref. 2 we add the partial-wave expansion of the Coulomb propagator, which we shall achieve in Sec. 4.2 directly. Applying the formula [20]

$$zI_0\left(z\cos\tfrac{\Theta}{2}\right) = \sum_{\ell=0}^{\infty} 2(2\ell+1)\,I_{2\ell+1}(z)\,P_\ell(\cos\Theta) \tag{58}$$

to Eq. (54), where P_ℓ denotes the ℓth Legendre polynomial, one obtains immediately

$$w(E;y,y',t) = \sum_{\ell=0}^{\infty} (2\ell+1)\, P_\ell(\cos\Theta)\, w_\ell(E;y,y',t) \quad \text{with}$$

$$w_\ell(E;y,y',t) = \exp\left[4e^2t - \sqrt{2iEm}\,\mathrm{cth}(t\sqrt{8iE/m})\,(|y|+|y'|)\right]\cdot \tag{59}$$

$$(|y|\,|y'|)^{-1/2}\ \frac{\sqrt{2iEm}}{\pi\,\mathrm{sh}(t\sqrt{8iE/m})}\ I_{2\ell+1}\left(\frac{\sqrt{8iEm}}{\mathrm{sh}(t\sqrt{8iE/m})}\sqrt{|y|\,|y'|}\right).$$

4.2. Evaluation based on $TK(Bes_x^d) = Bes_{x^2}^{1+d/2}$.

Since for d = 2, 3, ... the d-dimensional Bessel process can be obtained as the radial part of a d-dimensional Brownian motion, the following relation between Wiener and Bessel path integrals

$$\int_{C_x^d} L(|\psi|)\, d\,W_x^d(\psi) = \int_{C_{|x|}^+} L(\varphi)\, d\,Bes_{|x|}^d(\varphi) \tag{60}$$

is valid for every functional L on $C_{|x|}^+$, $x \in \mathbb{R}^d\backslash\{0\}$. The transition density of the Bessel process follows from that of the Brownian motion

$$p^{(d)}(a,b,u) = \left(\frac{m}{2\pi u}\right)^{d/2} \exp\left(-\frac{m}{2u}|a-b|^2\right) \tag{61}$$

by integrating over the sphere around the origin through b [22]:

$$bes^{(d)}(\alpha,\beta,u) = \frac{m}{u}\,\beta^{d-1}(\alpha\beta)^{1-d/2}\, e^{-\frac{m}{2u}(\alpha^2+\beta^2)}\, I_{d/2-1}\left(\frac{m}{u}\alpha\beta\right) \tag{62}$$

with $\alpha = |a|$, $\beta = |b|$.

A series of Bessel path integrals occurs in Eq. (49) for a rotationally invariant potential when expanding the initial distribution w(y) into partial waves. The general situation is as follows: Let y and y' be two fixed directions in $\mathbb{R}^3\backslash\{0\}$ forming the angle Θ, and denote by Θ_u the angle between y' and $\psi(u)$ for $\psi \in C_y^3$. Then

$$\int_{C_y^3} P_\ell(\cos\Theta_u)\, L(|\psi|)\, d\,W_y^3(\psi)$$

$$= P_\ell(\cos\Theta) \int_{C_{|y|}^+} \left(\frac{|y|}{\varphi(u)}\right)^\ell L(\varphi)\, d\,Bes_{|y|}^{2\ell+3}(\varphi) \tag{63}$$

is valid for every functional L depending only on $\varphi|[0,u]$ for $\varphi \in C^{+}_{|y|}$. The proof is straightforward using the polynomial approximation for path integrals. [23] Specializing to the case of the Coulomb propagator Eq. (48) one finds the partial wave expansion of the initial distribution

$$\delta^3(y-y') = \sum_{\ell=0}^{\infty}(2\ell+1)\, P_\ell(\cos\Theta)\, \frac{\delta(|y|-|y'|)}{4\pi\, y'^2} \tag{64}$$

and hence the functionals

$$L(\varphi) = (2\ell+1)\, \frac{\delta(\varphi(u)-|y'|)}{4\pi\, y'^2}\, \exp\left[\int_0^u \frac{e^2}{\varphi(s)}\, ds\right]. \tag{65}$$

Now the equality of measures Eq. (46) allows to convert the Bessel path integrals Eqs. (63), (65) to Bessel path integrals with a quadratic potential by quite the analogous steps leading from Eq. (49) to Eq. (50). One obtains, cf. Eq. (59)

$$w_\ell(E;y,y',t) = \pi^{-1}\, |y|^\ell\, |y'|^{-\ell-1} \exp(4e^2 t)\cdot$$
$$\cdot\left\{\int_{C_r^+} \delta(\varphi(t)^2-|y'|)\, \exp\left[-\int_0^t 4iE\, \varphi(s)^2\, ds\right] d\, \mathrm{Bes}_r^{4(\ell+1)}\right\}$$

with $r := \sqrt{|y|}$. The path integral within the curly brackets is equal to a $4(\ell+1)$-dimensional Gaussian path integral, see Eq. (60). The corresponding propagator is given by Eq. (53) for $d = 4(\ell+1)$. Thus it remains to compute [22]

$$\int \delta(x'^2-|y'|)\, \varrho_d(E;x,x',t)\, d^d x'$$

where x satisfies $x^2 = |y|$. The result coincides with Eq. (59).

As a final remark let us point out the relation of the computations in this section to those in the literature Refs. 7, 8, 9, concerning path integrals in polar coordinates and their application to the Coulomb propagator. This relation is brought about by the basic fact that the Bessel measures Bes_x^d, $x > 0$, for $d \geq 2$ are mutually absolutely continuous with a Radon-Nikodym derivative yielding the well-known additional radial potential proportional to x^{-2}. In particular

$$\frac{d\,\mathrm{Bes}_x^d}{d\,\mathrm{Bes}_x^3}(\varphi) = \left(\frac{\varphi(u)}{x}\right)^{(d-3)/2} \exp\left[-\frac{1}{m}\int_0^u \frac{(d-1)(d-3)}{8\,\varphi(s)^2}\, ds\right] \tag{66}$$

holds [17]. Hence combining this with Eq. (63) the partial wave expansion of

$$w(y,u) \equiv \int_{C_y^3} w(\psi(u)) \exp\left[-\int_0^u V(|\psi(s)|)\,ds\right] dW_y^3(\psi) \tag{67}$$

for some potential V on $\mathbb{R}$ reads

$$w(y,u) = \sum_{\ell=0}^{\infty} (2\ell+1)\, P_\ell(\cos\Theta) \cdot \tag{68}$$

$$\cdot \int_{C^+_{|y|}} w_\ell(\varphi(u)) \exp\left\{-\int_0^u \left[V(\varphi(s)) + \frac{\ell(\ell+1)}{2m}\varphi(s)^{-2}\right] ds\right\} d\,Bes^3_{|y|}(\varphi)$$

with $w(y) = \sum_{\ell=0}^{\infty}(2\ell+1)\,P_\ell(\cos\Theta)\,w_\ell(|y|)$ being the partial wave expansion of the initial distribution. Now, the transition density of the 3-dimensional Bessel process, see Eq. (62)

$$bes^{(3)}(\alpha,\beta,u) = \frac{\beta}{\alpha}\left[p^{(1)}(\alpha,\beta,u) - p^{(1)}(\alpha,-\beta,u)\right] \tag{69}$$

is equal to that of the Absorbing Brownian motion [24] times (β/α). Thus, formally, the 3-dimensional Bessel path integrals in Eq. (68) are written as 3-dimensional Wiener path integrals over $C^1_{|y|}$ (!) with an additional density $\varphi(u)/|y|$. This is the form of the radial path integrals one finds in Refs. 7, 8, 9. Finally, the passage from the $(2\ell+3)$-dimensional Bessel path integral in Eq. (63) to the $4(\ell+1)$-dimensional one by Eq. (66) gives rise, besides to an additional power of $\varphi(u)$ and $|y|$, to the new radial potential term $(4\ell+1)(4\ell+3)/(8mx^2)$.

References

1. Duru, I.H. and Kleinert, H., "Solution of the Path Integral for the H-Atom", Phys. Lett. 84 B, 185-188 (1979).
   —·—, "Quantum Mechanics of H-Atom from Path Integrals", Fortsch. Phys. 30, 401-435 (1982).

2. Ho, R. and Inomata, A., "Exact-Path-Integral Treatment of the Hydrogen Atom", Phys. Rev. Lett. 48, 231-234 (1982).

3. Blanchard, Ph. and Sirugue, M., "Treatment of some singular potentials by change of variables in Wiener integrals", J. Math. Phys. 22, 1372-1376 (1981).

4. Blanchard, Ph., "Transformations of Wiener integrals and the desingularization of Coulomb problem", in: Springer L.N. Phys. 173, Stochastic processes in quantum theory and statistical physics, Eds. S. Albeverio, Ph. Combe and M. Sirugue-Collin, 1982, p. 19-28.

5. Young, A. and DeWitt-Morette, C., "Time Substitutions in Stochastic Processes as a Tool in Path Integration", Ann. Phys. 169, 140-166 (1986).

6. Castrigiano, D.P.L. and Stärk, F., "New aspects of the path integrational treatment of the Coulomb potential", J. Math. Phys., to appear in Dec. 1989.

7. Peak, D. and Inomata, A., "Summation over Feynman Histories in Polar Coordinates", J. Math Phys. 16, 1422-1428 (1969).

8. Inomata, A., "Alternative Exact-Path-Integral Treatment of the Hydrogen Atom", Phys. Lett. 101A, 253-257 (1984).

9. Steiner, F., "Space-Time Transformations in Radial Path Integrals", Phys. Lett. 106A, 356-362 (1984).
—·— "Exact Path Integral Treatment of the Hydrogen Atom", Phys. Lett. 106A, 363-367 (1984).
—·— "Path Integrals in Polar Co-ordinates from eV to GeV", in: Path Integrals from mev to MeV, Eds. M.C. Gutzwiller, A. Inomata, I.R. Klauder, L. Streit (World Scientific, Singapore 1986).

10. Pak, N.K. and Sokmen, I., "General new-time formalism in path integral", Phys. Rev. A 30, 1629-1635 (1984).

11. Inomata, A., "Remarks on the time transformation technique for path integration", in: Path Integrals from meV to MeV, Eds. M.C. Gutzwiller, A. Inomata, I.R. Klauder, L. Streit (World Scientific, Singapore 1986), p. 433-447.

12. Carpio-Bernido, M.V. and Bernido, Ch. C., "A path integral solution of axially symmetric potentials", and Carpio-Bernido, M.V. and Inomata, A., "Exact path integral treatment of a class of axially symmetric potentials", in: Bielefeld Encounters in Mathematics and Physics VII, Eds. Hida, T. and Saito, K., 1988.

13. Karatzas, I. and Shreve, S.E., Brownian motion and stochastic calculus, Springer, New York, 1988.

14. Gihman, I.I. and Skohorod, A.V., Stochastic differential equations, Part I, Chap. 3.15, Springer, New York, 1972.

15. This is true, since for a d-dimensional ($d \geq 3$) Brownian motion every prescribed point (different from the starting point) is nonattainable and every compact subset is transient.

16. For simplicity the general considerations in Sec. 2 and elsewhere refer to diffusion constant 1, but the results, as e.g. the relations (36) and (46), of course are valid for every diffusion constant $1/\sqrt{m}$.

17. Pitman, J. and Yor, M., "Bessel processes and infinitely divisible laws", in: Springer L.N. Math. 851, Stochastic Integrals, Ed. D. Williams, 1981, p. 285–370, Chap. 2.

18. This is shown verifying the premises of the Levy Criterion: Since

$$\mathbb{E}\Big(\int_0^t (W_s^i/X_s)^2\, ds\Big) \leq \mathbb{E}\Big(\int_0^t 1\, ds\Big) = t \quad \text{is follows that} \quad C_t^i := \int_0^t (W_s^i/X_s)\, dW_s^i$$

is a square integrable martingale. For its mutual variation processes one obtains

$$\langle C^i, C^j\rangle_t = \int_0^t (W_s^i W_s^j/R_s^2)\, d\langle W^i, W^j\rangle = \delta_{ij} \int_0^t (W_s^i/X_s)^2\, ds, \quad \text{and hence}$$

$$\Big\langle \sum_{i=1}^d C^i \Big\rangle_t = \sum_{i=1}^d \int_0^t (W_s^i/X_s)^2\, ds = t\,.$$

19. Rogers, L.C.G. and Williams, D., Diffusions, Markov processes, and martingales, Vol. II, V. 48.

20. Formula (58) is an example for a Neumann series considered in Ref. 21, 7.10.1. It follows from the general formalism using Ref. 21, 7.2.2 (12) and 10.10 (23). In comparison with Eq. (58) a factor $2(-1)^\ell$ is missing for every term on the left side of Ref. 2, Eq. (23).

21. Erdėlyi, A., Ed., Higher transcendental functions, Vol. 2, McGraw-Hill, New York 1953.

22. Use $\int_0^\pi e^{z\cos\Theta}(\sin\Theta)^{2\nu}\, d\Theta = \sqrt{\pi}\, \Gamma\big(\nu+\tfrac{1}{2}\big)(z/2)^{-\nu}\, I_\nu(z)$, see Ref. 21, 7.8 (11), and the area $2\big(\pi^{d/2}\big)/\Gamma(d/2)$ of the d-dimensional unit sphere Ω^d, see Ref. 21, 11.1.1 (11), to obtain $\int e^{z\cos\Theta}\, d\Omega^d = (2\pi)^{d/2}\, z^{1-d/2}\, I_{d/2-1}(z)$.

23. It relies on $\int\limits_{-\pi}^{\pi}\int\limits_{0}^{\pi} P_\ell\,(\cos\Theta_2)\; e^{z\cos\vartheta}\,\sin\vartheta\; d\vartheta\; d\varphi = (2\pi)^{3/2}\, z^{-1}\, I_{\ell+1/2}(z)\; P_\ell\left(\cos\Theta_1\right)$,

where $\Theta_i = \sphericalangle\left(y_i, y'\right)$ and $\vartheta = \sphericalangle\left(y_1, y_2\right)$, which can be varified using Ref. 21, 10.10 (47) and 7.8 (11).

24. Williams, D., "Path Decomposition and Continuity of Local Time for one-dimensional Diffusions, I", Proc. London Math. Soc. (3), 28, 738-768 (1974).

EXPONENTIAL ESTIMATES OF LARGE DEVIATION TYPE FOR ITÔ PROCESSES IN HILBERT SPACES*

Pao-Liu Chow

Department of Mathematics, Wayne State University

Detroit, Michigan 48202, U.S.A.

ABSTRACT.

The article gives a brief exposition on a large deviation type of estimates for some Itô processes in a Hilbert space. Exponential tail probability estimates are established for such processes to leave a ball of radius r before a finite time $t > 0$. An application is given to show the almost sure convergence of discrete-time approximations to a stochastic integral.

1. Introduction

Let b_t be a standard Brownian motion (Wiener process) in d dimensions defined in $(\Omega, \mathcal{F}_t, P)$. Then it is well known that [1]

$$P\{\sup_{0\leq s\leq t} |b_s| \geq r\} \leq 2d\exp\{-\frac{r^2}{2dt}\}. \tag{1}$$

Such an estimate is of basic importance in the study of stochastic differential equations and the relataed Itô processes. For instance, consider the Itô process

$$x_t = \int_0^t a(s)ds + \int_0^t \sigma(s)db_s, \tag{2}$$

where $a(t)$ and $\sigma(t)$ are $\mathcal{F}_t$-progressively measurable processes denoting the drift vector in $\mathcal{R}^d$ and an $d \times d$ diffusion matrix, respectively. Suppose that $|a(t)| \leq k$ and the trace $tr\{\sigma(t)\sigma^*(t)\} \leq m \;\; \forall t \geq 0$. Then the following exponential estimate holds [2]:

$$P\{\sup_{0\leq s\leq t} |x_s| \geq r\} \leq 2d\exp\{-\frac{(r-\sqrt{d}kt)^2}{2dmt}\}. \tag{3}$$

*This work was supported by the National Science Foundation under grant # DMS-87-02236.

Referring to the inequalities (1) and (3), One notes that the estimates depend on the dimension d. In the infinite-dimensional stochastic analysis related to a Wiener process, One is naturally interested in generalizing the above estimates from a finite -dimensional case. However, because of the dimensional dependence, such estimates cannot be generalized in a straight-forward manner. Here we will give a brief account for how to resolve this problem. The results to be presented are based on a recent joint work with my colleague, J.L. Menaldi [3].

Let H be a real separable Hilbert space with inner product $(\cdot,\cdot)$ and norm $|\cdot|$. Let $\{W_t\}$ be a Wiener process over $(\Omega, \mathcal{F}_t, P)$ with $W_0 = 0$ and covariance operator R. That is , for any $g, h \in H$,

$$E(W_t, g) = 0$$

and

$$E(W_t, g)(W_s, h) = (t \wedge s)(Rg, h).$$

Recall that $R : H \to H$ is a self-adjoint, nuclear (trace class) operator. Suppose that $F(t)$ and $\Sigma(t)$ are $\mathcal{F}_t$-progressively measurable processes with values in H and $\mathcal{L}(H)$, respectively, where $\mathcal{L}(H)$ denotes the space of bounded linear operators in H. Assume that

$$E \int_0^T |F(s)|^2 ds < \infty$$

and

$$E \int_0^T tr.[R\Sigma(s)\Sigma^*(s)]ds < \infty,$$

where Σ^* means the adjoint of Σ. Then the following Itô process in H is well-defined [4]:

$$X_t = \int_0^t F(s)ds + \int_0^t \Sigma(s)dW_s. \tag{4}$$

Furthermore, suppose that $\varphi : H \to \mathcal{R}$ is continuous and twice H-differentiable such that the first two derivatives $\varphi'(h)$ and $\varphi''(h)$ are locally bounded, continuous from H into H and $\mathcal{L}(H)$, respectively. Then the following Itô formula holds:

$$\varphi(X_t) = \varphi(0) + \int_0^t L\varphi(X_s)ds + \int_0^t (\varphi'(X_s), \Sigma(s)dW_s), \tag{5}$$

where

$$L\varphi(X_s) = \frac{1}{2}tr[\Sigma^*(s)\varphi''(X_s)\Sigma(s)] + (\varphi'(X_s), F(s)). \tag{6}$$

In what follows, we will first present an exponential estimate for the Wiener process W_t and then indicate how a similar estimate can be obtained for the Itô process X_t.

2. Exponential Estimates for Itô Processes

Let us consider the Wiener process W_t in H with covariance operator R. The complete set of normalized eigenvectors of R is denoted by $\{e_k\}$ so that

$$Re_k = \lambda_k e_k, \quad k = 1, 2, \cdots, \tag{7}$$

where λ_k's are the corresponding eigenvalues with $\lambda_k > 0$ and $trR = \sum_{k=1}^{\infty} \lambda_k < \infty$. By the Paley-Wiener representation, we can write

$$W_t = \sum_{k=1}^{\infty} \sqrt{\lambda_k} b_t^k e_k, \tag{8}$$

where $\{b_t^k\}$ is a sequence of i.i.d. standard Brownian motions in one dimension. We set

$$W_t = W_t^n + R_t^n, \tag{9}$$

where

$$W_t^n = \sum_{k=1}^{n} \sqrt{\lambda_k} b_t^k e_k, \tag{10}$$

and

$$R_t^n = \sum_{k=n+1}^{\infty} \sqrt{\lambda_k} b_t^k e_k. \tag{11}$$

Then, for any $\beta > 0$, we have

$$E\{\exp[\beta |W_t^n|^2]\} = \prod_{k=1}^{n} E\{\exp[\beta \lambda_k |b_t^k|^2]\} = \prod_{k=1}^{n} (1 - 2\beta\lambda_k t)^{-\frac{1}{2}} , \tag{12}$$

since b_t^k is a Gaussian random variable with mean zero and variance t. In view of the fact that

$$\ln(1-\alpha) \geq -2\alpha \quad \text{for } \alpha \in [0, \frac{1}{2}],$$

Eq.(12) yields

$$E\{\exp[\beta |W_t^n|^2]\} \leq \exp\{2\beta t \sum_{k=1}^{n} \lambda_k\}$$

for $\beta \leq (4\lambda_k t)^{-1} \ \ \forall k,$ or

$$E \exp[\beta |W_t^n|^2] \leq \exp(2\beta \Lambda_1 t), \tag{13}$$

for $\beta \leq (4\Lambda_0 t)^{-1}$, where $\Lambda_1 = tr.R$ and $\Lambda_0 = \| R \| = \sup_k\{\lambda_k\}$. On the other hand, since $\{R_t^n\}$ is a sequence of symmetric H-valued Gaussian r.v.'s such that $E|R_t^n|^2 \to 0$ as $n \to \infty$, it can be shown that

$$\lim_{n\to\infty} E\exp[\,\beta\,|R_t^n|^2\,] = 1. \tag{14}$$

Now, by a submartingale inequality, we get

$$\begin{aligned} P\{\sup_{0\leq s\leq t}|W_s| \geq r\} &= P\{\sup_{0\leq s\leq t}\exp[\,\beta\,|W_s|^2\,] \geq e^{\beta r^2}\} \\ &\leq e^{-\beta r^2} E\exp[\beta|W_t|^2] \\ &\leq e^{-\beta r^2} E\{\exp[\beta|W_t^n|^2]\}E\{\exp[\beta|R_t^n|^2]\}, \end{aligned} \tag{15}$$

because W_t^n and R_t^n are independent and orthogonal. In view of (13) and (14), the inequality (15) implies that

$$P\{\sup_{0\leq s\leq t}|W_s| \geq r\} \leq \exp\{-\beta[r^2 - 2\Lambda_1 t]\}, \quad \text{for } \beta \leq \frac{1}{4\Lambda_0 t},$$

from which, by choosing $\beta = \frac{1}{4\Lambda_1 t}$, we obtain the exponential estimate:

$$P\{\sup_{0\leq s\leq t}|W_s| \geq r\} \leq 2\exp\{-\frac{r^2}{4\Lambda_1 t}\}, \quad t > 0, \tag{16}$$

with $\Lambda_1 = trR$. In case that $H = \mathcal{R}^d$, $\Lambda_1 = d$. In contrast, the estimate (16) is not as sharp as that of (1) in the decay rate but its multiplicative constant is independent of dimension d. Also, we note that the result depends heavily on the Gaussian properties of the random process. For a general Itô process, an alternative approach is needed. In what follows, we will turn our attention to this subject.

To obtain an exponential estimate for an Itô process, we shall make use of the Itô calculus and the associated semi-martingale property. To this end, let us introduce a real-valued function φ_λ on H, depending on a parameter $\lambda > 0$, as follows:

$$\varphi_\lambda(h) = (1 + \lambda|h|^2)^{\frac{1}{2}}, \quad h \in H. \tag{17}$$

For the Itô process X_t given by Eq.(4), the Itô formula (5) applied to $\varphi_\lambda(X_t)$ yields

$$\varphi_\lambda(X_t) = 1 + \lambda\int_0^t [\varphi_\lambda(X_s)]^{-1}(X_s, \Sigma(s)dW_s) + \int_0^t L_\lambda\varphi_\lambda(X_s)ds, \quad 0 \leq t \leq T, \tag{18}$$

where

$$\begin{aligned} L_\lambda\varphi_\lambda(X_s) &= \lambda[\varphi_\lambda(X_s)]^{-1}(X_s, F(s)) \\ &+ \frac{1}{2}\lambda[\varphi_\lambda(X_s)]^{-1}\{tr[R\Sigma^*(s)\Sigma(s)] - \lambda[\varphi_\lambda(X_s)]^{-2}|R^{\frac{1}{2}}\Sigma^*(s)X_s|^2\}. \end{aligned} \tag{19}$$

Let $Z_\lambda(t)$ be an exponential martingale defined by

$$Z_\lambda(t) = \exp\{\eta_\lambda(t)\} \tag{20}$$

where

$$\eta_\lambda(t) = \lambda \int_0^t [\varphi_\lambda(X_s)]^{-1}(X_s, \Sigma(s)dW_s) - \frac{1}{2}\lambda^2 \int_0^t [\varphi_\lambda(X_s)]^{-2}|R^{\frac{1}{2}}\Sigma^*(s)X_s|^2 ds. \tag{21}$$

Let us assume that

$$|F(t)| \leq K \tag{22}$$

and

$$\| \Sigma^*(t)\Sigma(t) \|_{\mathcal{L}(H)} \leq M \quad \text{a.s.} \quad \forall\, t \in [0,T] \tag{23}$$

Note that

$$0 < \Big[\frac{\max.(1, \lambda|h|)}{\varphi_\lambda(h)}\Big]^j < 1 \quad \text{for } j = 1,2 \text{ and } h \in H. \tag{24}$$

By invoking (21) to (24), it is not difficult to check that Eq.(18) gives rise to the following bound:

$$\varphi_\lambda(X_t) \leq 1 + (K + \lambda M \Lambda_1)t + \eta_\lambda(t). \tag{25}$$

Therefore, in view of (17), (20) and (25), we have

$$\begin{aligned} P\{\sup_{0\leq s\leq t} |X_s| \geq r\} &= P\{\sup_{0\leq s\leq t} \varphi_\lambda(X_s) \geq (1+\lambda r^2)^{\frac{1}{2}}\} \\ &\leq P\{\sup_{0\leq s\leq t} Z_\lambda(s) \geq \exp[(1+\lambda r^2)^{\frac{1}{2}} - 1 - (K + \lambda M \Lambda_1)t]\} \\ &\leq \exp[1 + (K + \lambda M \Lambda_1)t - (1+\lambda r^2)^{\frac{1}{2}}], \end{aligned} \tag{26}$$

as a result of Doob's martingale inequality. By choosing $\lambda = (\frac{r}{2\Lambda_1 M t})^2 - \frac{1}{r^2} > 0$, it follows from (26) that

$$P\{\sup_{0\leq s\leq t} |X_s| \geq r\} \leq \exp\{1 + Kt - \frac{r^2}{4\Lambda_1 M t}\}.$$

This, in turn, implies that, for $0 < t \leq T$,

$$P\{\sup_{0\leq s\leq t} |X_s| \geq r\} \leq 3\exp\{\frac{(r-ct)^2}{4M\Lambda_1 t}\} \tag{27}$$

with $c = 2(KM\Lambda_1)^{\frac{1}{2}}$, which generalizes the finite-dimensional inequality (2). In particular, if we set $F \equiv 0$ in Eq.(4), one obtains an exponential bound for the stochastic integral as a special case:

$$P\{\sup_{o\leq t\leq T} |\int_0^t \Sigma(s)dW_s| \geq r\} \leq 3\exp\{-\frac{r^2}{4\Lambda_1 MT}\}, \quad T > 0. \tag{28}$$

We will give an application of this inequality in the next section.

3. An Application

The derivation of exponential inequalities of the type (27) was first motivated by a large deviation problem for parabolic Ito equations [5]. Here we will give another application, namely, the proof of the a.s. convergence of a discrete-time approximation to the following stochastic integral in H:

$$\xi_t = \int_0^t \Sigma(s)dW_s, \quad 0 \leq t \leq T. \tag{29}$$

In addition to the previous assumptions, let $\Sigma : [0,T] \to \mathcal{L}(H)$ be Hölder-continuous with exponent $\alpha \in (0,1)$ so that

$$\| \Sigma(t) - \Sigma(s) \|_{\mathcal{L}(H)} \leq \nu|t-s|^\alpha , \tag{30}$$

for any $t, s \in [0,T]$ and some constant $\nu > 0$. Let ξ_t be approximated by a discrete-time version of Eq.(29) as follows:

$$\xi_t^n = \int_0^t \Sigma_n(s)dW_s, \tag{31}$$

where $\Sigma_n(s) = \Sigma(\frac{k-1}{n})$, for $\frac{k-1}{n} \leq s < \frac{k}{n}$, $k = 1,2,\cdots,[nt]$, and $[\alpha]$ denotes the integral part of α. By definition, it is easy to see that

$$\lim_{n\to\infty} E\{|\xi_t^n - \xi_t|^2\} = 0.$$

To show it actually converges almost surely, let

$$\eta_t^n = (\xi_t - \xi_t^n) = \int_0^t \Delta\Sigma_n(s)dW_s, \tag{32}$$

where

$$\Delta\Sigma_n(s) = [\Sigma(s) - \Sigma_n(s)] = [\Sigma(s) - \Sigma(\frac{k-1}{n})] \quad \text{for } \frac{k-1}{n} \leq s < \frac{k}{n}\,.$$

Since, by condition(30),

$$\sup_{\frac{k-1}{n} \leq s < \frac{k}{n}} \| \Delta\Sigma_n(s) \|_{\mathcal{L}(H)} \leq \frac{\nu}{n^\alpha} \quad \forall k,$$

it follows that

$$\sup_{0 \leq t \leq T} \| \Delta\Sigma_n(t) \|_{\mathcal{L}(H)} \leq \frac{\nu}{n^\alpha}. \tag{33}$$

Let $\{r_n\}$ be a sequence of positive numbers such that $r_n \downarrow 0$ as $n \to \infty$. For each n, consider the event

$$A_n = \{ \sup_{0 \leq t \leq T} |\eta_t^n| \geq r_n \}.$$

In view of (33), an application of the inequality (28) to η_t^n defined by (32) yields

$$P\{A_n\} \leq 3 \exp\{-r_n^2 / 4\nu n^{-\alpha} \Lambda_1 T\}.$$

If we choose $r_n = (\beta \ln n / n^\alpha)^{\frac{1}{2}}$ with $\beta > \rho = 4\nu\Lambda_1 T$, then

$$\sum_{n=1}^{\infty} P\{A_n\} \leq 3 \sum_{n=1}^{\infty} n^{-\beta/\rho} < \infty,$$

so that, by the Borel-Cantelli lemma,

$$P\{A_n \text{ i.o.}\} = 0.$$

Of course, this implies that $\lim_{n\to\infty} \xi_t^n = \xi_t$ uniformly in $t \in [0, T]$ with probability one. Therefore the stochastic integral (29) can be approximated uniformly over [0,T] by a discrete-time version (30) with certainty.

REFERENCES

1. D.W. Stroock and S.R.S. Varadhan, Multidimensional Diffusion Processes, Springer Verlag, New York, 1979.

2. D.W. Stroock, An Introduction to the Theory of Large Deviations, Springer-Verlag, New York, 1984.

3. P.L. Chow and J.L. Menaldi, Exponential Estimates in Exit Probability for Some Diffusion Processes in Hilbert Spaces, Stochastics, (to appear) 1990.

4. H.H. Kuo, Gaussian Measures in Banach Spaces, Springer-Verlag, Berlin, 1975.

5. P.L. Chow, Large Deviation Problem for Some Parabolic Itô Equations, (preprint) 1989.

NONSTANDARD REPRESENTATION OF GAUSSIAN MEASURES

Nigel J. Cutland

Department of Pure Mathematics, University of Hull,
HULL HU6 7RX, England

ABSTRACT

Nonstandard methods allow a 'flat' integral representation of certain Gaussian measures - Wiener measure, the measures associated with the Brownian bridge and Lévy Brownian motion, and Gaussian measures on separable Hilbert space. This paper surveys these representations and a number of applications.

INTRODUCTION

In this brief survey we shall discuss a particular kind of nonstandard representation of certain Gaussian measures - Wiener measure, the measures induced by the Brownian bridge, fractional Brownian motion (including Lévy Brownian motion), and Gaussian measures on Hilbert space. The common theme is a 'flat integral' representation for these measures, and the convenience of such a representation for applications,

In a short article such as this we cannot discuss in any detail the basics of nonstandard analysis: for a good introduction see [10]. The main features are as follows. To each standard object A there is a nonstandard extension $^*A$, so that we have, for example,

$$({}^*\mathbb{R},\ {}^*+,\ {}^*\times,\ {}^*\leq)$$

which is an ordered field extending $\mathbb{R}$. Usually * is dropped from *+, etc. The set $^*\mathbb{R}$ is the nonstandard reals, or *hyperreals* and contains both infinite and infinitesimal members. Similarly $^*\mathbb{N}$, the *hypernatural* numbers, is an extension of $\mathbb{N}$ that contains infinite members, and $^*\mathbb{N} \subseteq {}^*\mathbb{R}$.

The *standard part mapping* associates with each finite $x \in {}^*\mathbb{R}$ a unique real $r = {}^\circ x = st(x)$ such that $x \approx r$ (i.e. $x - r$ is *infinitesimal*).

Let $\mathscr{C} = C_0[0,1]$ = continuous functions with $f(0) = 0$. The *nearstandard* or S-*continuous* functions in $^*\mathscr{C}$ are those functions g with $g(s) \approx g(t)$ whenever $s \approx t$ (for $s, t \in {}^*[0,1]$). For each such g there is a unique (standard) $f \in \mathscr{C}$ that is infinitely close to g; i.e. $f(t) \approx g(t)$ for each $t \in [0,1]$; we write $f = {}^\circ g = st(g)$.

An alternative way to represent functions in $\mathscr{C}$ nonstandardly is as follows. Fix $N \in {}^*\mathbb{N}\backslash\mathbb{N}$, for convenience of the form $N = H!$,and let $\Delta t = 1/N$ (≈ 0). Define the discrete time line T by

$$T = \{\Delta t,\ 2\Delta t, \ldots N\Delta t = 1\}.$$

An internal (nonstandard) function $X\colon T \to {}^*\mathbb{R}$ is S-continuous if $s \approx t$ implies $X_s \approx X_t$ all $s,t \in T \cup \{0\}$, where we specify $X_0 = 0$. This is equivalent to $\overline{X} \in {}^*\mathscr{C}$ being S-continuous, where $\overline{X}$ is the polygonalisation of X, and we write ${}^\circ X = st(X) = {}^\circ\overline{X}$.

1. NONSTANDARD CONSTRUCTION OF BROWNIAN MOTION

In the paper [2] Anderson constructed Brownian motion from an internal (nonstandard) random walk as follows.

Let Ω^A be the collection of all paths $B\colon T \cup \{0\} \to {}^*\mathbb{R}$ with $B(0) = 0$, and with $\Delta B(t) = B(t+\Delta t)-B(t) = \pm\Delta t^{1/2}$, and let ν be the nonstandard counting probability on Ω^A. Thus the canonical process $B_A(t,\omega) = \omega(t)$ for $\omega \in \Omega^A$ is an internal symmetric random walk, with $\Delta B_A^2 = \Delta t$.

Let $P = \nu_L$, the Loeb measure corresponding to ν. This is the standard measure defined by setting $\nu_L(D) = {}^\circ\nu(D)$ for internal ('nonstandard') $D \subseteq \Omega^A$ and extending to the σ-algebra generated by such D. (Loeb [11] showed that this is possible.) Anderson showed in [2]:

1.1 Theorem (a) For P-almost all $\omega \in \Omega^A$, $B_A(\cdot,\omega)$ is S-continuous

(b) If we define $b_A(\cdot,\omega) = {}^\circ B_A(\cdot,\omega)$ then $b_A(t,\omega)$ is Brownian motion on [0,1].

The essence of Anderson's proof is to use the Central Limit theorem to show that ${}^\circ B_A(t,\cdot) \sim \mathcal{N}(0,{}^\circ t)$ for each t, and to use the reflection property of B_A to obtain the S-continuity.

We shall present a slight modification of Anderson's idea below, with $\Delta B_t \sim \mathcal{N}(0,\Delta t)$, independent, so that $B(t,\cdot) \sim \mathcal{N}(0,t)$ automatically.

First fix an internal probability space $\boldsymbol{\Omega} = (\Omega,\mathcal{A},\mu)$ carrying independent $\mathcal{N}(0,t)$ random variables $(\eta_t)_{t\in T}$. Define a process $B: T \cup \{0\} \times \Omega \to {}^*\mathbb{R}$ by

$$\begin{cases} B(0,\omega) = 0 \\ \Delta B(t,\omega) = B(t,\omega)-B(t-\Delta t,\omega) = \eta_t \quad (t \in T). \end{cases}$$

Let $P = \mu_L$. Then we have the following modification of Theorem 1.1.

1.2 Theorem (a) For P-a.a. ω, $B(\cdot,\omega)$ is S-continuous

(b) The process $b(\cdot,\omega) = {}^\circ B(\cdot,\omega)$ is Brownian motion.

Proof (After [4]). Note first that an alternative description of B is

$$B(t,\omega) = \sum_{s\le t} \eta_s$$

so that $B(t,\cdot)$ is $\mathcal{N}(0,t)$; so $E(B(t)^4) = 3t^2$ and

$$\mu(|B(t)|\ge\alpha) \le 3t^2/\alpha^4 \tag{1.3}$$

by Chebychev.

Now $B(\cdot,\omega)$ is S-continuous iff $B(\cdot,\omega) \in C$ where

$$C = \bigcap_k \bigcup_m \{X \in {}^*\mathbb{R}^T: |s-t| < \tfrac{1}{m} \Rightarrow |X_s-X_t| < \tfrac{1}{k}\}$$

$$= \bigcap_k \bigcup_m C_{km} \quad \text{say.}$$

(As noted before, we set $X_0 = 0$ for $X \in {}^*\mathbb{R}^T$)

We will show that $P(B(\cdot,\omega) \in C^c) = 0$. It is sufficient to show that $\mu(B(\cdot,\omega) \in C^c_{k,M}) \approx 0$ for all sufficiently small infinite M. Taking $M \le H$, since $M|N$ we have $\frac{1}{M} \in T$ and

$$C^c_{k,M} = \{X: \exists s,t \text{ with } |s-t| < \frac{1}{M} \text{ and } |X_s - X_t| \ge \frac{1}{k}\}$$

$$\subseteq \bigcup_{r=0}^{M-2} \{X: \exists u \in [\frac{r}{M}, \frac{r+2}{M}] \text{ with } |X_u - X_{r/M}| \ge \frac{1}{2k}\}.$$

Now, given the identical distribution of the increments of B, we have (writing B for $B(\cdot,\omega)$ and B_u for $B(u,\omega)$)

$$P(B \in C^c_{k,m}) \;\le\; (M-1)P(\exists u \le \frac{2}{M} \text{ with } |B_u| \ge \frac{1}{2k}).$$

The reflection property of B means that given $u \le \frac{2}{M}$ and $|B_u| \ge \frac{1}{2k}$), then with probability at least ½ we will have $|B_{2/M}| \ge \frac{1}{2k}$ (see the following diagram).

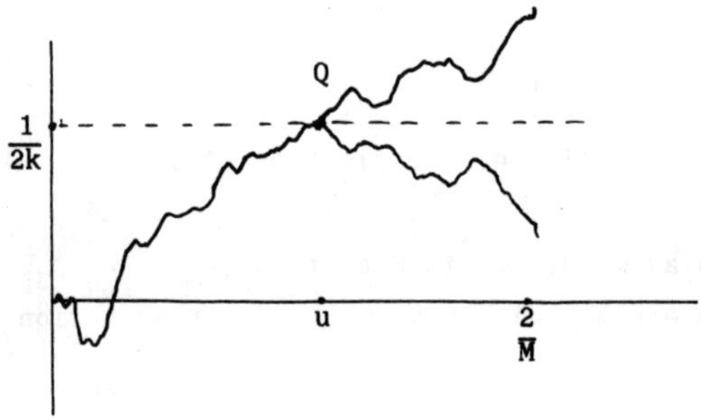

For every path starting from Q there is a corresponding reflected path.

Thus
$$P(B \in C^c_{k,m}) \le 2(M-1)\, P(|B_{2/m}| \ge \frac{1}{2k})$$
$$\le 2(M-1)3(\frac{2}{M})^2(2k)^4 \qquad \text{by (1.3)}$$
$$\approx 0.$$

Thus we can define a continuous process $b = {}^\circ B$; i.e.

$$b_{ot}(\omega) = {}^\circ B(t,\omega)$$

It is immediate from the construction that the increments of b are independent, and normally distributed with the correct variances. Hence b is Brownian motion.

Remarks on White Noise Analysis

1. This construction may provide a better nonstandard model for Hida's white noise calculus than that proposed by Takenaka [14]. What is required is a model for Brownian motion $(B_t)_{t\in T}$ such that $\dot{B}_t = \Delta B_t/\Delta t$ is $\mathcal{N}(0,\Delta t^{-1})$, which is precisely what we have. Takenaka's construction is essentially to modify Anderson's to sample the hyperfinite random walk B_A at a courser set of time points, so that increments are *approximately* $\mathcal{N}(0, \Delta t^{-1})$ (via CLT).

2. The strong law of large numbers implies that for $\dot{B} = (\dot{B}_t)_{t\in T}$

$$\frac{\|\dot{B}\|_2}{\sqrt{N}} \approx 1 \quad \text{almost surely}$$

(where $\|\dot{B}\|_2^2 = \Sigma\, \dot{B}_t^2\Delta t$) and, in the *Euclidean norm

$$\|\Delta B\| \approx 1 \quad \text{almost surely}$$

where $\Delta B = (\Delta B_t)_{t\in T}$. These remarks give some sense to the idea that white noise is uniform measure on $S^{\infty}(\sqrt{\infty})$. Alternatively, we can proceed as follows. Let $\mathcal{H}$ be the non-standard hull of $^*\mathbb{R}^T$ under the *Euclidean norm; i.e.

$$\mathcal{H} = \text{Finite }(^*\mathbb{R}^T)/\approx.$$

Then the last remark above shows that $^\circ\Delta B$ lies on the unit sphere $S(\mathcal{H})$ in $\mathcal{H}$ almost surely; and it is uniformly distributed on $S(\mathcal{H})$. This gives another representation of white noise that may prove useful.

The above construction of b gives a construction of Wiener measure that can be expressed as follows. Let Γ be the internal measure on $^*\mathbb{R}^T$ induced by B; i.e.

$$\Gamma(A) = \mu(B(\cdot,\omega) \in A)$$

$$= (2\pi\Delta t)^{-N/2}\int_A \prod_{t\in T} \exp\left[-\frac{(X_t-X_{t-\Delta t})^2}{2\Delta t}\right] dX_{\Delta t}dX_{2\Delta t}\ldots dX_1$$

where dX_t denotes *Lebesgue measure over $^*\mathbb{R}$. Writing dX for *Lebesgue measure on $^*\mathbb{R}^T$ and $\dot{X}_t = \dfrac{X_t-X_{t-\Delta t}}{\Delta t} = \Delta X_t/\Delta t$ we can rewrite this as

$$\Gamma(A) = \kappa \int_A \exp(-\tfrac{1}{2} \sum_{t\in T} \dot{X}_t^2 \Delta t)dX \qquad (1.4)$$

where $\kappa = (2\pi\Delta t)^{-N/2}$. Then Theorem 1.2 says

1.5 Theorem For almost all $X \in {}^*\mathbb{R}^T$ (with respect to Γ_L) X is S-continuous, and Wiener measure W is given by

$$W(D) = \Gamma_L(st^{-1}(D)) \qquad (D \subseteq \mathscr{C}). \qquad (1.6)$$

(where $st^{-1}(D) = \{X \in {}^*\mathbb{R}^T : {}^\circ X \in D\}$).

Remark The formulae (1.4), (1.6) may be viewed as a way to make precise Donsker's intuitive flat integral formula for W:

$$W(D) = \kappa \int_D \exp(-\tfrac{1}{2} \int_0^1 \dot{x}_t^2 dt)dx$$

where dx is a 'flat' (i.e. translation invariant) measure on $\mathscr{C}$.

We shall now give three applications of this construction - the Cameron-Martin formula, the Brownian Bridge and Schilder's Large Deviation principle.

2. THE WIENER INTEGRAL AND CAMERON-MARTIN FORMULA

Anderson [2] showed how to define the Wiener integral nonstandardly (in fact he did this for the Itô integral), and his approach is valid for our modification.

2.1. **Theorem** Let $y \in L^2[0,1]$ with SL^2 lifting $Y \in {}^*\mathbb{R}^T$. Then

$$\int_0^1 y_t db_t = {}^\circ\left[\sum_{t\in T} Y_t \Delta B_t\right] \qquad (a.s) \qquad (2.2)$$

Proof (Sketch)

Show that the right hand side is independent of the lifting Y, and verify (2.2) for step functions; then show that (2.2) is preserved under limits in L^2.

Remark Anderson showed that for $s \leq 1$,

$$^\circ\left[\sum_{t\leq s} Y_t \Delta B_t\right] = \int_0^s y_t db_t$$

defines a version of the Wiener integral that is continuous in s for almost all ω.

Our first application is

2.3 **Theorem** (Cameron-Martin)

Let $z \in \mathscr{C}$ with $\dot{z} \in L^2$, and define W^z by

$$W^z(A) = W(A-z). \qquad (A \subseteq \mathscr{C})$$

Then $W^z \ll W$ and

$$\frac{dW^z}{dW}(x) = \exp\left[\int_0^1 \dot{z}_t dx_t - \tfrac{1}{2}\int_0^1 \dot{z}_t^2 dt\right]$$

where the first integral on the right is the Wiener integral.

Proof (Sketch) Let $Z_t = {}^*z_t$ for $t \in T$, and then

$$\dot{Z}_t = \Delta t^{-1} \int_{t-\Delta t}^{t} {}^*\dot{z}_\tau d\tau.$$

$\dot{Z}$ is an SL^2 lifting of $\dot{z}$, (see [4] or [12]), which means that

$$\sum \dot{Z}_t^2 \Delta t \approx \int \dot{z}_t^2 dt. \tag{2.4}$$

Now $^\circ Z = z$, and so

$$\begin{aligned} W^z(A) = W(A-z) &= \Gamma_L(st^{-1}(A-z)) \\ &= \Gamma_L(st^{-1}(A) - Z)) \\ &= \Gamma_L^Z(st^{-1}(A)) \end{aligned}$$

where $\Gamma^Z(D) = \Gamma(D-Z)$ for $D \subseteq {}^*\mathbb{R}^T$. We have

$$\begin{aligned} \Gamma^Z(D) &= \kappa \int_{D\ Z} \exp\ (-\tfrac{1}{2} \sum_{t\in T} \dot{X}_t^2\ \Delta t) dX \\ &= \kappa \int_D \exp(-\tfrac{1}{2} \sum_{t\in T} (\dot{X}_t - \dot{Z}_t)^2 \Delta t) dX \end{aligned}$$

(since dX is translation invariant)

$$= \kappa \int_D \exp(\sum \dot{Z}_t \Delta X_t - \tfrac{1}{2} \sum \dot{Z}_t^2 \Delta t) \exp(-\tfrac{1}{2} \sum \dot{X}_t^2 \Delta t) dX$$

$$= \int_D R(X)d\Gamma(X)$$

so that nonstandardly $\Gamma^Z \ll \Gamma$ with $\frac{d\Gamma^Z}{d\Gamma} = R$

where
$$R(X) = \Sigma\, \dot{Z}_t \Delta X_t - \tfrac{1}{2} \Sigma\, \dot{Z}_t^2 \Delta t$$
$$\approx \int_0^1 \dot{z}_t dx_t - \tfrac{1}{2} \int \dot{z}_t^2 dt \quad \text{a.s.} \quad (\text{where } x = {}^\circ X)$$
$$= \rho(x), \text{ say.}$$

We can show that R is S-integrable, which implies that

$$\Gamma_L^Z \ll \Gamma_L \text{ with } \frac{d\Gamma_L^Z}{d\Gamma_L}(X) = {}^\circ R(X) = \rho({}^\circ X).$$

But $W(\cdot) = \Gamma_L(st^{-1}(\cdot))$ and $W^Z = \Gamma_L^Z(st^{-1}(\cdot))$ so this shows that $dW^Z/dW = \rho$ as required.

3. THE BROWNIAN BRIDGE

For a fixed $a \in \mathbb{R}$ the Brownian bridge b^a: $[0,1] \to \mathbb{R}$ is a continuous process that is Brownian motion conditioned to be at $a \in \mathbb{R}$ at time 1. The corresponding measure W^a is a probability on $\mathscr{C}^a = \mathscr{C} \cap \{x : x_1 = a\}$, so that

$$W^a(A) = P(b^a \in A) \qquad \text{for } A \subseteq \mathscr{C}^a.$$

The family of measures W^a has the property that for Borel $D \subseteq \mathscr{C}$,

$$W(D) = \int_{\mathbb{R}} \varphi(a)\, W^a(D \cap \mathscr{C}^a) da \tag{3.1}$$

where $\varphi(a)$ is the normal density $(2\pi)^{-\frac{1}{2}}\exp(-\frac{1}{2}a^2)$. Of course it is required that $W^a(D \cap \mathscr{C}^a)$ be measurable in a for any such D.

We will sketch below how the construction of Brownian motion above provides a natural construction of b^a and W^a, and provides the working characterisation of b^a as a Wiener integral that is often used in practice. For full details see [6].

Let $S = T \setminus \{1\}$ and for convenience write $J(X) = \frac{1}{2} \sum_{t\in T} \dot{X}_t^2 \Delta t$ for $X \in {}^*\mathbb{R}^T$.

For $a \in {}^*\mathbb{R}$ the natural nonstandard counterpart of W^a is the internal measure Γ^a on ${}^*\mathbb{R}^S \times \{a\}$ given by

$$\Gamma^a(D) = \kappa\, \varphi(a)^{-1} \int_D \exp(-\,J(Y))dY$$

where dY = *Lebesgue measure on ${}^*\mathbb{R}^S \times \{a\}$. Clearly we have

3.2 Proposition For *Borel $A \subseteq {}^*\mathbb{R}^T$,

$$\Gamma(A) = \int_{{}^*\mathbb{R}} \varphi(a)\Gamma^a(A^a)da$$

where $A^a = \{X \in A : X_1 = a\}$.

Proof Transfer standard finite dimensional theory.

It is easy to check that

$$\Gamma^a(D) = \Gamma^{a+d}(D+(dt)_{t\in T}). \tag{3.3}$$

This is used to establish

3.4 Theorem For finite $a \in {}^*\mathbb{R}$, the paths in ${}^*\mathbb{R}^S \times \{a\}$ are S-continuous a.a. (Γ^a_L).

Proof By (3.3) it is sufficient to prove this for $a = 0$. If $C \subseteq {}^*\mathbb{R}^S \times \{0\}$ is an internal set of S-discontinuous paths then $D = \bigcup_{|a|\le 1} (C+(at)_{t\in T})$ is also such a set, with $D^a = C+(at)_{t\in T}$ for $|a|\le 1$. So (from Theorem 1.5)

$$0 \approx \Gamma(D) = \int_{|a|\le 1} \varphi(a)\Gamma^a(C+(at)_{t\in T})da$$

$$= \int_{|a|\le 1} \varphi(a)\Gamma^0(C)da \qquad \text{(by 3.3)}$$

and so $\Gamma^0(C) \approx 0$.

We are now in a position to define measures $(W^a)_{a\in\mathbb{R}}$ by

$$W^a(A) = \Gamma_L^a(st^{-1}(A)) \qquad (A \subseteq \mathscr{C}^a)$$

and it is a routine to prove:

3.5 Theorem W^a is a regular conditional probability of W conditional on $x_1 = a$. i.e. (3.1) holds.

Proof Omitted.

The *Brownian bridge* b^a is now defined to be the canonical process given by W^a with paths in $\mathscr{C}^a$. In the literature b^a is often defined in one of the following ways:

$$\text{(3.6)}\quad \text{(a)}\qquad \begin{cases} b_t^a = (1-t)\displaystyle\int_0^t \frac{dw_s}{1-s} + ta & 0 \le t < 1 \\ b_1^a = a \end{cases}$$

where w is a Brownian motion,

or

(3.6) (b) b_t^a is a continuous process such that $b_t^a - at + \displaystyle\int_0^t \frac{b_t^a - at}{1-t}\,dt$ is a Brownian motion.

It is not immediately clear that these give conditional Brownian motion; we shall see that this is so from our construction. It is sufficient to restrict to the case a = 0.

We write Ω^0 for $^*\mathbb{R}^S \times \{0\}$. Consider J(X) for $X \in \Omega^0$:

$$\begin{aligned} J(X) &= \tfrac{1}{2} \sum_{t\in T} \dot{X}_t^2 \Delta t \\ &= \frac{1}{2\Delta t}(X_{\Delta t}^2 + (X_{2\Delta t} - X_{\Delta t})^2 + \ldots + (X_{1-\Delta t} - X_{1-2\Delta t})^2 + X_{1-\Delta t}^2). \end{aligned}$$

A simple completion of squares from the right shows

3.7 Lemma (a) $\sum_{t\in T} \dot{X}_t^2 = \sum_{t\in S} \hat{X}_t^2$

where $\hat{X}_t = (\alpha_t X_t - \frac{1}{\alpha_t} X_{t-\Delta t})/\Delta t$ for $t \in S$

and $\alpha_t = \left[\frac{1-t+\Delta t}{1-t}\right]^{1/2} = \left[1 + \frac{\Delta t}{1-t}\right]^{1/2}$ (so that $\alpha_t \approx 1$ provided $t \not\approx 1$).

(b) $\sum_{t\in T} \dot{X}_t \dot{Y}_t = \sum_{t\in S} \hat{X}_t \hat{Y}_t$.

Remark The whole point here is the reduction of the number of independent terms to $N-1 = |S|$, the dimension of Ω^0.

Routine calculation shows

3.8 Lemma

(a) $\dot{X}_t = \alpha_t \hat{X}_t - \frac{X_t}{1-t}$ $(t < 1)$

(b) $X_s = \sum_{t\leq s} \left[\frac{1-s}{1-t}\right] \alpha_t^{-1} \hat{X}_t \Delta t$ $(s < 1)$

Let B^0 be the canonical internal process on Ω^0 given by Γ^0, so that $°B^0 = b^0$ is the Brownian bridge, and define $\bar{\eta} \in {}^*\mathbb{R}^S$ by

$$\bar{\eta} = \widehat{B^0}\Delta t.$$

Then we have

3.9 Lemma The variables $(\bar{\eta}_s)_{s\in S}$ are $\mathcal{N}(0,\Delta t)$ and independent under Γ^0.

Proof For $D \subseteq {}^*\mathbb{R}^S$

$$\Gamma^0(\bar{\eta} \in D) = \kappa(2\pi)^{1/2} \int_{\substack{\hat{X}\Delta t\in D \\ X\in\Omega^0}} \exp(-\tfrac{1}{2} \Sigma \dot{X}_t^2 \Delta t)dX$$

Make the change of variable $Y = \hat{X}\Delta t \in {}^*\mathbb{R}^S$; by Lemma 3.7(a) we have

$$\Gamma^0(\bar{\eta} \in D) = \kappa' \int_D \exp(-\tfrac{1}{2} \sum_{t\in S} Y_t^2/\Delta t)dY$$

where the constant κ' must be $(2\pi\Delta t)^{-\frac{1}{2}(N-1)}$ to normalise. This is precisely the required distribution for $\overline{\eta}$.

As a consequence of Lemma 3.9, if we define

$$\beta_t = \sum_{s\leq t} \overline{\eta}_s \qquad (t\in S)$$

then (as in §1) this gives a Brownian motion w say, where

$$w = {}^\circ\beta.$$

We easily have

3.10 Theorem

(a) $$b^0_s = w_s - \int_0^s \frac{b^0_t}{1-t}\,dt \qquad (s < 1)$$

(b) $$b^0_s = \int_0^s \frac{(1-s)}{(1-t)}\,dw_t \qquad (s < 1)$$

Proof Combine Lemma 3.8(a,b) together with Theorem 2.1 and the fact that $\alpha_t \approx 1$ if $t \not\approx 1$, so that α_t is a lifting of the function that is constantly 1. Note that under Γ^0, ${}^\circ X = b^0$ and ${}^\circ(\Sigma\hat{X}\Delta t) = w$.

Another application of this approach to the Brownian bridge is to obtain a Cameron-Martin formula in the same way as for Wiener measure. Instead of the Wiener integral we have an integral

$$\int_0^1 y_t\,db^0_t$$

defined for those $y \in L^2$ with $\int y = 0$, and we consider translations of W^a by $z \in \mathscr{C}^0$ with $\dot{z} \in L^2$. For details see [6].

4. LARGE DEVIATIONS - SCHILDER'S THEOREM

Schilder's Large Deviation Principle (LDP) shows that the convergence to zero of $W_\delta(A) = W(\delta^{-1}A)$ is closely related to $\inf I(A)$ where, for $x \in \mathscr{C}$

$$I(x) = \begin{cases} \frac{1}{2}\int_0^1 \dot{x}_t^2\, dt & \text{if } x \text{ is absolutely continuous and } \dot{x} \in L^2 \\ \infty & \text{otherwise.} \end{cases}$$

I(x) is known as the *action* of the path x.

Recall our nonstandard flat integral formula for W in §1, which involved the nonstandard representation of I:

$$J(X) = \tfrac{1}{2} \sum_{t \in T} \dot{X}_t^2 \Delta t \qquad (X \in {}^*\mathbb{R}^T).$$

The flat integral representation of W provides a fairly simple and intuitive proof of Schilder's LDP, which we now sketch; for full details see [4].

First note

4.1. **Proposition** If J(X) is finite then X is S-continuous, and $I({}^\circ X) \le {}^\circ J(X)$.

Proof Omitted.

Also:

4.2 **Proposition** If $x \in \mathscr{C}$ and we define $J({}^*x) = J(({}^*x_t)_{t \in T})$ then $I(x) = {}^\circ J({}^*x)$.

Proof Omitted.

Schilder's Theorem has two parts, corresponding to the cases A open and A closed.

4.3. **Schilder's Theorem - Open Set.**

Let A be open, $A \subseteq \mathscr{C}$. Then

$$\underline{\lim}\ \delta^2 \log W_\delta(A) \ \ge\ -\inf I(A).$$

Proof Let $z \in A$ with $I(z) < \infty$; we show that

$$\underline{\lim}\ \delta^2 \log W_\delta(A) \ \ge\ -I(z).$$

Since A is open we can find standard r > 0 with $A \supseteq \{x \in \mathscr{C} : |x-z| \le r\} = A_1$ say. Let $D = \{X : |X-Z| \le r\}$ be the corresponding internal set, where $Z_t = ({}^*z_t)$ $(t \in T)$. (Here $|\cdot|$ denotes the uniform norm on $\mathscr{C}$ and on ${}^*\mathbb{R}^T$.)

Clearly, for standard $\delta > 0$, using Theorem 1.5 we have

$$\begin{aligned} W_\delta(A) &= W(\delta^{-1}A) = \Gamma_L(\delta^{-1}st^{-1}(A)) \\ &\geq \Gamma_L(\delta^{-1}D) = {}^\circ\Gamma(\delta^{-1}D) \\ &\approx \kappa \int_{\delta^{-1}D} \exp(-J(X))dX \qquad \text{(from (1.4))} \\ &= \kappa \int_C \exp(-J(Y + \delta^{-1}Z))dY \end{aligned}$$

(putting $Y = X - \delta^{-1}Z$, and putting $C = \{Y : |Y| \leq \delta^{-1}r\}$)

$$\begin{aligned} &= \kappa \int_C \exp\left(-\frac{J(Z)}{\delta^2} - \frac{1}{\delta}\Sigma\, \dot{Z}_t\Delta Y_t\right)\exp(-J(Y))dY \\ &= \int_C \exp\left(-\frac{J(Z)}{\delta^2} - \frac{1}{\delta}\Sigma\, \dot{Z}_t\Delta Y_t\right)d\Gamma(Y). \end{aligned}$$

Jensen's inequality gives

$$\delta^2 \log W_\delta(A) \geq -J(Z) - \frac{\delta}{\Gamma(C)}\int_C (\Sigma\, \dot{Z}_t\Delta Y_t)d\Gamma(Y) + \delta^2 \log \Gamma(C) \tag{4.4}$$

Now for $\delta \approx 0$ we have $C \supseteq$ S-continuous paths, so $\Gamma(C) \approx 1$; moreover

$$\begin{aligned} \left|\int_C (\Sigma\, \dot{Z}_t\Delta Y_t)d\Gamma(Y)\right|^2 &\leq E_\Gamma(\Sigma\, \dot{Z}_t\Delta Y_t)^2 = \Sigma\, \dot{Z}_t^2\Delta t \\ &= 2J(Z) \approx 2I(z) < \infty. \end{aligned}$$

Hence, for $\delta \approx 0$ the right side of (4.4) is $\gtrsim -J(Z)$, which shows that

$$\underline{\lim}\ \delta^2 \log W_\delta(A) \geq -I(z).$$

The closed set part of Schilder's Theorem is a little more delicate.

4.5 Schilder's Theorem - Closed Set

Let A be closed, $A \subseteq \mathscr{C}$. Then

$$\overline{\lim}\ \delta^2 \log W_\delta(A) \leq -\inf I(A).$$

Proof Let $\gamma < \inf I(A)$; it is sufficient to show that

$$\overline{\lim}\ \delta^2 \log W_\delta(A) \leq -\gamma.$$

Let $D = \{x \in {}^*\mathscr{C} : J(x) \geq \gamma\}$ where $J(x) = J((x_t)_{t\in T})$ for $x \in {}^*\mathscr{C}$; then ${}^*A \backslash D \subseteq NS^c$ where NS^c denotes the complement in ${}^*\mathscr{C}$ of the set of paths that are nearstandard in the uniform topology (equivalently, that are S-continuous).

We shall prove

$$\overline{{}^*\lim_{\delta^2\to 0}}\ \delta^2 \log {}^*W_\delta(D) \le -\gamma \tag{4.6}$$

and

$$\overline{{}^*\lim_{\delta^2\to 0}}\ \delta^2 \log {}^*W_\delta({}^*A\backslash D) \le -q \tag{4.7}$$

for any finite q.

These together imply that $\overline{{}^*\lim_{\delta\to 0}}\ \delta^2 \log {}^*W_\delta({}^*A) \le -\gamma$ which gives the result by transfer.

Proof of (4.6) Putting $C = \{X \in {}^*\mathbb{R}^T : J(X) \ge \gamma\}$ we have (for $0 < \delta < 1$ in ${}^*\mathbb{R}$):

$$\begin{aligned}
{}^*W_\delta(D) &= {}^*W(\delta^{-1}D) \\
&= \kappa \int_{\delta^{-1}C} \exp(-J(X))dX \\
&= \kappa\delta^{-N} \int_C \exp(-\delta^{-2}J(Y))dY \qquad \text{(putting } Y = \delta X) \\
&= \kappa\delta^{-N} \int_C \exp((1-\delta^{-2})J(Y))\exp(-J(Y))dY \\
&= \delta^{-N} \int_C \exp((1-\delta^{-2})J(Y))d\Gamma(Y) \\
&\le \delta^{-N} \exp((1-\delta^{-2})\gamma)\Gamma(C)
\end{aligned}$$

Thus

$$\begin{aligned}
\delta^2 \log {}^*W_\delta(D) &\le \delta^2 \log (\delta^{-N}) + (\delta^2-1)\gamma + \delta^2 \log \Gamma(C) \\
&\to -\gamma \quad \text{as } \delta \to 0 \text{ in } {}^*\mathbb{R}.
\end{aligned}$$

Proof of (4.7) We have

$$ {}^*A\backslash D \subseteq NS^c \subseteq \bigcup_{k\in\mathbb{N}} \bigcap_{m\in\mathbb{N}} D_{k,m}$$

where (as in the proof of Theorem 1.2)

$$D_{k,m} = \{x \in {}^*\mathscr{C} : \exists s,t \text{ with } |s-t| < \tfrac{1}{m} \text{ and } |x_s-x_t| \ge \tfrac{1}{k}\}$$

Now, for any standard sequence m_k (which we will choose later),

$$*A\backslash D \subseteq \bigcup_{k\leq K} D_{k,m_k} \qquad \text{all infinite } K$$

so by saturation there is finite k_0 with

$$*A\backslash D \subseteq \bigcup_{k\leq k_0} D_{k,m_k}$$

Note that, exactly as in the proof of Theorem 1.2, we have

$$*W_\delta(D_{k,m}) \leq 2(m-1)*W_\delta(|x_{2/m}| \geq \frac{1}{2k})$$

Now x_t under $*W_\delta$ is $N(0,\delta^2 t)$ and so using the well known standard estimate for a $\mathcal{N}(0,\sigma^2)$ random variable θ,

$$P(\theta \geq \gamma) \leq \exp\left(-\frac{\alpha^2}{2\sigma^2}\right),$$

we have

$$*W_\delta(D_{k,m}) \leq 2(m-1)2\exp\left[-\frac{m}{16k^2\delta^2}\right].$$

Now choose the sequence $m_k = 16k^2q$. Then

$$*W\delta(A\backslash D) \leq \sum_{k\leq k_0} *W_\delta(D_{k,m_k})$$

$$\leq 4(m_{k_0}-1)k_0 \exp\left(-\frac{q}{\delta^2}\right)$$

So $$\delta^2 \log *W_\delta(*A\backslash D) \leq \delta^2 \log(4(m_{k_0}-1)k_0) - q$$

$$\to -q \quad \text{as} \quad \delta \to 0 \text{ in } *\mathbb{R}.$$

So $\overline{\lim}\ \delta^2 \log *W_\delta(*A \setminus D) \leq -q$ as required.

5. LÉVY BROWNIAN MOTION AND FRACTIONAL BROWNIAN MOTION

The flat integral representation for Wiener measure was generalised in the paper [7] to the measure induced by Lévy Brownian motion (LBM) on $\mathbb{R}^d$, based on a modification of Stoll's nonstandard construction [13], which is expounded in the book [1]. The ideas extend naturally to fractional Brownian motion (FBM).

5.1 Definition (a) *Fractional Brownian motion* (FBM) on $\mathbb{R}^d$ with index γ (for $(0 < \gamma < 2)$ is a continuous random field $x: \mathbb{R}^d \to \mathbb{R}$ such that

(i) $x_0 = 0$

(ii) $x_t - x_s \sim \mathcal{N}(0, \|t-s\|^\gamma)$ $(s, t \in \mathbb{R}^d)$

(b) FBM with index $\gamma = 1$ is *Lévy Browninan motion* (LBM) on $\mathbb{R}^d$, denoted by $\ell: \mathbb{R}^d \to \mathbb{R}$.

Stoll's nonstandard construction of LBM [13] gave a new white noise integral representation

$$\ell_s = \int_{\mathbb{R}^d} f_{s,t}\xi(dt)$$

where ξ is a one-dimensional white noise measure on $\mathbb{R}^d$ and the kernels $f_{s,t}$ are given explicitly by

$$f_{s,t} = k_d(\|s-t\|^{\frac{1}{2}(1-d)} - \|t\|^{\frac{1}{2}(1-d)}).$$

To obtain FBM it is only necessary to replace the exponents $\frac{1}{2}(1-d)$ by $\frac{1}{2}(\gamma-d)$; this is valid for all $d \geq 1$ and $0 < \gamma < 2$ except $\gamma = d = 1$, which is ordinary Brownian motion.

For a flat integral representation for the measure ν on $\mathscr{C} = C_0(\mathbb{R}^d; \mathbb{R})$ induced by FBM it is necessary to replace the time set T in §1 by a lattice $S = \{-N, -N+\Delta t, \ldots, 0, \Delta t, 2\Delta t, \ldots N-\Delta t, N\}^d$. Functions in $\mathscr{C}$ will be represented nonstandardly by members of $^*\mathbb{R}^S$. Let F be the S×S matrix $(F_{s,t})_{s,t\in S}$ given by

$$F_{s,t} = {}^*f_{s,t} \qquad (s,t \in S)$$

and for $X \in {}^*\mathbb{R}^S$ let $\mathring{X} = F^{-1}X/\Delta s$.

Define an internal probability Γ on $^*\mathbb{R}^S$ by

$$\Gamma(D) = \kappa \int_D \exp(-\tfrac{1}{2} \sum_{s\in S} \mathring{X}_s^2 \Delta s)dX$$

where $\Delta s = \Delta t^d$ and dX is *Lebesgue measure on $^*\mathbb{R}^S$, with κ a normalising constant.

A modification of Stoll's proof shows:

5.2 Theorem Let x be FBM on $\mathbb{R}^d$ with index γ and let $\nu(A) = P(x \in A)$ for $A \subseteq \mathscr{C}$. Then almost all 'paths' in $^*\mathbb{R}^S$ are S-continuous under Γ_L, and

$$\nu(A) = \Gamma_L(st^{-1}(A)) \qquad (A \subseteq \mathscr{C}).$$

The internal functional $\hat{X}$ plays the same rôle for FBM as $\dot{X}$ for Brownian motion; and we have a corresponding *action functional* I as follows.

Define $\varphi: L^2(\mathbb{R}^d) \to \mathscr{C}$ by

$$(\varphi(y))_s = \int f_{s,t} y_t dt \tag{5.3}$$

$$= \langle f_s, y\rangle \qquad \text{(inner product in } L^2\text{)}.$$

Then φ is injective, with range $H = \varphi(\mathscr{C})$ say; H is the Cameron-Martin subspace of $\mathscr{C}$ (with respect to ν); the inner product for x, y $\in$ H is given by

$$(x,y)_H = (\varphi^{-1}(x), \varphi^{-1}(y))_{L^2} = \langle\hat{x}, \hat{y}\rangle$$

if we write $\hat{x} = \varphi^{-1}(x)$ for $x \in H$. Clearly $\hat{x}$ corresponds to the internal function $\hat{X}$. We now have

5.4 Definition The action functional I for FBM is defined by

$$I(x) = \begin{cases} \frac{1}{2}\langle\hat{x},\hat{x}\rangle = \frac{1}{2}(x,x)_H & \text{if } x \in H \\ \infty & \text{otherwise} \end{cases}$$

Now we can apply the flat integral representation for ν (Theorem 5.2) to obtain an explicit Cameron-Martin translation formula and a large deviation principle.

5.5 Theorem - Cameron Martin Formula. Let $z \in H$ and let $\nu^z(A) = \nu(A-z)$ for $A \subseteq \mathscr{C}$. Then $\nu^z \ll \nu$ with

$$\frac{d\nu^z}{d\nu}(x) = \exp(\langle\hat{z},\hat{x}\rangle) - I(z))$$

where the notation $\langle\hat{z},\hat{x}\rangle$ is used to denote a stochastic term that is similar to the Wiener integral in the case of Brownian motion. In fact, for any $y \in L^2$, $\langle y,\hat{x}\rangle$ takes the form

$$\langle y,\hat{x}\rangle = \int_{\mathbb{R}^d} y_t \xi^x(dt)$$

where ξ^x is a white noise constructed from the FBM x.

5.6 **Theorem - Large Deviation Principle** With I as defined above, the large deviation principle (4.3 and 4.5) for the measures $\nu^\delta(\cdot) = \nu(\delta^{-1}(\cdot))$ is valid.

The proofs of Theorems 5.5 and 5.6 follow lines that are similar to those for the corresponding results for Wiener measure (Theorems 2.3, 4.3 and 4.5) using in addition the nonstandard version of Kolmogorov's continuity theorem that is discussed in [1].

6. GAUSSIAN MEASURE ON A SEPARABLE HILBERT SPACE

If μ is a centred (mean 0) Gaussian measure on a Hilbert space H it is known that there is an orthonormal basis $(e_n)_{n\in\mathbb{N}}$ such that μ has covariance S_μ given by

$$S_\mu((x_n)_{n\in\mathbb{N}}) = (\alpha_n x_n)_{n\in\mathbb{N}}$$

with $\Sigma\alpha_n < \infty$, $\alpha_n \geq 0$. (Here we are identifying $x = \Sigma x_n e_n$ with $(x_n)_{n\in\mathbb{N}} \in \ell^2$.) The Cameron-Martin space is $H_0 = \{(x_n) : \Sigma x_n^2/\alpha_n < \infty\}$.

It is quite easy in this case to see that we have the following flat integral representation.

Let $N \in {}^*\mathbb{N}$, infinite as before and define Γ on ${}^*\mathbb{R}^N$ by

$$\Gamma(A) = \kappa \int_A \exp(-\tfrac{1}{2} \Sigma \frac{X_n^2}{\alpha_n}) dX$$

where $\kappa = \left[\prod_{n=1}^{N} 2\pi\alpha_n\right]^{-\frac{1}{2}}$. We have

6.1 Theorem For a.a. $X \in {}^*\mathbb{R}^N$, $({}^\circ X_n)_{n\in\mathbb{N}} \in \ell^2$ and for $D \subseteq \ell^2$

$$\mu(D) = \Gamma_L(st^{-1}(D))$$

where $st^{-1}(D) = \{X \in {}^*\mathbb{R}^N : ({}^\circ X_n)_{n\in N} \in D\}$.

From this follow routine proofs of the Cameron-Martin formula and the large deviation principle for μ, along the lines of earlier sections. The action functional for μ is $I(x) = \frac{1}{2}\sum x_n^2/\alpha_n$. Details are given in [8].

REFERENCES

[1] S. Albeverio, J. E. Fenstad, R. Høegh-Krohn, & T. Lindstrom, *Nonstandard Methods in Stochastic Analysis and Mathematical Physics*, Academic Press, New York 1986.

[2] R. M. Anderson, A non-standard representation for Brownian motion and Itô integration, *Israel J. Math.* 25 (1976) 15-46.

[3] N. J. Cutland, Nonstandard measure theory and its applications, *Bull. London Math. Soc.* 15 (1983) 529-589.

[4] N. J. Cutland, Infinitesimals in action, *J. London Math. Soc.* 35 (1987), 202-216.

[5] N. J. Cutland (Ed), *Nonstandard Analysis and its Applications*, CUP, 1988.

[6] N. J. Cutland, The Brownian bridge as a flat integral, *Math. Proc. Camb. Phil. Soc.* 106 (1989), 343-354.

[7] N. J. Cutland, An action functional for Lévy Brownian motion, submitted.

[8] N. J. Cutland, On Large deviations on Hilbert space, in preparation.

[9] Kuo, H.-H. *Gaussian Measures in Banach Spaces*, Lecture Notes in Mathematics, Vol. 463, 1975, Springer-Verlag, Berlin.

[10] T. Lindstrøm, An invitation to nonstandard analysis, in [5], pp.1-105.

[11] P. A. Loeb, Conversion from nonstandard to standard measure spaces and applications in probability theory, *Trans. Amer. Math. Soc.* 211 (1975), 113-122.

[12] H. Rodenhausen, A characterisation of nonstandard liftings of measurable functions and stochastic processes, *Israel J. Math.* 43 (1982) 1-22.

[13] A. Stoll, Nonstandard construction of Levy Brownian motion, *Prob. Th. and Related Fields* 71 (1986), 321-334.

[14] S. Takenaka, An approach to white noise calculus based on partitions of parameter space, in *Bielefeld Encounters in Mathematics and Physics VII*, BiBoS, University of Bielefeld, 1988.

MULTILEVEL BRANCHING SYSTEMS

D.A. DAWSON
Department of Mathematics and Statistics
Carleton University
Ottawa, Canada K1S 5B6

K.J. HOCHBERG
Department of Mathematics and Computer Science
Bar-Ilan University
52900 Ramat-Gan, Israel

Y. WU
Department of Mathematics and Statistics
Carleton University,
Ottawa, Canada K1S 5B6

ABSTRACT

We consider stochastic models of a multilevel poulation structure in which an individual at one population level consists of a collection of individuals at the next lower level. We first consider the existence problem for a general class of multilevel birth and death processes and obtain a non-explosion criterion. In the case of multilevel branching we consider the diffusion limit which is given by a measure-valued process. We do this for two cases, one in which the population at the lowest level has a spatial structure. The main tools which are developed are the Laplace functional and moment hierarchy. Finally we give two applications of the moment structure to the spatial structure and long-time behavior of the process.

1. Introduction

The notion of multilevel or hierarchical measure-valued process was introduced in Dawson (1985). Such processes arise naturally in many contexts, for example, a population model in which the successive levels in the hierarchy correspond to individuals, colonies, clans and so on. In this paper we consider models having two levels which can be described as follows. The first level consists of a finite number of branching particles and is described by a counting random variable. At level two, we assume that each "superparticle", that is, collection of level one particles, also undergoes a binary branching process in which it either dies or produces a copy (another level two particle (superparticle) having the same number of level one particles), each with probability one-half, after exponentially distributed holding times. Each of these replicas or superparticles then evolves independently and undergoes further branching, thus yielding a random number of copies. Thus this system can be described as an integer-valued random measure, $X(t)$, on Z^+. In particular $X(t,\{i\})$ denotes the number of superparticles at time t which contain exactly i level one particles.

2. The Two Level Birth and Death Process

Dynamic multilevel population systems can be studied from several different points of view. We will describe them as multilevel birth and death processes since this is the framework that most naturally extends to higher level situations. For the sake of simplicity we restrict our attention in this paper to the two level birth and death case.

We consider the two level birth and death process $\{X(t):t\geq 0\}$ as a $N(Z_+)$-valued pure jump Markov process where $N(Z_+)$ denotes the set of integer-valued measures on Z_+. A measure in $N(Z_+)$ can be represented by a vector $((n_i):i\in Z_+)$ where n_i denotes the number of level two particles consisting of exactly i level one particles.

The possible transitions of state in this process can be described as follows:

(a) the birth of a level 1 particle in a level 2particle of size i: $(n_i,n_{i+1}) \longrightarrow (n_i-1,n_{i+1}+1)$ at rate $\lambda_1(n_i,i)$,

(b) death of a level 1 particle in a level 2 particle of size i:

$(n_{i-1},n_i) \longrightarrow (n_{i-1}+1,n_i-1)$ at rate $\mu_1(n_i,i)$,

(c) birth of a level 2 particle of size i:

$n_i \longrightarrow n_i+1$ at rate $\lambda_2(n_i,i)$,

(d) death of a level 2 particle of size i:

$n_i \longrightarrow n_i - 1$ at rate $\mu_2(n_i, i)$.

To describe the generator G of this $N(Z_+)$-valued Markov process, we consider the class of test functions of the form

$F(\nu) = f(\Sigma\phi_i n_i) = f(\langle\phi,\nu\rangle)$, if $\nu = \Sigma\, n_i \delta_i$,

and where $\langle\phi,\nu\rangle = \int \phi d\nu$.

Applying (a)-(d) we obtain

$$Gf(N) = \lim_{t \downarrow 0} \frac{E\{F(X_t)\,|X_0 = \nu\} - F(X_0)}{t}$$

which is given by

$$\begin{aligned} GF(\nu) = & \sum_k \lambda_1(\nu(k),k)\left[f(\langle\phi,\nu\rangle - \phi(k) + \phi(k+1)) - f(\langle\phi,\nu\rangle)\right] \\ & + \sum_k \mu_1(\nu(k),k)\left[f(\langle\phi,\nu\rangle - \phi(k) + \phi(k-1)) - f(\langle\phi,\nu\rangle)\right] \\ & + \sum_k \lambda_2(\nu(k),k)\left[f(\langle\phi,\nu\rangle + \phi(k)) - f(\langle\phi,\nu\rangle)\right] \\ & + \sum_k \mu_2(\nu(k),k)\left[f(\langle\phi,\nu\rangle - \phi(k)) - f(\langle\phi,\nu\rangle)\right]. \end{aligned}$$

We can show that there exists a Markov process $\{X_t\}$ which is the unique solution of the $N(Z_+)$-valued martingale problem associated with G. To illustrate the ideas we will consider the two level pure birth process. To prove the existence of this process it suffices to demonstrate the absence of explosions. Non-explosion of a wide class of two level birth and death processes can then be obtained by a coupling argument.

Theorem 2.1

(a) Suppose that there exist functions $\lambda_i(.)$, $i=1,2$ such that for each $k \in Z_+$

(2.1) $\sup_j \lambda_1(n,k) \le \lambda_1(k)$, $\sup_k \lambda_2(n,k) \le \lambda_2(n)$,

and

(2.2) $\sum_k \frac{1}{\lambda_i(k)} = \infty, i=1,2.$

(b) Assume that $\sum_k \frac{1}{\lambda_1(n,k)} = \infty$ for each n, $\lambda_2(.)$

satisfies (2.2) and $\lambda_2(n_1)+\lambda_2(n_2) \le \lambda_2(n_1+n_2)$.

(c) $\lambda_1(n,k) \le c_1 nk$, and $\lambda_2(n,k) \le c_2 k$.

(This condition includes the case of multilevel branching.)

Under condition (a), (b), or (c) the two level pure birth process $\{X(t):t\ge 0\}$ has no explosion.

Proof. (a) Let us define the times between successive births as follows: $\tau_1 = \inf\{t:X(t)\ne X(0)\}$

$\tau_n = \inf\{t:X(t+\tau_1+\ldots+\tau_{n-1})\ne X(\tau_1+\ldots+\tau_{n-1})\}$

$T_n = \tau_1+\ldots+\tau_n$.

Because all jumps of the two level pure birth process are level 1 or level 2 births, therefore the first m jumps consist of $k\le m$ jumps of level 1 and $m-k$ jumps of level 2. Let $N^*(t) = \min\{k_0:X(t,k)=0\ \forall k\ge k_0\}$. We first prove that $N^*(t)$ has no explosion. Otherwise there is a subsequence of times τ_{m_k} at which there is a type one birth at $N^*(\tau_{m_k}-) = k$. But the Laplace transform of $\sum \tau_{m_k}$ is

$$(2.3)\qquad L(\theta) = \prod_{k=1}^{\infty}\left(1+\frac{\theta}{\lambda_1(n_k,k)}\right)^{-1} = 0\ ,\quad \text{by (2.2)},$$

yielding $\sum \tau_{m_k} = +\infty$, a.s., a contradiction.

On the other hand if $N^*(t)\le k_0$ for all $t\le T$ but $X(t)$ has explosion in $[0,T]$, then there must be countably many type two births at some $k\le k_0$. This also leads to a contradiction by a similar Laplace transform argument.

(b) The proof in this case is similar to that of (a). From the assumptions on $\lambda_2(.)$ we first prove that the number of type two particles remains bounded in bounded time intervals. Then using the assumption on $\lambda_1(.,.)$ we prove that the number of level one particles in each level two particle remains bounded on bounded time intervals.

(c) This follows immediately from (b). ∎

3. Multilevel Branching and its Continuous Limit

The two level branching process corresponds to the case in which $\lambda_1(n,i) = \lambda_1 ni = \gamma_1(1+\frac{1}{2}c_1)ni$, $\mu_1(n,i) = \mu_1 ni = \gamma_1(1-\frac{1}{2}c_1)ni$, $\lambda_2(n,i) = \lambda_2 n = \gamma_2(1+\frac{1}{2}c_2)n$, $\mu_2(n,i) = \mu_2 n = \gamma_2(1-\frac{1}{2}c_2)n$. It can be studied from several different points of view. The superparticle process can be considered as a multitype branching process in which the type corresponds to the number of particles it contains; thus the space Z^+ of non-negative integers serves as the type space. Another possibility is to view the superparticle process as a branching random walk on Z^+, where again, a superparticle consisting of k particles will be located at site k. However, we will adopt a third possible formulation, since it most naturally extends to higher level situations. We consider the two-level branching process $\{X_t\}$ as an $N(Z^+)$-valued pure jump Markov process, where $N(Z^+)$ denotes the set of integer-valued measures on Z^+. A neasure in $N(Z^+)$ can be represented by a vector $\{(n_i):i\in Z^+\}$ where n_i denotes the number of superparticles consisting of exactly i particles. The possible transitions of state in the process can be described as follows:

(a) birth of a particle in a superparticle of size i:

(3.1) $\mu \longrightarrow \mu - \delta_i + \delta_{i+1}$ at rate $i\gamma_1(1+ 1/2c_1)$

i.e. $(n_i,n_{i+1}) \longrightarrow (n_i-1,n_{i+1}+1)$ at rate proportional to in_i;

(b) death of a particle in a superparticle of size i:

(3.2) $\mu \longrightarrow \mu - \delta_i + \delta_{i-1}$ at rate $i\gamma_1(1- 1/2c_1)$,

i.e. $(n_{i-1},n_i) \longrightarrow (n_{i-1}+1,n_i-1)$ at rate proportional to in_i;

(c) copy of a superparticle of size i:

(3.3) $\mu \longrightarrow \mu + \delta_i$ at rate $\gamma_2(1+1/2c_2)$,

i.e. $n_i \longrightarrow n_i+1$ at rate proportional to n_i;

(d) death of a superparticle of size i:

(3.4) $\mu \longrightarrow \mu - \delta_i$ at rate $\gamma_2(1-1/2c_2)$,
i.e. $n_i \longrightarrow n_i-1$ at a rate proportional to n_i.
Here γ_1 and γ_2 are constants denoting the particle and superparticle branching rates, and the constants c_1 and c_2 are determined by the mean offspring sizes of the level one and level two branching mechanisms, so that $c_i = 0$ (>0, <0) correspond to critical (supercritical, subcritical) behavior at level i. Note that the total number of particles is $\sum_{i=1}^{\infty} in_i$, the number of null superparticles is n_0 and the number of nonnull superparticles is $\sum_{i=1}^{\infty} n_i$.

To describe the generator G of this $N(Z^+)$-valued Markov process, we consider the class of test functions of the form:

$$(3.5)\quad F(\mu) = f\left(\sum \phi_i n_i\right) = f\left(\int \phi d\mu\right) = f(\langle\phi,\mu\rangle)$$

where $\mu = \sum n_i\delta_i$ and $\langle\phi,\mu\rangle = \int\phi d\mu$. Applying (3.1)-(3.4) we obtain

$$(3.6)\quad GF(\mu) = \lim_{t\downarrow 0} \frac{E\{F(X_t)|X_0=\mu\} - F(\mu)}{t}$$

which is given by

$$\begin{aligned}(3.7)\quad GF(\mu) = {} & \gamma_2(1+1/2c_2)\int\Big[f(\langle\phi,\mu\rangle+\phi(x)) - f(\langle\phi,\mu\rangle)\Big]\mu(dx) \\ & + \gamma_2(1-1/2c_2)\int\Big[f(\langle\phi,\mu\rangle-\phi(x)) - f(\langle\phi,\mu\rangle)\Big]\mu(dx) \\ & +\gamma_1(1+1/2c_1)\int x\Big[f(\langle\phi,\mu\rangle-\phi(x)+\phi(x+1)) - f(\langle\phi,\mu\rangle)\Big]\mu(dx) \\ & + \gamma_1(1-1/2c_1)\int x\Big[f(\langle\phi,\mu\rangle-\phi(x)+\phi(x-1)) - \\ & \qquad f(\langle\phi,\mu\rangle)\Big]\mu(dx).\end{aligned}$$

The process $\{X_t\}$ can be characterized as the unique solution of the $N(Z^+)$-valued martingale problem associated with G. To prove the existence of this birth and death process, it suffices to demonstrate the absence of explosions; this follows from Theorem 2.1.

Remark. The uniqueness of the solution to this martingale problem can be proved by observing that the moment measures of X_t condititioned on X_0 are uniquely determined by the martingale problem and verifying that the corresponding moment problem is well-posed.

We next consider the continuous diffusion limit of the multilevel system. Let $X_t^{\varepsilon_1,\varepsilon_2}$ denote the two level process with $\gamma_2^{\varepsilon_2} = \gamma_2/\varepsilon_2^2$, $\gamma_1^{\varepsilon_1} = \gamma_1/\varepsilon_1^2\varepsilon_2$, $c_i^{\varepsilon_i} = c_i\varepsilon_i$, $i = 1,2$. Let

$$Y_t^{\varepsilon}(A) = Y_t^{\varepsilon_1,\varepsilon_2}(A) = \varepsilon_2 X_t^{\varepsilon_1,\varepsilon_2}(A/\varepsilon_1).$$

Theorem 3.1. Assume that $\varepsilon_1=\varepsilon_2=\varepsilon$ and $Y_0^{\varepsilon} \longrightarrow Y_0 \in M(R^+)$ as $\varepsilon \longrightarrow 0$. Then the $M(R^+)$-valued processes $\{Y_t^{\varepsilon}\}$ converge weakly to a $M(R^+)$-valued Markov process $\{Y_t\}$ which is characterized as the unique solution to the martingale problem associated with

$$(3.9)\qquad G_c F(\mu) = f'(\langle\phi,\mu\rangle)\langle L\phi,\mu\rangle + \gamma_2 f'(\langle\phi,\mu\rangle)\langle c_2\phi,\mu\rangle + \frac{1}{2}\gamma_2 f''(\langle\phi,\mu\rangle)\langle\phi^2,\mu\rangle,$$

$$(3.10)\qquad L\phi = \frac{1}{2}\gamma_1 x\phi'' + \gamma_1 c_1 x\phi'.$$

Moreover, the transition probabilities on $M(R^+)$ are determined by the Laplace functional

$$(3.11)\qquad E\left(e^{-\langle\phi,Y_t\rangle} \middle| Y_0=\mu\right) = e^{-\int u(t,x)\mu(dx)}$$

where the cumulant generating function $u(.,.)$ satisfies

$$(3.12)\qquad \frac{\partial u}{\partial t} = \frac{1}{2}\gamma_1 x\frac{\partial^2 u}{\partial x^2} + c_1\gamma_1 x\frac{\partial u}{\partial x} - \frac{1}{2}\gamma_2 u^2 + c_2\gamma_2 u,$$

$$u(0,x) = \phi(x).$$

Proof. The existence and uniqueness of the process with generator G_c follows from Dynkin (1988) or Fitzsimmons (1989). Hence it suffices to prove tightness and to verify that the limit satisfies the limiting martingale problem. The tightness is proved in Wu (1989) using a criterion of

Ethier and Kurtz (1986). To identify the limit we note that the generator of the rescaled process $Y^{\varepsilon_1,\varepsilon_2}$ is given by

$$G^{\varepsilon_1,\varepsilon_2}F(\mu)$$

$$= \sum_j [f(\langle\phi,\mu\rangle+\varepsilon_2\phi(j\varepsilon_1) - f(\langle\phi,\mu\rangle)]\frac{\gamma_2\mu(j\varepsilon_1)}{\varepsilon_2^2}(1+\varepsilon_2 c_2/2)$$

$$+ \sum_j [f(\langle\phi,\mu\rangle-\varepsilon_2\phi(j\varepsilon_1)) - f(\langle\phi,\mu\rangle)]\frac{\gamma_2\mu(j\varepsilon_1)}{\varepsilon_2^2}(1-\varepsilon_2 c_2/2)$$

$$+\sum_j [f(\langle\phi,\mu\rangle-\varepsilon_2\phi(j\varepsilon_1)+\varepsilon_2\phi((j+1)\varepsilon_1))-f(\langle\phi,\mu\rangle)]. \frac{j\varepsilon_1\gamma_1\mu(j\varepsilon_1)}{\varepsilon_1^2\varepsilon_2}(1+\varepsilon_1 c_1/2)$$

$$+\sum_j [f(\langle\phi,\mu\rangle-\varepsilon_2\phi(j\varepsilon_1)+\varepsilon_2\phi((j-1)\varepsilon_1))-f(\langle\phi,\mu\rangle)]. \frac{j\varepsilon_1\gamma_1\mu(j\varepsilon_1)}{\varepsilon_1^2\varepsilon_2}(1-\varepsilon_1 c_1/2)$$

and we obtain the limiting generator $G_c F(\mu)$ using a two term Taylor expansion. The equation (3.12) for the cumulant generating function is obtained as in Ethier and Kurtz (1986). ∎

In this continuous limit the total mass at time t is given by the real-valued process

$$x(t) = \langle x, Y_t\rangle.$$

It follows from the expression for the Laplace functional of $\langle x, Y_t\rangle$ that $x(t)$ is infinitely divisible but not multiplicative.

4. The Moment Structure

We first consider the moment measures for the particle system $\{X_t\}$. The nth moment measure is defined by

$$(4.1) \qquad m_n(t,\mu;j_1,\ldots,j_n) = E\left[\prod_{i=1}^{n} X_t(j_i) \mid X_0=\mu\right]$$

where $j_1,\ldots,j_n \in Z^+$ (not necessarily distinct).

We also define

$m_k^*(t) = E(\langle x^k, X_t\rangle)$.

Proposition 4.1.

(a) $m_1(t,\delta_i;j) = e^{\gamma_2 c_2 t} p(t,i,j)$

where $p(t,i,j)$ denotes the probability transition function for a linear birth and death process with birth rate $\gamma_1(1+1/2c_1)$ and death rate $\gamma_1(1-1/2c_1)$.

(b) The m_k^* satisfy the differential equations

$$(4.2) \qquad \frac{dm_k^*(t)}{dt} = 2\gamma_1 \sum_{\substack{\ell=1 \\ (k-\ell)\text{odd}}}^{k} \binom{k}{\ell-1} m_\ell^*(t) + \gamma_1 c_1 \sum_{\substack{\ell=1 \\ (k-\ell)\text{even}}}^{k} \binom{k}{\ell-1} m_\ell^*(t) + \gamma_2 c_2 m_k^*(t).$$

Proof of (a). From the martingale problem applied to $F(\mu) = \langle\phi,\mu\rangle$, we obtain

$$(4.3) \qquad \frac{dm_1(t;i)}{dt} = \{\gamma_1(1+1/2c_1)(i+1)m_1(t,i+1) + \gamma_1(1-1/2c_1)(i-1)m_1(t;i-1) - 2im_i(t,i)\} + \gamma_2 c_2 m_1(t;i).$$

The result then follows by comparison with the forward equation for the linear birth and death process and the Feynman-Kac formula.

(b) Substitute $F(\mu) = \langle f,\mu\rangle$ into the martingale problem for the process X_t. ∎

We next consider the moment equations for the measure-valued process Y_t. We define

$m_k(t) = E(\langle x^k, Y_t\rangle)$, *mean empirical moments*

$M_k(t) = E\left(\langle x, Y_t\rangle^k\right)$ *higher order moments of the empirical mean.*

Proposition 4.2.

(a) $m_k(t)$ satisfy the system of differential equations

$$(4.4)\qquad \frac{dm_k(t)}{dt} = \frac{1}{2}\gamma_1 k(k-1)m_{k-1}(t) + \gamma_1 c_1 k m_k(t) + \gamma_2 c_2 m_k(t).$$

(b) $m_1(t) = m_1(0)e^{(\gamma_1 c_1+\gamma_2 c_2)t}$, and

$$m_2(t) = e^{2(\gamma_1 c_1+\gamma_2 c_2)t}m_2(0) + \frac{1}{c_1}(1-e^{-\gamma_1 c_1 t})m_1(0)e^{(2\gamma_1 c_1+\gamma_2 c_2)t}$$

(c) If $(\gamma_1 c_1+\gamma_2 c_2) \neq 0$,

$$M_2(t) = M_2(0)e^{2(\gamma_1 c_1+\gamma_2 c_2)t}$$

$$+ \frac{m_1(0)}{(\gamma_1 c_1+\gamma_2 c_2)}\left\{\left[\frac{\gamma_2}{c_1} e^{(\gamma_1 c_1+\gamma_2 c_2)t} + \frac{\gamma_1}{c_2} e^{2(\gamma_1 c_1+\gamma_2 c_2)t}\right]\right\}$$

$$- \frac{m_1(0)}{c_1 c_2} e^{(2\gamma_1 c_1+\gamma_2 c_2)t}$$

$$+ \frac{m_2(0)}{c_2}\left\{ e^{2(\gamma_1 c_1+\gamma_2 c_2)t} - e^{(2\gamma_1 c_1+\gamma_2 c_2)t}\right\}.$$

Proof. Refer to Dawson and Hochberg (1989).

5. Multilevel Branching Brownian Motions

5.1 Characterization of the Process

Consider a finite system of branching random walks in $\mathbb{R}^d$. After an exponential time let the entire system which can be represented as a point in $M(\mathbb{R}^d)$ perform binary branching, that is, either die or produce a replica (each with probability 1/2) After this each of the two replicas (or "super-particles") develop independently and undergo further branching thus yielding a random number of such replicas which can then be represented as a random atomic measure on $M(\mathbb{R}^d)$,

$$(5.1)\quad Y_a(t) = \sum_{i=1}^{\infty}\sum_{k=1}^{n_i} \delta_{\sum_{r=1}^{i}\delta_{X_{i,k,r}(t)}} \in M_2 = M(M(\mathbb{R}^d)),$$

where $X_{i,k,r}$ denotes the location in $\mathbb{R}^d$ of the rth particle in the kth superparticle of size i at time t.

To obtain the continuous diffusion limit we let

$Y_n(t,A) = n^{-1}Y_a(nt,A_n)$

where $A_n = \{\mu:n^{-1}\mu \in A\}$. Then as $n\longrightarrow\infty$, $Y_n(t)$ converges weakly to a M_2-valued process $Y(t)$ which can be characterized as the unique solution of the martingale problem for the limiting generator G_c which is given by

$$(5.2)\quad G_cF(\nu) = \langle\langle \mathcal{L}F'(\nu,.),\nu\rangle\rangle + \frac{1}{2}\iint F''(\nu;\mu_1,\mu_2)\delta_{\mu_1}(d\mu_2)\nu(d\mu_1)$$

where if $H\in C(M(\mathbb{R}^d))$, the space of continuous functions on $M(\mathbb{R}^d)$, $\langle\langle H,\nu\rangle\rangle = \int H(\mu)\nu(d\mu)$. The test function $F(\nu)$ on M_2 has the form $F(\nu) = f(\langle\langle h_1(\langle h_2,.\rangle),\nu\rangle\rangle)$ where $\nu\in M_2$, $h_1,f \in C_b^2(\mathbb{R})$, $h_2\in C^2(\mathbb{R}^d)$, and

$$(5.3)\quad F'(\nu,\mu) = \frac{\delta F(\nu)}{\delta\nu(\mu)} = \frac{d}{d\varepsilon}[F(\nu+\varepsilon\delta_\mu)]_{\varepsilon=0}$$
$$= f'(\langle\langle h_1(\langle h_2,.\rangle),\nu\rangle\rangle)h_1(\langle h_2,\mu\rangle)$$

and $\mathcal{L}$ denotes the generator of the $M(\mathbb{R}^d)$-valued branching process.

(5.4)

$$\mathcal{L}F'(\nu,\mu) = f'(\langle\langle h_1(\langle h_2,.\rangle),\nu\rangle\rangle)h_1'(\langle h_2,\mu\rangle)\langle\Delta h_2,\mu\rangle$$
$$\frac{1}{2} f'(\langle\langle h_1(\langle h_2,.\rangle),\nu\rangle\rangle)h_1''(\langle h_2,\mu\rangle)\langle h_2^2,\mu\rangle$$

The Laplace functional of the M_2-valued process $Y(.,.)$ is given by: for $H\in C_+(M(\mathbb{R}^d))$,

$$(5.5)\qquad L_{t,\nu}(H) = E\left[\exp\left(-\int_{M(\mathbb{R}^d)} Y(t,d\mu)\right)\middle|Y(0)=\nu\right]$$
$$= \exp\left(-\int U(t,\mu)\nu(d\mu)\right)$$

where

$$(5.6)\qquad \frac{\partial U}{\partial t} = \mathcal{L}U - U^2.$$

Given the M_2-valued process Y we introduce the process which is the analogue of the measure-valued process that counts all the particles regardless of the superparticle to which they belong, namely

$$(5.7)\qquad \bar{Y}(t) = \int \mu Y(t,d\mu).$$

Note that $\bar{Y}(.)$ is infinitely divisible but not multiplicative.

The Laplace functional of $\bar{Y}$ is given by

$$(5.8)\qquad E\left(e^{-\int\theta\phi(x)\bar{Y}(t,dx)}\right) = E\left(e^{-\int\int\theta\phi(x)\mu(dx)X(t,d\mu)}\right)$$
$$=\sum_n \frac{(-\theta)^n}{n!}\int\ldots\int m_n(t;dx_1,\ldots,dx_n)\phi(x_1)\ldots\phi(x_n).$$

Hence we must find the solution of

$$(5.9)\qquad \frac{\partial U(t,\mu)}{\partial t} = \mathcal{L}U(t,\mu) - [U(t,\mu)]^2,\quad U(0,\mu) = \int\theta\phi(x)\mu(dx).$$

We next wish to determine the second moment measure $m_2(t;dx_1,dx_2)$ of $\bar{Y}(t)$. The coefficient of θ is given by $\int S_t\phi(x)X(0,dx)$. The coefficient of θ^2 in the solution of the above equation is

$$(5.10)\qquad \int_0^t T_{t-s}\left\{\iint S_s\phi(x_1)S_s\phi(x_2)\mu(dx_1)\mu(dx_2)\right\}ds$$

where $\{S_s:s\geq 0\}$ is the Brownian semigroup and $\{T_t:t\geq 0\}$ is the semigroup associated with the measure-valued process. We can use the covariance measure for the level 1 measure-valued process to obtain

$$(5.11)\qquad \int_0^t T_{t-s}\left\{\iint S_s\phi(x_1)S_s\phi(x_2)\mu(dx_1)\mu(dx_2)\right\}ds$$

$$= \int_0^t \left\{\iint S_s\phi(x_1)S_s\phi(x_2)m_2(t-s,dx_1,dx_2)\right\}ds$$

Moreover

$$(5.12)\qquad T_t\langle\phi,\mu\rangle^2 = \langle S_t\phi,\mu\rangle^2 + \int_0^t (S_{t-s}[S_s\phi]^2)(x)\mu(dx).$$

This yields the covariance density

$$(5.13)\qquad \iint V(t,\mu,x_1,x_2)\phi(x_1)\phi(x_2)dx_1dx_2$$

$$= \int_0^t\iint S_s\phi(x_1)S_s\phi(x_2)m_2(t-s,dx_1,dx_2)ds.$$

$$= \int_0^t\left[\langle S_t\phi,\mu\rangle^2 + \int_0^{t-s} (S_{t-s-u}[S_{u+s}\phi]^2)(x)\mu(dx)du\right]ds.$$

5.2 Application to the Hausdorff Dimension of Support

In the case $X(0,d\mu) = \delta_\mu$, $\mu(dx) = dx$,

$$(5.14)\qquad \iint V(t,\mu,x_1,x_2)\phi(x_1)\phi(x_2)dx_1dx_2$$

$$= \left[t\left(\int\phi dx\right)^2 + \int_0^t\int_s^t [S_v\phi]^2(x)dxdvds\right]$$

$$= t\left(\int\phi dx\right)^2 + \iiint_0^t vS_{2v}(x_1-x_2)dv\ \phi(x_1)\phi(x_2)dx_1dx_2.$$

It is then easy to verify that $V(t,\mu,x_1,x_2) \sim \text{const}|x_1-x_2|^{-(d-4)}$ for small $|x_1-x_2|$ provided that $d>4$. Applying the criterion of Zähle [12], this implies that the Hausdorff dimension of support ≥ 4 (in dimensions $d>4$).

This is in contrast to the one level super-Brownian motion which has dimension of support equal to 2.

5.3 Application to the Long-time Behavior

In the case $X(0,d\mu) = \delta_{\delta_x} dx$

$$(5.15)\quad \iint V(t,\mu,x_1,x_2)\phi(x_1)\phi(x_2)dx_1dx_2$$
$$= \left[\int t(T_t\phi(x))^2dx + \int_0^t\int_s^t [S_v\phi]^2(x)dxdvds\right]$$
$$= \int t(T_t\phi(x))^2dx$$
$$+ \iiint_0^t vS_{2v}(x_1-x_2)dv\ \phi(x_1)\phi(x_2)dx_1dx_2.$$

Hence in this case $E(\langle\bar{Y}(t),\phi\rangle)$ remains bounded if and only if $d>4$. The contrast between expressions (5.14) and (5.15) underlines the strong non-Markovian nature of $\bar{Y}(t)$ since in both cases $\bar{Y}(0) = dx$.

Acknowledgements. The research of the first author was partially supported by a Natural Sciences and Engineering Research Council of Canada grant. The research of the second author was supported by U.S. National Security Agency grant MDA 904-85-H-2044 and U.S. National Science Foundation grant DMS-8800289.

6. References

1. Athreya, K.B.and Ney, P.E. Branching Processes, Springer-Verlag, 1982.
2. Carmelli, D. and Cavalli-Sforza, L. (1976). Some models of population structure and evolution, Theoret. Popn. Biol. 9(1976) , 329-359.
3. Dawson, D.A.Stochastic ensembles and hierarchies, Lecture Notes in Math. 1203(1986) , 20-37, Springer-Verlag.
4. Dawson, D.A. and Hochberg, K.J. The carrying dimension of a stochastic measure diffusion, Annals of Probab. 7(1979) , 693-703.

5. Dawson, D.A. and Hochberg, K.J. A multilevel branching model, preprint, 1989.
6. Dynkin, E.B. Superprocesses and their linear additive functionals, Trans. Amer. Math. Soc. 1988.
7. Dynkin, E.B. Regular transition functions and regular superprocesses, Trans. Amer. Math. Soc. 1988.
8. Durrett, R. The genealogy of critical branching processes, Stoch. Proc. Appl. 8(1978), 101-116.
9. Ethier, S.N. and Kurtz, T.G. Markov Processes: Characterization and Convergence. New York: John Wiley, 1986.
10. Fitzsimmons, P.J. Construction and regularity of measure-valued Markov branching processes, Israel J. Math. 64(1988), 337-361.
11. Sawyer, S. and Felsenstein, J. Isolation by distance in a hierarchically clustered population, J. Appl. Prob. 20(1983), 1-10.
12. Zähle, U. Self-similar random measures, I. Notion, carying Hausdorff dimension and hyperbolic distribution, Probab. Th. Rel. Fields 80, 79-100.

Logarithmic Sobolev Inequalities for the Heat Kernel on a Lie Group

Leonard Gross
Department of Mathematics
Cornell University
Ithaca, NY 14853

ABSTRACT

Logarithmic Sobolev inequalities for the heat kernel measure on a Lie group G are derived from a G valued Brownian motion. The Brownian motion is represented as a limit of G valued random walks so as to permit an understanding of why the Gaussian constants are not best possible when G is compact.

§1. Introduction

Let $p_t(dx) = (2\pi t)^{-n/2}\exp[-|x|^2/2t]dx$ be Gauss measure on R^n with variance $t > 0$. The logarithmic Sobolev inequality for p_t is [Gr3]

$$\int_{R^n} |f(x)|^2\log|f(x)|p_t(dx) \tag{1.1}$$

$$\leq t\int_{R^n} |\nabla f(x)|^2 p_t(dx) + \|f\|^2\log\|f\|$$

where $\|f\|$ is the $L^2(p_t)$ norm. The constant t in front of the gradient term in 1.1) is the smallest possible.

Inequalities of this type arose as part of an effort to understand more deeply E. Nelson's method of proof of semiboundedness of Hamiltonians in quantum field theory. A first step in the direction of the inequality (1.1) was made by Paul Federbush [F] who showed that if one differentiates Nelson's family of hypercontractive inequalities for the Ornstein-Uhlenbeck semigroup with respect to the time parameter one gets an inequality similar to (1.1) and that the resulting relatively simple inequality is already sufficient to establish semiboundedness. Later the present author [Gr3] gave a direct proof of (1.1) and showed moreover that (1.1) implies Nelson's hypercontractive inequalities. Logarithmic Sobolev inequalities and hypercontractivity have been extensively explored and applied. For a recent survey see [DGS]. A bibliography of papers studying or using logarithmic Sobolev inequalities or hypercontractivity (updated to January 1990) is appended at the end of the present paper.

Since the coefficients in (1.1) are dimension independent it should not be surprising that (1.1) also holds in infinite dimensions for any Gaussian measure once ∇ is understood correctly. The appropriate notion of gradient has a long history.

For a real valued function f on a vector space V let us denote by $\partial_h f(x)$ the directional derivative defined by

$$\partial_h f(x) = df(x+th)/dt\,|_{t=0}\ , \qquad h \in V,\ x \in V$$

if it exists. Write W for the classical Wiener space $C_0([0,1])$ consisting of continuous real valued functions on $[0,1]$ which vanish at zero and denote by H the dense subspace consisting of absolutely continuous functions x in W such that $|x|^2 \equiv \int_0^1 \dot{x}(t)^2 dt < \infty$. In numerous papers concerned with topics in advanced calculus over Wiener space Cameron and Martin and their associates [C,CM1-3,CF,CH,CS] showed that although H is a subset of Wiener measure zero nevertheless the geometry of this Hilbert space (generally called the Cameron Martin space) controls the form of many of their identities. In particular Cameron [C] was the first to recognize that the derivative in H directions, $\partial_h f(x)$, $x \in W$, $h \in H$ relates well to the Wiener integral and in fact allows an integration by parts formula. (Although he considered only derivatives in W^* directions, with W^* properly contained in H by the usual injection $W^* \hookrightarrow H^* = H \hookrightarrow W$, the extension of his formula to H directions is immediate.) Many of the formulas of these early authors can be understood by thinking of Wiener measure as given by the heuristic expression $Z^{-1}\exp[-|x|^2/2]\mathcal{D}x$ where $\mathcal{D}x$ is some kind of infinite dimensional Lebesgue measure on H. It is well known that this expression, common in the physics literature, does not represent a countably additive measure on H but can be interpreted as a countably additive measure (Wiener measure) on any one of a number of extensions of H, such as W itself or even on the smaller space of Lip α functions in W if $0 < \alpha < 1/2$.

Gaussian measures on infinite dimensional space have for a long time provided an important tool for studying quantum field theory. One can see precursers of this in Von Neumann's 1932 book [V, Ch. III.6]. The directional derivative operator ∂_h has the physical interpretation of an annihilation operator for a particle in state h. Von Neumann used, implicitly, a product of real lines for his measure space and differentiated only in coordinate directions. However, in a more coordinate independent approach to the simplest quantum field theories in three space dimensions one would take $H = \text{Real } L^2(R^3)$ and one could realize the Gauss measure $Z^{-1}\exp[-|x|_H^2/2)\mathcal{D}x$ on $\mathcal{S}'(R^3)$ say, by Minlos' theorem. Recognizing the determining role that the Hilbert space H plays in these calculus and operator considerations, both in quantum field theory and the work of Cameron and associates, I. E. Segal [S1,S2] formulated, in the mid 1950's, the notion of a Gaussian measure associated with a real Hilbert space H, in a manner which dispensed with the need for a choice of a specific ambient vector space, such as W or $\mathcal{S}'(R^3)$ or R^Z, on which to interpret $Z^{-1}\exp[-|x|^2/2]\mathcal{D}x$ as a countably additive measure. In so doing he greatly clarified the central role of the Hilbert space H in the work of Cameron et al, sharpened some previous theorems, simplified their proofs and illuminated the isomorphism between L^2 (Gauss measure) and the symmetric tensor algebra over $\mathbb{C} \otimes H$ which had been described by Wiener [W] and Kakutani [K] and Ito [I] and was at the basis of the particle-wave duality in quantum field theory. The work described by Kallianpur [JK] at this meeting, in which all analysis is referred back to the controlling Hilbert space H, is a most recent example of Gaussian analysis in this spirit.

Nevertheless for many purposes it is essential to realize the countably additive Gaussian measure associated to a particular Hilbert space H as a measure on a specific space. For example in the case of Wiener space it is extremely important for probability theory that the points of W happen to be continuous functions. In order to develop a potential theory in infinite dimensions this author found it convenient [Gr2] to describe how one could complete a real separable Hilbert space H with respect to a sufficiently weak norm (a measurable norm) on H and obtain thereby a Banach space B, containing H as a dense subspace, and on which the informal expression $Z^{-1}\exp[-|x|_H^2/2]\mathcal{D}x$ could be interpreted as a countably additive measure. The point is that if one wishes to study the Laplacian for H, defined by $\Delta = \Sigma_{j=1}^{\infty} \partial^2/\partial x_j^2$ where $x_1, x_2, \ldots$ is an O.N. coordinate system on H, then one must expect that the semigroup $e^{t\Delta}$ will correspond to a Brownian motion with values in just such a Banach space B. It turns out in fact that the Dirichlet problem for the above Laplacian must be formulated for an open set in B (open in the B topology) rather than for an open set in H (open in the H topology). Moreover the way in which Hölder conditions enter classical potential theory reappears again in infinite dimensions but with the Hölder conditions defined in terms of the B norm rather than the H norm. Thus there appears in potential theory over a Hilbert space a need for two spaces - one, B, to carry the topology, the measurability and the Brownian motion process and the other, H, to define the relevant notion of differentiability. It was for this reason that this author developed [Gr1] the notion of H-derivative, ∇, which is the gradient in H directions defined by $((\nabla f)(x), h)_H = (\partial_h f)(x)$ for x in B, h in H. This notion, which is clearly weaker than the Gateaux or Frechet B-derivatives, seems to have been explored for the first time in this work of the mid 1960's [Gr1]. In the late 1960's and early 1970's the H-derivative was central in a number of works on partial differential operators in infinite dimensions. There was, for example, the work of M. A. Piech on parabolic equations on Hilbert space with variable coefficients [P1-4], the work of J. Alvarez [A] on the Riesz decomposition theorem, the work of C. Elson [E] on an infinite dimensional Weyl's lemma, the work of V. Goodman [G1-4] on harmonic functions on Hilbert space and related topics, and the early work of H. H. Kuo [Ku1-5] on Riemann-Wiener manifolds and related topics. Even through 1974 the preceding papers cannot be said to be near complete or even representative of work exploring the H-derivative because much of the work on mathematical quantum field theory can be interpreted as statements about the number operator $N = \nabla^* \nabla$ where ∇ is now the closure in L^2 (Gauss measure) of the H-derivative $\nabla : L^2 \to L^2 \otimes H$. Moreover there is significant work already in the 60's by Daletzky [D1, D2] and his students on infinite dimensional Brownian motions and potential theory. But after 1975 there was a rapid increase in the mathematics literature utilizing the H-derivative, too vast to attempt to survey here, inspired by Malliavin's approach to hypoelliptic equations.

In the present paper we first generalize the notion of H-derivative to the case in which Wiener space $C_0([0,1])$, is replaced by $W_G \equiv C_0([0,1],G)$, where G is a connected finite dimensional Lie group. We define what translation $h+g$ means when g is in W_G and h is in $L^2([0,1],\mathcal{G})$ where $\mathcal{G}$ is the Lie algebra of G with any inner product. This determines immediately the notion of H-derivative. We then use the H-derivative to formulate and prove a logarithmic Sobolev inequality with respect to a diffusion process measure P on W_G. P is the measure induced on W_G by the Ito development map, i.e., the solution to the stochastic differential equation [M]

$$dg(t) = g(t) \circ dB(t), \quad g(0) = I \qquad (1.2)$$

where B is a $\mathcal{G}$ valued Brownian motion and the equation (1.2) is interpreted in the sense of Fisk-Stratanowicz [IW]. Under the measurable map $B \to g$ the usual H-translations go over to our map $h, g \to h+g$ and the logarithmic Sobolev inequality, well known for Brownian motion goes over to our present inequality over (W_G, P).
This in turn induces a logarithmic Sobolev inequality on G itself. Although our translation operation $h+g$ is just the image of ordinary translation in Wiener space under the Ito development map its potential importance seems to make it worthwhile to

describe it directly in W_G. Moreover we describe P as a limit of G valued random walks rather than defining it through (1.2) in order to be able to use the approximation to P to understand why the Sobolev coefficient in the logarithmic Sobolev inequality over G is not best possible.

§2. Results

Choose an inner product on the Lie algebra $\mathcal{G}$ of a connected Lie group G and let w be Wiener measure on $W \equiv C_0([0,1],\mathcal{G})$. With $H = \{h \in W: |h|^2 \equiv \int_0^1 |\dot{h}(t)|^2 dt < \infty\}$ we let ∇ denote the H-derivative of functions on W. Then (1.1) (for $t = 1$) implies

$$\int_W |f(x)|^2 \log|f(x)| dw(x) \qquad (2.1)$$

$$\leq \int_W |\nabla f(x)|^2 dw(x) + \|f\|^2 \log\|f\|$$

for any function f in the L^2 domain of ∇.

If $\{\eta_a\}_{a=1}^N$ is an O.N. basis of $\mathcal{G}$ and $\hat{\eta}_a$ are the corresponding left invariant vector fields we put $\Delta = \Sigma_a \hat{\eta}_a^2$. Let P be the diffusion process measure on $W_G \equiv C_0([0,1];G)$ with generator $\Delta/2$ and $P(g(0) = e) = 1$. Then P is induced from w by the solution to (1.2). Translations in W by an element h in H leaves w quasiinvariant with the result that one can transfer such translations in $L^2(W,w)$ to translations in $L^2(W_G,P)$ and then transfer the H-derivative from $L^2(W,w)$ to $L^2(W_G,P)$. But the space W_G seems important enough to warrant understanding the translation operation at a more intrinsic level, independently of all the measure theory and stochastic analysis contained in the preceding sentence. To this end we define $h+g$ for h in H and g in W_G as follows. Let v be the element of W_G defined by the equation

$$(v(t)^{-1})_* \dot{v}(t) = (\mathrm{Ad}\, g(t))\dot{h}(t), \quad v(0) = e. \qquad (2.2)$$

Here $\dot{v}(t)$ is a tangent vector to G at $v(t)$ and the left side of (2.2) is its translate back to the identity of G. Put

$$(h+g)(t) = v(t)g(t). \qquad (2.3)$$

Here are some properties of this operation: a) $0+g = g$, b) $(h_1+h_2)+g = h_1+(h_2+g)$ for h_1 and h_2 in H and g in W_G, c) if $g \in W_G$ and $h_1+g = h_2+g$ then $h_1 = h_2$, d) if g is piecewise C^1, h is in H and $u = h+g$ then $u(t)$ is the unique solution to the initial value problem

$$(u(t)^{-1})_* \dot{u}(t) = (g(t)^{-1})_* \dot{g}(t)+\dot{h}(t), \quad u(0) = e. \qquad (2.4)$$

To gain further insight into the translation operation $h+g$ let $k(t)$ be the solution to

$$(k(t)^{-1})_* \dot{k}(t) = \dot{h}(t), \quad k(0) = e. \qquad (2.5)$$

Then k is an element of W_G. The correspondence $h \longleftrightarrow k$ may be used to identify H with the subset of elements k of W_G with finite energy: $\int_0^1 |(k(t)^{-1})_* \dot{k}(t)|^2 dt < \infty$. One does not have $(h+g)(t) = k(t)g(t)$ but rather $h+g$ "adds" the increments of k to the increments of g either in the sense of (2.4) or in the sense of the convergent representation

$$(h+g)(t) = \lim_{n\to\infty}(\Delta_k(1)\Delta_g(1))\cdots(\Delta_k(j)\Delta_g(j)), \quad j = [nt] \qquad (2.6)$$

where $\Delta_g(j) = g((j-1)/n)^{-1}g(j/n)$.

If $f: W_G \to R$ we define the H-derivative of f at g by the identity

$$((\nabla f)(g), h)_H = df(th+g)/dt\,|_{t=0}, \quad h \in H. \qquad (2.7)$$

In an appropriate measure theoretic sense the translation $x \to h+x$ in W goes over to $g \to h+g$ in W_G and the two H-derivatives correspond. In particular we have

<u>Theorem 1</u>.

$$\int_{W_G} f(g)^2 \log|f(g)|P(dg) \tag{2.8}$$

$$\le \int_{W_G} |\nabla f(g)|^2 P(dg) + \|f\|^2 \log\|f\|.$$

Moreover the coefficient (= one) in front of the gradient term is smallest possible.

In [Gr4] we prove (2.8) by representing P as a limit of G valued random walks rather than appealing to the stochastic differential equation (1.2). This allows us to make some comments later about the coefficient in the next inequality.

<u>Corollary 2</u>. Let ν_t be the heat kernel measure on G. That is, $e^{t\Delta}$ is convolution by ν_t. Assume that the given inner product on $\mathcal{G}$ is $\mathrm{Ad}\, G$ invariant. Then

$$\int_G u(x)^2 \log|u(x)|\nu_t(dx) \tag{2.9}$$

$$\le t \int_G |u'(x)|^2 \nu_t(dx) + \|u\|^2 \log\|u\|$$

where $u'(x)$ is the differential of u regarded as an element of $\mathcal{G}^*$. Moreover if the inequality (2.9) holds for all u with the coefficient t in front of the gradient term replaced by a smaller constant $c(t)$ then $\lim_{t \downarrow 0} c(t)/t = 1$.

There is no doubt that if G is compact then the smallest allowed coefficient $c(t)$ in front of the gradient term of (2.9) will satisfy $c(t) < t$, at least for large t because if $t \to \infty$ then ν_t converges to Haar measure for which existence of a logarithmic Sobolev inequality is known. In order to understand the mechanism responsible for making $c(t) < t$ we take $G = S^1$ and study the three step random walk approximation to υ_1 in [Gr4]. In this case a calculator assisted computation yields precise bounds for $c(1)$ which turns out to be less than .602 in this simplified model of a circle.

References

[A] J. Alvarez, The Riesz decomposition theorem for distributions on a Wiener space, Ph. D. Thesis, Cornell, Jan. 1973.

[C] R. H. Cameron, The first variation of an indefinite Wiener integral, Proc. Amer. Math. Soc. 2 (1951), 914-924.

[CM1] R. H. Cameron and W. T. Martin, Evaluation of various Wiener integrals by use of certain Sturm-Liouville differential equations, Bull. A.M.S. 51 (1945), 73-89.

[CM2] ______________________, Transformation of Wiener integrals under a general class of linear transformations, Trans. A.M.S. 58 (1945), 184-219.

[CM3] ______________________, The transformation of Wiener integrals by nonlinear transformations, Trans. A.M.S. 66 (1949), 253-283.

[CH] R. H. Cameron and C. Hatfield, On the summability of certain series for unbounded nonlinear functionals, Proc. A.M.S. 4 (1953), 375-387.

[CF] R. H. Cameron and R. E. Fagen, Nonlinear transformations of Volterra type in Wiener space, Trans. A.M.S. 75 (1953), 552-575.

[CS] R. H. Cameron and J. M. Shapiro, Nonlinear integral equations, Ann. of Math. 62 (1955), 472-497.

[D1] Yu. L. Deletzky, Differential equations with functional derivatives and stochastic equations for generalized random processes, Dokl. Akad. Nauk. SSSR 166 (1966), 1035-1038.

[D2] ________________, Infinite dimensional elliptic operators and the parabolic equations connected with them, English Transl. Russian Math. Surveys 22 (1967), #4, 1-53.

[DGS] E. B. Davies, L. Gross and B. Simon, Hypercontractivity: a bibliographic review, Proc. Hoegh-Krohn Memorial Conference, Ed. S. Albeverio, to appear.

[E] C. Elson, An extension of Weyl's lemma to infinite dimensions, Trans. A.M.S. 194 (1974), 301-324.

[F] P. Federbush, A partially alternate derivation of a result of Nelson, J. Math. Phys. 10 (1969), 50-52.

[G1] V. Goodman, Quasi-differentiable functions on Banach spaces, Proc. A.M.S. 30 (1971), 367-370.

[G2] __________, A divergence theorem for Hilbert space, Trans. A.M.S. 164 (1972), 411-426.

[G3] __________, Harmonic functions on Hilbert space, J. Funct. Anal. 10 (1972), 451-470.

[G4] __________, A Liouville theorem for abstract Wiener spaces, Amer. J. of Math. 45 (1973), 215-220.

[Gr1] L. Gross, Potential theory on Hilbert space, J. of Funct. Anal. 1 (1967), 123-181.

[Gr2] ________, Abstract Wiener spaces, Proc. Fifth Berkeley Symposium on Math. Stat. and Prob. 1965, U. of Calif. Press, 31-42.

[Gr3] ________, Logarithmic Sobolev inequalities, Amer. J. Math. 97 (1975), 1061-1083.

[Gr4] ________, Logarithmic Sobolev inequalities on Lie groups, Cornell preprint, June 1989.

[I] K. Ito, Multiplc Wiener integral, J. Math. Soc. of Japan, 3 (1951), 157-169.

[IW] N. Ikeda and S. Watanabe, Stochastic Differential Equations and Diffusion Processes, North Holland Pub. Co. 1981.

[JK] G. W. Johnson and G. Kallianpur, Homogeneous chaos, p-forms, scaling and the Feynman integral.

[K] S. Kakutani, Spectrum of the flow of Brownian motion, Proc. Nat. Acad. Sci. U.S.A. 36 (1950), 319-323.

[Ku1] H. H. Kuo, Integration by parts for abstract Wiener measures, Duke Math. J. 41 (1974), 373-379.

[Ku2] __________, On Gross differentiation on Banach spaces, Pacific J. Math. 59 (1975), 135-145.

[Ku3] __________, Gaussian measures in Banach spaces, Lecture Notes in Math. Vol. 463, Springer-Verlag, N.Y. 1975.

[Ku4] __________, Potential theory associated with Ornstein- Uhlenbeck process, J. Funct. Anal. 21 (1976), 63-75.

[Ku5] __________, Ornstein-Uhlenbeck process on a Riemann-Wiener manifold, Proc. Symp. on Stoch. Diff. Eq. Kyoto 1976, 187-193, K. Ito, Ed.

[M] H. P. McKean, Jr., Stochastic Integrals, Academic Press 1969, New York.

[P1] M. A. Piech, A fundamental solution of the parabolic equation on Hilbert space, J. Funct. Anal. 3 (1969), 85-114.

[P2] ____________, Diffusion semigroups on abstract Wiener space, Trans. A.M.S. 166 (1972), 411-430.

[P3] ____________, Regularity of the Green's operator on abstract Wiener space, J. Diff. Equs. 12 (1972), 353-360.

[P4] ____________, Smooth functions on Banach spaces, J. Math. Anal. and Applications 57 (1977), 56-67.

[S1] I. E. Segal, Tensor algebras over Hilbert spaces, Trans. Amer. Math. Soc. 81 (1956), 106-134.

[S2] ____________, Distributions in Hilbert space and canonical systems of operators, Trans. Amer. Math. Soc. 88 (1958), 12-41.

[V] J. von Neumann, Mathematical Foundations of Quantum Mechanics, Princeton Univ. Press 1955 (Originally published by Julius Springer, 1932.)

[W] N. Wiener, The homogeneous chaos, Amer. J. Math. 60 (1938), 897-936.

Bibliography on Logarithmic Sobolev Inequalities and Hypercontractivity

[1] R. A. Adams, General logarithmic Sobolev inequalities and Orlicz imbeddings, J. Funct. Anal. 34 (1979), 292-303.

[2] ______________ and F. H. Clarke, Gross's logarithmic Sobolev inequality: a simple proof, Amer. J. Math. 101 (1979), 1265-1270.

[3] Joel Avrim, Perturbation of Log. Sob. generators by electric and magnetic potentials, Berkeley Thesis 1982.

[4] D. Bakry, Une remarque sur les diffusions hypercontractives, Preprint Nov. 1987, 3 pp.

[5] D. Bakry and M. Emery, Hypercontractivite de semi-groupes de diffusion, Compte-Rendus 299 (1984), 775-778.

[6] ______________________________, Diffusions hypercontractives, pp. 177-207 in Sem. de Probabilites XIX, Lecture Notes in Mathematics #1123, eds. J. Azema and M. Yor, Springer- Verlag, New York, 1985.

[7] D. Bakry and P. A. Meyer, Sur les inequalities de Sobolev logarithmique I & II Preprints, Univ. de Strasbourg 1980/81.

[8] D. Bakry and D. Michel, Inegalites de Sobolev et minorations du semigroupe de la chaleur, Toulouse preprint, Oct. 1989, 37 pp.

[9] R. Banuelos, Intrinsic ultracontractivity and eigenfunction estimates for Schrödinger operators, Purdue preprint, Jan. 1990, 29 pp.

[10] W. Beckner, Inequalities in Fourier Analysis on R^n, Proc. Nat. Acad. Sci. U.S.A. 72 (1975), 638-641.

[11] ___________, Inequalities in Fourier Analysis, Ann. of Math. 102 (1975), 159-182.

[12] __________, S. Janson and D. Jerison, Convolution inequalities on the circle, Conference on harmonic analysis in honor of Antoni Zygmund, Wadsworth, Belmont 1983, vol. 1, 32-43.

[13] Iwo Bialynicki-Birula, Entropic uncertainty relations in quantum mechanics, in Quantum Probability and Applications II, eds. L. Accardi and W. von Waldenfels, Lect. Notes in Math. 1136, Springer 1985, p. 90.

[14] Iwo Bialynicki-Birula and Jerzy Mycielski, Uncertainty relations for information entropy in wave mechanics, Comm. Math. Phys. 44 (1975), 129-132.

[15] __________, Wave equations with logarithmic nonlinearities, Bull. Acad. Polon. Sci. Cl. III, 23 (1975), 461.

[16] __________, Ann. Phys. 100 (1976) 62.

[17] A. Bonami, Étude des coefficients de Fourier des fonctions de $L^p(G)$, Ann. Inst. Fourier 20:2 (1970), 335-402 [Contains "two point" inequalities, see p. 376-380].

[18] Christer Borell, Positivity improving operators and hypercontractivity, Math. Z. 180 (1982), 225-234.

[19] __________, A Gaussian correlation inequality for certain bodies in R^n, Math. Ann. 256 (1981), 569-573.

[20] __________, Convexity in Gauss space, Les aspects statistiques et les aspects physiques des processus gaussiens, CNRS, Paris, No. 307, (1981), 27-37.

[21] __________, On polynomial chaos and integrability, Preprint 1980, Univ. of Göteborg, Prob. & Math. Statistics.
(This prolific author has many other papers tangentially related to hypercontractivity.)

[22] H. J. Brascamp and E. H. Lieb, Best constants in Youngs inequality, its converse, and its generalization to more than three functions, Advances in Math. 20 (1976), 151-173.

[23] Henri Buchwalter et chen Dian Jie, Une liaison entre une inegalite logarithmique de Weissler sur le cercle et une question d'extrememum dans le probleme des moments de Stieltjes, Preprint U.E.R. de Mathématiques, Univ. Claude-Bernard-Lyon 1, 43, bd du 11 Nov. 1918; 69622 Villeurbanne Cedex.

[24] Eric Carlen, Superadditivity of Fisher's information and logarithmic Sobolev inequalities, Princeton Preprint July 1989 (to appear).

[25] Eric Carlen and Daniel Stroock, An application of the Bakry-Emery criterion to infinite dimensional diffusions, Strasbourg Sem. de Probabilite XX, 1986, Ed. Azima and Yor.

[26] ______________________________, Ultracontractivity, Sobolev inequalities, and all that, M.I.T. preprint, June 1985.

[27] Eric Carlen and Michael Loss, Extremals of functionals with competing symmetries, Princeton preprint, 1988.

[28] R. Carmona, Regularity properties of Schrödinger and Dirichlet semigroups, J. Funct. Anal. 33 (1979), 259-296.

[29] E. B. Davies, Spectral properties of metastable Markov semigroups, J. Funct. Anal. 52 (1983), 315-329 (and other references to Davies therein).

[30] ___________, Hypercontractive and related bounds for double well Schrödinger operators, Quart. J. Math. Oxford Ser. (2) 34 (1983), 407-421.

[31] ___________, Perturbations of ultracontractive semigroups, Quart. J. Math. Oxford (2) 37 (1986), 167-176.

[32] ___________, Explicit constants for Gaussian upper bounds on heat kernels, Amer. J. Math. 109 (1987), 319-334.

[33] ______________, The equivalence of certain heat kernel and Green function bounds, J. Funct. Anal. 71 (1987), 88-103.

[34] ______________, Heat kernel bounds for second order elliptic operators on Riemannian manifolds, Amer. J. of Math. 109 (1987), 545-570.

[35] ______________, Gaussian upper bounds for heat kernels of some second order operators on Riemannian manifolds, J of Funct. Anal. 80 (1988), 16-32.

[36] ______________, Heat kernels and spectral theory, Cambridge Univ. Press (1989), 197 pp.

[37] E. B. Davies, L. Gross, B. Simon, Hypercontractivity: a bibliographic review, Proc. Hoegh-Krohn Memorial Conference, Ed. S. Albeverio, to appear.

[38] E. B. Davies and N. Mandouvalos, Heat kernel bounds on manifolds with cusps, J. Funct. Anal. 75 (1987), 311-322.

[39] E. B. Davies and M. M. H. Pang, Sharp heat kernel bounds for some Laplace operators, King's College Preprint March 1988.

[40] E. B. Davies and O. S. Rothaus, Markov semigroups on C^*-bundles, J. Funct. Anal. 85 (1989), 264-286.

[41] E. B. Davies and B. Simon, Ultracontractive semigroups and some problems in analysis, in "Aspects of Mathematics and its Applications", North-Holland, Amsterdam 1984, 18 pp.

[42] E. B. Davies and B. Simon, Ultracontractivity and the heat kernel for Schrödinger operators and Dirichlet Laplacians, J. of Funct. Anal. 59 (1984), 335-395.

[43] E. B. Davies and B. Simon, L^1 properties of intrinsic Schrödinger semigroups, J. Funct. Anal. 65 (1986), 126-146.

[44] Burgess Davis, Intrinsic ultracontractivity and the Dirichlet Laplacian, Purdue preprint Jan. 1990, 24 pp.

[45] J. D. Deuschel, Logarithmic Sobolev inequalities of symmetric diffusions, Cornell preprint, July 1989.

[46] J. D. Deuschel and D. W. Stroock, Large deviations, Academic Press, Boston, 1989.

[47] ________________, Hypercontractivity and spectral gap of symmetric diffusions with applications to the stochastic Ising model, J. of Funct. Anal. to appear.

[48] J. P. Eckmann, Hypercontractivity for anharmonic oscillators, J. Funct. Anal. 16 (1974), 388-406.

[49] M. Emery and J. E. Yukich, A simple proof of the logarithmic Sobolev inequality on the circle, in "Seminaire de Probabilites XXI, Lecture Notes in Math. 1247, Springer, Berlin 1987, 173-176.

[50] J. Epperson, The hypercontractive approach to exactly bounding an operator with complex Gaussian kernel, J. of Funct. Anal. 86 (1989).

[51] W. Faris, Product spaces and Nelson's inequality, Helv. Phys. Acta, 48 (1975), 721-730.

[52] P. Federbush, A partially alternate derivation of a result of Nelson, J. Math. Phys. 10 (1969), 50-52.

[53] G. Feissner, Hypercontractive semigroups and Sobolev's inequality, T.A.M.S. 210 (1975), 51-62.

[54] E. Getzler, Dirichlet forms on loop space, Bull. Sc. math. 2^{e} ser. 113 (1989), 151-174.

[55] J. Glimm, Boson fields with nonlinear self-interaction in two dimensions, Comm. Math. Phys. 8 (1968), 12-25.

[56] L. Gross, Logarithmic Sobolev inequalities, Amer. J. of Math. 97 (1975), 1061-1083.

[57] ________, Existence and uniqueness of physical ground states, J. of Funct. Anal. 10 (1972), 52-109.

[58] ________, Hypercontractivity and logarithmic Sobolev inequalities for the Clifford-Dirichlet form, Duke Math. J. 42 (1975), 383-396.

[59] ________, Logarithmic Sobolev inequalities - a Survey in Vector Space Measures and Applications I, Springer Lect. Notes in Math. 644 (1977), Eds. R. Aron and S. Dineen, 196-203.

[60] O. Guennoun, Inéqalités logarithmiques de Gross-Sobolev et hypercontractivité, Thèse 3ème cycle, Université Lyon 1 (1980), 90 pages.

[61] Ira Herbst, Minimal decay rate for hypercontractive measures, Private Communication, November 1975.

[62] R. Holley and D. Stroock, Diffusions on an infinite dimensional torus, J. Funct. Anal. 42 (1981), 29-63.

[63] ____________________________, Logarithmic Sobolev inequalities and stochastic Ising models, J. of Statistical Physics 46 (1987), 1159-1194.

[64] ____________________________, Uniform and L^2 convergence in one dimensional stochastic Ising models, M.I.T. preprint, Sept. 1988, 8 pp.

[65] J. G. Hooton, Dirichlet forms associated with hypercontractive semigroups, Trans. A.M.S. 253 (1979), 237-256.

[66] ____________, Dirichlet semigroups on bounded domains, Rocky Mountain J. Math. 12 (1982), 283-297.

[67] ____________, Compact Sobolev imbeddings on finite measure spaces, J. Math. Anal. and Applications 83 (1981), 570-581.

[68] S. Janson, On hypercontractivity for multipliers on orthogonal polynomials [Best constant for Laplacian on S^2] Ark. Mat. 21 (1983), 97-110.

[69] Abel Klein, Selfadjointness of the locally correct generators of Lorentz transformations for $P(\varphi)_2$, in Math. of Contemporary Physics, Ed. R. F. Streater, Academic Press, London 1972, 227-236.

[70] Abel Klein and L. Landau, Construction of a unique self- adjoint generator for a symmetric local semigroup, J. Funct. Anal. 44 (1981), 121-137.

[71] ____________, Singular perturbations of positivity preserving semigroups via path space techniques, J. Funct. Anal. 20 (1975), 44-82.

[72] ____________ and D. Shucker, Decoupling inequalities for stationary Gaussian processes, Ann. of Probability 10 (1982), 702-708.

[73] Abel Klein and Bernard Russo, Sharp inequalities for Weyl operators and Heisenberg groups, Math. Ann. 235 (1978), 175-194.

[74] Andrzej Korzeniowski, On hypercontractive estimates for the Laguerre semigroup, Preprint, U. of Texas, Arlington, Feb. 1984.

[75] ____________, On logarithmic Sobolev constant for diffusion semigroups, J. Funct. Anal. 71 (1987), 363-370.

[76] Andrzej Korzeniowski and D. W. Stroock, An example in the theory of hypercontractive semigroups, Proc. A.M.S. 94 (1985), 87-90.

[77] W. Krakowiak and J. Szulga, Hypercontraction principle and random multilinear forms, Wroclaw Univ. Preprint #34 (1985), to appear in Prob. Theory and Related Fields.

[78] S. Kusuoka and D. Stroock, Some boundedness properties of certain stationary diffusion semigroups, J. Funct. Anal. 60 (1985), 243-264.

[79] S. Kwapien and J. Szulga, Hypercontraction methods for comparison of moments of random series in normed spaces, Case Western Reserve Preprint, Sept. 1, 1988, 13 pages.

[80] M. Ledoux, Isopérimétrie et inégalités de Sobolev logarithmiques Gaussiennes, C. R. Acad. Sci. Paris, ser. 1, v. 306 (1988), 79-82.

[81] E. H. Lieb, Proof of an entropy conjecture of Wehrl, Comm. Math. Phys. 62 (1978), 35-41.

[82] ____________, Gaussian kernels have only Gaussian maximizers, Princeton preprint 1989.

[83] Paul Meyer, Some analytical results on the Ornstein-Uhlenbeck semigroup in infinitely many dimensions, in IFIP Working Conference on Theory and Applications of Random Fields, Bangalore, Jan. 4-9, 1982 preprint (10 pages)

[84] P. H. Müller and A. Müllner, A representation formula in a space of functionals of the Wiener process and some applications, preprint, Tech. Univ. Dresden, May, 1987.

[85] E. Nelson, A quartic interaction in two dimensions, in Mathematical Theory of Elementary Particles, 69-73. (R. Goodman and I. Segal, eds.) M.I.T. Press, Cambridge, Mass. 1966.

[86] __________, Probability theory and Euclidean field theory, in Constructive quantum field theory. Ed. G. Velo and A. Wightman, Springer Lecture Notes in Physics, Vol. 25 (1973).

[87] __________, The free Markoff field, J. of Funct. Anal. 12 (1973), 211-227.

[88] J. Neveu, Sur l'espérance conditionnelle par rapport à un mouvement brownien, Ann. Inst. H. Poincare [Contains a probabilistic proof of hypercontractivity] 12 (1976), 105-109.

[89] M. M. H. Pang, L^1 properties of two classes of singular second order elliptic operators, J. London Math. Soc. (2) 38 (1988), 525-543.

[90] J. Peetre, A class of kernels related to the inequalities of Beckner and Nelson, "A tribute to Ake Pleijel" Uppsala 1980, 171-210.

[91] Ross Pinsky, Hypercontractivity estimates for nonselfadjoint diffusion semigroups, Proc. of A.M.S. 104 (1988), 532-536.

[92] J. Rosen, Sobolev inequalities for weight spaces and supercontractivity, Trans. A.M.S. 222 (1976), 367-376.

[93] O. S. Rothaus, Lower bounds for eigenvalues of regular Sturm- Liouville operators and the logarithmic Sobolev inequality, Duke Math. Journal 45 (1978), 351-362.

[94] ______________, Logarithmic Sobolev inequalities and the spectrum of Sturm-Liouville operators, J. Funct. Anal. 39 (1980), 42-56.

[95] ______________, Logarithmic Sobolev inequalities and the spectrum of Schrodinger operators, J. Funct. Anal. 42 (1981), 110-120.

[96] ______________, Diffusion on compact Riemannian manifolds and logarithmic Sobolev inequalities, J. Funct. Anal. 42 (1981), 102-109.

[97] ______________, Analytic inequalities, isoperimetric inequalities and logarithmic Sobolev inequalities, J. of Funct. Anal. 64 (1985), 296-313.

[98] ______________, Hypercontractivity and the Bakry-Emery criterion for compact Lie groups, J. Funct. Anal. 65 (1986), 358-367.

[99] I. E. Segal, Construction of nonlinear local quantum processes, I, Ann. of Math. 92 (1970), 462-481.

[100] ____________, Construction of nonlinear local quantum processes, II, Inventiones Math. 14 (1971), 211-241.

[101] R. Seneor, Logarithmic Sobolev inequalities from Sobolev inequalities, Private communication, November 1975.

[102] I. Shigekawa, Existence of invariant measures of diffusions on an abstract Wiener space, Osaka J. Math. 24 (1987), 37-59.

[103] B. Simon, Schrödinger semigroups, Bull. A.M.S. 7 (1982), 447-526.

[104] ________, A remark on Nelson's best hypercontractive estimates, Proc. Amer. Math. Soc. 55 (1976), 376-378.

[105] ________, The $P(\varphi)_2$ Euclidean (Quantum) field theory, Princeton Univ. Press, Princeton, N.J. 1974.

[106] B. Simon and R. Hoegh-Krohn, Hypercontractive semi-groups and two dimensional self-coupled Bose fields, J. Funct. Anal. 9 (1972), 121-180.

[107] D. W. Stroock, Some remarks about Beckner's inequality, Proc. of Symposia in Pure Math. 35 (1979) Part I, 149-157.

[108] ______________, An introduction to the theory of large deviations "Universitext", Springer-Verlag 1984, 196 pages.

[109] Jerzy Szulga, On hypercontractivity of α-stable random variables, $0 < \alpha < 2$, Preprint, July 1987, U. of North Carolina Statistics Dept.

[110] N. Th. Varopoulos, Iteration de J. Moser, Perturbation de semigroups sous-markoviens, C. R. Acad. Sci. Paris Ser. 1, t.300 (1985), 617-620.

[111] ____________________, Hardy-Littlewood theory for semigroups, J. Funct. Anal. 63 (1985), 240-260.

[112] F. Weissler, Logarithmic Sobolev inequalities for the heat- diffusion semigroup, Trans. A.M.S. 237 (1978), 255-269.

[113] _____________, Two-point inequalities, the Hermite semigroup, and the Gauss-Weierstrass semigroup, J. Funct. Anal. 32 (1979), 102-121.

[114] _____________, Logarithmic Sobolev inequalities and hypercontractive estimates on the circle, J. Funct. Anal. 37 (1980), 218-234.

[115] _____________ and C. E. Mueller, Hypercontractivity for the heat semigroup for ultraspherical polynomials and on the n-sphere, J. of Funct. Anal. 48 (1982), 252-283.

[116] I. Wilde, Hypercontractivity for Fermions, J. Math. Phys. 14 (1973), 791-792.

[117] B. Zegarlinski, On log-Sobolev inequalities for infinite lattice systems, preprint (1989).

[118] ________________, Log-Sobolev inequalities for infinite one- dimensional lattice systems, Bochum preprint, Sept. 1989.

[119] ________________, Dobrushin uniqueness theorem and logarithmic Sobolev inequalities, Bochum preprint, Dec. 1989, 34 pp.

SOLUTIONS OF QUANTUM MECHANICS AND FIELD THEORY IN TERMS OF STOCHASTIC (PARTIAL) DIFFERENTIAL EQUATIONS

Zbigniew Haba
Institute of Theoretical Physics, University of Wroclaw, Wroclaw, Poland

Abstract
We show that stochastic differential equations are a useful tool for a construction of stochastic processes and conformal invariant two-dimensional fields with values in non-compact cosets.

I. Introduction

One can approach the quantum mechanics on three main ways: i) Hamiltonian method, ii) functional integration, iii) stochastic equations. A relation between these three approaches can be symbolically expressed by the formula

$$(e^{tH}f)(q) = E_q[f(\xi_t(b))] = \int d\mu[\xi(\cdot)]\, f[\xi(t)] \tag{1}$$

In this formula the Hamiltonian H is the generator of a diffusion process ξ, which is a solution of the Ito stochastic equation

$$d\xi = \alpha(\xi)db + \beta(\xi)dt \tag{2}$$

The expectation $E_q[\ \cdot\ ]$ is with respect to the Wiener measure μ_W. Then, the "functional measure" μ on the r.h.s. of eq.(1) is the image of the Wiener measure μ_W under the transformation

$$T: b \rightarrow \xi \tag{3}$$

determined by eq.(2) i.e. $\mu = \mu_W T^{-1}$. The virtues of each of the three approaches depend on the model and on questions which we are asking about the model.

We are interested in extending the method of stochastic equations to quantum field theory. This involves a treatment of multidimensional time $x \in R^d$. There are two possibilities: a) to distinguish one direction and to treat the other components of x as indexing an infinite dimensional state space, b) to treat all components of $x \in R^d$ on an equal footing. The method

a) applies to all models. However, for a construction of random fields a knowledge of the ground state is required [1]. In this approach the Euclidean invariance pertinent to quantum field theory is not evident. We shall discuss here the approach b). In such an approach instead of ordinary differential equations (2) we shall have stochastic partial differential equations. We are interested here mainly in finding explicit solutions. The functional integral in field theory is far from beeing completly understood. Explicit solutions can elucidate problems involved in functional integration and in the stochastic analysis with multidimensional time.

II. Stochastic equations in quantum mechanics on a manifold

The usefulness of the method of stochastic differential equations for finding solutions can be illustrated in a discussion of the quantum mechanics on a manifold. In general, the stochastic equation for the Brownian motion on a Riemannian manifold M has the form [2] ($\circ d\xi$ denotes the Stratonovitch differential)

$$dq^{\mu} = e^{\mu}{}_{a}(q) \circ db^{a} \tag{4a}$$

$$de^{\mu}{}_{a} + \Gamma^{\mu}{}_{\nu\rho} e^{\rho}{}_{a} \circ dq^{\nu} = 0 \tag{4b}$$

where e is a frame on M ($e^{\mu}{}_{a} e^{\nu}{}_{a} = g^{\mu\nu}$, where g is the metric) and Γ is the Christoffel connection. The Laplace-Beltrami operator is the generator of the Brownian motion (4). In coordinates we can express the equation (4) in a non-covariant form (Ito differentials)

$$dq^{\mu} = e^{\mu}{}_{a}(q) db^{a} - \tfrac{1}{2}\Gamma^{\mu} dt \tag{5}$$

We may treat eq.(4) as a transformation (3) of paths. The Jacobian of this transformation is

$$J = \frac{\delta b^{a}}{\delta q^{\mu}} = e^{\mu}{}_{a} \frac{d}{dt} + \Gamma^{\mu}{}_{\nu\rho} e^{\rho}{}_{a} \frac{dq^{\nu}}{dt} \tag{6}$$

It can be shown that $\det J = 1$. Then, the (formal) functional measure μ resulting from the transformation (3) of the Wiener measure $d\mu_W$ is

$$d\mu(q) = d\mu_W(T^{-1}q) = dq \exp\left\{-\int \tfrac{1}{2}\left(e_{\mu a}(q)\frac{dq^{\mu}}{dt}\right)^2\right\} \tag{7}$$

A large symmetry group G can simplify the problem of finding solutions. In particular, there is a large class of non-compact cosets G/H such that with a proper choice of coordinates the Brownian motion on the coset can be explicitly solved in terms of the Wiener process. Moreover, using this solution an integral representation of the heat kernel can be obtained [3]-[5]. The soluble coset models include

$$SO(n,1)/SO(n),\quad SU(n,1)/SU(n),\quad SL(n,R)/SO(n),\quad SL(n,C)/SU(n),$$
$$Sp(n,C)/USp(n),\quad SU(n,n)/SU(n)\times SU(n) \tag{8}$$

As soon as we have solved the Brownian motion on the coset G/H we can find the Brownian motion on G if the Brownian motion on H is known.

Let us start with the simplest non-compact non-abelian group $SO(2,1)\sim SL(2,R)$. We introduce a left invariant Riemannian metric on SL(2,R)

$$ds^2 = Tr\left\{g^{-1}dg(g^{-1}dg)^+\right\} \tag{9}$$

If we parameterize $g\in SL(2,R)$ by (Iwasawa decomposition)

$$g = \begin{bmatrix} 1 & x \\ 0 & 1 \end{bmatrix}\begin{bmatrix} \sqrt{y} & 0 \\ 0 & \sqrt{y}^{-1} \end{bmatrix}\begin{bmatrix} \cos\theta & -\sin\theta \\ \sin\theta & \cos\theta \end{bmatrix} \qquad y>0$$

then we get from eq.(9)

$$ds^2 = \tfrac{1}{2}y^{-2}dy^2 + y^{-2}(dx + yd\theta)^2 + d\theta^2 \tag{10}$$

The Laplace-Beltrami operator corresponding to the metric (10) is

$$H = y^2\left(\frac{\partial^2}{\partial x^2} + \frac{\partial^2}{\partial y^2}\right) - y\frac{\partial}{\partial x}\frac{\partial}{\partial\theta} + \frac{1}{2}\frac{\partial^2}{\partial\theta^2} \tag{11}$$

H generates a stochastic process on SL(2,R) fulfilling the equations

$$dx = ydb_1 - ydb_3,\quad dy = \sqrt{2}\,ydb_2,\quad d\theta = db_3 \tag{12}$$

with the solution

$$y(t) = y_0\exp\left(\sqrt{2}b_2(t) - t\right),\quad x(t) = \int_0^t y(s)db_1(s) - \int_0^t y(s)db_3(s) + x_0$$
$$\theta(t) = b_3(t) + \theta_0 \tag{13}$$

In general, a stochastic equation for the left invariant (it will be biinvariant if and only if the group is compact) Brownian motion on a group G reads

$$g^{-1}\circ dg = db \tag{14}$$

where the Wiener process b takes values in the Lie algebra of G.

Let us apply the Iwasawa decomposition

$$g = NDU = SU \qquad (15)$$

where N is upper triangular, D is diagonal and $U \in H$ is compact. It follows from eq.(14) that

$$S^{-1} \circ dS = U(\circ db - U^{-1} \circ dU)U^{-1} \qquad (16)$$

The orthogonal Iwasawa decomposition in the algebra of G reads

$$db = db_s + db_c \qquad (17)$$

where db_c denotes the compact part. It follows that eq.(16) splits into two equations

$$U^{-1} \circ dU = db_c$$
$$S^{-1} \circ dS = U \circ db_s U^{-1} \qquad (18)$$

S has zeros below the diagonal. Therefore, eq.(18) can be solved in an elementary way (see ref.[3]). So, the stochastic equation on a non-compact group is solved explicitly in terms of the Brownian motion on its maximal compact subgroup. Whenever we have the Brownian motion on G we can get the Brownian motion on the coset G/H by means of a projection. In the case of SL(2,R)/O(2) the projection sets $b_3=0$ in eq.(13).

It is easy to compute correlation functions $E[g(t_1)\ldots\ldots\ldots g(t_n)]$ for the solution (13). However, in quantum mechanics we are interested in the evolution kernel (heat kernel at imaginary time). Let us consider the heat kernel for the upper half-plane SL(2,R)/0(2)

$$e^{tH}(x_0,y_0;x_1,y_1) = y_1^{\,2}\, E_{x_0,y_0}[\, \delta\big(x(t) - x_1\big)\, \delta\big(\, y(t) - y_1\big)\,] \equiv p_t(z_0,z_1) \qquad (19)$$

(here z=x+iy). It is not easy to compute $\delta(\, x(t) - x_1)$. However, this will not be necessary. Let us integrate p over x_1. Then, we get the equation

$$\int dx_1\; p_t(z_0,z_1) = y_1^{\,2}\int dx_1 E[\, \delta\big(x(t) - x_1\big)\, \delta\big(y(t) - y_1\big)] = y_1^{\,2} E_{y_0}[\delta\big(y(t) - y_1\big)]$$

$$= y_1^{\,2}(2\pi t)^{-\frac{1}{2}}\exp\Big\{-(\ln y_1 - \ln y_0 + \tfrac{t}{2})^2\Big\} \qquad (20)$$

We have got an integral equation for p. In order to solve it we use the invariance of the heat kernel under SL(2,R) transformations. The invariance implies that p is a function of the hyperbolic distance d

$$\mathrm{ch}d(z_0,z_1) = 1 + [(x_1-x_0)^2 + (y_1-y_0)^2]\,(2y_0y_1)^{-1}$$

The solution of eq.(20) is [5]

$$p_t(z_0,z_1) = e^{-\frac{t}{8}}\sqrt{2}(2\pi t)^{-\frac{3}{2}}\int_d^{\infty} dr\; (\mathrm{ch}r - \mathrm{ch}d)^{-\frac{1}{2}} r \exp(-\tfrac{r^2}{2t}) \qquad (21)$$

The method applied here to SL(2,R)/O(2) is successful in application to other cosets (8) as well as to some non-compact groups (this will be worked out in ref.[4]).

III. Two-dimensional Euclidean fields.

In order to generalize stochastic equations to two dimensions we need a generalization of the Brownian motion and its derivative (the white noise). It comes out that such a generalization does exist if the fields are complex. So, let ϕ be the Gaussian complex massless scalar field with the covariance

$$E[\ \phi(z)\overline{\phi(z')}\ ] = (-\Delta)^{-1}(z,z') = -\frac{1}{4\pi}\ln|z-z'|^2 \tag{22}$$

Then, its holomorphic derivative $\bar{\partial} = \partial_1 + i\partial_2$ is the noise i.e.

$$E[\bar{\partial}\phi(z)\ \partial\bar{\phi}(z')] = \delta(z-z') \tag{23}$$

Now, eq.(14) can be generalized to define a field with values in a *complex* group G

$$g^{-1}\bar{\partial}g = \bar{\partial}\phi \tag{24}$$

where ϕ takes values in the Lie algebra of G. In coordinates, eq.(24) takes the form

$$\bar{\partial}w^{\mu} = L^{\mu}{}_{a}(w)\bar{\partial}\phi^{a} \tag{25}$$

where ϕ^a are independent scalar fields (22) and $L^{\mu}{}_{a}$ are the left invariant vector fields on G. The equations (24) constitute a system of linear partial differential equations, which can be solved by iteration. We could also apply the Iwasawa decomposition (15). Then, eq.(24) reads

$$S^{-1}\bar{\partial}S = U(\bar{\partial}\phi - U^{-1}\bar{\partial}U)U^{-1} \tag{26}$$

or

$$N^{-1}\ \bar{\partial}N = DU(\bar{\partial}\phi - U^{-1}\bar{\partial}U)U^{-1}D^{-1} + \bar{\partial}DD^{-1} \tag{27}$$

However, the disentanglement of the compact and non-compact parts from eqs.(26)-(27) becomes a non-linear problem. For nilpotent groups N we can get the solution of eq.(24) in a compact form e.g. if N is a nilpotent group of order 3 (upper triangular matrix with entries Z_{12}, Z_{13}, Z_{23}) the solution is (see ref.[3])

$$Z_{12}=\phi_{12}\cdot \quad Z_{23}=\phi_{23}$$

$$Z_{13}= \phi_{13} + \bar{\partial}^{-1}(\phi_{12}\bar{\partial}\phi_{23}) \tag{28}$$

where

$$\bar{\partial}^{-1}(z,\varsigma) = \frac{1}{2\pi(z-\varsigma)}$$

A generalization of eq.(25) to a general Riemannian complex manifold has the form [6]

$$\bar{\partial}w^{\mu}= e^{\mu}{}_{a}(w)\bar{\partial}\phi^{a} \tag{29}$$

On a Kahler manifold $\Gamma^{\mu}=0$ in complex coordinates, hence the second term on the r.h.s. of eq.(5) is absent. Then, eq.(29) when reduced to one dimension $\bar{\partial} \rightarrow \frac{d}{dt}$ coincides with eq.(5). It is remarkable that on a Kahler manifold the stochastic integral appearing in the application of the Girsanov formula to the stochastic equation (29) interpreted as the Ito equation with the drift $\partial_{x}w^{\mu}$

$$\int \partial_{x}w^{\mu} \; g_{\mu\bar{\nu}}(w)\partial_{t}\bar{w}^{\nu}dxdt$$

is a constant (the topological charge). In general it is a torsion term [5].

As an example consider the sphere S^2. We can choose the complex coordinate w on S^2 by means of the stereographic projection of S^2 on the plane. Then, eq.(29) reads

$$\bar{\partial}w = (1 + w\bar{w})\bar{\partial}\phi \tag{30}$$

Let us consider eq.(30) in a region R_+ bounded by a curve γ. Denote the exterior of R_+ by R_- . The boundary value problem for equations of the type (30) has been discussed in ref.[7]. We require that on the boundary

$$w_{|R_-} - w_{|R_+}= \sigma \tag{31}$$

Then, the boundary value problem (30)-(31) has the unique solution in a bounded region R_+ (for regular functions $\bar{\partial}\phi$) . The solution can be obtained by iteration of the integral equation

$$w(z) = \frac{1}{2\pi}\int_{\gamma} d\varsigma\; \sigma(\varsigma)(z-\varsigma)^{-1} - \frac{1}{2\pi}\int_{R_+} d\varsigma\wedge d\bar{\varsigma}\; (z-\varsigma)^{-1}(1 + w(\varsigma)\bar{w}(\varsigma)\,)\,\bar{\partial}\phi \tag{32}$$

In the case of the pseudosphere

$$x_1^2 + x_2^2 - x_3^2 = 1$$

realized as the upper half-plane with the Poincare metric

$$ds^2 = -\left(\frac{2}{w-\bar{w}}\right)^2 dw d\bar{w} \tag{33}$$

the stochastic equations (29) become linear [5]

$$\bar{\partial} w = \frac{1}{2i}(w - \bar{w})\bar{\partial}\phi \tag{34}$$

Eq.(34) is equivalent to a system of two equations (for w and $\bar{w}$). A solution can be obtained in a form of an infinite series.

The measure corresponding to the solution of the stochastic equation (29) can be derived as a transformation (3) of the Gaussian measure with the covariance (22). In a formal way it can be written as

$$d\mu = dw \, |\det J(w)| \exp\left\{-\int |e_{\mu a}\bar{\partial} w^{\mu}|^2\right\} \tag{35}$$

However, in contradistinction to one dimension (see eq.(6)) the Jacobian $\det J \neq 1$. The Jacobian can be expressed as an integral over (anticommuting) Fermi fields. So, the Fermi fields are intrinsically connected with a description of Euclidean fields in terms of stochastic differential equations.

We shall discuss now a somewhat different situation than (29), when we can treat the frame e in eq.(29) as an *external* random field. It comes out that the problem of a construction of random fields on the coset spaces (8) reduces to such a linear external field problem.

On SO(2n+1,1)/SO(2n+1) the invariant metric is

$$L = \int y^{-2}\left(\partial y \bar{\partial} y + \sum_1^n \partial \bar{w}_k \bar{\partial} w_k\right) \tag{36}$$

Sp(r,C)/USp(r,C) is a generalization of the half-space (36) with the invariant metric

$$L = \mathrm{Tr}\int\left[Y^{-1}\partial Y(Y^{-1}\partial Y)^+ + Y^{-1}\bar{\partial} W(Y^{-1}\bar{\partial} W)^+\right] \tag{37}$$

where Y is a positive $r \times r$ matrix and $W \in GL(r,C)$. SL(n,C)/SU(n) can be realized as a subset of positive matrices in SL(n,C) with the invariant metric defined by

$$L = \mathrm{Tr}\int Y^{-1}\bar{\partial} Y(Y^{-1}\bar{\partial} Y)^+ \tag{38}$$

The stochastic equation determining solutions of the quantum field theory with the Lagrangian (36) has the form [10] (we set $y = \exp\phi$)

$$\bar{\partial} w_k = e^{\phi}\bar{\partial}\Phi_k \tag{39}$$

where ϕ is a real Gaussian massless scalar field and Φ_k are independent Gaussian complex massless scalar fields with the covariance (22).

In order to solve the model (38) we express Y in the form

$$Y = NDN^{+} \tag{40}$$

where N is upper triangular and D is diagonal $D=\{\exp\phi_r\}$. Then the Lagrangian (38) has the form

$$L = Tr\int\left(D^{-1}\bar{\partial}D\, D^{-1}\partial D + (D^{-1}N^{-1}\bar{\partial}N)^{+}N^{-1}\bar{\partial}N\, D\right) \tag{41}$$

Now, the stochastic equations describing the model (41) have the form [11]

$$(N^{-1}\bar{\partial}N)_{km} = \exp\tfrac{1}{2}(\phi_k-\phi_m)\, \bar{\partial}\Phi_{km} \tag{42}$$

where ϕ_k are real Gaussian fields and Φ_{km} are independent complex Gaussian fields with the covariance (22). The equation (42) for the nilpotent matrix N can easily be solved (see ref.[3]).

It can be seen that the solution of the model (37) can be obtained as soon as we solve the model (38) (eq.(42)). In fact, Y is expressed by eqs.(40) and (42), whereas $W = \bar{\partial}^{-1}(Y\bar{\partial}\Phi)$, where Φ is a matrix-valued free field.

The models (36)-(38) are conformal invariant i.e. invariant under the transformations

$$z \rightarrow (\alpha z+\beta)(\gamma z+\delta)^{-1} \qquad \text{where} \begin{bmatrix} \alpha & \beta \\ \gamma & \delta \end{bmatrix} \in SL(2,C)$$

The simplicity of stochastic equations describing the models (36)-(38) reflects the simplicity of the conformal field theory itself, which is closely related to a free field theory.

Let us discuss briefly a possible extension of the method of stochastic equations to four dimensional quantum field theory. The proper counterpart to the holomorphic derivative is the quaternionic derivative

$$\nabla = \sigma^{\mu}\partial_{\mu} \qquad \mu=1,..,4$$

where the σ's form the basis of quaternions. An analog of the massless scalar field is the electromagnetic field $A = A_{\mu}\sigma^{\mu}$ (see refs.[8]-[9]). Then,

$$\nabla A^{+}=\eta \tag{43}$$

is the noise. A is conformal invariant. However, we do not know how to built non-Gaussian conformal fields out of A. Perhaps, a proper building block for conformal field theory in four dimensions is a quaternion-valued field Φ with the covariance

$$E[\Phi(x)\Phi(y)^{+}] = -\frac{1}{(4\pi)^2}\ln(x-y)^2 \tag{44}$$

An analog of the stochastic equation (29) is

$$g^{-1}\nabla g = \nabla\Phi \tag{45}$$

There is a remarkable analogy between the two dimensional conformal field theory and the four dimensional one if only the complex groups in (8) are replaced by quaternionic ones.

Acknowledgements

The studies reported here have been done in the framework of a research project SFB 237 in collaboration with Professor Sergio Albeverio during author's stay at Bochum University and at BiBoS Research Center of Bielefeld University. The author would like to thank the organizers of this conference and organizers of SFB 237 for stimulating atmosphere and hospitality.

References

[1] S. Albeverio and R. Hoegh-Krohn, Z.Wahr.verv.Gebiete,40,1(1977)

[2] N. Ikeda and S. Watanabe, Stochastic Differential Equations and Diffusion Processes, North Holland, 1981

[3] S. Albeverio, T. Arede and Z. Haba, BiBoS preprint No.330, 1988, to appear in Journ.Math.Phys.

[4] S. Albeverio, T. Arede and Z. Haba, in preparation

[5] Z. Haba, Phys.Rev.D38,647(1988)

[6] Z. Haba, Phys.Rev.D33,2428(1986)

[7] I.N. Vekua, Generalized Analytic Functions, Pergamon Press

[8] Z. Haba, in Lect.Notes in Phys.,No.262,Springer,1986

[9] S. Albeverio, K. Iwata and T. Kolsrud, BiBoS preprint No.377, 1989

[10] Z. Haba, Int.Journ.Mod.Phys.A4,467(1989)

[11] Z. Haba, BiBoS preprint No.382,1989

White Noise Analysis — An Overview

T. Hida
Department of Mathematics, Nagoya University
Nagoya 464, Japan

and

J. Potthoff
Department of Mathematics, Louisiana St. University
Baton Rouge, LA 70803, USA

1. Introduction.

Let $\{B(t),\, t \in I\!R\}$ be an ordinary Brownian motion. We are interested in functionals of Brownian motion:

$$f(B(t),\, t \in T), \quad T: \text{ an interval of } I\!R \tag{1.1}$$

In order to analyze such functionals, we propose to consider instead the expression

$$\varphi(\dot{B}(t),\, t \in T), \tag{1.2}$$

where $\{\dot{B}(t),\, t \in I\!R\}$ is a white noise: $\dot{B}(t) = \frac{d}{dt}B(t)$. The main reason why like to deal with Eq. (1.2) instead of (1.1) is that $\{\dot{B}(t),\, t \in I\!R\}$ is a continuous analogue of i.i.d. (independent identically distributed) random variables. Therefore, we may think of $\{\dot{B}(t),\, t \in I\!R\}$ as a system of "coordinates" and we shall call φ in Eq. (1.2) a *white noise functional.* In fact, expression (1.1) is equivalent to (1.2): cf. Section 2. The price we have to pay for using white noise instead of Brownian motion as the argument of functionals is, that we have to deal with functionals of distributions $\dot{B}$ instead of functionals of continuous functions B.

There are several important classes of white noise functionals φ which depend on $\dot{B}$ through variables like $< \dot{B},\, \xi >$, where ξ is a sufficiently smooth test function. For example, the class $\mathcal{P}$ of polynomials: these are functionals p of the form

$$p(\dot{B}) = P(< \dot{B},\, \xi_1 >,\, \dots\, ,\, < \dot{B},\, \xi_n >), \tag{1.3}$$

where $\xi_1, \ldots, \xi_n \in \mathcal{S}(\mathbb{R})$ and P is a polynomial on $\mathbb{R}^n$. However, for many applications (such as in stochastic analysis or in quantum field theory) it is important to consider $\dot{B}(t)$ "without smearing" by a test function. For instance, considering stochastic integration with respect to Brownian motion we have to think of $\dot{B}(t)dt$ as the increment of the Brownian path at time t.

Once we accept this point of view, it is natural to try to formulate simple functionals of the basic variables $\dot{B}(t)$, such as monomials or polynomials. This leads immediately to the construction of spaces of generalized functionals of white noise $\dot{B}$. In fact, an expression like $\dot{B}(t)^2$, $t \in \mathbb{R}$, does not have any meaning as a random variable. But after a suitable "renormalization", it can be given a meaning as a continuous linear functional on a space of smooth white noise functionals. It turns out that generalized functionals of white noise appear in many applications. In Sections 3 and 4 we shall sketch the theory, properties and some examples of generalized white noise functionals.

The next step in the analysis of white noise functionals is to establish the calculus of differential operators. According to the above, we are interested to introduce a "partial differentiation"

$$\frac{\partial}{\partial \dot{B}(t)} \equiv \partial_t, \quad t \in \mathbb{R} \tag{1.4}$$

with respect to the basic variable $\dot{B}(t)$. In Section 5 we describe the resulting calculus, which we shall also refer to as the *causal calculus*.

An important tool for the analysis of white noise functionals is the *infinite dimensional rotation group*. Like in finite dimensional harmonic analysis, it describes the behaviour of functionals and operations acting on functionals under "coordinate transformations". The infinite dimensional rotation group will be discussed in Section 6. Also, we mention there the relation between certain subgroups of the infinite dimensional rotation group and various Laplacians.

The applications of white noise analysis are manifold: in (non-linear) filtering, in prediction theory, in quantum physics, in stochastic analysis, in molecular biology and so on. Some of them are sketched in Section 7.

Problems in many domains of applications suggest the extension of white noise analysis to the analysis of (Gaussian) random fields, i.e. the variational calculus of functionals depending on random fields indexed by a manifold. Here, the "state of art" is still far from a general theory, but there are indications that white noise analysis provides the intuition needed to develop such a theory. We shall discuss this aspect shortly in the concluding remarks in Section 8.

The present paper is based on the talks given by the authors at the Meeting in Bielefeld. However, it owes very much to the manuscript of a monograph in preparation with H.-H. Kuo and L. Streit [HKPS].

2. L^2 – Theory.

As we have mentioned in the introduction, considering the time derivative $\{\dot{B}(t),\, t \in I\!R\}$ of Brownian motion as a coordinate system, we have to deal with functionals of generalized functions, because the paths of Brownian motion are with probability one non-differentiable functions. It is convenient to choose the triple $(\mathcal{S}'(I\!R), \mathcal{B}, \mu)$ as the probability space of $\dot{B}$, where $\mathcal{S}'(I\!R)$[1] is the Schwartz space of tempered distributions over $I\!R$, $\mathcal{B}$ the σ-algebra over $\mathcal{S}'(I\!R)$ generated by the cylinder sets, and μ is the white noise measure with characteristic functional given by

$$\int_{\mathcal{S}'(I\!R)} \exp(i < x,\, \xi >)\, d\mu(x) = \exp(-\frac{1}{2}\, \|\xi\|_2^2), \quad \xi \in \mathcal{S}(I\!R). \tag{2.1}$$

In (2.1), $< \cdot\,, \cdot >$ denotes the pairing between $\mathcal{S}'(I\!R)$ and $\mathcal{S}(I\!R)$, and $\|\cdot\|_2$ is the norm of $L^2(I\!R, dt)$. Informally speaking, the "path" $\{x(t),\, t \in I\!R\}$ for $x \in \mathcal{S}'(I\!R)$, is a "path" of $\dot{B}$ under μ. We shall make this more precise below. Another informal remark is that considering $x(t)$ as a random variable on $(\mathcal{S}'(I\!R), \mathcal{B}, \mu)$, we conclude immediately from (2.1) that $x(t)$ is centered Gaussian with covariance given by

$$I\!E\, x(t)x(s) = \delta(t - s),$$

which reflects the fact noted in the introduction that $\dot{B}(t)$, $t \in I\!R$, are continuously indexed i.i.d. random variables.

We denote by $X.$ the canonical coordinate process on $(\mathcal{S}'(I\!R), \mathcal{B}, \mu)$ indexed by $\mathcal{S}(I\!R)$, i.e.

$$X : \xi \longmapsto X_\xi, \quad \xi \in \mathcal{S}(I\!R)$$

with

$$X_\xi(x) = < x,\, \xi >, \quad x \in \mathcal{S}'(I\!R).$$

It easy to see that X maps $\mathcal{S}(I\!R)$ into $L^p(\mathcal{S}'(I\!R), d\mu)$ for every $p \in [1, +\infty)$. Clearly, $X.$ is centered Gaussian with covariance

$$I\!E X_\xi X_\eta = (\xi,\, \eta),$$

where $(\cdot\,, \cdot)$ is the inner product of $L^2(I\!R)$. Observe that $X.$ has a (continuous) extension from $\mathcal{S}(I\!R)$ to $L^2(I\!R)$. We shall continue to denote this extension as $X_f(x) = < x, f >$, $f \in L^2(I\!R)$. In particular, we may form the random variables $\{B(t,x) := < x,\, 1_{[0,t)} >,\, t \in I\!R_+\}$. A straightforward computation shows that $\{B(t),\, t \in I\!R_+\}$ is a Brownian motion. Therefore, the relation $B(t,x) = < x,\, 1_{[0,t)} >$

---

[1] here, and throughout unless stated otherwise, we consider $\mathcal{S}'(I\!R)$ as equipped with the weak topology

expresses rigorously the above mentioned fact that under μ the "paths" of $x \in \mathcal{S}'(\mathbb{R})$ are those of a white noise $\dot{B}$ and it establishes the equivalence of (1.1) and (1.2).

It is well-known that the measure space $(\mathcal{S}'(\mathbb{R}), \mu)$ can be approximated by $(S^n(\sqrt{n}), \sigma_n)$, where σ_n is the uniform probability measure on the n-dimensional sphere S^n of radius $\sqrt{n}$. This suggests to introduce a group of rotations acting on the "support" of μ, such that it is kept invariant under these rotations.

Following H. Yoshizawa, the rotation group $O(\mathcal{S}(\mathbb{R}))$ of $\mathcal{S}(\mathbb{R})$ is defined by

$$O(\mathcal{S}(\mathbb{R})) = \{g \text{ lin. homeom. of } \mathcal{S}(\mathbb{R}), \text{ s.t. } \|g\xi\|_2 = \|\xi\|_2 \text{ f. a. } \xi \in \mathcal{S}(\mathbb{R})\}.$$

If we denote by g^* the adjoint action of $g \in O(\mathcal{S}(\mathbb{R}))$ on $\mathcal{S}'(\mathbb{R})$, then we obtain as expected

$$g^*\mu = \mu. \tag{2.2}$$

The last remark lays the ground for the *harmonic analysis* arising from the infinite dimensional rotation group [Hi 80, Hi 85, HKPS], which will later be discussed for spaces of generalized functionals. Also, there are the related concepts of *Fourier transformation* and *Laplacians*, cf. Sections 6 and 7.

The complex Hilbert space $L^2(\mathcal{S}'(\mathbb{R}), d\mu)$ which we simply denote as (L^2) admits the direct sum decomposition (see e.g. [Hi 80, HKPS, Itô 51, Ne 73])

$$(L^2) = \bigoplus_{n=0}^{\infty} \mathcal{H}_n, \tag{2.3}$$

where $\mathcal{H}_n$ is the space of multiple Wiener integrals of degree n.

To make the decomposition (2.3) more concrete, we introduce the following two (related) transformations on (L^2): let $\varphi \in (L^2)$, $\xi \in \mathcal{S}(\mathbb{R})$

1) $\mathcal{T}$-transform

$$(\mathcal{T}\varphi)(\xi) = \int_{\mathcal{S}'(\mathbb{R})} \exp(i < x, \xi >)\varphi(x)d\mu(x) \tag{2.4}$$

2) $\mathcal{S}$-transform

$$(\mathcal{S}\varphi)(\xi) = \int_{\mathcal{S}'(\mathbb{R})} \exp\left(< x, \xi >) - \frac{1}{2}\|\xi\|_2^2\right)\varphi(x)d\mu(x), \tag{2.5a}$$

$$= \int_{\mathcal{S}'(\mathbb{R})} \varphi(x+\xi)d\mu(x) \tag{2.5b}$$

$(\mathcal{S}\varphi)(\xi)$ is often called the *U-functional* of φ. (In Section 4 we shall give a slightly more general definition of U-functionals.)

With the help of $\mathcal{T}$ or $\mathcal{S}$ one can establish the well-known isomorphism between (L^2) and the symmetric Fock space over $L^2(\mathbb{R}, du)$ [Hi 80, HKPS, Itô 51, Ne 73]:

$$(L^2) \simeq \bigoplus_{n=0}^{\infty} L^2(\widehat{\mathbb{R}^n}, du) \tag{2.6a}$$

since

$$\mathcal{H}_n \simeq L^2(\widehat{\mathbb{R}^n}, du), \tag{2.6b}$$

where $\widehat{\cdot}$ means symmetrization.

According to (2.6), every φ in (L^2) is in 1–1 correspondence with a sequence $(f^{(n)} \in L^2(\widehat{\mathbb{R}^n}), n \in \mathbb{N}_0)$. Moreover, it is not hard to see that for such a φ

$$\|\varphi\|_2^2 = \sum_{n=0}^{\infty} n! \left\| f^{(n)} \right\|_{L^2(\mathbb{R}^n)}^2 \tag{2.7}$$

and

$$(\mathcal{S}\varphi)(\xi) = \sum_{n=0}^{\infty} \int_{\mathbb{R}^n} f^{(n)}(u) \xi^{\otimes n}(u) du. \tag{2.8}$$

We can represent $\varphi \in (L^2)$ corresponding to $(f^{(n)}, n \in \mathbb{N}_0)$ through the $f^{(n)}$'s directly in the following way:

$$\varphi(x) = \sum_{n=0}^{\infty} <: x^{\otimes n} :, f^{(n)} >, \quad x \in \mathcal{S}'(\mathbb{R}) \tag{2.9}$$

In (2.9) $: x^{\otimes n} :$ is the *Wick-ordering* of the distribution $x^{\otimes n} \in \mathcal{S}'(\widehat{\mathbb{R}^n})$, $x \in \mathcal{S}'(\mathbb{R})$, which can be defined recursively as follows.

$$: x^{\otimes 0} : = 1 \tag{2.10a}$$

$$: x^{\otimes 1} : = x \tag{2.10b}$$

$$: x^{\otimes n} : =: x^{\otimes (n-1)} : \widehat{\otimes} x - n : x^{\otimes (n-2)} : \widehat{\otimes} Tr, \tag{2.10c}$$

and $Tr \in \mathcal{S}'(\mathbb{R}^2)$ is given by

$$Tr\, f = \int_{\mathbb{R}} f(u,u) du, \quad f \in \mathcal{S}(\mathbb{R}^2).$$

For example, $: x^{\otimes 2} := x^{\otimes 2} - Tr$.

3. Generalized White Noise Functionals.

Generalized functionals of white noise have proved to be extremely useful in many domains of applications (cf. e.g. [AHP 89, Ch 86, Hi 83b, HKPS, Kuo 83a, Kuo 83b, KP 89a, KP 89b, KS 87, NZ 88, PR 89, PY 89, SH 83]) and are of considerable theoretical interest. Just as in finite dimensional (functional) analysis, the choice of the space of test or generalized functionals will depend very much on the application one has in mind.

In this section we shall discuss three classes of Gel'fand triples

$$(E)^* \supset (L^2) \supset (E) \tag{3.1}$$

where – as usual – we have identified (L^2) with its dual.

3.1 Classical $(L^2)^\pm$ Spaces.

The construction of the spaces $(L^2)^\pm$ of test and generalized functionals of white noise, carried out in [Hi 75], was firstly motivated by the same remark as in Section 1: to formulate monomials and polynomials of $\dot{B}(t)$. However, the triple

$$(L^2)^- \supset (L^2) \supset (L^2)^+ \tag{3.2}$$

has found various applications: for example in quantum mechanics [SH 82,SH 83], or in stochastic partial differential equations [Ch 86].

Fix $n \in \mathbb{N}$ and consider the diagram

$$\begin{array}{ccccc} \mathcal{H}_n^- & \supset & \mathcal{H}_n & \supset & \mathcal{H}_n^+ \\ \downarrow & & \downarrow & & \downarrow \\ \widehat{H_{-\frac{n+1}{2}}(\mathbb{R}^n)} & \supset & L^2(\mathbb{R}^n) & \supset & \widehat{H_{\frac{n+1}{2}}(\mathbb{R}^n)} \end{array} \tag{3.3}$$

where the middle arrow represents the unitary mapping given by the $\mathcal{S}$–transform restricted to $\mathcal{H}_n$ (cf. (2.6b)) and $H_\alpha(\mathbb{R}^n)$ denotes the Sobolev space of order $\alpha \in \mathbb{R}$ over $\mathbb{R}^n$. The right exterior arrow in (3.3) is the restriction of the $\mathcal{S}$–transform (considered as an isomorphism of vector spaces) to those functionals in $\mathcal{H}_n$ whose kernel functions $f^{(n)}$ belong to $\widehat{H_{\frac{n+1}{2}}(\mathbb{R}^n)}$. $\mathcal{H}_n^+$ is topologized so that this arrow too becomes a unitary mapping, i.e. if $\varphi^{(n)} \in \mathcal{H}_n^+$ with

$$\varphi^{(n)}(x) = <: x^{\otimes n} :, f^{(n)} >, \tag{3.4}$$

then

$$\left\| \varphi^{(n)} \right\|_{\mathcal{H}_n^+} = \sqrt{n!} \left\| f^{(n)} \right\|_{H_{\frac{n+1}{2}}(\mathbb{R}^n)} . \tag{3.5}$$

The dual of $\mathcal{H}_n^+$ is denoted by $\mathcal{H}_n^-$ and this space is isomorphic to $H_{-\frac{n+1}{2}}\widehat{(\mathbb{R}^n)}$.

Set

$$(L^2)^+ := \bigoplus_{n=0}^{\infty} c_n \mathcal{H}_n^+ \tag{3.6}$$

and

$$(L^2)^- := \bigoplus_{n=0}^{\infty} c_n^{-1} \mathcal{H}_n^-, \tag{3.7}$$

where $\{c_n,\, n \in \mathbb{N}_0\}$ is a positive, non-decreasing sequence. If $\varphi \in (L^2)^+$ corresponding to $(f^{(n)},\, n \in \mathbb{N}_0)$, then

$$\begin{aligned} \|\varphi\|^2_{(L^2)^+} &= \sum_{n=0}^{\infty} c_n^2 \left\|\varphi^{(n)}\right\|^2_{\mathcal{H}_n^+} \\ &= \sum_{n=0}^{\infty} c_n^2 n! \left\|f^{(n)}\right\|^2_{H_{\frac{n+1}{2}}(\mathbb{R}^n)} . \end{aligned} \tag{3.8}$$

Examples. The following examples are informal expressions which can be given a rigorous meaning as generalized functionals in $(L^2)^-$, for details we refer the reader to [Hi 75, Kuo 83b, HKPS]. Here we choose $c_n = 1$, for all $n \in \mathbb{N}_0$.

1) *Wick monomials of white noise*

$$: \dot{B}(t)^n :=: x(t)^n :\in \mathcal{H}_n^-, \quad t \in \mathbb{R}$$

2) *Donsker's delta function*

$$\delta(B(t) - a) \in (L^2)^-, \quad t \in \mathbb{R}_+,\, a \in \mathbb{R}$$

3) *Gauss kernel*

$$\mathcal{N} \exp\left(c \int_0^T x(u)^2 du\right), \quad c \in \mathbb{C},\, c \neq \frac{1}{2},\, T \in \mathbb{R}_+$$

Here $\mathcal{N}$ stands for a multiplicative renormalization which – informally speaking – is division by the expectation of the above exponential.

3.2 $d\Gamma(A)$ – Spaces

This class of spaces has been investigated in [HKPS, Po 87, Po 88a, Po 88b]. It is most conveniently described in terms of second quantized operators, cf. e.g. [Co 53, Ne 73, Si 74].

Let A be a self-adjoint, non-negative operator on $L^2(\mathbb{R})$. Then

$$A \otimes 1 \otimes \cdots \otimes 1 + 1 \otimes A \otimes 1 \otimes \cdots \otimes 1 + \ldots + 1 \otimes \cdots \otimes 1 \otimes A \tag{3.9}$$

(each term in the above sum having n factors) acts on $L^2(\mathbb{R})^{\otimes n}$, therefore on $L^2(\mathbb{R}^n) \simeq L^2(\mathbb{R})^{\otimes n}$ and by restriction on $L^2\widehat{(\mathbb{R}^n)}$. The direct sum of (3.9) over n acts on Fock space and its $\mathcal{S}^{-1}$-transform on (L^2): we denote the corresponding operator by $d\Gamma(A)$. (A domain on which $d\Gamma(A)$ is essentially self-adjoint can be constructed explicitly from $\mathcal{D}(A)$ – we denote by $d\Gamma(A)$ also the self-adjoint extension of $d\Gamma(A)$.) Note that $d\Gamma(A)$ is non-negative too. Now introduce the norms

$$|||\, \varphi \,|||_{p,k} := \left\| (1 + d\Gamma(A))^k \varphi \right\|_p, \quad p \in [1, +\infty),\, k \in \mathbb{N}_0, \tag{3.10}$$

where $\|\cdot\|_p$ denotes the norm of $L^p(\mathcal{S}'(\mathbb{R}), \mu)$. Let $\mathcal{D}_A^{p,k}$ be the completion of $\mathcal{P}$ (cf. Sect. 1) under $|||\cdot|||_{p,k}$.

Set

$$\mathcal{D}_A := \text{proj} - \lim_{p \in \mathbb{N},\, k \in \mathbb{N}_0} \mathcal{D}_A^{p,k}. \tag{3.11}$$

(See e.g. [GV 64] for the definition of projective and inductive limits.) The dual of $\mathcal{D}_A$ is denoted by $\mathcal{D}_A^*$ and we have the triple

$$\mathcal{D}_A^* \supset (L^2) \supset \mathcal{D}_A. \tag{3.12}$$

A special, important choice is given by $A =$ identity on $L^2(\mathbb{R})$, and in this case we simply use the notation $\mathcal{D}$ and $\mathcal{D}^*$. For this choice, $d\Gamma(A)$ is equal to the number operator N (and hence equal to the negative of the Ornstein–Uhlenbeck operator), and the triple (3.12) is the white noise version of the spaces of test and generalized functionals introduced by Meyer [Me 83]. In this case, $\mathcal{D}$ can be shown to be an algebra (e.g. [Me 83, Po 88b]), which turns out to be a very essential property. The triple (3.12) with its special properties was used in [Wa 83] for a reformulation of the Malliavin calculus (cf. also [Po 88b, Su 85]).

Examples. Let X be a random variable in $\mathcal{D}$ with values in $\mathbb{R}^d$, such that its Malliavin covariance matrix is non-degenerate (cf. e.g. [Wa 83, Po 88b]). Furthermore let $T \in \mathcal{S}'(\mathbb{R}^d)$. Then

$$T \circ X \in \mathcal{D}^*.$$

3.3 $\Gamma(A)$ – Spaces.

Consider an operator A as in Sect. 3.2 and the action of

$$A \otimes A \otimes \cdots \otimes A \quad ,\text{(n factors)} \tag{3.13}$$

on $L^2(\widehat{\mathbb{R}^n})$ (cf. above). Then

$$\Gamma(A) := \mathcal{S}^{-1}\left(\bigoplus_{n=0}^{\infty} A^{\otimes n}\right)\mathcal{S} \tag{3.14}$$

acts on (L^2) and is essentially self-adjoint. Again, we denote the closure by the same symbol.

The following two well-known formulae are useful for calculations involving $\Gamma(A)$.

(i) Let $\xi \in \mathcal{D}(A) \subset L^2(\mathbb{R})$, and define

$$: \exp \lambda < \cdot, \xi >:= \exp\left(\lambda < \cdot, \xi > -\frac{\lambda^2}{2}\|\xi\|_2^2\right), \quad \lambda \in \mathbb{C}.$$

Then

$$\Gamma(A) : \exp \lambda < \cdot, \xi >: = : \exp \lambda < \cdot, A\xi >: . \tag{3.15}$$

(ii) If $\varphi \in \mathcal{D}(\Gamma(A)) \subset (L^2)$ with chaos decomposition

$$\varphi(x) = \sum_{n=0}^{\infty} <: x^{\otimes n} :, f^{(n)} >$$

then

$$(\Gamma(A)\varphi)(x) = \sum_{n=0}^{\infty} <: x^{\otimes n} :, A^{\otimes n} f^{(n)} > \tag{3.16}$$

(Clearly, (i) is a special case of (ii).)

Of special importance is the choice where A is the self-adjoint extension of

$$Af(u) = -f''(u) + (1 + u^2)f(u), \quad f \in \mathcal{S}(\mathbb{R}), \tag{3.17}$$

and we shall make this choice from now on. The resulting spaces of test and generalized functionals will be denoted by $(\mathcal{S})$ and $(\mathcal{S})^*$, respectively. These spaces have a number of interesting properties which will be described in the next section. We conclude this section with some examples of elements in these spaces.

Examples.

1) By construction, $\mathcal{P} \subset (\mathcal{S})$, but also $\exp(\lambda < \cdot, \xi >) \in (\mathcal{S})$, for all $\lambda \in \mathbb{C}, \xi \in \mathcal{S}(\mathbb{R})$.
2) Suppose $\xi_1, \dots, \xi_n \in \mathcal{S}(\mathbb{R})$, and let X_ξ denote $(X_{\xi_1}, \dots, X_{\xi_n})$, X_{ξ_k} being the canonical coordinate process as in Sect. 2. Assume furthermore that f is a function on $\mathbb{R}^n$, such that f is the Fourier transform of a distribution with compact support. Then $f \circ X_\xi$ belongs to $(\mathcal{S})$.
3) Assume that K is a self-adjoint operator on $L^2(\mathbb{R})$, with $\mathcal{D}(K) \supset \mathcal{S}(\mathbb{R})$. Suppose that K leaves $\mathcal{S}(\mathbb{R})$ invariant and that inf spec$K > -1$. Then the informal expression
$$\mathcal{N} \exp\left(-\frac{1}{2} < x,\, Kx >\right)$$
has a well-defined meaning as an element in $(\mathcal{S})^*$.
4) The measures defined by the Gibbs' states and the measures defined by the groundstates of the following quantum field theories (cf. e.g. [AH 84] and references quoted there) are represented by elements in $(\mathcal{S})^*$ ([AHP 89] and [PS 89]):
   i) Høegh-Krohn model
   ii) Sine-Gordon model
   iii) $P(\varphi)_2$ models with unique ground state
5) One has the inclusions $(\mathcal{S})^* \supset \mathcal{D}^*$ and $(\mathcal{S})^* \supset (L^2)^-$, cf. [PY 89]. Therefore, all examples of generalized functionals given in Sections 3.1 and 3.2 are also examples of generalized functionals in $(\mathcal{S})^*$.

4. Properties of $(\mathcal{S})$ and $(\mathcal{S})^*$.

In this section we shall review some properties of the spaces $(\mathcal{S})$ and $(\mathcal{S})^*$ of test and generalized white noise functionals defined in Section 3.3. For more details about these spaces we refer the interested reader to the works [AHP 89b, AHP 89c, HPS 88, HKPS, IKT 89, KT 80, KY 87, Lee 89, MY 88, PS 89, PY 89, Yo 87].

First we want to list the more basic properties of $(\mathcal{S})$ in the following theorem.

Theorem 4.1.

a) $(\mathcal{S})$ is a nuclear Fréchet algebra;
b) If $\varphi \in (\mathcal{S})$ and φ has chaos decomposition given by the sequence $(f^{(n)}, n \in \mathbb{N}_0)$ (cf. Sect. 2), then for each $n \in \mathbb{N}$, $f^{(n)}$ has a version $\widetilde{f^{(n)}}$ belonging to $\widehat{\mathcal{S}(\mathbb{R}^n)}$;
c) Every $\varphi \in (\mathcal{S})$ has a unique version $\widetilde{\varphi}$ which is pointwise defined and continuous with respect the strong topology on $\mathcal{S}'(\mathbb{R})$;
d) Let N denote the number operator. For all $\lambda \in \mathbb{C}$, $\exp(\lambda N)$ acts continuously on $(\mathcal{S})$.

For proofs of the variuos statements in Theorem 4.1 we refer to the articles quoted above.

In order to compare $(\mathcal{S})$ with the space $\mathcal{D}$ in Sect. 3.2, one applies Nelson's hypercontractivity theorem (e.g. [Ne 73]) to obtain the following result [HKPS, PY 89].

Theorem 4.2. $(\mathcal{S})$ is continuously embedded in $\mathcal{D}$.

As a consequence one has the fact already mentioned in Sect. 3.3: $\mathcal{D}^* \subset (\mathcal{S})^*$ – providing a large number of examples of generalized white noise functionals, given by the composition of (non-degenerate) elements in $\mathcal{D}$ with elements in $\mathcal{S}'(\mathbb{R}^n)$.

Very interesting consequences for measures given by positive functionals in $(\mathcal{S})^*$ (see below) are implied by the next two results (for proofs cf. [HKPS, PY 89]).

For $y \in \mathcal{S}'(\mathbb{R})$, $\varphi \in (\mathcal{S})$ with continuous version $\widetilde{\varphi}$ set

$$\widetilde{\varphi_y}(x) := \widetilde{\varphi}(x+y), \quad x \in \mathcal{S}'(\mathbb{R}). \tag{4.1}$$

Let φ_y denote the μ–class of $\widetilde{\varphi_y}$ and denote the mapping $\varphi \longmapsto \varphi_y$ by τ_y.

Theorem 4.3. For every $y \in \mathcal{S}'(\mathbb{R})$, τ_y maps $(\mathcal{S})$ continuously into itself.

In Section 5 we shall give the Taylor series of $\tau_y \varphi$.

Similarly as for the translations in $\mathcal{S}'(\mathbb{R})$, one can discuss scaling transformations (and in general affine transformations) (cf. also the article [IKT 89]). Let $\lambda \in \mathbb{R}$, $\varphi \in (\mathcal{S})$ and set

$$\widetilde{\varphi^{(\lambda)}}(x) := \widetilde{\varphi}(\lambda x), \quad x \in \mathcal{S}'(\mathbb{R}). \tag{4.2}$$

$\varphi^{(\lambda)}$ denotes the μ–class of $\widetilde{\varphi^{(\lambda)}}$.

Theorem 4.4. For every $\lambda \in \mathbb{R}$, the mapping $\varphi \longmapsto \varphi^{(\lambda)}$ is continuous from $(\mathcal{S})$ into itself.

The proofs of both theorems are quite straightforward: they involve some (algebraic) properties of the Wick-ordered distributions, combined with standard estimates of the $\|\cdot\|_{2,p}$–norms (cf. Sect. 3.3). Moreover, it is shown in [KPY 89] that $\tau_y \varphi$ and $\varphi^{(\lambda)}$ are strongly continuous in $y \in \mathbb{R}$ and $\lambda \in \mathbb{R}$, respectively.

$(\mathcal{S})$ has also interesting properties with respect to differential operators — this will be discussed in Section 5.

Next we want to quote several results concerning $(\mathcal{S})^*$. The first is a result in [PS 89], which characterizes the elements of $(\mathcal{S})^*$ in terms of analytic properties of their U–functionals. Let us introduce the following definition.

Definition 4.5. Let F be a functional on $\mathcal{S}(\mathbb{R})$. F is called a *U–functional*, if and only if it admits the following two properties:

(i) For every $\xi \in \mathcal{S}(\mathbb{R})$, the mapping $\lambda \mapsto F(\lambda\xi)$, $\lambda \in \mathbb{R}$, has an entire analytic extension, denoted by $f(z, \xi)$;

(ii) There exists a $p \in \mathbb{N}_0$, so that the entire function $z \mapsto f(z, \xi)$ is of order less than or equal to 2, uniformly on the set $\{\xi \in \mathcal{S}(\mathbb{R});\ \|A^p\xi\|_2 \leq 1\}$.

In Section 3 we have stated that $\exp(\lambda < \cdot\,, \xi >) \in (\mathcal{S})$ for all $\lambda \in \mathbb{C}$, $\xi \in \mathcal{S}(\mathbb{R})$. Therefore, the $\mathcal{S}$– transform (2.5) extends to $(\mathcal{S})^*$ as

$$\mathcal{S}\Phi(\xi) := \exp\left(-\frac{1}{2}\|\xi\|_2^2\right) < \Phi, \exp < \cdot\,, \xi >>, \tag{4.3}$$

and similarly one can extend the $\mathcal{T}$–transformation (2.4).

Theorem 4.6. If $\Phi \in (\mathcal{S})^*$ then $\mathcal{S}\Phi$ is a U–functional. Conversely, if F is a U–functional then there is a $\Phi \in (\mathcal{S})^*$ so that $\mathcal{S}\Phi = F$.

Next we mention Yokoi's theorem [Yo 87], which is a version of the Bochner–Schwartz theorem for $(\mathcal{S})^*$. (For similar results about $\mathcal{D}^*$ cf. also [Po 87, Su 88].)

Theorem 4.7. If $\Phi \in (\mathcal{S})^*$ is positive (i.e. if $< \Phi, \varphi > \geq 0$ for all $\varphi \in (\mathcal{S})$ with $\varphi \geq 0$ (μ-a.e.)), then there is a finite measure ν on $\mathcal{B}$ with

$$< \Phi, \varphi >= \int_{\mathcal{S}'(\mathbb{R})} \widetilde{\varphi}(x)\, d\nu(x), \tag{4.4}$$

for all $\varphi \in (\mathcal{S})$.

For the following it will be useful to know whether the product of two $\mathcal{S}$–transforms is again an $\mathcal{S}$–transform, i.e. whether for given Φ, $\Psi \in (\mathcal{S})^*$ there exists an element $\Phi \diamond \Psi \in (\mathcal{S})^*$, so that

$$\mathcal{S}\Phi \diamond \Psi(\xi) = \mathcal{S}\Phi(\zeta) \diamond \mathcal{S}\Psi(\xi). \tag{4.5}$$

The answer is given in the following theorem due to Meyer and Yan [MY 88] (cf. also [PS 89] for another proof).

Theorem 4.8. If Φ, $\Psi \in (\mathcal{S})^*$, then $\Phi \diamond \Psi \in (\mathcal{S})^*$. Furthermore, if Φ and Ψ have chaos decomposition given by $(F^{(n)}, n \in \mathbb{N}_0)$ and $(G^{(n)}, n \in \mathbb{N}_0)$, respectively, where $F^{(n)}, G^{(n)} \in \widehat{\mathcal{S}'(\mathbb{R}^n)}$, $n \in \mathbb{N}_0$, then $\Phi \diamond \Psi$ has chaos decomposition given by $(H^{(n)}, n \in \mathbb{N}_0)$ with

$$H^{(n)} = \sum_{m=0}^{n} F^{(n-m)} \hat{\otimes} G^{(m)} \quad \in \widehat{\mathcal{S}'(\mathbb{R}^n)}. \tag{4.6}$$

$\Phi \diamond \Psi$ is called the *Wick-product* of Φ and Ψ.

From now on let us consider a measure ν on $\mathcal{B}$ which is given by a positive element Φ in $(\mathcal{S})^*$ as in (4.4). In this case, we shall use the notation

$$\Phi = \frac{d\nu}{d\mu} \tag{4.7}$$

and call $\frac{d\nu}{d\mu}$ the *generalized Radon–Nikodym derivative of ν with respect to μ.*

As a consequence of Theorems 4.3 and 4.4, we have the following result [PY 89].

Theorem 4.9. Assume that $\frac{d\nu}{d\mu} \in (\mathcal{S})^*$.

a) For $y \in \mathcal{S}'(\mathbb{R})$, denote by ν_y the translation of ν by y. Then also $\frac{d\nu_y}{d\mu} \in (\mathcal{S})^*$, and

$$\frac{d\nu_y}{d\mu} = \frac{d\mu_y}{d\mu} \diamond \frac{d\nu}{d\mu}, \tag{4.8}$$

and, explicitly, $\frac{d\mu_y}{d\mu}$ has $\mathcal{S}$–transform

$$\mathcal{S}\left(\frac{d\mu_y}{d\mu}\right)(\xi) = \exp < y, \xi >, \quad \xi \in \mathcal{S}(\mathbb{R}). \tag{4.9}$$

b) For $\lambda \in \mathbb{R}$, $\lambda \neq 0$, let $\nu^{(\lambda)}$ denote the measure obtained by scaling ν, i.e. $\nu^{(\lambda)}(E) = \nu(\lambda^{-1}E)$, $E \in \mathcal{B}$. Then also $\frac{d\nu^{(\lambda)}}{d\mu} \in (\mathcal{S})^*$ and

$$\frac{d\nu^{(\lambda)}}{d\mu} = \left(\frac{d\nu}{d\mu}\right)_{(\lambda)} \diamond \frac{d\mu^{(\lambda)}}{d\mu}, \tag{4.10}$$

where

$$\mathcal{S}\left(\frac{d\nu}{d\mu}\right)_{(\lambda)}(\xi) = \mathcal{S}\left(\frac{d\nu}{d\mu}\right)(\lambda\xi),$$

(which defines a unique element $\left(\frac{d\nu}{d\mu}\right)_{(\lambda)}$ in $(\mathcal{S})^*$), and

$$\mathcal{S}\left(\frac{d\mu^{(\lambda)}}{d\mu}\right)(\xi) = \exp\left(-\frac{1}{2}(1-\lambda^2)\,\|\xi\|_2^2\right). \tag{4.11}$$

5. Causal Calculus

For $y \in \mathcal{S}'(\mathbb{R})$, let D_y denote the Gâteaux derivative of functions on $\mathcal{S}'(\mathbb{R})$ in direction y, i.e. if $\varphi : \mathcal{S}'(\mathbb{R}) \longrightarrow \mathbb{R}$, then

$$D_y\varphi(x) = \lim_{s\to 0} s^{-1}\left(\varphi(x+sy) - \varphi(x)\right), \quad x \in \mathcal{S}'(\mathbb{R}) \tag{5.1}$$

$$= \frac{d}{ds}\varphi(x+sy)|_{s=0}, \tag{5.2}$$

whenever the limit (5.1) (or the derivative (5.2)) exists.

It is not hard to see that if φ is a polynomial in $\mathcal{P}$ of the form (1.3), then φ is Gâteaux differentiable in every direction y and

$$D_y\varphi(x) = \sum_{k=1}^{n} < y,\, \xi_k > P_{,k}(< x,\, \xi_1 >, \ldots, < x,\, \xi_n >), \tag{5.3}$$

where $P_{,k}$ denotes the k-th partial derivative of P. On the other hand, we may represent $\varphi \in \mathcal{P}$ as in (2.9), with $f^{(n)} \in \mathcal{S}(\widehat{\mathbb{R}^n})$ ($f^{(n)}$ being a linear combination of symmetric tensor products of the ξ_k's). Then one can show that

$$D_y\varphi(x) = \sum_{n=1}^{\infty} n <: x^{\otimes(n-1)} :,\, < y,\, f^{(n)} >>, \tag{5.4}$$

where $< y,\, f^{(n)} > \in \mathcal{S}(\widehat{\mathbb{R}^{n-1}})$ is the function which is defined by pairing $f^{(n)}$ in its first argument with y. In particular, if $y \in L^2(\mathbb{R})$ then

$$< y,\, f^{(n)} > (u) = \int_{\mathbb{R}} y(v) f(v,u)\, dv, \quad u \in \mathbb{R}^{n-1} \tag{5.5}$$

or if $y = \delta_t$, δ_t being the Dirac distribution at $t \in \mathbb{R}$,

$$< y,\, f^{(n)} > (u) = f(t,u), \quad u \in \mathbb{R}^{n-1}. \tag{5.6}$$

The last choice of y is of special importance and we denote

$$D_{\delta_t}\varphi(x) = \partial_t\varphi(x). \tag{5.7}$$

Thus we have (at least on polynomials φ)

$$\partial_t\varphi(x) = \sum_{n=1}^{\infty} n <: x^{\otimes(n-1)} :,\, f^{(n)}(t,\,\cdot) > . \tag{5.8}$$

Also, it is straightforward to show that for $h \in L^2(\mathbb{R})$

$$D_h\varphi(x) = \int_{\mathbb{R}} h(t)\partial_t\varphi(x)dt. \tag{5.9}$$

Combining (5.8) with (2.8), we find

$$\partial_t = \mathcal{S}^{-1}\frac{\delta}{\delta\xi(t)}\mathcal{S}, \tag{5.10}$$

where $\frac{\delta}{\delta\xi(t)}$ is the Fréchet functional derivative. The last remark shows that we have found another formulation of the derivative defined in [Hi 75] (cf. also [KT 80] and [Kuo 83b]). Equation (5.10) is very useful, since it is sometimes easier to deal with U-functionals rather than with white noise functionals directly.

Next, let us discuss the action of D_y, and in particular of ∂_t, on $(\mathcal{S})$. Using formula (5.4), density of $\mathcal{P}$ in $(\mathcal{S})$ and some straightforward estimates, we obtain the following result [PY 89] (cf. also [HKPS]).

Theorem 5.1. For every $y \in \mathcal{S}'(\mathbb{R})$, D_y maps $(\mathcal{S})$ continuously into itself. In particular, every $\varphi \in (\mathcal{S})$ is C^∞ in every direction of $\mathcal{S}'(\mathbb{R})$. Moreover, if $\widetilde{\varphi}$ is the pointwise defined continuous version of φ, then $\widetilde{D_y\varphi} = D_y\widetilde{\varphi}$.

Remark. Strictly speaking, in the preceding theorem D_y has to be considered as the continuous extension of D_y from $\mathcal{P}$ to $(\mathcal{S})$ — the existence of this extension being a part of the statement.

Due to Theorem 5.1 formulae (5.4) and (5.8) continue to hold for $\varphi \in (\mathcal{S})$. Also, for D_y we have the product rule on $(\mathcal{S})$, and the chain rule for $f \circ \varphi$, when $\varphi \in (\mathcal{S})$ and f is differentiable, such that $f \circ \varphi \in (\mathcal{S})$ and $f' \circ \varphi \in (\mathcal{S})$.

It is interesting to note Taylor's formula on $(\mathcal{S})$ [PY 89].

Theorem 5.2. For every $y \in \mathcal{S}'(\mathbb{R})$, $\varphi \in (\mathcal{S})$

$$\tau_y\varphi = \sum_{n=0}^{\infty}\frac{1}{n!}D_y^n\varphi \tag{5.11}$$

$$\equiv \exp D_y\,\varphi, \tag{5.12}$$

where the sum in (5.11) converges in $(\mathcal{S})$.

In the special case of $y = \delta_t$, $t \in \mathbb{R}$, we are interested to consider ∂_t as an operator on (L^2). The next theorem (cf. [PY 89]) provides a domain for ∂_t which is larger than $(\mathcal{S})$.

Theorem 5.3. For every $p > \frac{1}{4}$ and $t \in \mathbb{R}$, $(\mathcal{S})_p$ is an (L^2)–domain of ∂_t.

Let us now consider "$\partial_t\varphi$ as a function of t". For $\varphi \in (\mathcal{S})$ set

$$\nabla\varphi(t,x) = (\partial_t\varphi)(x), \quad t \in \mathbb{R},\, x \in \mathcal{S}'(\mathbb{R}) \tag{5.13}$$

We have

Theorem 5.4.

a) ∇ maps $(\mathcal{S})$ continuously into $\mathcal{S}(\mathbb{R}) \otimes (\mathcal{S})$. More generally, for every $k \in \mathbb{N}$, ∇^k maps $(\mathcal{S})$ continuously into $\widehat{\mathcal{S}(\mathbb{R}^k)} \otimes (\mathcal{S})$.

b) For every $k \in \mathbb{N}$, ∇^k maps $\mathcal{D}(N^{k/2})$ continuously into $\widehat{L^2(\mathbb{R}^k)} \otimes (L^2)$. Moreover, for $h \in L^2(\mathbb{R})$, $\varphi \in \mathcal{D}(N^{1/2})$

$$(h,\, \nabla\varphi)_{L^2(\mathbb{R})} = D_h\varphi. \tag{5.14}$$

(5.14) is a consequence of (5.9) and it shows that ∇ is a gradient.

The adjoint ∂_t^* of ∂_t (∂_t defined on $(\mathcal{S})_p$, $p > \frac{1}{4}$), $t \in \mathbb{R}$, maps (L^2) into $(\mathcal{S})_{-p}$. If φ has chaos decomposition given by $(f^{(n)};\, n \in \mathbb{N}_0)$, then $\partial_t^*\varphi$ is given by $(0, f^{(n)} \otimes \delta_t;\, n \in \mathbb{N}_0)$. Also, we note that

$$\partial_t^* = \mathcal{S}^{-1}\xi(t)\mathcal{S}, \tag{5.15}$$

i.e. the $\mathcal{S}$–transform of ∂_t^* acts as multiplication by $\xi(t)$.

On the other hand, multiplication by white noise $\dot{B}(t)$ is given by [KT 80]

$$\dot{B}(t)\varphi = (\partial_t^* + \partial_t)\varphi, \tag{5.16}$$

for example as an application from $(\mathcal{S})$ into $(\mathcal{S})^*$, or from $(\mathcal{S})_p$, $p > \frac{1}{4}$ into its dual $(\mathcal{S})_{-p}$.

Equation (5.16) is the starting point of the theory of stochastic integration within white noise analysis, cf. [AP 88, KR 88, KP 89, Po 88c].

Generally, one can also discuss the adjoint of D_y^*, $y \in \mathcal{S}'(\mathbb{R})$, from $(\mathcal{S})^*$ into itself. If $\eta \in \mathcal{S}(\mathbb{R})$, we can prove that D_η^* acts continuously on $(\mathcal{S})$. As a consequence, one has the following result [HKPS].

Theorem 5.5. ∇ is well-defined on $(\mathcal{S})^*$ as a continuous mapping

$$\nabla : (\mathcal{S})^* \longrightarrow \mathcal{S}'(\mathbb{R}) \otimes (\mathcal{S})^*, \tag{5.17}$$

and more generally, for every $k \in \mathbb{N}$

$$\nabla^k : (\mathcal{S})^* \longrightarrow \mathcal{S}'\widehat{(\mathbb{R}^k)} \otimes (\mathcal{S})^*. \tag{5.18}$$

We conclude this section by mentioning in an informal manner the important *canonical commutation relations* (e.g. [HKPS, KT 80]):

$$[\partial_t, \partial_s] = [\partial_t^*, \partial_s^*] = 0 \tag{5.19a}$$
$$[\partial_t, \partial_s^*] = \delta(t-s) \tag{5.19b}$$

6. Rotation Group and Laplacians.

For simplicity, let us denote the Schwartz space $\mathcal{S}(\mathbb{R})$ of test functions in this section by E.

With the flavour of "Harmonic Analysis" we shall make further considerations of the infinite dimensional rotation group $O(E)$ already mentioned at the end of Section 2. Intuitively speaking, the group is quite large — however, we can find a number of interesting subgroups that can be investigated to some extent by classical tools.

I. Inductive Limit of Finite Dimensional Rotations.

The first subgroup we consider consists of finite dimensional rotations. Obviously, there is a subgroup G_n of $O(E)$ which is isomorphic to the group $SO(n)$. The inductive limit of G_n is again a subgroup of $O(E)$ and is denoted by G_∞. A few important remarks are now in order.

1) The group G_∞ can be defined with a choice of a complete orthonormal system (CONS) $\{\xi_n, n \in \mathbb{N}\}$ in $L^2(\mathbb{R})$, such that $\xi_n \in \mathcal{S}(\mathbb{R})$, for every $n \in \mathbb{N}$.
2) An irreducible unitary representation of G_n is given on each subspace $\mathcal{H}_n$ of (L^2), cf. (2.3).
3) We can characterize (uniquely up to constant multiples) an infinite dimensional Laplace–Beltrami operator Δ_∞ by G_∞ (just as the Laplace–Beltrami operator Δ_n on the n-dimensional unit sphere is characterized by $SO(n)$). It turns out that each $\mathcal{H}_n$ is an eigenspace of Δ_∞ of eigenvalue $-n$. In other words, Δ_∞ is

the Ornstein–Uhlenbeck operator $-N$. Note that Δ_∞ may be expressed in the form

$$\Delta_\infty = \sum_{n=0}^{\infty} \left(D^2_{\xi_n} - < x, \xi_n > D_{\xi_n}\right). \tag{6.1}$$

4) In terms of the operators ∂_t and ∂_t^*, we can express Δ_∞ as follows.

$$\Delta_\infty = -\int_{\mathbb{R}} \partial_t^* \partial_t \, dt. \tag{6.2}$$

II. The Lévy Group $\mathcal{G}$.

Choose again a CONS $\{\xi_n, n \in \mathbb{N}\}$ and consider transformations of E that come from change of infinitely many coordinates. More precisely, $g \in O(E)$ belongs to $\mathcal{G}$ if for $\xi = \sum_n a_n \xi_n$

$$g\xi = \sum_{n=0}^{\infty} a_n \xi_{\pi(n)}, \tag{6.3}$$

where π is an automorphism of the positive integers, such that

$$\lim_{N\to\infty} \frac{1}{N} \#\{n \le N;\ \pi(n) \ge N\} = 0.$$

This group has a close connection with the Lévy Laplacian, which is defined by

$$\Delta_L = \lim_{N\to\infty} \frac{1}{N} \sum_{n=1}^{N} D^2_{\xi_n}. \tag{6.4}$$

Incidentally, this operator vanishes on (L^2), but it acts effectively on the spaces $(L^2)^-$ and $(\mathcal{S})^*$ of generalized white noise functionals.

III. Whiskers.

We are interested in one-parameter subgroups of $O(E)$ which come from diffeomorphisms of the parameter space $\mathbb{R}$. Namely, $\{g(t), t \in \mathbb{R}\}$ is a *whisker* if there exists a family of diffeomorphisms $\{\psi_t, t \in \mathbb{R}\}$ of $\mathbb{R}$, such that $g(t)$ is continuous in t, and

$$g_t\xi(u) = \xi(\psi_t(u))|\psi_t'(u)|^{1/2}, \tag{6.5}$$

$$\psi_t \circ \psi_s = \psi_{t+s}. \tag{6.6}$$

Example. The following *shift* $\{S_t,\ t \in \mathbb{R}\}$ is a whisker:

$$S_t\xi(u) = \xi(u-t).$$

The shift describes the propagation of time and plays therefore an important role in white noise analysis.

It is interesting to investigate various type of Laplacians in infinite dimensions from point of view of the causal calculus. Above, we have discussed the infinite dimensional Laplace–Beltrami operator Δ_∞ which is characterized by the inductive limit $\mathcal{G}_\infty$ of finite dimensional rotation groups and which turned out to be equal to the Ornstein–Uhlenbeck operator. Δ_∞ acts on the space (L^2) of ordinary white noise functionals. Lévy's Laplacian Δ_L acts on the spaces of generalized functionals $(L^2)^-$ and $(\mathcal{S})^*$. Δ_L has a close connection with the Lévy group.

We now make the following remarks.

If we consider a function space over $\mathbb{R}^n$, we can implement a representation of the n-dimensional rotation group $SO(n)$ with the differential operators $x_j \frac{\partial}{\partial x_k} - x_k \frac{\partial}{\partial x_j}$, $j, k = 1, 2, \ldots, n$, which are the generators of rotations. Since we think of white noise $\{\dot{B}(t),\ t \in \mathbb{R}\}$ as a coordinate system for infinite dimensional calculus, it is natural to consider the following class of operators

$$\gamma_{s,t} = \dot{B}(s)\partial_t - \dot{B}(t)\partial_s \tag{6.7a}$$

$$= \partial_s^*\partial_t - \partial_t^*\partial_s \tag{6.7b}$$

where we used (5.16).

The operators $\gamma_{s,t}$ are important and basic, as one can see in many situations. For example, in [HSa 89] it is shown that Δ_∞ can be characterized as a quadratic form of the ∂_t^*, ∂_t, $s, t \in \mathbb{R}$, which commutes with all $\gamma_{s,t}$. Further discussion in this line may be found in [HOS 89].

One may hope that some good relation will be found between the Fourier transform defined by Kuo [Kuo 83b, Kuo 87, Kuo 89] and other operators characterized by the $\gamma_{s,t}$.

7. Applications

In this section we shall sketch some of the applications of white noise analysis. For other domains of application, we refer the interested reader to [Hi 80, Hi 83a, Hi 83b, HKPS, HKP 88b, Kuo 83b] and to the articles quoted there.

7.1 Feynman Integrals.

Consider a classical particle, whose motion in $\mathbb{R}^d$ is described by a Lagrange function $L(q(t), \dot{q}(t))$, $t \in T \subset \mathbb{R}$, where $q(t)$ is the position of the particle at time $t \in T$, while $\dot{q}(t)$ is its velocity. The physical trajectories $(q(t),\ t \in T)$ of the particle are the solutions of the Euler–Lagrange equations

$$\frac{\partial}{\partial q_k} L(q, \dot{q}) - \frac{d}{dt} \frac{\partial}{\partial \dot{q}_k} L(q, \dot{q}) = 0, \quad k = 1, \ldots, d \tag{7.1}$$

Most typically L is of the form $L = L_0 + L_I$ with $L_0(q(t), \dot{q}(t)) = \frac{1}{2}(\dot{q}(t))^2$.

In his celebrated article [Fe 48], Feynman proposed to "quantize" the particle by averaging over fluctuations around the classical trajectory given by the solution of (7.1) (with appropriate initial conditions) with the complex weight

$$\exp\left(\frac{i}{\hbar} \int_T L(q(t), \dot{q}(t))dt\right). \tag{7.2}$$

In order to find a mathematically rigorous formulation of the above averaging process (which is called the "Feynman integral"), it was proposed in [HS 82] to describe the fluctuations by the paths of a (pinned) Brownian motion. Consequently, L_0 contributes a term $\exp(i/\hbar \int_T \dot{B}(t)^2 dt)$ to (7.2). In view of the examples mentioned in Section 3.1 it is natural to express Feynman integrals as expectations of properly defined generalized white noise functionals. This has been carried out in [HS 82, SH 83] for a number of examples (cf. also [Kuo 83b]).

It is still an interesting open problem to prove (or disprove) that Lagrangians of the above type with $L_I(q, \dot{q}) = -V(q)$ and a potential $V(q)$ which has stronger growth at infinity than q^2 can be treated in this way.

For further developments about Feynman integrals in white noise analysis see also the article of D.C. Khandekar [Kh 89] in these proceedings.

7.2 Dirichlet Forms.

Generalizations of the classical Dirichlet form

$$\mathcal{E}(f) = \int_D |\nabla f|^2(x)dx, \quad f \in C_0^1(D), \tag{7.3}$$

where D is a suitable domain in $\mathbb{R}^d$, to infinite dimensional spaces offer an attractive strategy for a mathematical formulation and study of quantum field theory (see e.g. [AH 84] and references quoted there).

In [HPS 88a] white noise analysis was used to begin the investigation of forms over $\mathcal{S}'(\mathbb{R})$ of the type

$$\mathcal{E}_\Phi(\varphi) = <\Phi, |\nabla\varphi|^2> \tag{7.4a}$$

$$= \int_{\mathcal{S}'(\mathbb{R})} \widetilde{|\nabla\varphi|^2}(x) d\nu_\Phi(x), \tag{7.4b}$$

where $\nabla\varphi$ is defined in (5.13), $\varphi \in (\mathcal{S})$, and $\Phi \in (\mathcal{S})^*$ is positive. In (7.4b), ν_Φ is the measure associated with Φ via Yokoi's Theorem.

It can be shown [HPS 88a], that if the form (7.4) is closable on $L^2(\mathcal{S}'(\mathbb{R}), \nu_\Phi)$, then it is indeed a Dirichlet form (e.g. [Fu 80]).

Recently [AHP 89c], it was established that the form (7.4) coincides with those constructed by other authors, whenever the corresponding measure is given by a positive generalized white noise functional in $(\mathcal{S})^*$. The last condition is not a very serious restriction as far as applications in quantum field theory are concerned: the results in [AHP 89, PS 89] show that all so far constructed models in quantum field theory in two space-time dimensions have a description in terms of generalized white noise functionals in $(\mathcal{S})^*$ (cf. also the examples in Section 3.3).

For more details about Dirichlet forms in white noise analysis, we refer to the article [AR 90] of S. Albeverio and M. Röckner in these proceedings.

7.3 Stochastic Analysis.

In [KR 88] H.-H. Kuo and A. Russek used Equation (5.16) for the multiplication by the increment $\dot{B}(t)dt$ at time $t \in \mathbb{R}$, as the starting point for the definition of stochastic integrals in white noise analysis. They showed that for suitable processes $\varphi : [0,1] \longrightarrow (L^2)$, the stochastic integrals

$$\int_0^t (\partial_s^* + \partial_{s\pm})\varphi(s) ds \tag{7.5a}$$

with

$$\partial_{s\pm}\varphi(s) = \lim_{\varepsilon \downarrow 0} \partial_{s\pm\varepsilon}\varphi(s), \tag{7.5b}$$

are extensions of Itô's forward ("+") and backward ("−") integrals to non-anticipating φ. Moreover, these integrals are *different* extensions of Itô's integral than the Hitsuda–Skorokhod integral, cf. e.g. [Hit 72, NP 88, Sk 75]. In fact, the latter corresponds to the first term in Eq. (7.5a).

In [AP 88] it was proved that for the integrals (7.5), Itô's Lemma holds true in its usual form (cf. also [Po 88c]).

Naturally, the next step is to consider (anticipating) stochastic differential equations in this set-up. An investigation in this line was begun in [KP 89a] (cf. also the contribution [KP 89b] to these proceedings).

Applications in biology suggest to consider equations of the following type (cf. e.g. [OTK 87]). Define the operator

$$E_t(f) := \exp\left(\int_0^t f(t-u)\partial_u^* du\right), \tag{7.6}$$

where $f \in L^2_{loc}(\mathbb{R})$, $t \geq 0$. $E_t(f)$ acts naturally on $(\mathcal{S})^*$ and can be used to describe fluctuation effects in molecular biology. For instance, the *Oosawa Equation* for the movement of paramecium with fluctuations becomes

$$\frac{d}{dt}X(t) = -k_+ E(-f)X(t) + k_- E_t(f)(1 - X(t)), \tag{7.7}$$

and can be studied within white noise calculus.

8. Concluding Remarks.

This section is devoted to a short discussion of an important direction of generalization of the white noise analysis described so far.

In line with the harmonic analysis arising from the infinite dimensional rotation group, we can develop white noise analysis for the case of random fields with parameter space $\mathbb{R}^d$, $d \geq 1$, or with parameter set consisting of smooth manifolds. In the last case, the introduction of a non-trivial geometrical structure into the parameter space seems to provide a very interesting field of research.

Here is a simple, instructive example. Consider a Gaussian random field indexed by a curve. If the curve is deformed by diffeomorphisms, then the way the random field depends on these transformations exhibits special features of the random field. In particular, if the parameter is taken to be a circle in $\mathbb{R}^2$, then the family of whiskers plays a very important role. We refer to the papers [Le 89] and [HSi 88] for further discussions of this problem.

Finally, we should like to mention that classical functional analysis turns out to be very helpful in this approach. For instance, the Hadamard equation provides a tool for the investigation of random fields defined by a Green's function.

Acknowledgement. We would like to thank S. Albeverio, I. Kubo, M. Röckner and J.-A. Yan for fruitful discussions. We are particularly grateful to H.-H. Kuo and L. Streit for the joy of our collaboration.

References

[AHP 89a] Albeverio, S., Hida, T., Potthoff, J. and Streit, L.: The vacuum of the Høegh-Krohn model as a generalized white noise functional; *Phys. Lett. B* **217** (1989) 511–514

[AHP 89b] Albeverio, S., Hida, T., Potthoff, J., Röckner, M. and Streit, L.: Dirichlet forms in terms of white noise analysis I – Construction and QFT examples; *BiBoS Preprint* (1989)

[AHP 89c] Albeverio, S., Hida, T., Potthoff, J., Röckner, M. and Streit, L.: Dirichlet forms in terms of white noise analysis II – Construction of infinite dimensional diffusions and QFT examples; *BiBoS Preprint* (1989)

[AH 84] Albeverio, S. and Høegh-Krohn, R.: Diffusion fields, quantum fields and fields with values in Lie groups; in: *Stochastic Analysis and Applications*; ed. by M. Pinsky, New York and Basel: Marcel Dekker Inc. (1984)

[AR 90] Albeverio, S. and Röckner M.: Infinite dimensional diffusions connected with positive generalized white noise functionals; these proceedings

[AP 88a] Asch, J., Potthoff, J.: Itô's lemma without non-anticipatory conditions; *BiBoS Publications* **316** (1988), to appear in *Probab. Th. Rel. Fields*

[Ch 86] Chow, P.L.: Generalized solution of some parabolic equations with a random drift; to appear in *J. Appl. Math. Optimization*

[Ch 89] Chow, P.L.: Stochastic differential equations in infinite dimensions; in: *Mathematics + Physics*, Vol. 3; L. Streit (ed.). Singapore: World Scientific (1989)

[dF 87] de Falco, D.: Lévy's stochastic area in stochastic mechanics; *Lett. Math. Phys.* **14** (1987) 41–45

[Fe 48] Feynman, R.: Space-time approach to non-relativistic quantum mechanics; *Rev. Mod. Phys.* **20** (1948) 367–387

[Fu 80] Fukushima, M.: *Dirichlet Forms and Markov Processes.* Amsterdam: Kodansha and North-Holland (1980)

[GV 64] Gel'fand, I.M., Vilenkin, N.Y.: *Generalized Functions* IV. New York and London: Academic Press (1964)

[Hi 75] Hida, T.: *Analysis of Brownian Functionals.* Carleton Mathematical Lecture Notes no. **13** (1975)

[Hi 80a] Hida, T.: *Brownian Motion.* Berlin, Heidelberg, New York: Springer (1980)

[Hi 83a] Hida, T.: Causal analysis and an application to prediction theory; in: *Prediction Theory and Harmonic Analysis*; V. Mandrekar and H. Salehi (eds.). (1983)

[Hi 83b] Hida, T.: Generalized Brownian functionals; in: *Theory and Applications of Random Fields*; G. Kallianpur (ed.). Berlin, Heidelberg, New York: Springer-Verlag (1983)

[HKPS] Hida, T., Kuo, H.-H., Potthoff, J. and Streit, L.: *White Noise: An Infinite Dimensional Calculus*. Monograph in preparation

[HOS 89] Hida, T., Obata, N. and Saito, K.: White noise analysis and rotation group; *Preprint* (1989)

[HPS 88] Hida, T., Potthoff, J. and Streit, L.: Dirichlet forms and white noise analysis; *Commun. Math. Phys.* **116** (1988) 235–245

[HSa 89] Hida, T. and Saito, K.: Introduction to white noise analysis; in: *Bielefeld Encounters in Mathematics and Physics, VII. BiBoS Publications* **400** (1989)

[HSi 88] Hida, T. and Si Si: Variational calculus for Gaussian random fields; in: *Proc. IFIP Warsaw* (1988)

[HS 82] Hida, T. and Streit, L.: White noise analysis and its application to Feynman integral; *Lecture Notes in Math.* **1033** 219-226

[Hit 72] Hitsuda, M.: Formula for Brownian partial derivatives; in: *Proc. Japan-USSR Symp. on Probab. Th., Kyoto* (1972)

[Itô 51] Itô, K.: Multiple Wiener integral; *J. Math. Soc. Japan* **3** (1951) 157–169

[Kh 89] Khandekar, D.C.: White – noise analysis as a tool in computing Feynman integrals; these proceedings

[KT 80a] Kubo, I. and Takenaka, S.: Calculus on Gaussian white noise I; *Proc. Japan Acad.* **56A** (1980) 376–380

[KT 80b] Kubo, I. and Takenaka, S.: Calculus on Gaussian white noise II; *Proc. Japan Acad.* **56A** (1980) 411–416

[KY 87] Kubo, I. and Yokoi, Y.: A remark on the space of testing random variables in the white noise calculus; *Preprint* (1987)

[Kuo 83a] Kuo, H.-H.: Donsker's delta function as a generalized Brownian functional and its application; in: *Theory and Application of Random Fields*, G. Kallianpur (ed.). Berlin, Heidelberg, New York: Springer-Verlag (1983)

[Kuo 83b] Kuo, H.-H.: Brownian functionals and applications; *Acta Appl. Math.* **1** (1983) 175-188

[Kuo 86] Kuo, H.-H.: On Laplacian operators of generalized Brownian functionals; *Lecture Notes in Math.* **1203** 119–128. Springer-Verlag (1986)

[Kuo 87] Kuo, H.-H.: The heat equation and the Fourier transform of generalized Brownian functionals; *Lecture Notes in Math.* **1236** 154–163. Springer-Verlag (1987)

[Kuo 89] Kuo, H.-H.: The Fourier transform in white noise calculus; *Preprint* (1989)

[KP 89a] Kuo, H.-H. and Potthoff, J.: Anticipating stochastic differential equations; *Preprint* (1989), to appear in: *Proc. 5th Int. Vilnius Conf. on Probab. Th. and Math. Stat.*, Vilnius (1989)

[KP 89b] Kuo, H.-H. and Potthoff, J.: Stochastic differential equations in white noise analysis; these proceedings

[KPY 89] Kuo, H.-H., Potthoff, J. and Yan, J.-A.: Smoothing semigroups in infinite dimensions; in preparation

[KR 88] Kuo, H.-H. and Russek, A.: White noise approach to stochastic integration; *J. Multivariate Analysis* **24** (1988) 218–236

[KS 87] Kuo, H.-H. and Shieh, N. R.: A generalized Itô's formula for multidimensional Brownian motions and its applications; *Chinese J. Math.* **15** (1987) 163–174

[Le 89] Lee, K.-S.: White noise approach to Gaussian random fields; *Preprint* (1989), to appear in *Nagoya Math. J.*

[Lee 89] Lee, Y.-L.: A reformulation of white noise calculus; these proceedings

[Me 83] Meyer, P.A.: Quelques résultats analytiques sur le semigroupe d'Ornstein-Uhlenbeck en dimension infinie; in: *Theory and Application of Random Fields*; G. Kallianpur (ed.). Berlin, Heidelberg, New York: Springer (1983)

[MY 87] Meyer, P.A. and Yan, J.A.: A propos des distributions sur l'espace de Wiener; *Séminaire de Probabilités* **XXI**; J. Azéma and M. Yor (ed.s). Berlin, Heidelberg, New York: Springer (1987)

[MY 88] Meyer, P.A. and Yan, J.A.: Distributions sur l'espace de Wiener (suite); *Preprint* (1988), to appear in *Séminaire de Probabilités*

[Ne 73] Nelson, E.: Probability theory and Euclidean quantum field theory; in: *Constructive Quantum Field Theory*; G. Velo and A. Wightman (ed.s). Berlin, Heidelberg, New York: Springer (1973)

[NP 88] Nualart, D. and Pardoux, E.: Stochastic calculus with anticipating integrands; *Probab. Th. Rel. Fields* **78** (1988) 535–581

[NZ 88b] Nualart, D. and Zakai, M.: Generalized Brownian functionals and the solution to a stochastic partial differential equation; *Preprint* (1988), to appear in *J. Funct. Anal.*

[Ob 88a] Obata, N.: A note on certain permutation groups in the infinite dimensional rotation group; *Nagoya Math. J.* **109** (1988) 91–107

[Ob 88b] Obata, N.: Analysis of the Lévy Laplacian; *Soochow J. Math.* **14** (1988) 105–109

[Ob 88c] Obata, N.: Density of Natural Numbers and the Lévy Group; *J. Number Th.* **30** (1988) 288–97

[Oo 75] Oosawa, F.: The effect of field fluctuation on a macromolecular system; *J. Th. Biol.* **52** (1975) 175–186

[OTK 87] Oosawa, F., Tsuchiya, M. and Kubori, T.: Fluctuation in living cells: effect of field fluctuation and asymmetry of fluctuation; in: *Lecture Notes in Biomathematics* **70**. Berlin, Heidelberg, New York: Springer (1987)

[Po 87b] Potthoff, J.: On positive generalized functionals; *J. Funct. Anal.* **74** (1987) 81–95

[Po 88a] Potthoff, J.: Littlewood-Paley theory on Gaussian spaces; *Nagoya Math. J.* **109** (1988) 47–61

[Po 88b] Potthoff, J.: On Meyer's equivalence; *Nagoya Math. J.* **111** (1988) 99–109

[Po 88c] Potthoff, J.: Stochastic integration in Hida's white noise analysis; *BiBoS Preprint* (1988), to appear in: *Proc. of the 2nd Int. Conference on Stoch. Processes, Physics and Geometry*, Ascona/Locarno, S. Albeverio and D. Merlini (ed.s)

[PR 89] Potthoff, J. and Röckner, M.: On the contraction property of energy forms in infinite dimensions, *BiBoS Preprint* (1989), to appear in *J. Funct. Anal.*

[PS 89a] Potthoff, J. and Streit, L.: A characterization of Hida distributions; *Preprint* (1989)

[PS 89b] Potthoff, J. and Streit, L.: Hida distributions as vacua of quantum fields; in preparation

[PY 89] Potthoff, J. and Yan J.A.: Some results about test and generalized functionals of white noise, *BiBoS Preprint* (1989), to appear in: *Proc. Singapore Probab. Conf.* (1989), L.Y. Chen (ed.)

[Sk 75] Skorokhod, A. V.: On a generalization of a stochastic integral; *Theory Probab. Appl.* **20** (1975) 219–233

[SH 83] Streit, L. and Hida T.: Generalized Brownian functionals and the Feynman integral; *Stoch. Processes Appl.* **16** (1983) 55–69

[Su 85] Sugita, H.: Sobolev spaces of Wiener functionals and Malliavin's calculus; *J. Math. Kyoto Univ.* **25** (1985) 31–48

[Su 88] Sugita, H.: Positive generalized Wiener functionals and potential theory over abstract Wiener spaces; *Osaka J. Math.* **25** (1988) 665–696

[Wa 83] Watanabe, S.: Malliavin's calculus in terms of generalized Wiener functionals; in: *Theory and Application of Random Fields*; G. Kallianpur (ed.). Berlin, Heidelberg, New York: Springer (1983)

[Yo 87] Yokoi, Y.: Positive generalized functionals, Preprint, to appear in *Hiroshima Math. J.*

POSITIVE GENERALIZED FUNCTIONALS ON $(\mathbb{R}^\infty, \mathcal{B}^\infty, N^\infty)$

Kiyosi ITÔ

RIMS, Kyoto University
Kyoto, JAPAN

1. Introduction.

In our paper [1] we defined the Malliavin infinitely many times differentiable functionals $\mathcal{D}$ and the Watanabe generalized functionals $\mathcal{D}'$ on a Segal system without referring to the linear topological structure. In fact the base probability space has only a measurable structure and a probability measure. Several mathematicians [2],[3],[4] have recently proved that positive generalized functionals are measures on the Wiener space (or the abstract Wiener space) on which the Malliavin calculus is built. However, this is not possible in our general framework, because the measures representing positive generalized functionals turn out to be no longer absolutely continuous with respect to the base probability measure. To cope with this difficulty we have to take a nice version of the Segal system under consideration.

Let S be a Segal system on a probability space $(\Omega, \mathcal{F}, P)$ and $\{e_n\}$ an orthonormal base of S. Then $\underline{e}(\omega) = (e_1(\omega), e_2(\omega), \cdots)$ is an $\mathbb{R}^\infty$-valued random variable that is N^∞-distributed. Since every statement on S is P-equivalent to a statement on $\underline{e}(\omega)$, it suffices to observe all problems on $(\mathbb{R}^\infty, \mathcal{B}^\infty, N^\infty)$.

In Section 2 of this note we outline the Malliavin calculus on $(\mathbb{R}^\infty, \mathcal{B}^\infty, N^\infty)$. In Section 3 we introduce *strictly null sets, regular functions* and *regular versions.* These concepts play the role of slim sets, quasi-continuous functions and quasi-continuous versions introduced by P. Malliavin respectively, though we cannot clarify the relationship between these two groups of concepts. In any case our concepts work in the same way as Malliavin's. In Section 4 we will prove that every positive generalized functional is represented by a unique finite measure.

It is customary to choose an abstract Wiener space as a representing space of a Segal space and establish Malliavin theory on the space. However, the abstract Wiener spaces that have so far appeared in practical applications are the classical Wiener space for Brownian motions and the countably Hilbertian space with a centered Gauss measure for generalized random functions. These special cases can be treated on $(\mathbb{R}^\infty, \mathcal{B}^\infty, N^\infty)$ by taking nice versions in a way common to all representing measures. We will discuss these problems in a separate paper.

2. Generalized functions on $(\mathbb{R}^\infty, \mathcal{B}^\infty, N^\infty)$.

The product probability space of a countably infinite number of copies of $(\mathbb{R}, \mathcal{B}(\mathbb{R}), N)$ $(N = N_{0,1})$ is denoted by $(\mathbb{R}^\infty, \mathcal{B}^\infty, N^\infty)$. The space $\mathbb{R}^\infty$ endowed with the product topology is a Polish space. $\mathcal{B}^\infty$ is the Kolmogorov σ-algebra on $\mathbf{R}^\infty$ which in turn coincides with the topological σ-algebra on $\mathbf{R}^\infty$.

We denote a point of $\mathbf{R}^\infty$ by $\underline{x} = (x_1, x_2, \cdots)$. Unless there is no possibility of confusions, we regard x_i the i-th component function of $\underline{x}$. Then $x_1, x_2, \cdots$ are i.i.d. (N).

To define generalized functions on $\mathbf{R}^\infty = (\mathbf{R}^\infty, \mathcal{B}^\infty, N^\infty)$ there are many ways of choosing the class of testing functions . The simplest one is the set of all polynomials denoted by $\mathcal{P} = \mathcal{P}(\mathbf{R}^\infty)$. Here a *polynomial* on $\mathbf{R}^\infty$ is defined to be a polynomial of a finite number of components.

The Ornstein-Uhlenbeck operator L on $\mathcal{P}$ is defined by

$$(Lf)(\underline{x}) := \sum_{i=1}^{\infty}(\partial_i^2 f(\underline{x}) - x_i \partial_i f(\underline{x})). \tag{1}$$

Since $f \in \mathcal{P}$, the summands vanish for almost all i's (i.e. except for a finite number of i's), so there occurs no problem of convergence. Let

$$\mathcal{H}_k = \mathcal{H}_k(\mathbf{R}^\infty) := \{f \in \mathcal{P} : Lf = -kf\}, \quad k = 0, 1, 2, \cdots.$$

$\mathcal{H}_k$, $k = 0, 1, 2, \cdots$ are orthogonal to each other with respect to the inner product in the Hilbert space $L_2 = L_2(\mathbf{R}^\infty, \mathcal{B}^\infty, N^\infty)$, i.e. $< f, g >= E(f \cdot g)$. Every $f \in \mathcal{P}$ is uniquely represented in the form:

$$f = \sum_k f_k, \qquad f_k \in \mathcal{H}_k, \ k = 1, 2, \cdots,$$

where $f_k = 0$ for $k > d(f) :=$ the degree of f. The orthogonal projection of $\mathcal{P}$ to $\mathcal{H}_k$ is denoted by J_k. $J_k f = 0$ for $k > d(f)$, and

$$-L = \sum_k k J_k.$$

For $\varphi : \{0, 1, 2, \cdots\} \longrightarrow \mathbf{R}$ we define $\varphi(-L)$ by

$$\varphi(-L) = \sum_k \varphi(k) J_k.$$

Hence

$$(I - \alpha L)^{s/2} = \sum_k (1 + \alpha k)^{s/2} J_k \ (\alpha > 0).$$

It is known that if $s \geq 0$, then

$$(I - \alpha L)^{-s/2} f \geq 0 \quad \text{for} \ \ f \geq 0,$$

$$\|(I - \alpha L)^{-s/2} f\|_p \leq \|f\|_p,$$

where $\| \ \|_p$ is the usual *p-norm* in $L_p = L_p(\mathbf{R}^\infty, \mathcal{B}^\infty, N^\infty)$.

It is also known that $\mathcal{P}$ is $\| \ \|_p$-dense in L_p for $p \in [1, \infty)$. The *Sobolev* (s,p)-*norm* $\| \ \|_{s,p}$ is defined by

$$\|f\|_{s,p} := \|\{(I-L)^{s/2} f\|_p, \ s \in \mathbf{R}, p \in (1, \infty).$$

We denote the linear functionals on $\mathbf{R}^\infty$ by $\tilde{\mathcal{D}} = \tilde{\mathcal{D}}(\mathbf{R}^\infty)$. Define the *Sobolev* (s,p)-*norm* in $\tilde{\mathcal{D}}$ by

$$\|F\|_{s,p} := \sup\{|F(f)| : \ \|f\|_{-s,q} \leq 1\},$$

where q is conjugate to p. $\|F\|_{s,p}$ may be ∞. Let

$$\mathcal{D}_{s,p} := \{F \in \tilde{\mathcal{D}} : \|F\|_{s,p} < \infty\}.$$

$\mathcal{D}_{s,p}$ decreases as s and p increase. Let

$$\mathcal{D} := \bigcap_{s,p} \mathcal{D}_{s,p} \ \text{ and } \ \mathcal{D}' := \bigcup_{s,p} \mathcal{D}_{s,p}.$$

Elements of $\mathcal{D}$ and $\mathcal{D}'$ are called *Malliavin's infinitely differentiable functionals* and *Watanabes's generalized functionals*, respectively.

$(L_p, \| \ \|_p)$ is isomorphic to $(\mathcal{D}_{0,p}, \| \ \|_{0,p})$ under the map $f \mapsto F_f(g) \equiv < f, g >$, so f is often identified with F_f.

A linear map $V : \mathcal{P} \longrightarrow \mathcal{P}$ of the form $V = \sum_{i=1}^n f_i \partial_i$, $n = 1, 2, \cdots$, $f_i \in \mathcal{P}$ is called a *vector field*. The set of all vector fields is denoted by $\mathcal{T} = \mathcal{T}(\mathcal{P})$. Every $V \in \mathcal{T}$ has the *derivation property* : $V(f \cdot g) = V(f)g + V(g)f$. $\mathcal{T}$ is a $\mathcal{P}$-linear space (or $\mathcal{P}$-module).

For $f \in \mathcal{P}$ we define $\nabla f : \mathcal{T} \longrightarrow \mathcal{P}$ by

$$\nabla f(V) := V(f).$$

$\nabla f : \mathcal{T} \longrightarrow \mathcal{P}$ is obviously $\mathcal{P}$-linear. A $\mathcal{P}$-linear map $\alpha : \mathcal{T} \longrightarrow \mathcal{P}$ of the form $\alpha = \sum_{i=1}^n f_i \nabla x_i$ is called a *1-form*. The set of all 1-forms is denoted by $\mathcal{P}^{(1)}$. $\mathcal{P}^{(1)}$ is a $\mathcal{P}$-linear space.

We define $(\nabla \chi_i \otimes \nabla x_j)(V_1, V_2)$ to be $\nabla x_i(V_1) \nabla x_j(V_2)$. $\nabla x_i \otimes \nabla x_j : \mathcal{T} \times \mathcal{T} \longrightarrow \mathcal{P}$ is $\mathcal{P}$-bilinear. A $\mathcal{P}$-bilinear map α of the form :

$$\alpha = \sum_{i,j} f_{ij} \nabla x_i \otimes \nabla x_j, \ f_{ij} \in \mathcal{P}$$

is called a *2-form*. The set of all 2-forms is denoted by $\mathcal{P}^{(2)}$. Similarly we define *r-forms* and $\mathcal{P}^{(r)}$. $\mathcal{P}^{(r)}$ is a $\mathcal{P}$-linear space.

The *tensor product* $\alpha \otimes \beta \in \mathcal{P}^{(r+s)}$ of $\alpha \in \mathcal{P}^{(r)}$ and $\beta \in \mathcal{P}^{(s)}$ is defined in the obvious way. The scalar product $\alpha \cdot \beta \in \mathcal{P}$ and the inner product $< \alpha, \beta > \in \mathbf{R}$ of $\alpha, \beta \in \mathcal{P}^{(r)}$ are defined as follows. If $\alpha = \sum f_i \nabla x_i$, $\beta = \sum g_i \nabla x_i \in \mathcal{P}^{(1)}$, then

$$\alpha \cdot \beta := \sum f_i g_i \in \mathcal{P}, \qquad < \alpha, \beta > = E(\alpha \cdot \beta).$$

Similarly for the case where $r > 1$. The *length* $|\alpha|$ of $\alpha \in \mathcal{P}^{(r)}$ is defined to be $(\alpha \cdot \alpha)^{1/2}$ and the *p-norm* of α is defined to be $E(|\alpha|^p)^{1/p}$.

The *gradient operator* $\nabla : \mathcal{P}^{(r)} \longrightarrow \mathcal{P}^{(r+1)}$ was defined above for $r = 0$ $(\mathcal{P}^{(0)} := \mathcal{P})$. For $r = 1$ we define

$$\nabla\alpha := \sum_i \nabla f_i \otimes \nabla x_i \quad \text{for} \quad \alpha = \sum_i f_i \nabla x_i.$$

Similarly for $r > 1$. The *divergent operator* $\nabla^* : \mathcal{P}^{(r+1)} \longrightarrow \mathcal{P}^{(r)}$ is defined as follows:

$$\nabla^*\alpha = -\sum_i (\nabla f_i \cdot \nabla x_i - f_i x_i) \quad \text{for} \quad \alpha = \sum_i f_i \nabla x_i \in \mathcal{P}^{(1)},$$

$$\nabla^*\alpha = \sum_i \nabla^*\beta_i \otimes \gamma_i \quad \text{for} \quad \alpha = \sum_i \beta_i \otimes \gamma_i, \beta \in \mathcal{P}^{(1)}, \gamma \in \mathcal{P}^{(r)},$$

$-\nabla^* \nabla : \mathcal{P}^{(r)} \longrightarrow \mathcal{P}^{(r)}$ is called the *Ornstein-Uhlenbeck operator*, denoted by L. For the case $r = 0$ we obtain

$$\begin{aligned} Lf &= -\nabla^* \Big(\sum_i \partial_i f \cdot \nabla x_i\Big) = \sum_i (\nabla \partial_i f \cdot \nabla x_i - \partial_i f \cdot x_i) \\ &= \sum_{i,j} \partial_j \partial_i f \cdot \nabla x_j \cdot \nabla x_i - \sum_i x_i \partial_i f \\ &= \sum_i (\partial_i^2 f - x_i \partial_i f) \qquad\qquad (\nabla x_j \cdot \nabla x_i = \delta_{ji}). \end{aligned}$$

This matches the above defined operator $L : \mathcal{P} \longrightarrow \mathcal{P}$ for $r = 0$.

Then we can define $\|\alpha\|_{s,p}$ for $\alpha \in \mathcal{P}^{(r)}$. The following comparison theorem due to P. Meyer is very important:

$$c_1\|\alpha\|_{k,p} \le \sum_{i=0}^{k} \| \nabla^i \alpha\|_p \le c_2\|\alpha\|_{k,p}, \qquad \alpha \in \mathcal{P}^{(r)},$$

where $c_i = c_i(k,p)$, $i = 1, 2$, are constants independent of α.

As we defined $\tilde{\mathcal{D}}$, $\mathcal{D}_{s,p}$, etc., we can define the spaces of generalized forms $\tilde{\mathcal{D}}^{(r)}$, $\mathcal{D}^{(r)}_{s,p}, \cdots, \tilde{\mathcal{D}}^{(0)} := \tilde{\mathcal{D}}$, $\mathcal{D}^{(0)}_{s,p} := \mathcal{D}_{s,p}, \cdots$. The operators $\nabla, \nabla^*, L, (I - \alpha L)^{-s/2}, \cdots$ can be defined for generalized functionals and generalized forms.

$$(\nabla F)(f) := F(\nabla^* f), \quad (\nabla^* F)(f) := F(\nabla f), \cdots$$

$$(LF)(f) := F(Lf), \qquad (I - \alpha L)^{s/2} F(f) := F((I - \alpha L)^{s/2} f), \cdots.$$

3. Regular versions.

$\mathcal{D} = \mathcal{D}(\mathbb{R}^\infty)$ is topologized by the Sobolev norms :

$$\|\ \|_{s,p}, \qquad s \in \mathbb{R},\ p \in (1, \infty)$$

or equivalently by a subsystem :

$$\|\ \|_{k,k}, \quad k = 2, 3, \cdots,$$

since $\| \ \|_{s,p}$ is increasing in (s,p). This topology is equivalent to the topology determined by the pseudonorm :

$$\|f\|_{\mathcal{D}} = \sum_{k=2}^{\infty} 2^{-k}(\|f\|_{k,k} \wedge 1). \tag{1}$$

$(\mathcal{D}, \| \ \|_{\mathcal{D}})$ is a Fréchet space.

For any sequence $\underline{g} = \{g_n\} \subset \mathcal{P}$ satisfying

$$\sum_n \|g_n\|_{s,p} < \infty, \quad \forall (s,p) \in \mathbf{R} \times (1,\infty)$$

(namely $\sum_n \|g_n\|_{k,k} < \infty$, $\forall k = 2, 3, \cdots$), we define

$$N(\underline{g}) := \{\underline{x} \in \mathbf{R}^\infty : \sum_n |g_n(\underline{x})| = \infty\}. \tag{2}$$

Definition 3.1 A set $B \in \mathcal{B}^\infty$ is called *strictly null*, if B is a subset of a countable union of sets of the type $N(\underline{g})$. We denote the strictly null sets by $\mathcal{N}$.

If an assertion $A(\underline{x})$ holds for every $\underline{x} \in \mathbf{R}^\infty$ outside of a strictly null set, then we say that A holds *a.e.* $(\mathcal{N})$.

Proposition 3. 1

(*i*) *Every strictly null set is an N^∞-null set.* (See Proposition 4.8 for a set which is N^∞-null but not strictly null.)

(*ii*) *Every countable union of strictly null sets is strictly null.*

(*iii*) *Every subset of a strictly null set is strictly null.*

(*iv*) *If A holds* a.e. $(\mathcal{N})$, *then it holds* a.e. (N^∞).

(*v*) *If $f_i = g_i$* a.e. $(\mathcal{N})$, $i = 1, 2$, *then*

$$a_1 f_1 + a_2 f_2 = a_1 g_1 + a_2 g_2 \quad \text{a.e.} \quad (\mathcal{N}).$$

(*vi*) *If $f_n = g_n$* a.e. $(\mathcal{N})$, $n = 1, 2, \cdots$, *then*

$$\text{Lim } f_n = \text{Lim } g_n \quad \text{a.e. } (\mathcal{N}),$$

where Lim a_n *is defined to be* $\lim a_n$ *or* 0 *according to whether* $\{a_n\}$ *is convergent or not.*

Proof The only assertion that needs proof is (i). To prove this it suffices to check that $N(\underline{g})$ is N^∞-null. But this can be verified by observing that

$$E(\sum_n |g_n|) = \sum_n \|g_n\|_1 \leq \sum_n \|g_n\|_{0,2} < \infty.$$

Definition 3.2 $f \in \mathcal{D}$ is called *regular*, if there exists a sequence $\{g_n\} \subset \mathcal{P}$ such that

$$\|g_n - f\|_{\mathcal{D}} \longrightarrow 0 \quad \text{and} \quad g_n \longrightarrow f \quad \text{a.e.} \quad (\mathcal{N}).$$

Proposition 3.2
(*i*) *Every* $f \in \mathcal{P}$ *is regular.*
(*ii*) *If* $g_n \in \mathcal{P}, n = 1, 2, \cdots$ *and if* $\sum_n \|g_{n+1} - g_n\|_{s,p} < \infty, \forall\, (s,p)$, *then* $f := \text{Lim}\, g_n$ *is regular and* $\|g_n - f\|_{\mathcal{D}} \longrightarrow 0$.
(*iii*) *If* $f(\underline{x}) = \varphi(x_1, x_2, \cdots, x_m)$ *where* $\varphi : \mathbf{R}^m \longrightarrow \mathbf{R}$ *is a* C^∞*-function of compact support, then* f *is regular.*
(*iv*) *Suppose that* f_1 *and* f_2 *are regular. Then*
  (*a*) $f_1 = f_2$ a.e. (N^∞) *if and only if* $f_1 = f_2$ a.e. $(\mathcal{N})$.
  (*b*) $a_1 f_1 + a_2 f_2$ $(a_1, a_2 \in \mathbf{R})$ *is regular.*
  (*c*) $f_1 f_2$ *is regular.*

Proof (*i*) is obvious.
(*ii*) Since $\{g_n\}$ is $\| \;\|_{s,p}$-Cauchy for every (s,p) by the assumption, it is $\| \;\|_{\mathcal{D}}$-Cauchy. Hence we can find f' such that

$$\|g_n - f'\|_{\mathcal{D}} \longrightarrow 0. \tag{3}$$

Let $N := \{\underline{x} \in \mathbf{R}^\infty : \sum_n |g_{n+1} - g_n| = \infty\}$. Then $N \in \mathcal{N}$ by the definition of $\mathcal{N}$, and $\{g_n\}$ is convergent outside of N. Let $f := \text{Lim}\, g_n$. Then $g_n \longrightarrow f$ outside of N, so $g_n \longrightarrow f$ a.e. $(\mathcal{N})$. It remains to prove that $\|g_n - f\|_{\mathcal{D}} \longrightarrow 0$. It is obvious by (3) that $\|g_n - f\|_2 \longrightarrow 0$. Hence we can find a subsequence $\{g'_n\} \subset \{g_n\}$ such that $g'_n \longrightarrow f'$ a.e. (N^∞). Since $g_n \longrightarrow f$ a.e. $(\mathcal{N})$,

$$g'_n \longrightarrow f \text{ a.e. } (\mathcal{N}) \quad \text{and so} \quad g'_n \longrightarrow f \text{ a.e. } (N^\infty).$$

Hence $f = f'$ a.e. (N^∞), so

$$\|g_n - f\|_{\mathcal{D}} = \|g_n - f'\|_{\mathcal{D}} \longrightarrow 0.$$

(*iii*) Since $\mathcal{P}(\mathbf{R}^m)$ is dense in $\mathcal{D}(\mathbf{R}^m) := \mathcal{D}(\mathbf{R}^m, \mathcal{B}^m, N^m)$, we can find a sequence $\varphi_n \in \mathcal{P}(\mathbf{R}^m)$ such that $\|\varphi_n - \varphi\|_{\mathcal{D}(\mathbf{R}^m)} \longrightarrow 0$. Applying the Meyer comparison theorem, we obtain

$$\| \nabla^k \varphi_n - \nabla^k \varphi\|_p \longrightarrow 0, \quad \forall p > 1, \; k = 0, 1, 2, \cdots.$$

This implies that φ_n converges to φ uniformly on every compact subset of $\mathbf{R}^m$, so $\varphi_n \longrightarrow \varphi$ everywhere on $\mathbf{R}^m$. Let $g_n(\underline{x}) = \varphi_n(x_1, x_2, \cdots, x_m)$. Then $g_n \in \mathcal{P}$ and $g_n \longrightarrow f$ everywhere on $\mathbf{R}^\infty$. Hence $f = \text{Lim}\, g_n$ obviously. By taking a subsequence we can assume that $\{g_n\}$ satisfies the assumption of (*ii*). This proves that f is regular.
(*iv*) (*b*) is obvious. We will prove (*a*). $f := f_1 - f_2$ is regular by (*b*). It suffices to prove that

$$f = 0 \text{ a.e. } (N^\infty) \quad \text{if and only if} \quad f = 0 \text{ a.e. } (\mathcal{N}).$$

The 'if' part is ovbious. We will prove the 'only if' part. Since f is reglular, we can find a sequence $\{g_n\} \subset \mathcal{P}$ such that

$$\|g_n - f\|_{\mathcal{D}} \longrightarrow 0 \text{ and } g_n \longrightarrow f \text{ a.e. } (\mathcal{N}).$$

Since $f = 0$ a.e. (N^∞), the first condition implies

$$\|g_n\|_{\mathcal{D}} \longrightarrow 0.$$

By taking a subsequence we can assume that

$$\|g_n\|_{\mathcal{D}} < 2^{-n}, \quad \forall n.$$

Noting that $1 \wedge \|g_n\|_{k,k} \leq 2^k \|g_n\|_{\mathcal{D}} \leq 2^{k-n}$, we obtain that $\|g_n\|_{k,k} \leq 2^{k-n}$ for $n \geq k$. This implies that

$$\sum_n \|g_n\|_{k,k} < \infty, \quad \forall k,$$

so $N := \{\underline{x} : \sum |g_n| = \infty\} \in \mathcal{N}$. Hence $g_n \longrightarrow 0$ a.e. $(\mathcal{N})$. Since $g_n \longrightarrow f$ a.e. $(\mathcal{N})$ by the assumption, we obtain that $f = 0$ a.e. $(\mathcal{N})$. This completes the proof of (a).

We will prove (c). Noting the relations

$$\nabla(f_1 f_2) = f_2 \nabla f_1 + f_1 \nabla f_2,$$

$$\nabla^2(f_1 f_2) = \nabla f_2 \otimes \nabla f_1 + f_2 \nabla^2 f_1 + \nabla f_1 \otimes \nabla f_2 + f_1 \nabla^2 f_2, \quad etc.$$

and using the Meyer comparison theorem, we can choose $\ell = \ell(k) > k$ and $c_k > 0$ such that

$$\|f_1 f_2\|_{k,k} \leq c_k \|f_1\|_{\ell,\ell} \|f_2\|_{\ell,\ell}. \tag{3}$$

Since f_1 and f_2 are regular, we can find $\{g_{i,n}\} \subset \mathcal{P}, i = 1, 2$ such that

$$\lim_n \|f_i - g_{i,n}\|_{\mathcal{D}} = 0 \text{ and } \lim_n g_{i,n} = f_i \text{ a.e. } (\mathcal{N}), \quad i = 1, 2.$$

Using (3) we obtain

$$\lim_n \|f_1 f_2 - g_{1,n} g_{2,n}\|_{\mathcal{D}} = 0.$$

It is obvious that $\lim_n g_{1,n} g_{2,n} = f_1 f_2$ a.e. $(\mathcal{N})$. Hence $f_1 f_2$ is regular.

Definition 3.3 A *regular version* of $f \in \mathcal{D}$ is defined to be a regular function equal to f a.e. (N^∞).

Proposition 3.3

(i) *Every $f \in \mathcal{D}$ has at least one regular version.*

(ii) *If f_1 and f_2 are any regular versions of $f \in \mathcal{D}$, then $f_1 = f_2$ a.e. $(\mathcal{N})$.*

(iii) *If f_i' is a regular version of f_i for $i = 1, 2$, then $a_1 f'_1 + a_2 f'_2$ and $f'_1 f'_2$ are regular versions of $a_1 f_1 + a_2 f_2$ and $f_1 f_2$ respectively.*

Proof

(*i*) Take a sequence $\{g_n\} \subset \mathcal{P}$ such that

$$\|g_n - f\|_{\mathcal{D}} < 2^{-n}, \quad n = 1, 2, \cdots$$

By the argument used in the proof of (*iv*) of the last proposition we obtain

$$\sum_n \|g_n - f\|_{k,k} < \infty, \quad \forall k,$$

so

$$\sum_n \|g_{n+1} - g_n\|_{k,k} < \infty, \quad \forall k.$$

Using (*ii*) of the last proposition, we see that $f' := \mathrm{Lim}\; g_n$ is regular and $\|g_n - f'\|_{\mathcal{D}} \longrightarrow 0$. Since $\|g_n - f\|_{\mathcal{D}} \longrightarrow 0$, we obtain that $f' = f$ a.e. (N^∞). Hence f' is a regular version of f.

(*ii*) $f_1 = f_2 = f$ a.e. (N^∞). Since f_1 and f_2 are regular, we can use (*iv*) (*a*) of the last proposition to conclude that $f_1 = f_2$ a.e. $(\mathcal{N})$. This completes the proof of (*ii*). We can easily derive (*iii*) from (*iv*) (*b*) and (*c*) of the last proposition.

4. Positive generalized functionals.

$f \in L_0 = L_0(\mathbf{R}^\infty)$ is called *positive*, written

$$f \in L_0^+$$

if $f \geq 0$ a.e. (N^∞). Define

$$\mathcal{P}^+ := \mathcal{P} \cap L_0^+, \quad L_p^+ := L_p \cap L_0^+.$$

$F \in \tilde{\mathcal{D}}$ is called *positive*, written

$$F \in \mathcal{D}^+$$

if $F(f) \geq 0$ for $F \in \mathcal{P}^+$. Define

$$\mathcal{D}_{s,p}^+ := \mathcal{D}_{s,p} \cap \tilde{\mathcal{D}}^+, \quad \mathcal{D}^+ := \mathcal{D} \cap \tilde{\mathcal{D}}^+, \quad \mathcal{D}'^+ := \mathcal{D}' \cap \tilde{\mathcal{D}}^+.$$

The map $\theta : L_p \longrightarrow \mathcal{D}_{0,p}, (\theta f)(g) =< f, g >, \;\forall g \in \mathcal{P}$ gives an isomorphism of $(L_p, \|\;\|_p)$ to $(\mathcal{D}_{0,p}, \|\;\|_{0,p})$.

Proposition 4.1 $\theta(L_p^+) = \mathcal{D}_{0,p}^+$.

Proof If $f \in L_p^+$, then $(\theta f)(g) =< f, g > \geq 0$, $\forall g \in \mathcal{P}^+$. Suppose that $F \in \mathcal{D}_{0,p}^+$ and $f := \theta^{-1} F$. Then $f \in L_p$ and

$$< f, g >= F(g) \geq 0, \quad \forall g \in \mathcal{P}^+.$$

We will prove that $f \in L_p^+$. Let $g \in L_q^+$. Then $g^{1/2} \in L_{2q}$ and $\|g^{1/2}\|_{2q} = \|g\|_q^{1/2}$. Hence we can find a sequence $\{g_n\} \in \mathcal{P}$ such that

$$\|g_n - g^{1/2}\|_{2q} \longrightarrow 0, \quad n \longrightarrow \infty.$$

Hence

$$\|g_n^2 - g\|_q \le \|g_n - g^{1/2}\|_{2q}\|g_n + g^{1/2}\|_{2q} \longrightarrow 0, \quad n \longrightarrow \infty,$$

so

$$F(g) = \lim_n F(g_n^2) \ge 0.$$

Thus we have proved that $F(g) \ge 0, \forall g \in L_q^+$. Let $B \in \mathcal{B}^\infty$. Then $1_B \in L_q^+$. Hence

$$F(1_B) \ge 0, \quad \text{i.e.} \quad < f, 1_B > \ge 0$$

and so

$$\int_B f dN^\infty \ge 0, \quad \forall B \in \mathcal{B}^\infty.$$

This implies that $f \ge 0$ a.e. (N^∞), so $f \in L_p^+$.

Proposition 4.2 *Suppose that* $F \in \mathcal{D}_{-s,p}(s > 0)$. *Then* $F_n := (I - \frac{1}{n}L)^{-s/2}F \in D_{0,p}^+$. *Hence there exists* $f_n \in L_p^+$ *such that*

$$F_n(g) = < f_n, g > = \int_{\mathbb{R}^n} g f_n dN^\infty, \quad \forall g \in \mathcal{P}$$

by the last proposition.

Proof

$$|F_n(g)| = |F((I - \frac{1}{n}L)^{-s/2}g)|$$

$$\le c\|(I - \frac{1}{n}L)^{-s/2}g\|_{s,q}, \quad c = \|F\|_{-s,p}$$

$$\le c\|(I - L)^{s/2}(I - \frac{1}{n}L)^{-s/2}g\|_q$$

$$\le cc'\|g\|_q$$

by the L_p-multiplier theorem. Hence $F_n \in \mathcal{D}_{0,p}$. If $g \in \mathcal{P}^+$, then it is known that $(I - \frac{1}{n}L)^{-s/2}f \in \mathcal{P}^+$, so $F_n(g) \ge 0$. Hence $F_n \in \mathcal{D}_{0,p}^+$.

Proposition 4.3

$$F_n(g) \longrightarrow F(g) \quad (n \longrightarrow \infty), \quad \forall g \in \mathcal{P}.$$

Proof

$$g = \sum_{k=0}^{d} J_k g \qquad (d = \text{degree of } g),$$

$$(I - \frac{1}{n}L)^{-s/2}g = \sum_{k=0}^{d}(1 + \frac{k}{n})^{-s/2}J_k g,$$

$$F_n(g) = \sum_{k=0}^{d}(1 + \frac{k}{n})^{-s/2}F(J_k g)$$

$$\longrightarrow F(g). \quad \text{Q.E.D.}$$

Let ν_n be a measure on $\mathbf{R}^\infty$ defined by

$$d\nu_n = f_n dN^\infty.$$

Then

$$\nu_n(\mathbf{R}^\infty) = < f_n, 1 > = F_n(1) = F(1).$$

Hence $\{\nu_n\}$ is a sequence of measures on $\mathbf{R}^\infty$ with the same total measure. Thus we see that

$$\int_{\mathbf{R}^\infty} g d\nu_n = F_n(g) \longrightarrow F(g), \quad \forall g \in \mathcal{P}.$$

Proposition 4.4 *$\{\nu_n\}$ is tight.*

Proof It suffices to find a compact set $K_\varepsilon \subset \mathbf{R}^\infty$ for every $\varepsilon > 0$ such that

$$\nu_n(K_\varepsilon^c) < \varepsilon, \quad \forall n. \tag{1}$$

Observing that

$$a_{m,n} := \int (x_1^2 + x_2^2 + \cdots + x_m^2) d\nu_n = F_n(x_1^2 + \cdots + x_m^2)$$

$$\longrightarrow F(x_1^2 + x_2^2 + \cdots + x_m^2), \quad n \longrightarrow \infty,$$

we see that

$$a_m = \sup_n a_{mn} < \infty.$$

Then

$$\nu_n\{\underline{x}; x_1^2 + x_2^2 + \cdots + x_m^2 > b^2\} \le a_m/b^2.$$

Take $b = b_{m,\varepsilon} > 0$ such that $a_m/b^2 < 2^{-m}\varepsilon$. Then

$$\nu_n(A_{m,\varepsilon}) < 2^{-m}\varepsilon \quad \text{for} \quad A_{m,\varepsilon} := \{\underline{x}; x_1^2 + x_2^2 + \cdots + x_m^2 > b_{m,\varepsilon}^2\}.$$

Let $K_{m,\varepsilon} := A_{m,\varepsilon}^c$. Then $K_\varepsilon := \cap_m K_{m,\varepsilon}$ is closed and is included in $\Pi_m[-b_{m,\varepsilon}, b_{m,\varepsilon}]$. Hence K_ε is compact. Also

$$\nu_n(K_\varepsilon^c) \le \sum_m \nu_n(A_{m,\varepsilon}) < \sum_m 2^{-m}\varepsilon = \varepsilon. \quad \text{Q.E.D.}$$

Let $\psi_{m,k}(\underline{x}) := (1 + x_1^2 + x_2^2 + \cdots + x_m^2)^k$ and

$$d\nu_{m,k,n} := \psi_{m,k} d\nu_n, \quad m, k = 0, 1, 2, \cdots.$$

In the same way as above we can prove that $\{\nu_{m,k,n}, \ n = 1, 2, \cdots\}$ is tight. Hence for every (m, k) we can choose a subsequence $\{n'\} \subset \{n\}$ such that $\nu_{m,k,n'}$ converges to a finite measure $\nu_{m,k}$. Using the diagonal method, we can choose $\{n'\}$ independent of m, k. Then

$$\int \gamma d\nu_{m,k,n'} \longrightarrow \int \gamma d\nu_{m,k}, \quad \forall (k,m)$$

for every $\gamma \in C_b = C_b(\mathbf{R}^\infty)$, where C_b denotes the bounded continuous functions. If $\gamma \in C_b$, then $\gamma\psi_{m,k}^{-1} \in C_b$. Hence

$$\int \gamma\psi_{m,k}^{-1} d\nu_{m,k,n'} \longrightarrow \int \gamma\psi_{m,k}^{-1} d\nu_{m,k}.$$

The left hand side equals $\int \gamma d\nu_{n'}$, which converges to $\int \gamma d\nu$ ($\nu := \nu_{0,0}$). Hence

$$\int \gamma d\nu = \int \gamma\psi_{m,k}^{-1} d\nu_{m,k}, \quad \text{i.e.} \quad d\nu = \psi_{m.k}^{-1} d\nu_{m,k}.$$

Proposition 4.5 *Every $g \in \mathcal{P}$ is ν-integrable and*

$$\int_{\mathbf{R}^\infty} g d\nu = F(g).$$

Proof Let $g \in \mathcal{P}$ be a polynomial of $x_1, x_2, \cdots, x_m$ with degree k. Then $g\psi_{m,k}^{-1} \in C_b$. Hence $|g| \leq c\psi_{m,k}$ (c : const). But

$$\int_{\mathbf{R}^\infty} \psi_{m,k} d\nu = \int_{\mathbf{R}^\infty} d\nu_{m.k} = \nu_{m,k}(\mathbf{R}^\infty) < \infty.$$

Hence g is ν-integrable. Also

$$\int g d\nu = \int g\psi_{m,k}^{-1} d\nu_{m,k} = \lim_n \int g\psi_{m,k}^{-1} d\nu_{m,k,n'}$$

$$= \lim_n \int g d\nu_{n'} = \lim_n F_{n'}(g) = F(g).$$

Proposition 4.6 *Suppose that $\{g_n\} \subset \mathcal{P}$ satisfies*

$$\sum_n \|g_n\|_{k,k} < \infty, \quad \forall k = 2, 3, \cdots.$$

Then

$$\int_{\mathbf{R}^\infty} \sum_n |g_n| d\nu < \infty.$$

Proof

$$\int \sum_n |g_n| d\nu \leq \sum_n \left(\int g_n^2 d\nu\right)^{1/2} \nu(\mathbf{R}^\infty)^{1/2},$$

$$\int g_n^2 d\nu = F(g_n^2) \leq c\|g_n^2\|_{s,q}, \quad c: \text{ const}.$$

By choosing $k = k(s,q)$ and $c' = c'(s,q)$ sufficiently large we obtain

$$\|g_n^2\|_{s,q} \le c'\|g_n\|_{k,k}^2, \qquad n = 1, 2, \cdots.$$

(See (3) in the proof of (*iv*) (*c*) of Proposition 3.2.) Hence

$$\int \sum_n |g_n| d\nu \le c'' \sum_n \|g_n\|_{k,k}, \quad c'' : \text{ const.}$$
$$< \infty.$$

Proposition 4.7

$$\nu(A) = 0, \quad \forall A \in \mathcal{N}.$$

Proof It suffices to observe A of the form $N(\underline{g})$, i.e.

$$A = \{\underline{x} \in \mathbf{R}^\infty : \sum_n |g_n(\underline{x})| = \infty\},$$

where $\{g_n\} \subset \mathcal{P}$ and $\sum_n \|g_n\|_{k,k} < \infty$, $\forall k = 2, 3, \cdots$. By the last proposition we obtain

$$\int_{\mathbf{R}^\infty} \sum_n |g_n| d\nu < \infty,$$

which implies $\nu(A) = 0$.

Proposition 4.8 *The set* $A := \{\underline{x} \in \mathbf{R} : x_1 = 0\}$ *is not strictly null but* N^∞*-null set.*

Proof. A is obviously N^∞-null. To prove that it is not strictly null, consider the following functional :

$$F(g) = E(g(0, x_2, x_3, \cdots)), \quad g \in \mathcal{P}.$$

This is obviously linear and positive.

$$g(0, x_2, x_3, \cdots) = \int_{-\infty}^{0} \partial_1(e^{-x_1^2/2} g) dx_1$$

$$= c \int_{-\infty}^{0} (\partial_1 g - x_1 g) dN(x_1), \quad c : \text{const.}$$

$$|g(0, x_2, x_3, \cdots)| \le c \int_{-\infty}^{\infty} |\partial_1 g - x_1 g| dN(x_1).$$

Hence

$$|F(g)| \le E(|g(0, x_2, x_3, \cdots)|)$$
$$\le c\|\partial_1 g - x_1 g\|_1 \le c\|\partial_1 g - x_1 g\|_2.$$

Since $\partial_1 g - x_1 g = -\nabla^* (g \nabla x_1)$, we obtain

$$\|\partial_1 g - x_1 g\|_2 \le c'\|g\|_{1,2}, \quad c' : \text{ const.}$$

Hence we have that

$$|F(g)| \leq cc'\|g\|_{1,2},$$

so $F \in \mathcal{D}_{1,2}^{+}$. Therefore we can find a finite measure ν on $\mathbf{R}^\infty$ such that

$$F(g) = \int_{\mathbf{R}^\infty} g d\nu.$$

Let $g_1(\underline{x}) := x_1^2$. Since $g_1(0, x_2, x_3, \cdots) = 0$,

$$F(g_1) = E(g_1(0, x_2, x_3, \cdots)) = 0.$$

Since $g_1 = 0$ on A by the definition of A,

$$\int_{A^c} g_1 d\nu = \int_{\mathbf{R}^\infty} g_1 d\nu = F(g_1) = 0.$$

Since $g_1 > 0$ on $A^c, \nu(A^c)$ must be 0. Hence $\nu(A) = \nu(\mathbf{R}^\infty) = F(1) = 1 > 0$. This, combined with the last proposition, implies that A is not strictly null. Q.E.D.

Suppose that $F \in \mathcal{D}'^{+}$. Then $F \in \mathcal{D}_{-s,p}^{+}$ for some $s > 0$ and $p \in (1, \infty)$, so Proposition 4.5 ensures that

$$F(g) = \int_{\mathbf{R}^\infty} g d\nu, \quad g \in \mathcal{P}. \tag{1}$$

Since $F \in \mathcal{D}'^{+}, F : \mathcal{P} \longrightarrow \mathbf{R}$ can be extended to a continuous linear functional on $\mathcal{D}$.

If $g_1, g_2 \in \mathcal{D}$ and $g_1 = g_2$ a.e. (N^∞), then $F(g_1) = F(g_2)$. But we are not sure that $g_1 = g_2$ a.e. $(\mathcal{N})$. Hence we cannot expect that (1) holds for $g_1, g_2 \in \mathcal{D}$ in general, because the set $\{\underline{x} : g_1 \neq g_2\}$ may have positive ν-measure. Fortunately,if g_1, g_2 are regular, then $g_1 = g_2$ a.e. (N^∞) implies $g_1 = g_2$ a.e. $(\mathcal{N})$. Then

$$\int g_1 d\nu = \int g_2 d\nu,$$

if both exist.

Proposition 4.9 *If $f \in \mathcal{D}$, then any regular version $\tilde{f}$ of f is ν-integrable and*

$$F(f) = \int_{\mathbf{R}^\infty} \tilde{f} d\nu. \tag{2}$$

Proof Take a sequence $\{g_n\} \subset \mathcal{P}$ such that $\|g_n - f\|_{\mathcal{D}} < 2^{-n}$. Then

$$\sum_n \|g_{n+1} - g_n\|_{k,k} < \infty,$$

and $\tilde{f} = \text{Lim}\, g_n$ is a regular version of f (see the proof of Proposition 3.3 (i)). By Proposition 4.6 we obtain

$$\int_{\mathbf{R}^\infty} \sum_n |g_{n+1} - g_n| d\nu < \infty$$

Hence g_n converges to $\tilde{f}$ a.e. (ν), $\tilde{f}$ is ν-integrable and

$$\int_{\mathbf{R}^\infty} \tilde{f} d\nu = \lim_n \int_{\mathbf{R}^\infty} g_n d\nu = \lim_n F(g_n).$$

Since $F \in \mathcal{D}'$, we have $F \in \mathcal{D}_{-s,p}$ for some (s, p). Hence

$$|F(g_n) - F(f)| \leq \|F\|_{-s,p} \|g_n - f\|_{s,q} \longrightarrow 0 \quad (n \longrightarrow \infty).$$

Thus we obtain

$$F(f) = \int_{\mathbf{R}^\infty} \tilde{f} d\nu.$$

Remark If f_1, f_2 are regular versions of f, then $f_1 = f_2$ a.e. $(\mathcal{N})$ by Proposition 3.3 (ii), and so $f_1 = f_2$ a.e. (ν) by Proposition 4.7. Hence $\int f_1 d\nu = \int f_2 d\nu$.

Proposition 4.10 *The measure ν of the last proposition is uniquely determined by F.*

Proof By the last proposition it holds that

$$\int_{\mathbf{R}^\infty} f d\nu_1 = \int_{\mathbf{R}^\infty} f d\nu_2 = F(f)$$

for every regular function f. Hence this holds for every function f of the form $\varphi(x_1, x_2, \cdots, x_n)$, φ being a C^∞-function of compact support by Proposition 3.2 (ii). Therefore $\nu_1(B) = \nu_2(B)$ for every cylinder set B, so $\nu_1 = \nu_2$. Q.E.D.

If $f \in \mathcal{D}^+$, then $\tilde{f} \in \mathcal{D}^+$ because $f = \tilde{f}$ a.e. (N^∞). Hence

$$F(f) = F(\tilde{f}) = \int \tilde{f} d\nu \geq 0, \ \forall f \in \mathcal{D}^+.$$

References

[1] K.Itô: On Malliavin calculus, to appear in Proceedings of the Singapore Probability Conference, (1989).

[2] H.Airault - P.Malliavin : Intégration geometrique sur l'espace de Wiener, Bull. Sc. math. **2**, 112 (1988), 3-52.

[3] D.Naulari - A.S.Üstünel : Mesure cylindrique et distributions sur l'espace de Wiener, Preprint.

[4] H.Sugita : Positive generalized Wiener functions and potential theory over abstract Wiener spaces, Osaka J. Math. **25** (1989), 665-696.

CALCULUS ON GAUSSIAN WHITE NOISE AND KUO'S FOURIER TRANSFORM

Yoshifusa ITO, Izumi KUBO and Shigeo TAKENAKA

Nagoya University College of Medical Technology, Nagoya, JAPAN,
Faculty of Integrated Arts and Sciences, Hiroshima University, Hiroshima, JAPAN
and
Department of Mathematics, Faculty of Sciences Nagoya University, Nagoya, JAPAN

§1. Introduction.

Along the idea of generalized Brownian functionals by Hida [1], the authors established a calculus on Gaussian white noises and Poisson white noises based on Gel'fand triples in [5], [7] $\sim$ [10]. However the calculus is not enough to analyze Kuo's Fourier transform [13]. In the connection with the analysis, we are interested in the theory of variable changes of testing and generalized functionals on Gaussian white noise. In §2, the basic properties of the space $\mathcal{H}$ of testing functionals and $\mathcal{H}^*$ of generalized functionals will be referred. In §3, variable changes of affine type of $\varphi(x) \in \mathcal{H}$ will be discussed. The Fréchet derivative of φ will also be discussed. In §4, the algebraic properties of $\mathcal{F} = \mathcal{SH}$ and $\mathcal{F}^* = \mathcal{SH}^*$ will be observed. In §5, some examples of positive functionals which correspond to Gaussian measures will be remarked. In §6, it will be shown that the space $\mathcal{H}^*$ is invariant under Kuo's Fourier transform and that the Fourier transform of $h(< x, \zeta >)$ with $h \in \mathcal{S}^*(\boldsymbol{R})$ corresponds to the usual Fourier transform of h in $\mathcal{S}^*(\boldsymbol{R})$. Several examples will be given, too. In §7, some supplemental discussions will be given.

§2. Gel'fand triplets.

In [5] and [7], we discussed only real valued generalized functionals. Here we treat complex valued generalized functionals and Gel'fand triplets over $\boldsymbol{C}$.

For any topological vector space E, we denote by E^* the topological dual of E and by $<,>$ bilinear (not skew linear) form between E and E^*. When E is a real vector space, we define the complexification $E_c = E + iE$ as usual way. Then the dual space $E_c^* = (E_c)^*$ is equivalent to $(E^*)_c = E^* + iE^*$. The duality of E_c and E_c^* is given by

$$< \xi + i\eta, x + iy > = < \xi, x > - < \eta, y > + i(< \xi, y > + < \eta, x >).$$

for $\xi, \eta \in E$ and $x, y \in E^*$. When E has an inner product $(,)_E$, the inner product of

E_c is given by

$$(\xi_1 + i\eta_1, \xi_2 + i\eta_2)_E \equiv (\xi_1, \xi_2)_E + (\eta_1, \eta_2)_E - i(\xi_1, \eta_2)_E + i(\xi_2, \eta_1)_E$$

for ξ_1, ξ_2, η_1 and $\eta_2 \in E$. Then the norm is

$$\|\xi + i\eta\|_E^2 \ \equiv \|\xi\|_E^2 \ + \|\eta\|_E^2,$$

for $\xi, \eta \in \mathcal{E}$. For convenience we introduce

$$\ll \xi + i\eta \gg_E^2 \equiv (\xi,\xi)_E - (\eta,\eta)_E + 2i(\xi,\eta)_E. \tag{2.1}$$

Then for any $\zeta = \xi + i\eta \in E_c$ and any $\gamma = \alpha + i\beta \in C$, we have that

$$\ll \gamma\zeta \gg_E^2 = \gamma^2 \ll \zeta \gg_E^2 . \tag{2.2}$$

It is easily seen that the n-fold tensor product $(E_c)^{\otimes n}$ is naturally isomorphic to $(E^{\otimes n})_c$.

We consider a real Gel'fand triplet $\mathcal{E} \subset E_0 \subset \mathcal{E}^*$ which is topologized by a sequence of real Hilbert spaces $\{E_p\}$ with inclusions

$$E_{p+1} \subset E_p, \ \ p \in Z$$

with consistent inner products $\{(,)_p\}_{p \in Z}$. We assume the following:

(A. 1) $\|\xi\|_{p+1} \le \rho\|\xi\|_p$ for any $\xi \in \mathcal{E}$ and $p \in Z$ with $\rho < 1$.

(A. 2) $E_{-p} = E_p^*$ for $p \in Z$.

(A. 3) For any p there exists $q(> p)$ such that the injection mapping $\iota_{p,q} : E_q \to E_p$ is of Hilbert-Schmidt type. Particularly,$\iota_{1,0}$ is of Hilbert-Schmidt type.

Of course we associate to $\mathcal{E} \equiv \cap_p E_p$ the projective limit topology and to $\mathcal{E}^* \equiv \cup_p E_{-p}$ the inductive limit (convex) topology. By (A. 2), we see that

$$< \xi, \eta > = (\xi, \eta)_0 \quad \text{for} \quad \xi, \eta \in \mathcal{E}.$$

In [5] and [7], we treated more special case as folows :

Case (G). Let T be a topological space with the topological Borel field $\mathcal{B}$ and ν be a σ-finite measure on T without atoms. The intermediate space E_0 is equal to $L_r^2(T, \mathcal{B}, \nu)$, where L_r^2 (resp. L_c^2) means the real L^2-space (resp. the complex L^2-space). The evaluation map $\delta_t : \xi \to \xi(t)$ gives a continuous map $t \to \delta_t$ from T into E_{-1} with

$$\|\delta\|^2 \equiv \int_T \|\delta_t\|_{-1}^2 d\nu(t) < \infty.$$

In this case, each testing function $\xi(t) \in \mathcal{E}$ is continuous and the injection $\iota_{0,1}$ is of Hilbert-Schmidt type with norm

$$\|\iota\| \equiv \|\iota_{0,1}\|_{H.S.} = \|\delta\|. \tag{2.3}$$

Let $(E_p)_c^{\hat{\otimes} n}$ be the n-fold symmetric tensor product of $(E_p)_c$. By virtue of (A. 1), we have natural inclusions $(E_{p+1})_c^{\hat{\otimes} n} \subset (E_p)_c^{\hat{\otimes} n}$. Let us denote by $\mathcal{E}_c^{\hat{\otimes} n}$ the projective limit of $(E_p)_c^{\hat{\otimes}n}$ and by $\mathcal{E}_c^{*\hat{\otimes}n}$ the inductive limit of $(E_{-p})_c^{\hat{\otimes}n}$ as $p \to \infty$. We always associate *the inductive limit convex topology* with the inductive limit space.

The orthogonal direct sum

$$\exp[\hat{\otimes}(E_p)_c] \equiv \sum_{n=0}^{\infty} \oplus (n!)^{1/2} (E_p)_c^{\hat{\otimes}n}$$

with inner product

$$((f_n)_{n\geq 0}, (g_n)_{n\geq 0})_{\exp[\hat{\otimes}(E_p)_c]} = \sum_{n=0}^{\infty} n!(f_n, g_n)_{(E_p)_c^{\hat{\otimes}n}}$$

is called *a Fock space.* Its dual space is $\exp[\hat{\otimes}(E_{-p})_c]$ under the canonical bilinear form

$$< (G_n)_{n\geq 0}, (f_n)_{n\geq 0} >= \sum_{n=0}^{\infty} n! < G_n, f_n >$$

for $(G_n)_{n\geq 0} \in \exp[\hat{\otimes}(E_{-p})_c]$ and $(f_n)_{n\geq 0} \in \exp[\hat{\otimes}(E_p)_c], (p \geq 0)$. Again by virtue of (A. 1), we have natural inclusions $\exp[\hat{\otimes}(E_{p+1})_c] \subset \exp[\hat{\otimes}(E_p)_c]$ for $p \in \boldsymbol{Z}$. We denote by $\exp[\hat{\otimes}\mathcal{E}_c]$ the projective limit of $\exp[\hat{\otimes}(E_p)_c]$ and by $\exp[\hat{\otimes}\mathcal{E}_c^*]$ the inductive limit of $\exp[\hat{\otimes}(E_{-p})_c]$ as $p \to \infty$. The triplet $\exp[\hat{\otimes}\mathcal{E}_c] \subset \exp[\hat{\otimes}(E_0)_c] \subset \exp[\hat{\otimes}\mathcal{E}_c^*]$ is a Gel'fand triplet (cf. [5],[7]). The following lemma is proved similarly to Kubo-Yokoi [11], Yokoi [18].

Lemma 2.1 *The mappings*

$$\zeta \in \mathcal{E}_c \longrightarrow \exp[\hat{\otimes}\zeta] \equiv \sum_{n=0}^{\infty} \frac{1}{n!} \zeta^{\hat{\otimes}n} \in \exp[\hat{\otimes}\mathcal{E}_c]$$

and

$$z \in \mathcal{E}_c^* \longrightarrow \exp[\hat{\otimes}z] \equiv \sum_{n=0}^{\infty} \frac{1}{n!} z^{\hat{\otimes}} \in \exp[\hat{\otimes}\mathcal{E}_c^*]$$

are continuous.

Let $\mathcal{F}$ (resp. $\mathcal{F}^{(p)}, \mathcal{F}^*$) be the functional space which is the image space of $\exp[\hat{\otimes}\mathcal{E}_c]$ (resp. $\exp[\hat{\otimes}(E_p)]$, $\exp[\hat{\otimes}\mathcal{E}_c^*]$) under the mapping

$$(G_n)_{n\geq 0} \in \exp[\hat{\otimes}\mathcal{E}_c^*] \longrightarrow \sum_{n=0}^{\infty} < G_n, \xi^{\hat{\otimes}n} > . \tag{2.4}$$

and introduce a topology from the original space. Then $\mathcal{F}^{(p)}$ is the reproducing kernel Hilbert space with the reproducing kernel $\exp[(\xi, \eta)_{-p}]$. As shown in [5], [7],

Proposition 2.2 *(a)* $\mathcal{F} \subset \mathcal{F}^{(0)} \subset \mathcal{F}^*$ *is a Gel'fand triplet.*
(b) Let $\xi \in \mathcal{E}, \eta, \zeta \in \mathcal{E}_c$ *and* n, m *be non-negative integers. Then* $< \xi, \eta >^m$ *and* $< \xi, \zeta >^n$ *belong to* $\mathcal{F}^{(p)}$ *and satisfy the equality*

$$(< \xi, \eta >^m, < \xi, \zeta >^n)_{\mathcal{F}^{(p)}} = \delta_{m,n} n! (\eta, \zeta)_p^n \quad \textit{for any } p \in \boldsymbol{Z}.$$

Now let us introduce a Gaussian measure μ. Since $\mathcal{E} \subset E_0 \subset \mathcal{E}^*$ is a Gel'fand triplet, by Bochner-Minlos' theorem, we can find a probability measure μ on $\mathcal{E}^*$ such that

$$\int_{\mathcal{E}^*} \exp[i < x, \xi >] d\mu(x) = \exp[-\frac{1}{2}\|\xi\|_0^2].$$

Then the measure μ is full on E_{-1}, i.e. $\mu(E_{-1}) = 1$, since $\|\iota_{1,0}\|_{H.S.} < \infty$. Let us denote $L_c^2(\mathcal{E}^*, \mu)$ simply by (L^2). It is useful to notice

$$\int_{\mathcal{E}^*} \exp[< x, \zeta >] d\mu(x) = \exp[\frac{1}{2} \ll \zeta \gg_0^2] \quad \text{for} \quad \zeta \in \mathcal{E}_c.$$

Proposition 2.3 *For each fixed $\zeta \in \mathcal{E}_c$, write*

$$f(\zeta; x) \equiv \exp[< x, \zeta > -\frac{1}{2} \ll \zeta \gg_0^2]. \tag{2.5}$$

The mapping $\mathcal{S}$ defined by

$$(\mathcal{S}\varphi)(\xi) \equiv \int_{\mathcal{E}^*} \varphi(x) f(\xi; x) d\mu(x) = \int_{\mathcal{E}^*} \varphi(x+\xi) d\mu(x) \tag{2.6}$$

for $\xi \in \mathcal{E}$, is an isomorphism from (L^2) onto $\mathcal{F}^{(0)}$. Especially,

$$(\mathcal{S}f(\zeta; \cdot))(\xi) = \exp[< \zeta, \xi >] \quad \textit{for any} \quad \xi \in \mathcal{E}, \ \zeta \in \mathcal{E}_c$$

and

$$\mathcal{S} : H_n(< x, \zeta >; \ll \zeta \gg^2) \longrightarrow < \xi, \zeta >^n \quad \textit{for} \quad \zeta \in \mathcal{E}_c,$$

where $H_n(z; \gamma)(n = 0, 1, 2, \cdots)$ are the Hermite polynomials with parameter γ defined by the generating function $\exp[\omega z - \frac{\gamma}{2}\omega^2]$;

$$\sum_{n=0}^{\infty} \frac{1}{n!} \omega^n H_n(z; \gamma) \equiv \exp[\omega z - \frac{\gamma}{2}\omega^2]. \tag{2.7}$$

Put $\mathcal{H}^{(p)} \equiv \mathcal{S}^{-1}(\mathcal{F}^{(p)})$ for $p \geq 0$ and $\mathcal{H} \equiv \mathcal{S}^{-1}(\mathcal{F})$ and introduce inner products by

$$(\varphi, \psi)_{\mathcal{H}^{(p)}} \equiv (\mathcal{S}\varphi, \mathcal{S}\psi)_{\mathcal{F}^{(p)}}. \tag{2.8}$$

in $\mathcal{H}^{(p)}$. Let be $\mathcal{H}^{(-p)}$ the dual of $\mathcal{H}^{(p)}$ for $p \geq 1$,$\mathcal{H}$ and (resp. $\mathcal{H}^*$) be the projective (resp. inductive) limit of $\mathcal{H}^{(p)}$. We call $\mathcal{H}$ the space of *complex valued testing functionals* and $\mathcal{H}^*$ the space of *complex valued generalized functionals.*

Proposition 2.4 *Let $f(\zeta; x)$ be the functional defined by (2. 5). Then the mapping $\zeta \in \mathcal{E}_c \to f(\zeta; \cdot) \in \mathcal{H}$ is continuous and*

$$(f(\zeta_1; \cdot), f(\zeta_2; \cdot))_{\mathcal{H}^{(p)}}$$

$$= \exp[(\zeta_1, \zeta_2)_p + \frac{1}{2}(\ll \zeta_1 \gg_p^2 + \ll \zeta_2 \gg_p^2 - \ll \zeta_1 \gg_0^2 - \ll \zeta_2 \gg_0^2)]$$

Proposition 2.5 *The mapping $\mathcal{S}$ is extended to $\mathcal{H}^*$ by*

$$(\mathcal{S}\Psi)(\xi) \equiv < \Psi(x), f(\xi; x) > . \tag{2.9}$$

Then $\mathcal{S}$ gives the isomorphism from $\mathcal{H} \subset (L^2) \subset \mathcal{H}^*$ to $\mathcal{F} \subset \mathcal{F}^{(0)} \subset \mathcal{F}^*$. Namely, $\mathcal{H}^{(p)}$ is isomorphic to $\mathcal{F}^{(p)}$ through $\mathcal{S}$ for any $p \in Z$. Moreover the functional $(\mathcal{S}\Psi)(\xi)$ can be extended to a continuous functional on $\mathcal{E}_c$ by

$$(\mathcal{S}\Psi)(\zeta) \equiv < \Psi(x), f(\zeta; x) > \quad \text{for } \zeta \in \mathcal{E}_c. \tag{2.10}$$

Generally, for a given $F_n \in (\mathcal{E}^*)_c^{\hat{\otimes} n}$, we denote by $I(F_n)$ the element in $\mathcal{H}^*$ which satisfies

$$(\mathcal{S}I(F_n))(\xi) \equiv < F_n, \xi^{\hat{\otimes} n} > \tag{2.11}$$

and call it *the multiple Wiener-Ito integral with kernel F_n.*

Proposition 2.6 *For $\varphi \in \mathcal{H}^{(p)}$, there exists $\boldsymbol{f} = \{f_n\}_{n=0}^{\infty} \in \exp[\hat{\otimes}(E_p)_c]$ such that*

$$\varphi = \sum_{n=0}^{\infty} I(f_n) \quad \text{with} \quad (I(f_n), I(f_m))_{\mathcal{H}^{(p)}} = \delta_{n,m} n! (f_n, f_m)_{(E_p)_c^{\hat{\otimes} n}}.$$

For $\Psi \in \mathcal{H}^*$, there exists $(F_n)_{n \geq 0} \in \exp[\hat{\otimes}\mathcal{E}_c^*]$ such that

$$(\mathcal{S}\Psi)(\zeta) = \sum_{n=0}^{\infty} < F_n, \zeta^{\hat{\otimes} n} > \quad \text{for } \zeta \in \mathcal{E}_c.$$

Proposition 2.7 *In Case (G), for $p \geq 0$, $I(f_n)$ is the multiple Wiener-Itô integral*

$$I(f_n) = \int \cdots \int_{T^n} f_n(t_1, \cdots, t_n) W(dt_1) \cdots W(dt_n)$$

with respect to the Gaussian white noise $W(dt)$ given by the relation

$$< x, \xi > \equiv \int_T \xi(t) W(dt, x) \ a.s. \ x \in \mathcal{E}^*(\mu).$$

By Proposition 2. 5, Hida's $\mathcal{T}$-transformation is given by

$$(\mathcal{T}\Phi)(\xi) \equiv < \exp[i < x, \xi >], \Phi > = < f(i\xi; x) \exp[-\frac{1}{2} \ll \xi \gg_0^2], \Phi > \tag{2.12}$$

$$= \exp[-\frac{1}{2} \ll \xi \gg_0^2](\mathcal{S}\Phi)(i\xi)$$

for $\xi \in \mathcal{E}$ and $\mathcal{S}^{-1}\mathcal{T}\Phi \in \mathcal{H}^*$.

Lastly, we remark that the space $\mathcal{H}^*$ is richer than the space of Malliavin calculus.

Proposition 2.8 *The space $\mathcal{H}$ is an algebra and included in the Sobolev space $\mathcal{D}$ of Malliavin calculus.*

Proof. The continuity of the product in $\mathcal{H}$ was proved in [8], [18]. Hence $\mathcal{H}$ is an algebra. On the other hand, it is included in $L^2(E^*, \mu)$ by definition. Therefore it's elemsnts belong to $L^p(E^*, \mu)$ for any $p > 0$. Let N be the number operator (= − Ornstein-Uhlenbeck operator). Then for any $\varphi \in \mathcal{H}$,

$$\|\exp[tN]\varphi\|_{\mathcal{H}^{(p)}} \leq \|\varphi\|_{\mathcal{H}^{(p+q)}} \quad \text{for } q \geq -\frac{t}{log\rho}.$$

This implies the assertion. □

§3. Complexification.

For $G_k \in (E_{-p})_c^{*\hat{\otimes}k}$, the operator G_k* from $(E_p)_c^{\hat{\otimes}n}$ to $(E_p)_c^{\hat{\otimes}(n-k)}$ $(n \geq k)$ is defined by

$$(3.1) \qquad < G_k * f_n, F_{n-k} > \equiv < f_n, G_k \hat{\otimes} F_{n-k} > \quad \text{for any } F_{n-k} \in \mathcal{E}_c^{*\hat{\otimes}(n-k)}.$$

Then we can show that

$$(3.2) \qquad \|G_k * f_n\|_{(E_p)_c^{\hat{\otimes}(n-k)}} \leq \|G_k\|_{(E_{-p})_c^{\hat{\otimes}k}} \|f_n\|_{(E_p)_c^{\hat{\otimes}n}}$$

in the same way as Lemma 2. 2 in Kubo-Takenaka [7]. Define an operator T from $(E_p)_c^{\hat{\otimes}n}$ to $(E_p)_c^{\hat{\otimes}(n-2)}$ $(p \geq 1)$ by

$$(3.3) \qquad Tf_n \equiv \sum_{k=1}^{\infty} (e_k \hat{\otimes} e_k) * f_n$$

with an complete orthonormal system $\{e_k\}_{k=1}^{\infty}$ in E_0. The operator T does not depend on choices of complete orthonormal basis. The norm of Tf_n is estimated as

$$\|Tf_n\|_{(E_p)_c^{\hat{\otimes}(n-2)}} \leq \|\iota_{p,0}\|_{H.S.}^2 \|f_n\|_{(E_p)_c^{\hat{\otimes}n}}.$$

by (3. 2). By (A. 1), the norm $\|\iota_{p,0}\|_{H.S.}$ is evaluated as

$$\|\iota_{p,0}\|_{H.S.} \leq \rho^{p-1}\|\iota\| \qquad \text{with } \|\iota\| = \|\iota_{1,0}\|_{H.S.}.$$

For given $f_n \in (E_p)_c^{\hat{\otimes}n}, p \geq 1$ and $n \geq 2k$, define

$$(3.4) \qquad f_{n|n-2k} \equiv T^k f.$$

Here the norm of $f_{n|n-2k}$ is estimated as

$$(3.5) \qquad \|f_{n|n-2k}\|_{(E_p)_c^{\hat{\otimes}(n-2k)}} \leq (\|\iota\|\rho^{p-1})^{2k} \|f_n\|_{(E_p)_c^{\hat{\otimes}n}}.$$

In Case (G), for $f_n \in \mathcal{E}_c^{\hat{\otimes} n}$, the relation between the multiple Wiener-Ito integral $I(f_n)(x)$ and the nonlinear functional $< f_n, x^{\hat{\otimes} n} >$ are given in Remark 6. 7 of Ito-Kubo [5] and Theorem 3. 1 of Kubo-Yokoi [11] as follows:

$$(3.6)\qquad I(f_n)(x) = \sum_{k=0}^{[n/2]} (-1)^k \frac{n!2^{-k}}{(n-2k)!k!} < f_{n|n-2k}, x^{\hat{\otimes}(n-2k)} >$$

and

$$(3.7)\qquad < f_n, x^{\hat{\otimes} n} >= \sum_{k=0}^{[n/2]} \frac{n!2^{-k}}{(n-2k)!k!} I(f_{n|n-2k})(x).$$

In this case, the functions $f_{n|n-2k} \in \mathcal{E}_c^{\hat{\otimes}(n-2k)}$ are given directly by

$$f_{n|n-2k}(t_1, \cdots, t_{n-2k})$$

$$\equiv \int \cdots \int_{T^k} f_n(t_1, \cdots, t_{n-2k}, u_1, u_1, \cdots, u_k, u_k) d\nu(u_1) \cdots d\nu(u_k).$$

In general cases also, we have the expressions (3. 6) and (3. 7). Therefore the norm of $< f_n, x^{\hat{\otimes} n} >$ is estimated as

$$(3.8)\qquad \| < f_n, x^{\hat{\otimes} n} > \|_{\mathcal{H}^{(p)}} \leq (n!)^{1/2} (\|\iota\|^2 \rho^{2p-2} + 1)^{n/2} \|f_n\|_{(E_p)_c^{\hat{\otimes} n}}.$$

by (3. 5), (3. 7) and the inequality $(2k)! \leq (2^k k!)^2$. By (3. 6), we can extend the function $I(f_n)(x)$ to $\mathcal{E}_c^*$ by

$$(3.9)\qquad I(f_n)(z) \equiv \sum_{k=0}^{[n/2]} (-1)^k \frac{n!2^{-k}}{(n-2k)!k!} < f_{n|n-2k}, z^{\hat{\otimes}(n-2k)} >$$

for $z \in \mathcal{E}_c^*$. By (3. 5), we have:

Lemma 3.1 *For $f_n \in \mathcal{E}_c^{\hat{\otimes} n}, I(f_n)(z)$ defined by (3. 9) is a continuous functional in $z \in \mathcal{E}_c^*$ and estimated as*

$$|I(f_n)(z)| \leq (n!)^{1/2} (\|z\|_{-p} + \|\iota\| \rho^{p-1})^n \|f_n\|_{(E_p)_c^{\hat{\otimes} n}}.$$

Let η_k be in $\mathcal{E}_c$ and z_k be in $\mathcal{E}_c^*$ for $k = 1, \cdots, d$. For $g_m \in \mathcal{E}_c^{\hat{\otimes} m}$ and $\gamma \in \mathbf{C}$, we have that

$$(3.10)\qquad < g_m, (\gamma x + \sum_{k=1}^{d} < \eta_k, x > z_k)^{\hat{\otimes} m} >$$

$$= \sum_{j_0+j_1+\cdots j_d=m} \frac{m!}{j_0!j_1! \cdots j_d!} < g_m, \gamma^{j_0} \prod_{i=1}^{d} < \eta_i, x >^{j_i} x^{\hat{\otimes} j_0} \hat{\otimes} \prod_{i=1}^{d} \hat{\otimes} z_i^{\hat{\otimes} j_i} >$$

$$= \sum_{j_0+j_1+\cdots+j_d=m} \frac{m!}{j_0!j_1! \cdots j_d!} \gamma^{j_0} < (\prod_{i=1}^{d} \hat{\otimes} \eta_i^{\hat{\otimes} j_i}) \hat{\otimes} (\prod_{i=1}^{d} \hat{\otimes} z_i^{\hat{\otimes} j_i} * g_m), x^{\hat{\otimes} m} > .$$

Lemma 3.2 *Let η_k be in $\mathcal{E}_c$ and z_k be in $\mathcal{E}_c^*$ for $k = 1, \cdots, d$. For $f_n \in \mathcal{E}_c^{\hat{\otimes} n}$ and $\gamma \in \boldsymbol{C}$, $\psi(x) = I(f_n)(\gamma x + \sum_{k=1}^d < \eta_k, x > z_k)$ belongs to $\mathcal{H}$ and satisfies*

$$\|\psi(x)\|_{\mathcal{H}(p)} \leq ((n+1)!)^{1/2}(|\gamma| + \sum_{i=1}^d \|\eta_i\|_p \|z_i\|_{-p} + 1)^n (\|\iota\|^2 \rho^{2p-2} + 1)^{n/2} \|f_n\|_{(E_p)_c^{\hat{\otimes} n}}.$$

Proof. By (3. 8), (3. 9) and (3. 10), the norm of $\psi(x)$ is estimated as follows:

$$\|\psi(x)\|_{\mathcal{H}(p)} \leq \sum_{k=0}^{[n/2]} \sum_{j_0+j_1+\cdots j_d = n-2k} \frac{n! 2^{-k}}{(n-2k)! k!} \frac{(n-2k)!}{j_0! j_1! \cdots j_d!} |\gamma|^{j_0}$$

$$\times \| < (\prod_{i=1}^d \hat{\otimes} \eta_i^{\hat{\otimes} j_i}) \hat{\otimes} (\prod_{i=1}^d \hat{\otimes} z_i^{\hat{\otimes} j_i} * f_{n|n-2k}), x^{\hat{\otimes}(n-2k)} > \|_{\mathcal{H}(p)}$$

$$\leq \sum_{k=0}^{[n/2]} \sum_{j_0+j_1+\cdots j_d=n-2k} \frac{n! 2^{-k}}{k! j_0! j_1! \cdots j_d!} |\gamma|^{j_0} \prod_{i=1}^d (\|\eta_i\|_p \|z_i\|_{-p})^{j_i} ((n-2k)!)^{1/2}$$

$$\times (\|\iota\|^2 \rho^{2p-2} + 1)^{(n-2k)/2} \|f_{n|n-2k}\|_{(E_p)_c^{\hat{\otimes} n-2k}}$$

$$\leq \sum_{k=0}^{[n/2]} \frac{n! 2^{-k}}{k!(n-2k)!} (|\gamma| + \sum_{i=1}^d \|\eta_i\|_p \|z_i\|_{-p})^{n-2k} ((n-2k)!)^{1/2}$$

$$\times (\|\iota\|^2 \rho^{2p-2} + 1)^{(n-2k)/2} \|f_{n|n-2k}\|_{(E_p)_c^{\hat{\otimes} n-2k}}$$

$$\leq \sum_{k=0}^{n} \left(\frac{(n!)^2}{(n-k)! k!} \right)^{1/2} (|\gamma| + \sum_{i=1}^d \|\eta_i\|_p \|z_i\|_{-p})^{n-k}$$

$$\times (\|\iota\|^2 \rho^{2p-2} + 1)^{(n-k)/2} (\|\iota\| \rho^{p-1})^k \|f_n\|_{(E_p)_c^{\hat{\otimes} n}}$$

$$\leq ((n+1)!)^{1/2} (|\gamma| + \sum_{i=1}^d \|\eta_i\|_p \|z_i\|_{-p} + 1)^n (\|\iota\|^2 \rho^{2p-2} + 1)^{n/2} \|f_n\|_{(E_p)_c^{\hat{\otimes} n}}. \quad \square$$

Theroem 3.3 *Any $\varphi(x) \in \mathcal{H}$ has the expression*

$$\varphi(x) = \sum_{n=0}^{\infty} I(f_n)(x)$$

with $\boldsymbol{f} = \{f_n\}_{n=0}^{\infty} \in \exp[\hat{\otimes} \mathcal{E}_c]$ and it is extended to a continuous functional in $z \in \mathcal{E}_c^*$ by

$$\varphi(z) \equiv \sum_{n=0}^{\infty} I(f_n)(z). \tag{3.11}$$

The value $\varphi(z)$ is estimated as

$$|\varphi(z)| \leq (1 - (\|z\|_{-p} + \|\iota\| \rho^{p-1}) \rho^q)^{-1} \|\varphi\|_{\mathcal{H}(p+q)},$$

if $z \in (E_{-p})_c$ and $(\|z\|_{-p} + \|\iota\| \rho^{p-1}) \rho^q < 1$. Hence the evaluation map

$$\delta_z : \varphi \in \mathcal{H} \longrightarrow \varphi(z) \in \boldsymbol{C}$$

belongs to $\mathcal{H}^*$ (and is called the delta functional at *z).*

Proof. The first part is seen by Proposition 2. 6. By Lemma 3. 1, the absolute convergence of the series (3. 11) and the estimation of $|\varphi(z)|$ are seen. The continuity in z can be shown similarly to [11] and [13]. The linearity of δ_z is obvious and the continuity on $\mathcal{H}$ is clear by the estimation of $|\varphi(z)|$. □

By Lemma 3. 2 and Theorem 3. 3, we have :

Theroem 3.4 *Let η_k be in $\mathcal{E}_c$ and z_k be in $\mathcal{E}_c^*$ for $k = 1, \cdots, d$. For $\varphi(x) \in \mathcal{H}$ and $\gamma \in \boldsymbol{C}$, $\varphi(\sum_{k=1}^d < \eta_k, x > z_k + \gamma x)$ belongs to $\mathcal{H}$ and its norm is estimated as*

$$\|\varphi(\sum_{k=1}^{d} < \eta_k, x > z_k + \gamma x)\|_{\mathcal{H}^{(p)}}$$

$$\leq \left(1 - \rho^q(|\gamma| + \sum_{i=1}^{d} \|\eta_i\|_p \|z_i\|_{-p} + 1)(\|\iota\|^2 \rho^{2p-2} + 1)^{1/2}\right)^{-1} \|\varphi\|_{\mathcal{H}^{(p+q)}},$$

if q is sufficiently large as $(|\gamma| + \sum_{i=1}^d \|\eta_i\|_p \|z_i\|_{-p} + 1)(\|\iota\|^2 \rho^{2p-2} + 1)^{1/2} < \rho^{-q}$.

By (3. 7) and (3. 9), we have :

$$(3.12) \qquad < f_n, (x+z)^{\hat{\otimes} n} >$$

$$= \sum_{k=0}^{n} \binom{n}{k} < (z^{\hat{\otimes}(n-k)}) * f_n, x^{\hat{\otimes} k} >$$

$$= \sum_{m=0}^{n} \binom{n}{m} \sum_{k=0}^{[m/2]} \frac{m! 2^{-k}}{(m-2k)! k!} I(z^{\hat{\otimes}(n-m)} * f_{n|n-2k})(x),$$

for any $f_n \in \mathcal{E}^{\hat{\otimes} n}$ and $z \in \mathcal{E}_c^*$.

Lemma 3.5 *For any $f_n \in \mathcal{E}^{\hat{\otimes} n}$ and $z \in \mathcal{E}_c^*, < f_n, (x+z)^{\hat{\otimes} n} >$ belongs to $\mathcal{H}$. Its norm is estimated as*

$$\| < f_n, (x+z)^{\hat{\otimes} n} > \|_{\mathcal{H}^{(p)}} \leq (n!)^{1/2} \left((\|\iota\|^2 \rho^{2p-2} + 1)^{1/2} + \|z\|_{-p} \right)^n \|f_n\|_{(E_p)^{\hat{\otimes} n}}$$

for $z \in E_{-p}$ and its $\mathcal{S}$-transformation is given by

$$(3.13) \qquad (\mathcal{S} < f_n, (x+z)^{\hat{\otimes} n} >)(\xi) = (\mathcal{S} < f_n, x^{\hat{\otimes} n} >)(\xi + z).$$

Moreover $I(f_n)(\cdot + z)$ belongs to $\mathcal{H}$ and its norm is estimated as

$$\|I(f_n)(\cdot + z)\|_{\mathcal{H}^{(p)}} \leq \left(3\|\iota\|^2 \rho^{2p-2} + 2\|z\|_{-p}^2 + 2\right)^{n/2} ((n+1)!)^{1/2} \|f_n\|_{(E_p)^{\hat{\otimes} n}}$$

and its $\mathcal{S}$-transformation is given by

$$(3.14) \qquad (\mathcal{S}I(f_n)(\cdot + z))(\xi) = (\mathcal{S}I(f_n))(\xi + z) = < f_n, (\xi + z)^{\hat{\otimes} n} > .$$

Proof. By (3. 2), (3. 5) and (3. 12), we see that

$$\| < f_n, (x+z)^{\hat{\otimes} n} > \|_{\mathcal{H}(p)}$$

$$\leq \sum_{k=0}^{n} \binom{n}{k} (k!)^{1/2} (\|\iota\|^2 \rho^{2p-2} + 1)^{k/2} \|z\|_{-p}^{n-k} \|f_n\|_{(E_p)_c^{\hat{\otimes} n}}$$

$$\leq (n!)^{1/2} \left((\|\iota\|^2 \rho^{2p-2} + 1)^{1/2} + \|z\|_{-p} \right)^n \|f_n\|_{(E_p)_c^{\hat{\otimes} n}}.$$

By (3. 12) and Proposition 2. 6, we have that

$$(\mathcal{S} < f_n, (x+z)^{\hat{\otimes} n} >)(\xi)$$

$$= \sum_{m=0}^{n} \sum_{k=0}^{[m/2]} \frac{n!}{m!(n-m)!} \frac{m! 2^{-k}}{(m-2k)!k!} < z^{\hat{\otimes}(n-m)} * f_{n|n-2k}, \xi^{\hat{\otimes}(m-2k)} >$$

$$= \sum_{m=0}^{n} \sum_{k=0}^{[m/2]} \frac{n! 2^{-k}}{(n-2k)!k!} \frac{(n-2k)!}{(n-m)!(m-2k)!} < f_{n|n-2k}, z^{\hat{\otimes}(n-m)} \hat{\otimes} \xi^{\hat{\otimes}(m-2k)} >$$

$$= \sum_{k=0}^{[n/2]} \sum_{m=0}^{n-2k} \frac{n! 2^{-k}}{(n-2k)!k!} \frac{(n-2k)!}{(n-m-2k)!m!} < f_{n|n-2k}, z^{\hat{\otimes}(n-2k-m)} \hat{\otimes} \xi^{\hat{\otimes} m} >$$

$$= \sum_{k=0}^{[n/2]} \frac{n! 2^{-k}}{(n-2k)!k!} < f_{n|n-2k}, (\xi+z)^{\hat{\otimes}(n-2k)} >$$

$$= (\mathcal{S} < f_n, x^n >)(\xi + z).$$

By (3. 5), (3. 6) and Lemma 3. 5, we have that

$$\|I(f_n)(\cdot + z)\|_{\mathcal{H}(p)}$$

$$\leq \sum_{k=0}^{[n/2]} \frac{n! 2^{-k}}{(n-2k)!k!} \| < f_{n|n-2k}, (\cdot + z)^{\hat{\otimes}(n-k)} > \|_{\mathcal{H}(p)}$$

$$\leq \sum_{k=0}^{[n/2]} \frac{n! 2^{-k}}{((n-2k)!)^{1/2} k!} \left((\|\iota\|^2 \rho^{2p-2} + 1)^{1/2} + \|z\|_{-p} \right)^{n-2k} (\|\iota\| \rho^{p-1})^{2k} \|f_n\|_{(E_p)_c^{\hat{\otimes} n}}$$

$$\leq \left(3\|\iota\|^2 \rho^{2p-2} + 2\|z\|_{-p}^2 + 2 \right)^{n/2} ((n+1)!)^{1/2} \|f_n\|_{(E_p)_c^{\hat{\otimes} n}}$$

By (3. 6) and (3. 13), we have the last formula (3. 14). □

Theroem 3.6 *For $\varphi(x) \in \mathcal{H}$ and $z \in \mathcal{E}_c^*, \varphi(x+z)$ belongs to $\mathcal{H}$. If $z \in (E_{-p})_c$ and $(3\|\iota\|^2 \rho^{2p-2} + 2\|z\|_{-p}^2 + 2)\rho^{2q} < 1$, then*

$$\|\varphi(\cdot + z)\|_{\mathcal{H}(p)} \leq \left(1 - (3\|\iota\|^2 \rho^{2p-2} + 2\|z\|_{-p}^2 + 2)\rho^{2q}\right)^{-1} \|\varphi\|_{\mathcal{H}(p+q)}.$$

Moreover, its $\mathcal{S}$-transformation is given by

$$(\mathcal{S}\varphi(\cdot + z))(\xi) = (\mathcal{S}\varphi)(\xi + z).$$

Proof. By Theorem 3. 3, $\varphi(x+z)$ is expressed in the form

$$\varphi(x+z)=\sum_{n=0}^{\infty} I(f_n)(x+z) \text{ with } \boldsymbol{f}=(f_n)_{n\geq 0}\in \exp[\hat{\otimes}\mathcal{E}_c].$$

Therefore by Lemma 3. 5,

$$\|\varphi(\cdot+z)\|_{\mathcal{H}^{(p)}} \leq \sum_{n=0}^{\infty} \|I(f_n)(\cdot+z)\|_{\mathcal{H}^{(p)}}$$

$$\leq \sum_{n=0}^{\infty} \left(3\|\iota\|^2\rho^{2p-2}+2\|z\|_{-p}^2+2\right)^{n/2}((n+1)!)^{1/2}\|f_n\|_{(E_p)_c^{\hat{\otimes} n}}$$

$$\leq \left(1-(3\|\iota\|^2\rho^{2p-2}+2\|z\|_{-p}^2+2)\rho^{2q}\right)^{-1}\|\varphi\|_{\mathcal{H}^{(p+q)}},$$

The second formula is derived from (3. 14) in Lemma 3. 5. □

Theroem 3.7 *For $\varphi(x)\in\mathcal{H}^*$ and $z\in(E_{-p})_c$, $(E_{-p})_c$-Fréchet derivative $D_z\varphi(x)=\frac{\delta}{\delta x}\varphi(x;z)$ exists; that is,*

$$\|\varphi(x+z)-\varphi(x)-D_z\varphi(x)\|_{\mathcal{H}^{(p)}} \leq C\|z\|_{-p}^2\|\varphi\|_{\mathcal{H}^{(p+q)}}.$$

for sufficiently large q. In particular,

$$D_z\varphi(x)=\lim_{\lambda\to 0}\frac{1}{\lambda}(\varphi(x+\lambda z)-\varphi(x)) \text{ for each } x \in\mathcal{E}^*.$$

The $\mathcal{S}$-transformation of the derivative $D_z\varphi(x)$ is

$$(\mathcal{S}D_z\varphi(x))(\xi)=\frac{\delta}{\delta\xi}(\mathcal{S}\varphi)(\xi;z).$$

Proof. For $f_n\in\mathcal{E}^{\hat{\otimes} n}$ and $z\in(E_{-p})_c$, similarly to (3. 12), we have

$$(3.15)\qquad < f_n,(x+z)^{\hat{\otimes} n}> - < f_n, x^{\hat{\otimes} n}> -n < z * f_n, x^{\hat{\otimes}(n-1)}>$$

$$=\sum_{m=0}^{n-2}\binom{n}{m} < (z^{\hat{\otimes}(n-m)}) * f_n, x^{\hat{\otimes} m}>$$

$$=\sum_{m=0}^{n-2}\binom{n}{m}\sum_{k=0}^{[m/2]}\frac{m!2^{-k}}{(m-2k)!k!}I(z^{\hat{\otimes}(n-m)} * f_{n|n-2k})(x).$$

Hence by (3. 2) and (3. 5)

$$\| < f_n,(x+z)^{\hat{\otimes} n}> - < f_n, x^{\hat{\otimes} n}> -n < z * f_n, x^{\hat{\otimes}(n-1)}> \|_{\mathcal{H}^{(p)}}$$

$$\leq \sum_{m=0}^{n-2}\binom{n}{m}\|z\|_{-p}^{n-m}\left(\sum_{k=0}^{[m/2]}\frac{(m!)^2 2^{-k}}{(m-2k)!(2k)!}(\|\iota\|\rho^{p-1})^{4k}\right)^{1/2}\|f_n\|_{(E_p)_c^{\hat{\otimes} n}}$$

$$\leq (n!)^{1/2} \sum_{m=0}^{n-2} \binom{n}{m} \|z\|_{-p}^{n-m} (\|\iota\|^2 \rho^{2p-2} + 1)^{n/2} \|f_n\|_{(E_p)_c^{\hat{\otimes} n}}$$

$$\leq (n!)^{1/2} n(n-1) \|z\|_{-p}^2 \left(\|z\|_{-p} + (\|\iota\|^2 \rho^{2p-2} + 1)^{1/2} \right)^{n-2} \|f_n\|_{(E_p)_c^{\hat{\otimes} n}}.$$

Therefore by (3. 6),

$$\|I(f_n)(\cdot + z) - I(f_n)(x) - nI(z * f_n)\|_{\mathcal{H}^{(p)}}$$

$$\leq \sum_{k<2n} \frac{n! 2^{-k}}{(n-2k)! k!} \| < f_{n|n-2k}, (x+z)^{\hat{\otimes}(n-2k)} - x^{\hat{\otimes}(n-2k)} - nz \hat{\otimes} x^{\hat{\otimes}(n-2k-1)} > \|_{\mathcal{H}^{(p)}}$$

$$\leq \sum_{k<2n} \left(\frac{(n!)^2}{(n-2k)! k!} \right)^{1/2} n(n-1) \|z\|_{-p}^2$$

$$\times \left((\|\iota\|^2 \rho^{2p-2} + 1)^{1/2} + \|z\|_{-p} \right)^{n-2k-1} \|f_{n|n-2k}\|_{(E_p)_c^{\hat{\otimes}(n-k)}}$$

$$\leq \sum_{k<2n} \left(\frac{(n!)^2}{(n-2k)! k!} \right)^{1/2} n(n-1) \|z\|_{-p}^2$$

$$\times \left((\|\iota\|^2 \rho^{2p-2} + 1)^{1/2} + \|z\|_{-p} \right)^{n-2k-1} (\|\iota\| \rho^{p-1})^{2k} \|f_n\|_{(E_p)_c^{\hat{\otimes} n}}$$

$$\leq \left(3\|\iota\|^2 \rho^{2p-2} + 2\|z\|_{-p}^2 + 2 \right)^{n/2} ((n+1)!)^{1/2} n(n-1) \|z\|_{-p}^2 \|f_n\|_{(E_p)_c^{\hat{\otimes} n}}.$$

Put

$$D_z \varphi(x) = \frac{\delta}{\delta x} \varphi(x; z) \equiv \sum_{n=1}^{\infty} nI(z * f_n). \tag{3.16}$$

Then

$$\|\varphi(x+z) - \varphi(x) - D_z \varphi(x)\|_{\mathcal{H}^{(p)}}$$

$$\leq \sum_{n=0}^{\infty} \left(3\|\iota\|^2 \rho^{2p-2} + 2\|z\|_{-p}^2 + 2 \right)^{n/2} ((n+1)!)^{1/2} n(n-1) \|z\|_{-p}^2 \|f_n\|_{(E_p)_c^{\hat{\otimes} n}}$$

$$\leq 120 \left(1 - (3\|\iota\|^2 \rho^{2p-2} + 2\|z\|_{-p}^2 + 2) \rho^{2q} \right)^{-3} \|z\|_{-p}^2 \|\varphi\|_{\mathcal{H}^{(p+q)}}.$$

From the existence of Fréchet derivative Gateux derivative exists in $\mathcal{H}^{(p)}$-norm if $z \in (E_{-p})_c$. Since the evaluation map is given by the delta functional δ_z in $\mathcal{H}^*$, the derivative exists for each $x \in \mathcal{E}$. Since $\mathcal{S}$-transformation of $\varphi(x+z)$ is given in Theorem 3. 6, we have the last equality. □

The delta functional $\delta_z \in \mathcal{H}^*$ defined in Theorem 3. 3 satisfies

$$\varphi(z) = < \varphi, \delta_z > \qquad \text{for any } \varphi \in \mathcal{H}.$$

Its $\mathcal{S}$-transformation is

$$(\mathcal{S}\delta_z)(\xi) = < \exp[< x, \xi > - \frac{1}{2} \ll \xi \gg_0^2], \delta_z(x) > \tag{3.17}$$

$$= \exp[< z, \xi > -\frac{1}{2} \ll \xi \gg_0^2]$$
$$= f(\xi; z).$$

Therefore

$$\delta_z(x) =: \exp[< z, x > -\frac{1}{2} \ll x \gg_0^2] \cdot :_x 1 \tag{3.18}$$

by the notation of additive renormalization defined in [7]. By Theorem 3. 7, the derivative D_z is a continuous operator on $\mathcal{H}$. Define the dual operator D_z^* of D_z by

$$< \varphi, D_z^* \Psi > \equiv < D_z \varphi, \Psi > \quad \text{for} \quad \varphi \in \mathcal{H}, \Psi \in \mathcal{H}^*.$$

Then the $\mathcal{S}$-transformation of $D_z^* \Psi$ is

$$(\mathcal{S} D_z^* \Psi)(\xi) = < \exp[< x, \xi > -\frac{1}{2} \ll \xi \gg_0^2], D_z^* \Psi(x) > \tag{3.19}$$

$$= < D_z \exp[< x, \xi > -\frac{1}{2} \ll \xi \gg_0^2], \Psi(x) >$$
$$= < z, \xi >< \exp[< x, \xi > -\frac{1}{2} \ll \xi \gg_0^2], \Psi(x) >$$
$$= < z, \xi > (\mathcal{S}\Psi)(\xi).$$

We see that D_z is an annihilation operator and D_z^* is a creation operator (cf. [8]). Now we can assert

Proposition 3.8 *For* $\varphi \in \mathcal{H}$, $z \in \mathcal{E}^*$,

$$(\mathcal{S} D_z \varphi)(\xi) = \frac{\delta}{\delta \xi} (\mathcal{S}\varphi)(\xi; z).$$

For $\Psi \in \mathcal{H}^*, z \in \mathcal{E}^*$,

$$(\mathcal{S} D_z^* \Psi)(\xi) = < z, \xi > (\mathcal{S}\varphi)(\xi).$$

In particular

$$(\mathcal{S}\delta_z)(\xi) = f(\xi; z) = \exp[< z, \xi > -\frac{1}{2} \ll \xi \gg_0^2],$$
$$(\mathcal{S} D_z^* \delta_y)(\xi) = < z, \xi > f(\xi; z).$$

§4. Algebraic structure of $\mathcal{F}$ and $\mathcal{F}^*$.

The algebraic properties of $\mathcal{H}$ is referred in Proposition 2. 8. Those of $\mathcal{F}$ and $\mathcal{F}^*$ are easy to see and valuable to mention.

Proposition 4.1 *Let $U(\xi)$ and $V(\xi)$ be in $\mathcal{F}$ then $U(\xi)V(\xi)$ is also in $\mathcal{F}$. Moreover, the multiplication UV is a continuous mapping from $\mathcal{F}\times\mathcal{F}$ to $\mathcal{F}$.*

Proof. There exist $(f_n)_{n\geq 0}$ and $(g_n)_{n\geq 0}\in \exp[\hat{\otimes}\mathcal{E}_c]$ such that

$$U(\xi)=\sum_{n=0}^{\infty}< f_n,\xi^{\hat{\otimes}n}> \quad\text{and}\quad V(\xi)=\sum_{n=0}^{\infty}< g_n,\xi^{\hat{\otimes}n}>$$

by definition. Obviously,

$$U(\xi)V(\xi)=\sum_{n=0}^{\infty}< h_n,\xi^{\hat{\otimes}n}> \quad\text{for } h_n\equiv\sum_{k=0}^{n} f_k\hat{\otimes}g_{n-k}.$$

The norm of $\mathcal{UV}$ is estimated through $(h_n)_{n\geq 0}$ as

$$\begin{aligned}\|UV\|^2_{\mathcal{F}_c^{(p)}} &= \sum_{n=0}^{\infty} n!\|h_n\|^2_{(E_p)_c^{\hat{\otimes}n}} \leq \sum_{n=0}^{\infty} n!\rho^{2qn}\|h_n\|^2_{(E_{p+q})_c^{\hat{\otimes}n}}\\
&\leq \sum_{n=0}^{\infty} n!\rho^{2qn}\left(\sum_{k=0}^{n}\|f_k\|_{(E_{p+q})_c^{\hat{\otimes}k}}\|g_{n-k}\|_{(E_{p+q})_c^{\hat{\otimes}(n-k)}}\right)^2\\
&\leq \sum_{n=0}^{\infty} 2^n\rho^{2qn}\left(\sum_{k=0}^{n}((n-k)!k!)^{1/2}\|f_k\|_{(E_{p+q})_c^{\hat{\otimes}k}}\|g_{n-k}\|_{(E_{p+q})_c^{\hat{\otimes}(n-k)}}\right)^2\\
&\leq \sum_{n=0}^{\infty} 2^n\rho^{2qn}\left(\sum_{k=0}^{n}(k!)\|f_k\|^2_{(E_{p+q})_c^{\hat{\otimes}k}}\right)\left(\sum_{k=0}^{n}(k!)\|g_k\|^2_{(E_{p+q})_c^{\hat{\otimes}k}}\right)\\
&\leq (1-2\rho^{2q})^{-1}\|U\|^2_{\mathcal{F}^{(p+q)}}\|V\|^2_{\mathcal{F}^{(p+q)}},\end{aligned}$$

for q with $2\rho^{2q}<1$. □

Proposition 4.2 *Let $U(\xi)$ and $V(\xi)$ be in $\mathcal{F}^*$ then $U(\xi)V(\xi)$ is also in $\mathcal{F}^*$. Moreover, the multiplication UV is a continuous mapping from $\mathcal{F}^*\times\mathcal{F}^*$ to $\mathcal{F}^*$.*

Proof. There exist $(F_n)_{n\geq 0}$ and $(G_n)_{n\geq 0}\in \exp[\hat{\otimes}\mathcal{E}_c^*]$ such that

$$U(\xi)=\sum_{n=0}^{\infty}< F_n,\xi^{\hat{\otimes}n}> \quad\text{and}\quad V(\xi)=\sum_{n=0}^{\infty}< G_n,\xi^{\hat{\otimes}n}>$$

by definition. Obviously,

$$U(\xi)V(\xi)=\sum_{n=0}^{\infty}< H_n,\xi^{\hat{\otimes}n}> \quad\text{for } H_n\equiv\sum_{k=0}^{n} F_k\hat{\otimes}G_{n-k}.$$

The norm of UV is estimated through $(H_n)_{n\geq 0}$ as

$$\|UV\|^2_{\mathcal{F}^{(p)}} = \sum_{n=0}^{\infty} n!\|H_n\|^2_{(E_p)_c^{\hat{\otimes} n}} \leq \sum_{n=0}^{\infty} n!\rho^{2qn}\|H_n\|^2_{(E_{p+q})_c^{\hat{\otimes} n}}$$

$$\leq \sum_{n=0}^{\infty} n!\rho^{2qn}\left(\sum_{k=0}^{n}\|F_k\|_{(E_{p+q})_c^{\hat{\otimes} k}}\|G_{n-k}\|_{(E_{p+q})_c^{\hat{\otimes}(n-k)}}\right)^2$$

$$\leq (1-2\rho^{2q})^{-1}\|U\|^2_{\mathcal{F}^{(p+q)}}\|V\|^2_{\mathcal{F}^{(p+q)}},$$

if $U, V \in \mathcal{F}^{(-r)}, p \leq -r$ and $2\rho^{2q} < 1$. □

By (2. 10), we see that $U(\xi)$ in $\mathcal{F}$ and $V(\xi)$ in $\mathcal{F}^*$ are extended to functionals on $\mathcal{E}_c$. Actually, they are represented as

(4.1) $$U(\zeta) = \sum_{n=0}^{\infty} < f_n, \zeta^{\hat{\otimes} n} > \quad \text{and} \quad V(\zeta) = \sum_{n=0}^{\infty} < G_n, \zeta^{\hat{\otimes} n} >$$

with $(f_n)_{n\geq 0} \in \exp[\hat{\otimes}\mathcal{E}_c]$ and $(G_n)_{n\geq 0} \in \exp[\hat{\otimes}\mathcal{E}_c^*]$.

Proposition 4.3 *For $U(\xi) \in \mathcal{F}$, $V(\xi) \in \mathcal{F}^*$ and $\gamma \in \mathbf{C}$, $U(\gamma\xi) \in \mathcal{F}$ and $V(\gamma\xi) \in \mathcal{F}^*$. Furthermore, for q with $|\gamma|\rho^q < 1$, their norms are estimated as*

$$\|U(\gamma\cdot)\|^2_{\mathcal{F}^{(p)}} \leq \|U\|^2_{\mathcal{F}^{(p+q)}} \quad \textit{and} \quad \|V(\gamma\cdot)\|^2_{\mathcal{F}^{(p-q)}} \leq \|V\|^2_{\mathcal{F}^{(p)}},$$

if $V \in \mathcal{F}^{(p)}$.

Proof. By (4. 1),

$$U(\gamma\xi) = \sum_{n=0}^{\infty} < \gamma^n f_n, \xi^{\hat{\otimes} n} > \quad \textit{and } V(\gamma\xi) = \sum_{n=0}^{\infty} < \gamma^n G_n, \xi^{\hat{\otimes} n} >$$

hold. Therefore

(4.2) $$\|U(\gamma\cdot)\|^2_{\mathcal{F}^{(p)}} = \sum_{n=0}^{\infty} n!|\gamma|^{2n}\|f_n\|^2_{(E_p)_c^{\hat{\otimes} n}}$$

$$\leq \sum_{n=0}^{\infty} n!|\gamma|^{2n}\rho^{2qn}\|f_n\|^2_{(E_{p+q})_c^{\hat{\otimes} n}} \leq \|U\|^2_{\mathcal{F}^{(p+q)}}.$$

Similarly,

(4.3) $$\|V(\gamma\cdot)\|^2_{\mathcal{F}^{(p-q)}} = \sum_{n=0}^{\infty} n!|\gamma|^{2n}\|F_n\|^2_{(E_{p-q})_c^{\hat{\otimes} n}}$$

$$\leq \sum_{n=0}^{\infty} n!|\gamma|^{2n}\rho^{2qn}\|F_n\|^2_{(E_p)_c^{\hat{\otimes} n}} \leq \|V\|^2_{\mathcal{F}^{(p)}}.$$

Thus we see the assertions. □

Definition :

(4.4) $$(T_\gamma\varphi)(x) \equiv \varphi(\gamma x) \ \text{ on } \ \mathcal{H} \ \text{ and } (T_\gamma U)(\xi) \equiv U(\gamma\xi) \ \text{ on } \ \mathcal{F}^*.$$

By Theorem 3. 4, (4. 2) and (4. 3), we see easily that

Proposition 4.4 *The opertor T_γ is continuous on $\mathcal{H}$. The operator T_γ is continuous on $\mathcal{F}$ and also on $\mathcal{F}^*$. In particular, T_γ is isometric on both $\mathcal{F}$ and $\mathcal{F}^*$ if $|\gamma| = 1$.*

Define the reflection operator R on $\mathcal{H}^*$ by

$$(4.5) \qquad R\Psi \equiv \sum_{n=0}^{\infty}(-1)^n I(G_n) \qquad \text{if } \Psi = \sum_{n=0}^{\infty} I(G_n).$$

Then we see

Proposition 4.5 *R is isometric on $\mathcal{H}$ and also on $\mathcal{H}^*$. For $\varphi \in \mathcal{H}$,*

$$R\varphi(x) = \varphi(-x) \quad \textit{and} \quad (\mathcal{S}R\varphi)(\xi) = T_{-1}(\mathcal{S}\varphi)(\xi)$$

hold. Moreover

$$< R\varphi, \Psi > = < \varphi, R\Psi > \quad \textit{for any} \quad \varphi \in \mathcal{H} \textit{ and } \Psi \in \mathcal{H}^*.$$

Proof. Since $I(f_n)(-x) = (-1)^n I(f_n)(x)$ holds by (3. 6), the first equality is true. Since $(\mathcal{S}(-1)^n I(f_n))(\xi) = < f_n, (-\xi)^{\hat{\otimes} n} > = (\mathcal{S}I(f_n))(-\xi)$, the second equality is seen. By the duality of $\mathcal{H}$ and $\mathcal{H}^*$ we see the last equality. □

Symbolically we will write $\Psi(-x) = R\Psi(x)$.

§5. Generalized random variables corresponding to Gaussian measures.

In Kubo-Takenaka [10], they defined Volterra's Laplacian Δ_V, which is equivalent to

$$(5.1) \qquad \Delta_V \varphi(x) \equiv \sum_{k=1}^{\infty} D_{\zeta_k} D_{\zeta_k} \varphi(x)$$

with a complete orthonormal system $\{\zeta_k\}$. It acts on $\mathcal{H}$ continuously:

$$(5.2) \qquad \|\Delta_V^k \varphi\|_{\mathcal{H}(p)} \leq ((2k)!)^{1/2}(\|\iota\|\rho^q)^{2k}(1-\rho^{2q})^{-k+1/2}\|\varphi\|_{\mathcal{H}(p+q)}$$

by Theorem 12. 2 in [10].

Proposition 5.1 *The one parameter group*

$$\{\exp[\tau\Delta_V] = \sum_{k=0}^{\infty} \frac{1}{k!}\tau^k \Delta_V^k ; \tau \in C\}$$

acts on $\mathcal{H}$ continuously. The one parameter group of dual operators

$$\{\exp[\tau\Delta_V^*] = \sum_{k=0}^{\infty} \frac{1}{k!}\tau^k \Delta_V^{*k} ; \tau \in C\}$$

acts on $\mathcal{H}^*$ continuously. The $\mathcal{S}$-transformation of $\exp[\tau\Delta_V^*]$ *is*

$$(\mathcal{S}\exp[\tau\Delta_V^*]\Psi)(\xi) = \exp[\tau{<}\xi{>}_0^2](\mathcal{S}\Psi)(\xi).$$

Proof. By (5. 2), we have the estimation

$$\|\exp[\tau\Delta_V]\varphi\|_{\mathcal{H}^{(p)}} \le \sum_{k=0}^{\infty} \frac{((2k)!)^{1/2}}{k!}|\tau|^k(\|\iota\|\rho^q)^{2k}(1-\rho^{2q})^{-k+1/2}\|\varphi\|_{\mathcal{H}^{(p+q)}}$$

$$\le \sum_{k=0}^{\infty}\left(2|\tau|(\|\iota\|\rho^q)^2(1-\rho^{2q})^{-1}\right)^k (1-\rho^{2q})^{1/2}\|\varphi\|_{\mathcal{H}^{(p+q)}}$$

$$\le \left(1-2|\tau|(\|\iota\|\rho^q)^2(1-\rho^{2q})^{-1}\right)^{-1}(1-\rho^{2q})^{1/2}\|\varphi\|_{\mathcal{H}^{(p+q)}}$$

if q is so large as $2|\tau|(\|\iota\|\rho^q)^2(1-\rho^{2q})^{-1} < 1$. This means that the dual operator $\exp[\tau\Delta_V^*]$ acts on $\mathcal{H}^*$ continuously :

$$\|\exp[\tau\Delta_V^*]\Psi\|_{\mathcal{H}^{(-p-q)}} \le \left(1-2|\tau|(\|\iota\|\rho^q)^2(1-\rho^{2q})^{-1}\right)^{-1}(1-\rho^{2q})^{1/2}\|\Psi\|_{\mathcal{H}^{(-p)}}$$

for $\Psi \in \mathcal{H}^{(-p)}$. By Theorem 12.2 in [10], we see the last equality. □

Taking $\Psi = 1$, we have that $\exp[\tau \ll \xi \gg_0^2]$ belongs to $\mathcal{F}^*$ for any $\tau \in C$. Put $\Phi_\gamma \equiv \mathcal{S}^{-1}\exp[\gamma{<}\xi{>}_0^2]$ for $\gamma \in C$ and $\mathcal{N}_\sigma \equiv \Phi_{(\sigma^2-1)/2} = \mathcal{S}^{-1}\exp[\frac{\sigma^2-1}{2} \ll \xi \gg_0^2]$ for $\sigma > 0$. Then, for $\xi \in \mathcal{E}$,

$$(5.3)\qquad (\mathcal{T}\Phi_\gamma)(\xi) =< \exp[i < x,\xi >], \Phi_\gamma > =< f(i\xi;x)\cdot\exp[\frac{1}{2} \ll i\xi \gg_0^2], \Phi_\gamma >$$

$$= \exp[\frac{1}{2} \ll i\xi \gg_0^2]\cdot\exp[\gamma \ll i\xi \gg_0^2] = \exp[-\frac{1}{2}(1+2\gamma) \ll \xi \gg_0^2].$$

By Yokoi [15], Φ_γ is the generalized functional which is equivalent to the measure of Gaussian white noise with variance $1 + 2\gamma$, if γ is real and $1 + 2\gamma > 0$. This suggests that the space of testing random variables $\mathcal{H}^*$ does not depend on variances. Let us fix a positive constant $\alpha > 0$. Put $\nu_\alpha \equiv \alpha\nu$, $(\ ,\ )_{\alpha,p} \equiv \alpha(\ ,\ )_p, p \in N$. Let $\mathcal{E}_\alpha \subset L^2_r(T,\nu_\alpha) \subset \mathcal{E}^*_\alpha$ be the Gel'fand triplet derived from those inner products. Let us denote by $<\ ,\ >_\alpha$ the duality corresponding to

$$(5.4)\qquad <\xi,\eta>_\alpha = (\xi,\eta)_{\alpha,0}.$$

For given $\boldsymbol{f} = (f_n)_{n\ge 0} \in \exp[\hat{\otimes}(\mathcal{E}_\alpha)_c]$, let us estimate its norms. Obviously, if $1 \ge \alpha > 0$,

$$\|f\|_{\exp[\hat{\otimes}(E_p)_c]} \le \|f\|_{\exp[\hat{\otimes}(E_{\alpha,p})_c]}$$

holds. We now assume that $\alpha > 1$. Then,

$$\|f\|^2_{\exp[\hat{\otimes}(E_p)_c]} = \sum_{n=0}^{\infty} n!\|f_n\|_p^2$$

$$\le \sum_{n=0}^{\infty} n!\rho^{2qn}\|f_n\|^2_{p+q} = \sum_{n=0}^{\infty} n!\rho^{2qn}\alpha^{2n}\|f_n\|^2_{\alpha,p+q}$$

$$\leq \sum_{n=0}^{\infty} n!\|f_n\|_{\alpha,p+q}^2 = \|f\|^2_{\exp[\hat{\otimes}(E_{\alpha,p+q})_c]},$$

if q is so large as $\alpha\rho^q < 1$.

Now we give some remarks related to delta functionals and Gaussian measures. Let $\vec{\zeta} = \{\zeta_k\}_{k=1}^d$ be a set of linearly independent elements in $\mathcal{E}$ and put

$$\Gamma(\vec{\zeta}) \equiv (<\zeta_i, \zeta_j>)_{i,j=1}^d, \tag{5.5}$$

which is the covariance matrix of $\{<x, \zeta_j>\}_{j=1}^d$. Let us denote by $\pi_{\vec{\zeta}}x$ the orthogonal projection of x to the linear space spanned by $\vec{\zeta}$; that is, $<x - \pi_{\vec{\zeta}}x, \zeta_k>= 0$ for $1 \leq k \leq$ d. Put $\pi_{\vec{\zeta}}^{\perp}x \equiv x - \pi_{\vec{\zeta}}x$. In particular, if $\{\zeta_k\}_{k=1}^d$ are orthonormal, i.e. $<\zeta_i, \zeta_j>= \delta_{i,j}$,

$$\pi_{\vec{\zeta}}x \equiv \sum_{k=1}^{d} <\zeta_k, x> \zeta_k \text{ and } \quad \pi_{\vec{\zeta}}^{\perp}x \equiv x - \pi_{\vec{\zeta}}x. \tag{5.6}$$

By Theorem 3. 4 and Theorem 3. 6, for η_k in in $\mathcal{E}_c$, z_k in $\mathcal{E}_c^*$, $k = 1, \cdots, d, z \in \mathcal{E}_c^*$ and $\gamma \in C$,

$$\varphi(\sum_{j=1}^{d} <x, \eta_j> \zeta_j + \gamma x + z)$$

belongs to $\mathcal{H}$, if $\varphi \in \mathcal{H}$. Therefore $\varphi(\pi_{\vec{\zeta}}x + z)$ and $\varphi(\pi_{\vec{\zeta}}^{\perp}x + z)$ belong to $\mathcal{H}$, too. Since

$$\int \exp[i <\pi_{\vec{\zeta}}^{\perp}x, \xi> + i <\pi_{\vec{\zeta}}x, \eta>]d\mu(x)$$

$$=< \exp[i < x, \pi_{\vec{\zeta}}^{\perp}\xi + \pi_{\vec{\zeta}}\eta>], 1>= \exp[-\frac{1}{2}<\pi_{\vec{\zeta}}^{\perp}\xi>_0^2 - \frac{1}{2}<\pi_{\vec{\zeta}}\eta>_0^2],$$

$\{<\pi_{\vec{\zeta}}^{\perp}x, \xi>; \xi \in \mathcal{E}_c\}$ and $\{<\pi_{\vec{\zeta}}x, \eta>; \eta \in \mathcal{E}_c\}$ are mutually independent to each other. By Proposition 5. 2, functionals $\delta_z(\pi_{\vec{\zeta}}^{\perp}x)$ and $\delta_\eta(\pi_{\vec{\zeta}}x)$ can be defined by

$$< \varphi(x), \delta_z(\pi_{\vec{\zeta}}^{\perp}x) > \equiv \int \varphi(z + \pi_{\vec{\zeta}}x)d\mu(x) \tag{5.7}$$

$$= \int_R \varphi(z + \sum_{k=1}^{d} u_k\zeta_k)(2\pi)^{-n/2}det\Gamma(\vec{\zeta}))^{-1/2} \exp[-\frac{1}{2} < \Gamma(\vec{\zeta})^{-1}\vec{u}, \vec{u}>]d\vec{u}.$$

and

$$< \varphi(x), \delta_\eta(\pi_{\vec{\zeta}}x) > \equiv \int \varphi(\eta + \pi_{\vec{\zeta}}^{\perp}x)d\mu(x) \tag{5.8}$$

for $z \in \mathcal{E}^*$ with $<z_k, \zeta>= 0$, $k = 1, \cdots, d$, and $\eta \in \mathcal{E}$ belonging to the linear hull of $\vec{\zeta}$. Then $\delta_z(\pi_{\vec{\zeta}}^{\perp}x)$ is a degenerated Gaussian measure with the characteristic functional

$$(T\delta_z(\pi_{\vec{\zeta}}^{\perp}x))(\xi) =< \exp[i <\pi_{\vec{\zeta}}x, \xi> + i <\pi_{\vec{\zeta}}^{\perp}x, \xi>], \delta_z(\pi_{\vec{\zeta}}^{\perp}x) > \tag{5.9}$$

$$= \exp[i < z, \xi >] < \exp[i < x, \pi_{\zeta}\xi >], 1 >$$
$$= \exp[i < z, \xi > -\frac{1}{2} < \pi_{\zeta}\xi >_0^2].$$

More exactly, we may say that $\delta_z(\pi_{\zeta}^{\perp} x)$ is the conditional probability conditioned by $\pi_{\zeta}^{\perp} x = z$. The generalized functional $\delta_\eta(\pi_{\zeta} x)$ is the conditional probability conditioned by $\pi_{\zeta} x = \eta$ with the characteristic functional

$$(5.10) \qquad (\mathcal{T}\delta_\eta(\pi_{\zeta} x))(\xi) =< \exp[i < \pi_{\zeta} x, \xi > +i < \pi_{\zeta}^{\perp} x, \xi >], \delta_\eta(\pi_{\zeta} x) >$$
$$= \exp[i < \eta, \xi >] < \exp[i < x, \pi_{\zeta}^{\perp} \xi >], 1 >$$
$$= \exp[i < \eta, \xi > -\frac{1}{2} < \pi_{\zeta}^{\perp} \xi >_0^2].$$

§6. Fourier transform.

Kuo's Fourier transform (see [13]) is defined by

$$(6.1) \qquad \boldsymbol{F}\Phi(x) = \tilde{\Phi}(x) \equiv \mathcal{S}^{-1}\mathcal{T}\Phi(x)$$
$$= \mathcal{S}^{-1}T_i\mathcal{S}\exp[\frac{1}{2}\Delta_V^*]\tilde{\Phi}(x)$$
$$= \mathcal{S}^{-1}\left(\exp[-\frac{1}{2} \ll \xi \gg_0^2](\mathcal{S}\Phi)(i\xi)\right)(x)$$

for any $\Phi \in \mathcal{H}^*$.

Theroem 6.1 *The Fourier transform $\boldsymbol{F}$ is an isomorphic operator on $\mathcal{H}^*$, whose inverse formula is given by*

$$(6.2) \qquad \Phi(x) = \mathcal{S}^{-1}\mathcal{T}R\tilde{\Phi}(x)$$
$$= \mathcal{S}^{-1}\left(\exp[-\frac{1}{2} \ll \xi \gg_0^2](\mathcal{S}\tilde{\Phi})(-i\xi)\right)(x)$$

for $\Phi \in \mathcal{H}^*$. The relation

$$(6.3) \qquad \tilde{\tilde{\Phi}}(x) = \Phi(-x) \quad i.e.\ \boldsymbol{F}\boldsymbol{F} = R.$$

holds.

Proof. As discussed in §5, the operator $\exp[\frac{1}{2}\Delta_V^*]$ acts on $\mathcal{H}^*$ continuously and T_i acts on $\mathcal{F}^*$ isometrically. Hence $\boldsymbol{F}$ acts continuously. By definition, we see

$$(\mathcal{S}\tilde{\Phi})(-i\xi) = (\mathcal{T}\Phi)(-i\xi) =< \exp[< x, \xi >], \Phi >= \exp[\frac{1}{2} \ll \xi \gg_0^2](\mathcal{S}\Phi)(\xi)$$

and

$$(\mathcal{S}\Phi)(\xi) = \exp[-\frac{1}{2} \ll \xi \gg_0^2](\mathcal{S}\tilde{\Phi})(-i\xi).$$

Hence we have the inverse formula (6. 2), which assures the continuity of $\boldsymbol{F}^{-1}$. By (6. 1) and (6. 2), we have that

$$\boldsymbol{F}^{-1}R = \mathcal{S}^{-1}\mathcal{T}RR = \boldsymbol{F}.$$

Hence we see the last equality. □

Put $P_t(u)$ be the heat kernel

$$P_t(u) \equiv (2\pi t)^{-1/2}\exp[-\frac{1}{2t}u^2]. \tag{6.4}$$

For a given $h \in \mathcal{S}^*(\boldsymbol{R})$($a$ temper distribution), its Fourier transform is defined by

$$< \psi, \bar{\tilde{h}} > = < \tilde{\psi}, \bar{h} > \qquad \text{for any } \psi \in \mathcal{S}(\boldsymbol{R}),$$

where

$$\tilde{\psi}(v) = (2\pi)^{-1/2}\int \psi(u)\exp[ivu]du.$$

We see that

$$h_t(u) \equiv \exp[-\frac{t}{2}u^2](P_t * h)(u) \longrightarrow h \ \text{ in } \ \mathcal{S}^*(\boldsymbol{R}) \ \text{ as } \ t \downarrow 0. \tag{6.5}$$

Notice that $h_t(u)$ belongs to $\mathcal{S}(\boldsymbol{R})$. Furthermore the Fourier transform of h_t is given by

$$\tilde{h}_t(v) = \exp[-\frac{t}{2}v^2](P_t * \tilde{h})(v). \tag{6.6}$$

By Theorem 1 in [6] and (6. 5), we can define $h(<\zeta, x>)$ for $\zeta \in \mathcal{E}$ by

$$h(<\zeta, x>) = \mathcal{H}^* - \lim_{t\downarrow 0}\exp[-\frac{t}{2} <\zeta, x>^2](P_t * h)(<\zeta, x>).$$

and we know that

$$(Sh(<\zeta, \cdot>))(\xi) = P_{\ll\zeta\gg_0^2} * h(<\zeta, \xi>).$$

Theroem 6.2 *Let h be in $\mathcal{S}^*(\boldsymbol{R})$ and $\tilde{h}$ be its Fourier transform in $\mathcal{S}^*(\boldsymbol{R})$. For $\zeta \in \mathcal{E}$, the Fourier transform of $\Phi(x) \equiv h(<x, \zeta>)$ is*

$$\tilde{\Phi}(x) = \delta_0(\pi_\zeta^\perp x)\frac{1}{\ll \zeta \gg_0}\tilde{h}(\frac{1}{\ll \zeta \gg_0^2} < x, \zeta >) \tag{6.7}$$

and its $\mathcal{S}$-transformation (which is equal to $\mathcal{T}$-transformation of Φ) is

$$\begin{aligned}(\mathcal{S}\tilde{\Phi})(\xi) = (\mathcal{T}\Phi)(\xi) &= \exp[-\frac{1}{2} \ll \xi \gg_0^2]P_{\ll\zeta\gg_0^2} * h(i < \zeta, \xi >). \\ &= \exp[-\frac{1}{2} \ll \pi_\zeta^\perp\xi \gg_0^2] \ll \zeta \gg_0^{-1} P_{\ll\zeta\gg_0^{-2}} * \tilde{h}(\frac{<\zeta, \xi>}{\ll \zeta \gg_0^2}).\end{aligned} \tag{6.8}$$

Proof. Put $\alpha \equiv <\zeta,\xi> / \ll \zeta \gg_0^2$. Since h_t belongs to $\mathcal{S}(\boldsymbol{R})$, the $\mathcal{T}$-transformation of $h_t(<x,\zeta>)$ is

(6.9)
$$(\mathcal{T}h_t(<x,\zeta>))(\xi) = \int \exp[i<x,\xi>]h_t(<x,\zeta>)d\mu$$
$$= \int \exp[i<\pi_\zeta^\perp x, \pi_\zeta^\perp \xi> + i<\pi_\zeta x, \pi_\zeta \xi>]h_t(<x,\zeta>)d\mu$$
$$= \exp[-\frac{1}{2} \ll \pi_\zeta^\perp \xi \gg_0^2] \int \exp[i\alpha <x,\zeta>]h_t(<x,\zeta>)d\mu$$
$$= \exp[-\frac{1}{2} \ll \pi_\zeta^\perp \xi \gg_0^2](2\pi \ll \zeta \gg_0^2)^{-1/2} \int \exp[i\alpha u - \frac{1}{2 \ll \zeta \gg_0^2} u^2]h_t(u)du$$
$$= \exp[-\frac{1}{2} \ll \pi_\zeta^\perp \xi \gg_0^2 - \frac{1}{2}\alpha^2 \ll \zeta \gg_0^2](2\pi \ll \zeta \gg_0^2)^{-1/2}$$
$$\times \int \exp[-\frac{1}{2 \ll \zeta \gg_0^2}(u - i\alpha \ll \zeta \gg_0^2)^2]h_t(u)du$$
$$= \exp[-\frac{1}{2} \ll \xi \gg_0^2]P_{\ll\zeta\gg_0^2} * h_t(i\alpha \ll \zeta \gg_0^2)$$

Let $\tilde{h}_t(v)$ be the Fourier transform of h_t. Put

$$\Psi_t(x) \equiv \delta_0(\pi_\zeta^\perp x) \frac{1}{\ll \zeta \gg_0} \tilde{h}_t(\frac{1}{\ll \zeta \gg_0^2} <x,\zeta>).$$

Then

(6.10)
$$(\mathcal{S}\Psi_t)(\xi)$$
$$=< \exp[<\pi_\zeta^\perp x, \xi> - \frac{1}{2} \ll \pi_\zeta^\perp \xi \gg^2 + <\pi_\zeta x, \xi> - \frac{1}{2} \ll \pi_\zeta \xi \gg^2],$$
$$\delta_0(\pi_\zeta^\perp x) \frac{1}{\ll \zeta \gg_0} \tilde{h}_t(\frac{1}{\ll \zeta \gg_0^2} <x,\zeta>) > .$$
$$= \exp[-\frac{1}{2} \ll \pi_\zeta^\perp \xi \gg^2] \frac{1}{\ll \zeta \gg_0}$$
$$\times < \exp[<\pi_\zeta x, \pi_\zeta \xi> - \frac{1}{2} \ll \pi_\zeta \xi \gg^2], \tilde{h}_t(\frac{1}{\ll \zeta \gg_0^2} <x,\zeta>) >$$
$$= \exp[-\frac{1}{2} \ll \xi \gg_0^2](2\pi \ll \zeta \gg_0^4)^{-1/2} \int \exp[\alpha u - \frac{u^2}{2 \ll \zeta \gg_0^2}]\tilde{h}_t(\frac{u}{\ll \zeta \gg_0^2})du$$
$$= \exp[-\frac{1}{2} \ll \xi \gg_0^2](2\pi \ll \zeta \gg_0^2)^{-1}$$
$$\times \int\int \exp[\alpha u - \frac{u^2}{2 \ll \zeta \gg_0^2}] \exp[i\frac{uv}{\ll \zeta \gg_0^2}]h_t(v)dudv$$
$$= \exp[-\frac{1}{2} \ll \xi \gg_0^2](2\pi \ll \zeta \gg_0^2)^{-1} \int\int \exp[-\frac{1}{2 \ll \zeta \gg_0^2}(\alpha \ll \zeta \gg_0^2 + iv)^2]$$

$$\times \exp[-\frac{1}{2 \ll \zeta \gg_0^2}(u - \alpha \ll \zeta \gg_0^2 - iv)^2] h_t(v) du dv$$

$$= \exp[-\frac{1}{2} \ll \xi \gg_0^2](2\pi \ll \zeta \gg_0^2)^{-1/2} \int \exp[-\frac{1}{2 \ll \zeta \gg_0^2}(\alpha \ll \zeta \gg_0^2 + iv)^2] h_t(v) dv$$

$$= \exp[-\frac{1}{2} \ll \xi \gg_0^2](2\pi \ll \zeta \gg_0^2)^{-1/2} \int \exp[-\frac{1}{2 \ll \zeta \gg_0^2}(v - i < \zeta, \xi >)^2] h_t(v) dv$$

$$= \exp[-\frac{1}{2} \ll \xi \gg_0^2] P_{\ll \zeta \gg_0^2} * h_t(i < \zeta, \xi >).$$

Therefore the Fourier transform of $h_t(< x, \xi >)$ is

$$\delta_0(\pi_\zeta^\perp x) \frac{1}{\ll \zeta \gg_0} \tilde{h}_t(\frac{1}{\ll \zeta \gg_0^2} < x, \zeta >).$$

Now we remember that the mapping $h \in \mathcal{S}^*(\boldsymbol{R}) \longrightarrow h(< x, \xi >) \in \mathcal{H}^*$ is continuous by Theorem 2 in [6]. Since Fourier transform $\boldsymbol{F}$ is continuous operator on $\mathcal{H}^*$ and since h_t (resp. $\tilde{h}_t$) converges to h (resp. $\tilde{h}$) in $\mathcal{S}^*(\boldsymbol{R})$ as $t \downarrow 0$, we have that the Fourier transform of $h(< x, \zeta >)$ is

$$\delta_0(\pi_\zeta^\perp x) \frac{1}{\ll \zeta \gg_0} \tilde{h}(\frac{1}{\ll \zeta \gg_0^2} < x, \zeta >)$$

and its $\mathcal{S}$-transformation is given by (6. 8). Since

$$P_t(v - ia) = \int \frac{1}{2\pi} \exp[-\frac{t}{2}\{w - \frac{1}{t}(a + iv)\}^2 - \frac{1}{2}(v - ia)^2] dw$$

$$= \int \frac{1}{2\pi} \exp[-\frac{t}{2} w^2] \exp[i(v - ia)w] dw$$

$$= \int \frac{1}{2\pi} \exp[-\frac{t}{2}(w - \frac{1}{t}a)^2 + \frac{1}{2t}a^2] \exp[ivw] dw,$$

the relation

(6.11) $$P_t * h(iu) = < P_t(\cdot - iu), h(\cdot) >$$

$$= < (2\pi)^{-1/2} \exp[-\frac{t}{2}(\cdot - \frac{1}{t}u)^2 + \frac{1}{2t}u^2], \tilde{h}(\cdot) >$$

$$= t^{-1/2} \exp[\frac{1}{2t}u^2] P_{1/t} * \tilde{h}(\frac{1}{t}u)$$

holds. Therefore the $\mathcal{S}$-transformation of $\tilde{\Phi}$ can be rewritten

$$\exp[-\frac{1}{2} \ll \xi \gg_0^2 + \frac{1}{2} \ll \pi_\zeta \xi \gg_0^2] \ll \zeta \gg_0^{-1} P_{\ll \zeta \gg_0^2} * \tilde{h}(\frac{< \zeta, \xi >}{\ll \zeta \gg_0^2})$$

$$= \exp[-\frac{1}{2} \ll \pi_\zeta^\perp \xi \gg_0^2] \ll \zeta \gg_0^{-1} P_{\ll \zeta \gg_0^{-2}} * \tilde{h}(\frac{< \zeta, \xi >}{\ll \zeta \gg_0^2}). \quad \square$$

Let Γ be a regular symmetric positive $d\times d$-matrix. Define the kernel $P_\Gamma(u_1,\cdots,u_d)$ by

$$(6.12)\qquad P_\Gamma \equiv (2\pi)^{-d/2}(det\Gamma)^{-1/2}\exp[-\frac{1}{2} < \vec{u}, \Gamma^{-1}\vec{u} >]$$

with $\vec{u} = (u_1,\cdots,u_d)$. Let I be the $d\times d$ identity matrix. For a temper distribution $h(\vec{u}) \in \mathcal{S}^*(\boldsymbol{R}^d)$, define

$$(6.13)\qquad h_\Gamma(\vec{u}) \equiv \exp[-\frac{1}{2} < \vec{u}, \Gamma\vec{u} >] < P_\Gamma(\vec{u}-\vec{v}), h(\vec{v}) > .$$

Then h_Γ belongs to $\mathcal{S}(\boldsymbol{R}^d)$ and converges to h in $\mathcal{S}^*(\boldsymbol{R}^d)$ as $\Gamma \downarrow 0$. For any linearly independent elements $\{\zeta_1,\cdots,\zeta_d\}$ of $\mathcal{E}$, let $\Gamma(\vec{\zeta})$ be the covariance matrix of $< x, \vec{\zeta} >= \{< x, \zeta_j >\}_{j=1}^d$ given in (5. 5). Let $\pi(\vec{\zeta}) = \pi_{\vec{\zeta}}$ and $\pi(\vec{\zeta})^\perp$ be the projections defined in §5. We can define $h(< x,\zeta_1 >,\cdots,< x,\zeta_d >)$ by

$$(6.14)\qquad \begin{aligned} h(< x, \vec{\zeta} >) &= h(< x,\zeta_1 >,\cdots,< x,\zeta_d >) \\ &\equiv \mathcal{H}^*\text{-}\lim_{\gamma\downarrow 0} h_{\gamma I}(< x,\zeta_1 >,\cdots,< x,\zeta_d >). \end{aligned}$$

The existence of the limit (6. 14) is seen similarly to [6]. Its $\mathcal{S}$-transformation is given by

$$\begin{aligned} (\mathcal{S}h(< \cdot, \vec{\zeta} >))(\xi) &= P_{\Gamma(\vec{\zeta})} * h(< \xi, \zeta >) \\ &=< P_{\Gamma(\vec{\zeta})}(< \xi, \vec{\zeta} > -\vec{v}), h(\vec{v}) > . \end{aligned}$$

The following theorem can be shown similarly to Theorem 6. 1.

Theroem 6.3 *For any given $h \in \mathcal{S}^*(\boldsymbol{R}^d)$, let $\tilde{h}(\vec{v})$ be the Fourier transform of h. For any linearly independent elements $\vec{\zeta} = \{\zeta_j\}_{j=1}^d$ of $\in \mathcal{E}$, the Fourier transform of $\Phi(x) \equiv h(< x, \vec{\zeta} >)$ is given by*

$$\tilde{\Phi}(x) = \delta_0(\pi_{\vec{\zeta}}^{\perp} x)(det\Gamma(\vec{\zeta}))^{-1/2}\tilde{h}(\Gamma(\vec{\zeta})^{-1} < x, \vec{\zeta} >).$$

and its $\mathcal{S}$-transformation (which is equal to $\mathcal{T}$-transformation of Φ) is

$$\begin{aligned} (\mathcal{S}\tilde{\Phi})(\xi) &= (\mathcal{T}\Phi)(\xi) \\ &= \exp[-\frac{1}{2} \ll \xi \gg_0^2] P_{\Gamma(\vec{\zeta})} * h(i < \vec{\zeta}, \xi >) \\ &= \exp[-\frac{1}{2} \ll \pi_{\vec{\zeta}}^{\perp}\xi \gg_0^2](det\Gamma(\vec{\zeta}))^{-1/2} P_{\Gamma(\vec{\zeta})^{-1}} * \tilde{h}(\Gamma(\vec{\zeta})^{-1} < \vec{\zeta}, \xi >). \end{aligned}$$

Example (1) (i) For $\gamma > 0$, $h(u) = \exp[-\gamma u^2]$ belongs to $\mathcal{S}(\boldsymbol{R})$ ($\subset \mathcal{S}^*(\boldsymbol{R})$) and its Fourier transform is

$$\tilde{h}(v) = (2\gamma)^{-1/2}\exp[-\frac{1}{4\gamma}v^2].$$

Applying Theorem 6. 1,

$$\boldsymbol{F}: \exp[-\gamma < x,\zeta >^2] \longrightarrow (2\gamma \ll \zeta \gg_0^2)^{-1/2} \exp[-\frac{1}{4\gamma \ll \zeta \gg_0^4} < x,\zeta >^2]\delta_0(\pi_\zeta^\perp x).$$

The $\mathcal{T}$-transformation is also calculated as

$$\mathcal{T}: \exp[-\gamma < x,\zeta >^2] \longrightarrow$$

$$\left(1+2\gamma \ll \zeta \gg_0^2\right)^{-1/2} \exp[-\frac{1}{2}\left(-\frac{2\gamma < \xi,\zeta >^2}{1+2\gamma \ll \zeta \gg_0^2} + \ll \xi \gg_0^2\right)].$$

(ii) For $h(u) = \delta^{(n)}(u), \tilde{h}(v) = (2\pi)^{-1/2}(iv)^n$. Hence

$$\boldsymbol{F}: \delta^{(n)}(< x,\zeta >) \longrightarrow \delta_0(\pi_\zeta^\perp x)(2\pi \ll \zeta \gg_0^2)^{-1/2}\left(i\frac{< x,\zeta >}{\ll \zeta \gg_0^2}\right)^n,$$

$$\mathcal{S}: \boldsymbol{F}\delta^{(n)}(< x,\zeta >) \longrightarrow$$

$$(2\pi \ll \zeta \gg_0^2)^{-1/2} \exp[-\frac{1}{2} \ll \pi_\zeta^\perp \xi \gg_0^2] H_n(i\frac{< \xi,\zeta >}{\ll \zeta \gg_0^2}; \frac{1}{\ll \zeta \gg_0^2})$$

and

$$\mathcal{S}: \delta^{(n)}(< x,\zeta >) \longrightarrow$$

$$(2\pi \ll \zeta \gg_0^2)^{-1/2} \exp[-\frac{1}{2}\frac{< \xi,\zeta >^2}{\ll \zeta \gg_0^2}] H_n(\frac{< \xi,\zeta >}{\ll \zeta \gg_0^2}; \frac{1}{\ll \zeta \gg_0^2}).$$

(iii) By (ii),

$$\begin{array}{lcll} \boldsymbol{F} & : & < x,\zeta >^n & \longrightarrow & (2\pi \ll \zeta \gg_0^2)^{1/2}(i \ll \zeta \gg_0^2)^n \delta_0(\pi_\zeta^\perp x)\delta^{(n)}(< x,\zeta >), \\ \mathcal{S} & : & \boldsymbol{F} < x,\zeta >^n & \longrightarrow & \exp[-\frac{1}{2} \ll \xi \gg_0^2 H_n(i < \xi,\zeta >; - \ll \zeta \gg_0^2), \\ \mathcal{S} & : & < x,\zeta >^n & \longrightarrow & H_n(< \xi,\zeta >; - \ll \zeta \gg_0^2). \end{array}$$

Example (2) Put $\Phi_\gamma \equiv \mathcal{S}^{-1}\exp[\gamma \ll \xi \gg_0^2]$. Then

$$\tilde{\Phi}_\gamma(x) = \Phi_{-(1+2\gamma)/2}(x).$$

In particular, $\Phi_{-1/4}(x)$ is invariant under the Fourier transform.

Proof. Since

$$(\mathcal{S}\Phi_\omega)(\xi) = \exp[\omega \ll \xi \gg_0^2],$$

$$(\mathcal{T}\Phi_\gamma)(\xi) = \exp[-\frac{1}{2}(1+2\gamma) \ll \xi \gg_0^2].$$

Taking $\omega = -\frac{1}{2}(1+2\gamma)$, we see that

$$\tilde{\Phi}_\gamma(x) = \Phi_{-(1+2\gamma)/2}(x). \qquad \square$$

Example (3) Let $\delta_y(x)$ be the delta functional defined in Theorem 3. 3. Then we see the followings:

$$\begin{array}{lcl} \mathcal{S} : \delta_y(x) & \longrightarrow & f(\xi; y) = \exp[-\frac{1}{2} \ll \xi \gg_0^2 + < y, \xi >], \\ \mathcal{T} : \delta_y(x) & \longrightarrow & \exp[i < y, \xi >], \\ \boldsymbol{F} : \delta_y(x) & \longrightarrow & : \exp[i < y, x >] :_x . \end{array}$$

Example (4) By Theorem 3. 7 and (3. 19), for $y \in \mathcal{E}^*$,

$$\begin{array}{lcl} \mathcal{S} : :< y, x > \cdot :_x 1 & \longrightarrow & < y, \xi >, \\ \mathcal{T} : :< y, x > \cdot :_x 1 & \longrightarrow & i < y, \xi > \exp[-\frac{1}{2} \ll \xi \gg_0^2], \\ \boldsymbol{F} : :< y, x > \cdot :_x 1 & \longrightarrow & i D_y^* \delta_0(x), \\ \mathcal{S} : D_y^* \delta_0 & \longrightarrow & < y, \xi > \exp[-\frac{1}{2} \ll \xi \gg_0^2], \\ \mathcal{T} : D_y^* \delta_0 & \longrightarrow & i < y, \xi >, \\ \boldsymbol{F} : D_y^* \delta_0 & \longrightarrow & i :< y, x > \cdot :_x 1. \end{array}$$

Furthermore

$$\begin{array}{lcl} \mathcal{S} : (D_y^*)^n \delta_0 & \longrightarrow & < y, \xi >^n \exp[-\frac{1}{2} \ll \xi \gg_0^2], \\ \mathcal{T} : (D_y^*)^n \delta_0 & \longrightarrow & i^n < y, \xi >^n \\ \boldsymbol{F} : (D_y^*)^n \delta_0 & \longrightarrow & i^n :< y, x >^n \cdot :_x 1. \end{array}$$

The last one is equal to $i^n H_n(< \eta, x >, \ll \eta \gg_0^2)$, if $y = \eta \in \mathcal{E}$.

Example (5) The Fourier transform of the multiplication operator $< x, \zeta > \cdot$ on $\mathcal{H}$ is given by

$$\boldsymbol{F} < x, \zeta > \cdot \boldsymbol{F}^{-1} = \frac{1}{i} D_\zeta .$$

for $\zeta \in \mathcal{E}$. But both operators $< x, \zeta > \cdot$ and D_ζ can be defined not on whole $\mathcal{H}^*$. To see the equality, put

$$U(\eta) \equiv \exp[\frac{1}{2} \ll \eta \gg_0^2](\mathcal{S}\varphi)(\eta).$$

Then by Theorem 7. 2 in [8], we have that

$$\mathcal{T}(< \zeta, x > \cdot \varphi)(\xi) = \exp[-\frac{1}{2} \ll \xi \gg_0^2](i < \zeta, \xi > (\mathcal{S}\varphi)(\xi) + (\mathcal{S}\varphi)^{(1)}(i\xi; \zeta))$$

$$= \left(\frac{\delta}{\delta \xi} \exp[\frac{1}{2} \ll \eta \gg_0^2](\mathcal{S}\varphi)(\eta; \zeta) \right) |_{\eta = i\xi}$$

$$= U^{(1)}(\eta; \zeta)|_{\eta = i\xi} .$$

Since

$$U^{(1)}(i\xi; \zeta) = \lim_{\lambda \to 0} \frac{1}{\lambda}(U(i\xi + \lambda\zeta) - U(i\xi)) = \lim_{\lambda \to 0} \frac{1}{\lambda}(U(i(\xi - i\lambda\zeta)) - U(i\xi)),$$

we see that

$$\boldsymbol{F}(< \zeta, x > \cdot \varphi(x)) = \mathcal{S}^{-1} U^{(1)}(i\xi; \zeta)(x)$$

$$= \lim_{\lambda\to 0} \frac{1}{\lambda}(\mathcal{S}^{-1}U(i\xi)(x - i\lambda\zeta) - \mathcal{S}^{-1}U(i\xi)(x))$$

$$= \lim_{\lambda\to 0} \frac{1}{\lambda}(\boldsymbol{F}\varphi(x - i\lambda\zeta) - \boldsymbol{F}\varphi(x)) = D_{-i\zeta}\boldsymbol{F}\varphi(x) = \frac{1}{i}D_\zeta \boldsymbol{F}\varphi(x).$$

§7. Supplements.

In the working shop, several authors presented results related to previous sections. We point out some of them here. Proposition 2. 8 and affine transformations of variables were also discussed by Pottohof and Yan [16]. Lee [14], [15] introduced a space of analytic testing functionals of exponential type (similar to Schwartz' $\mathbf{Z}(\boldsymbol{R})$). He defined an operator $\boldsymbol{F}_{\alpha,\beta}$ by

$$\boldsymbol{F}_{\alpha,\beta}\varphi(y) \equiv < \varphi(\alpha x + \beta y), 1 > \tag{7.1}$$

on his space (a group of similar operators were discussed in [14] for one dimensional case). On our space $\mathcal{H}$, the mapping $\varphi(x) \to \varphi(\alpha x + \beta y)$ is continuous for any $\alpha, \beta \in \boldsymbol{C}$ and $y \in \mathcal{E}_c^*$, as seen in Theorem 3. 4 (also given by Pottohof and Yan [16] for reals α, β). Therefore the operator $\boldsymbol{F}_{\alpha,\beta}$ is well defined on $\mathcal{H}$. By Theorem 12. 5 of [10] and (3. 14), we see that

$$\boldsymbol{F}_{1,1}\varphi(y) = \sum_{n=0}^{\infty} < f_n, y^{\hat{\otimes} n} > = (\mathcal{S}\varphi)(y) = \exp[\frac{1}{2}\Delta_V]\varphi(y),$$

for $\varphi = \sum I(f_n) \in \mathcal{H}$. Therefore we have :

Proposition 7.1 *The operator $\boldsymbol{F}_{\alpha,\beta}$ has the following expression on $\mathcal{H}$*

$$\boldsymbol{F}_{\alpha,\beta}\varphi = T_\beta \exp[\frac{1}{2}\Delta_V]T_\alpha\varphi \tag{7.2}$$

and hence it acts on $\mathcal{H}$ continuously for any $\alpha, \beta \in \boldsymbol{C}$. Further it is homeomorphic on $\mathcal{H}$ for $\alpha, \beta \neq 0$.

Lee introduced a Fourier transform $\boldsymbol{F}_{Lee}$ on his space, which can be carried in our space $\mathcal{H}^*$ as follows:

$$< \boldsymbol{F}_{Lee}\Phi, \varphi > \equiv < \Phi, \boldsymbol{F}_{1,i}\varphi > \text{ for any } \varphi \in \mathcal{H}. \tag{7.3}$$

Theroem 7.2 *Kuo's Fourier transform $\boldsymbol{F}$ coincides with Lee's Fourier transform $\boldsymbol{F}_{Lee}$ on $\mathcal{H}^*$.*

Proof. Since

$$\boldsymbol{F}_{1,i}f(\eta, y) = < \exp[< x + iy, \eta > -\frac{1}{2} \ll \eta \gg^2], 1 > = \exp[i < y, \eta >],$$

$$\begin{aligned}(\mathcal{S}F_{Lee}\Phi)(\eta) &= < F_{Lee}\Phi, f(\eta; x) > &&= < \Phi, F_{1,i}f(\eta;\cdot) > \\ &= < \Phi, \exp[i < x, \eta >] > &&= (\mathcal{T}\Phi)(\eta) = (\mathcal{S}F\Phi)(\eta)\end{aligned}$$

for any $\eta \in \mathcal{E}$. Therefore $F_{Lee} = F$. □

Thus we confirm that the Fourier transform F defined by Kuo and Lee is really infinite dimensional Fourier transform. The relation (6. 3) was discussed by Kuo himself in the lecture at Nagoya University for his previous definition of Fourier transform.

Acknowledgement: The authors thank to Dr. Hirotake Yaguchi for his converter to T_EX.

References

[1] Hida, T.: Analysis of Brownian functionals. Carleton Math. Lec. Notes, No.13, 2nd Ed. (1978).

[2] Hida, T.: Generalized Brownian functionals, Theory and application of random fields, Proc. IFIP-WG 7/1 Working Conf. Lecture Notes in Control and Information Sciences, Springer, (1982), 89-95.

[3] Hida, T., Pottohoff, J. & Streit, L. : Dirichlet forms and white noise analysis, Commun. Math. Physics, 116 (1988), 235-245.

[4] Ito,Y.: Generalized Poisson functionals. Prob. Th. Rel. Fields, 77 (1988), 1-28.

[5] Ito, Y. & Kubo, I.:Calculus on Gaussian and Poisson white noises. Nagoya Math. J. Vol. 111 (1988), 41-84.

[6] Kubo, I.: Ito formula for generalized Brownian functionals. Theory and application of random fields, Proc. IFIP-WG 7/1 Working Conf. Lecture Notes in Control and Information Sciences, Springer, (1982), 156-166.

[7] Kubo, I. & Takenaka, S.: Calculus on Gaussian white noises I. Proc. Japan Acad. 56 Ser. A, No.8, (1980), 376-380.

[8] Kubo, I. & Takenaka, S.: Calculus on Gaussian white noises II. Proc. Japan Acad. 56 Ser. A, No.9, (1980), 411-416.

[9] Kubo, I. & Takenaka, S.: Calculus on Gaussian white noises III. Proc. Japan Acad. 57 Ser. A, No.9 (1981), 433-437.

[10] Kubo, I. & Takenaka, S.: Calculus on Gaussian white noises IV. Proc. Japan Acad. 58 Ser. A, No.9 (1982), 186-189.

[11] Kubo, I. & Yokoi, T. : Some remarks on Gaussian white noises, to appear in Nagoya Math. J.)

[12] Kuo, H.-H.: Gaussian measures in Banach spaces. Lect. Notes in Math. Vol. 463, Springer-Verlag, (1975).

[13] Kuo, H.-H.: The Fourier transform in white noise calculus, Bielefeld Encounters in Mathematics and Physics VII, Editted by T.Hida & K.Sito, (1989), 19 pages.

[14] Lee, Y.-J. Generalization on infinite dimensional spaces and its Application to white noise calculus, J. Functional Analysis 82, (1988), 429-464.

[15] Lee, Y.-J. A reformation of white noise calculus, Lecture at White Noise Analysis-Mathematics and Physics (held at ZiF in University of Bielefeld), (1989).

[16] Pottohoff, J. & Yan, J.-a. : Some results about test and generalized functional of white noise, (preprint), (1989).

[17] Takenaka, S.: Invitation to white noise calculus. Theory and application of random fields, Proc. IFIP-WG 7/1 Working Conf. Lecture Notes in Control and Information Sciences, Springer, (1982), 249-257.

[18] Yokoi, Y.: Positivity of generalized Brownian functionals. (submitted to Hiroshima Math. J.)

MULTIPLE WIENER INTEGRALS ON ABSTRACT WIENER SPACES AND LIFTINGS OF p-LINEAR FORMS

G.W. JOHNSON
Dept. of Mathematics and Statistics, University of Nebraska
Lincoln, NE 68588, USA

and

G. KALLIANPUR
Dept. of Statistics, University of North Carolina
Chapel Hill, NC 27599-3260 USA

ABSTRACT

The theory of lifting maps and accessible random variables is used to derive formulas relating liftings of p-linear forms on H to multiple Wiener Itô integrals on an abstract Wiener space (H,B,P).

1. Introduction

In recent work which is yet to be published in its entirety, we have attempted to make a systematic and rigorous investigation of some of the fruitful ideas and suggestions put forward by Hu and Meyer in a fascinating recent paper [1,4]. A major motivation for [1], it seems to us, is to provide a distributional approach to Feynman integrals. _En route_ to their main goal, the authors of [1] have raised questions which are important for their own sake and which form the major concern of [4].

Let F(x) be a square integrable Wiener functional defined for $x \in \Omega_1$ where Ω_1 is the support of the standard Wiener measure in $C_0(\mathbb{R}_+)$. A natural problem that arises in defining the Feynman integral is one of extending the definition of F in a natural way to Ω_σ, the support of the Wiener measure with variance parameter σ^2 $(\sigma\neq 1)$. In other words, the problem is to define $F_\sigma(Y)$ for $y \in \Omega_\sigma$ such that F_σ is the "same" random variable on Ω_σ that F is on Ω_1. It has been suggested in [1] that it is perhaps more natural to consider the following question: The p-linear form on $L^2(\mathbb{R}_+)$,

$$I_p^o(m) := \int\ldots\int f_p(t_1,\ldots,t_p)m(t_1)\ldots m(t_p)dt_1\ldots dt_p$$

where $f_p \in L^2(\mathbb{R}_+^p)$ is symmetric, may be regarded as a multiple integral on the Cameron-Martin space Ω_0. The problem is to find a natural way to extend $I_p^o(m)$ to Ω_σ for all $\sigma>0$. The circle of ideas connected with the extension problems of Hu and Meyer is closely linked to an infinite dimensional calculus on Hilbert spaces developed, with quite a different purpose in mind in [6]. Since I_p^o is not a random variable in the usual sense, the above question of extension is naturally resolved by using the theory of accessible random variables of the above quoted monograph. In order to do this it has also been necessary to give the various definitions of traces of symmetric kernels in $L^2(\mathbb{R}_+)$ and examine their interrelationships. In [4] which contains a detailed exposition of scale invariant liftings leading to the Hu-Meyer definition of the Feynman integral based on the techniques of [6], some of the main results are also extended to abstract Wiener spaces. The full generalization of the results will be undertaken in a later paper. (It may be mentioned in this connection that an alternative approach to Feynman distributions on abstract Wiener spaces will be presented in a forthcoming paper by Ustunel and the second author [7]). In the present article, however, we confine ourselves to outlining the derivation of the main results of [4] for abstract Wiener spaces. The latter setting is of interest since it applies to a wide class of Gaussian processes including multiparameter Wiener processes and more general Gaussian random fields.

Section 2 introduces definitions and properties of four kinds of traces of elements f_p belonging to the p-fold symmetric tensor product of H. In Section 3, the basic definitions of the finitely additive theory needed for our purposes are presented. Section 4 contains the principal results.

Let (H,B,P) be an AWS. We then have the triplet

$$B^* \subset H^* = H \subset B$$

where $*$ denotes the dual and H^* is identified with H. Suppose that P is a centered Gaussian measure on $\mathcal{B}(B)$, the σ-field of Borel sets of B with covariance

$$E(\langle y,x\rangle\langle y',x\rangle) = (y,y'), \quad y,y' \in B^* \subset H^* = H.$$

Here $\langle\ ,\ \rangle$ is the (B^*,B)-canonical form and $(\ ,\ )$ denotes the inner product in H.

The norm on B restricted to H is a measurable norm in the sense of Gross [6]. Furthermore if γ denotes the injection of H into B and μ is the finitely additive Gauss measure on H, a basic result of Gross says that the induced (finitely additive) measure $\mu.\gamma^{-1}$ has a countably additive extension to $\mathcal{B}(B)$ which coincides with P. It is clear that for a given H, one obtains different abstract Wiener spaces (H,B,P), each AWS corresponding to a particular choice of measurable norm. From this point of view, it is the Hilbert space H that is central to the infinite dimensional calculus developed on an AWS.

When $B = C_o[0,1]$, the Banach space with sup norm of continuous real valued functions vanishing at 0 and P is the usual standard Wiener measure (H,B,P) is an AWS if $H = \{F: [0,1] \longrightarrow \mathbb{R}^1, F(\cdot) = \int_0^{(\cdot)} f(s)ds, f \in L^2[0,1]\}$ with $|F|_H = |f|_{L^2}$.

2. Hilbert Space-Valued Traces.

Let H be a real, separable Hilbert space and let $f_p \in H^{\hat{\otimes}p}$, the p-fold symmetric tensor product of H and let $0 \le k \le [p/2]$. We need to define the notion of a k-trace of f_p. There are various ways of doing so and we begin with the first (and possibly, the most natural) of these.

For any CONS (e_i) in $H^{\otimes k}$, denote by $f_p[e_i \otimes e_i]$ the unique element in $H^{\hat{\otimes}p-2k}$ such that $(f_p[e_i \otimes e_i], g) = (f_p, e_i \otimes e_i \otimes g)$ for all $g \in H^{\hat{\otimes}p-2k}$.

<u>Definition</u>. For $k \ge 1$, we say that $Tr^k f_p$ exists and equals $h \in H^{\hat{\otimes}p-2k}$ if for every CONS (e_i) in $H^{\otimes k}$, the series $\Sigma_{i=1}^{\infty} f_p[e_i \otimes e_i]$ converges to h in $H^{\hat{\otimes}p-2k}$. We define $Tr^o f_p = f_p$. The definition of the tensorial trace $Tr_t^k f_p$ is given in the same way as above except that (e_i) is any enumeration of the tensorial complete orthonormal system (CONS) $\{\varphi_{i_1}^{(1)} \otimes \ldots \otimes \varphi_{i_k}^{(k)}\}$ in $H^{\otimes k}$ where $\{\varphi_{i_1}^{(1)}\}, \ldots \{\varphi_{i_k}^{(k)}\}$ are arbitrarily chosen CONS's in H. Obviously, the existence of $Tr^k f_p$ implies that of $Tr_t^k f$ and their equality.

<u>Proposition 2.1</u> Suppose $Tr_t^k f_p$ exists. If $\{\varphi_{i_1}^{(1)}\}, \ldots, \{\varphi_{i_k}^{(k)}\}$ are any CONS's in H, then $Tr_t^k f_p$ can be expressed as the norm convergent multiple series

$$Tr_t^k f_p = \sum_{i_1, \ldots, i_k = 1} f_p[(\varphi_{i_1}^{(1)} \otimes \ldots \otimes \varphi_{i_k}^{(k)}) \otimes (\varphi_{i_1}^{(1)} \otimes \ldots \otimes \varphi_{i_k}^{(k)})]$$

or as an iterated series where the iteration can be done in any desired order.

Finally, we turn to the definition of the limiting k trace of f_p denoted by $\overrightarrow{Tr}^k f_p$. If f_p has a finite expansion in terms of some CONS $\{\varphi_{i_1}\otimes\ldots\otimes\varphi_{i_p}\}$ in $H^{\hat{\otimes}p}$,

$$f_p = \sum_{i_1,\ldots,i_p=1}^{N} a_{i_1,\ldots,i_p}\phi_{i_1}\otimes\ldots\otimes\phi_{i_p} \tag{1}$$

then $Tr^k f_p \in H^{\hat{\otimes}p-2k}$ exists and is given by

$$\sum_{i_{2k+1},\ldots,i_p=1}^{N}\Big(\sum_{j_1,\ldots,j_k=1}^{N} a_{j_1,j_1,\ldots,j_k,j_k,i_{2k+1},\ldots,i_p}\Big)\phi_{i_{2k+1}}\otimes\ldots\otimes\phi_{i_p}$$

$$= \sum_{j_1,\ldots,j_k=1}^{N} f_p[(\varphi_{j_1}\otimes\ldots\otimes\phi_{j_k})\otimes(\phi_{j_1}\otimes\ldots\otimes\phi_{j_k})].$$

Now suppose $f_p \in H^{\hat{\otimes}p}$ is arbitrary and given any CONS (ϕ_i) in H let $f^N_{p,(\phi_i)}$ have the expansion (1) where $a_{i_1,\ldots,i_p} = (f_p,\phi_{i_1}\otimes\ldots\otimes\phi_{i_p})$. For any k, $1\leq k\leq[p/2]$ suppose there exists $h\in H^{\hat{\otimes}p-2k}$ such that

$$\|Tr^k f^N_{p,(\phi_i)} - h\| \longrightarrow 0 \quad \text{as } N\longrightarrow\infty$$

for every CONS (ϕ_i) in H. We then define $\overrightarrow{Tr}^k f_p = h$. $\overrightarrow{Tr}^0 f_p = f_p$ by definition. When $\overrightarrow{Tr}^k f_p$ exists it can be shown that

$$\overrightarrow{Tr}^k f_p \tag{2}$$

$$= \sum_{i_{2k+1},\ldots,i_p=1}^{\infty}\Big(\sum_{j_1,\ldots,j_k=1}^{\infty} a_{j_1,j_1,\ldots,j_k,j_k,i_{2k+1},\ldots,i_p}\Big)\phi_{i_{2k+1}}\otimes\ldots\otimes\phi_{i_p}$$

where the inner sum in (2) is the limit as $N \longrightarrow \infty$ of $\sum_{j_1,\ldots,j_k=1}^{N} a_{j_1,j_1,\ldots,j_k,j_k,i_{2k+1},\ldots,i_p}$ and the outer sum is the limit in $H^{\otimes p-2k}$ norm.

The definition of the iterated k-trace, $Tr_i^k f_p$ is given as follows: $Tr_i^o f_p := f_p$ and $Tr_i^1 f := Tr^1 f_p$ provided the latter exists. Define $Tr_i^2 f_p = Tr^1(Tr^1 f_p)$ if the RHS quantity exists. This procedure is continued inductively and for $1 \leq k \leq [p/2]$.

The k-traces $Tr^k f_p$ so far defined for $k=0,\ldots,[p/2]$ may be called first order traces. For $\nu=0,1,\ldots,[(p-2k)/2]$, the ν-trace $Tr^\nu[Tr^k f_p]$ may exist. If all of these traces exist we say that f_p has all of its second order traces. The latter are said to be consistent with the first order traces if $Tr^\nu[Tr^k f_p] = Tr^{k+\nu} f_p$ for $k=0,1,\ldots,[p/2]$ and $\nu=0,1,\ldots,[(p-2k)/2]$.

The interrelationships among the various traces are described in the following result.

Proposition 2.2.

Assume that $Tr^k f_p$ exists for every k, $1 \leq k \leq [p/2]$. Then

(a) $Tr_t^k f_p$, $Tr_i^k f_p$ and $\overrightarrow{Tr}^k f_p$ exist for all k and $\overrightarrow{Tr}^k f_p = Tr_i^k f_p = Tr_t^k f_p = Tr^k f_p$. Further if f_p possesses all its second order traces, then

(b) $Tr^k[Tr^\nu f_p] = Tr^{k+\nu} f_p$ for all k,ν such that $0 \leq k+\nu \leq [p/2]$.

3. Accessible Random Variables on H.

As explained in the Introduction, we are interested in defining random variables on H regarded as a finitely additive (f.a.), Gauss measure space. The general definitions and terminology given below are taken, with slight changes, from [6] where the reader will find further details.

Let H be a real infinite dimensional, separable Hilbert space and $\mathscr{B}(H)$, the σ-field of Borel sets in H. Let $\mathscr{C}$ be the field of finite dimensional Borel cylinder sets and μ the f.a. canonical Gauss measure on H, i.e., the measure with characteristic functional $e^{-\frac{1}{2}\|h\|^2}$, $(h \in H)$. $(H,\mathscr{C},\mu)$ is our f.a. canonical Hilbert space.

Let $\mathscr{P}$ denote the class of all orthogonal projections on H with finite dimensional ranges. For $\Pi \in \mathscr{P}$ let $\mathscr{C}_\Pi := \{\Pi B^{-1}$: B a Borel set in the range of $\Pi\}$. Then the restriction of μ to $\mathscr{C}_\Pi$ is σ-additive and $\mathscr{C} = \bigcup_{\Pi\in\mathscr{P}} \mathscr{C}_\Pi$.

A representation of μ is a pair (L,P) where P is a σ-additive probability measure on some measurable space $(\Omega,\mathscr{A})$ and L is a mapping from H into the space of real valued Borel measurable functions on $(\Omega,\mathscr{A})$ such that for all $C \in \mathscr{C}$,

$$\mu(C) = P\{\omega: (L(h_1)(\omega),\ldots,L(h_j)(\omega)) \in B\}$$

where

$$C = \{h \in H: ((h,h_1),\ldots,(h,h_j)) \in B\} \quad \text{for } h_1,\ldots,h_j \in H$$

and B is a Borel set in $\mathbb{R}^j$.

L (or rather, an equivalence class) is also called a weak distribution. $(\Omega,\mathscr{A},P)$ is called the representation space of μ corresponding to L. In [6] it is shown that a representation of μ always exists. For our present purposes we take the probability space $(B,\mathscr{B}(B),P)$ as a representation space for $(H,\mathscr{C},\mu)$. That this can be done is shown in [6]. In [6] a representation of μ is given by (L,P) where for h∈H,

$$L(h)(x) := \sum_{j=1}^{\infty} (h,e_j)f_j[x], \quad (x\in B)$$

if the series converges in B and =0 otherwise. $(e_j)\subset\gamma^*(B^*)$ is a fixed CONS in H, given by $f_j\in B^*$ and $e_j = \gamma^*(f_j)$ where γ^* is the adjoint of the imbedding γ of H into B.

Let S be a separable Hilbert or Banach space with norm $|\cdot|$. A function f: H→S is a Borel cylinder function if it can be written as

$$f(h) = f_1((h,h_1),\ldots,(h,h_k)) \text{ for some } h_1,\ldots,h_k \in H$$

where $f_1: \mathbb{R}^k \to S$ is measurable w.r.t. the Borel σ-fields $\mathscr{B}(\mathbb{R}^k)$ and $\mathscr{B}(S)$. Define the <u>lifting</u> of f to be

$$Rf := f_1(L(h_1),\ldots,L(h_k)).$$

Let $\mathscr{L}^o(H,\mu;S)$ be the class of functions f: H→S such that for all $\Pi \in \mathscr{P}$, $f\circ\Pi(h)$:= is a Borel cylinder function and for all sequences $\{\Pi^N\} \subset \mathscr{P}$, converging strongly to the identity I ($\Pi^N \uparrow I$), $\{R(f\circ\Pi^N)\}$ is a $|\cdot|$- Cauchy sequence in P-probability. Then R(f) := lim in P-prob. of $R(f\circ\Pi^N)$ is called the S-lifting of f. Thus $f \in \mathscr{L}^o(h,\mu;S)$ if and only if f has an S-lifting. If $S=\mathbb{R}^1$ we drop the qualification S and simply write $\mathscr{L}^o(H,\mu)$. For $q\geq 1$ let $\mathscr{L}^q(H,\mu;S)=\{f \in \mathscr{L}^o(h,\mu;S)$: $E_P|R(f\circ\Pi^N)-R(f\circ\Pi^{N'})|^q \to \infty$ as N, N' → ∞. Note that if $f \in \mathscr{L}^q(H,\mu;H)$ then $E_P\|R(f)\|^q < \infty$. Since we will be dealing only with $\mathscr{L}^2$-liftings we make the following

<u>Definition:</u> f: H → S is an <u>accessible</u> r.v. if $f \in \mathscr{L}^2(H,\mu;S)$.

4. p-Linear Forms on H and their Liftings to an AWS.

We begin with some definitions and auxiliary results leading up to it. The following lemma is an immediate consequence of the Riesz representation theorem for Hilbert spaces.

Lemma 4.1. Let $F_m \in H^{\otimes m}$ and $g_k \in H^{\otimes k}$, $k \leq m$. Then there is a unique element $F_m^{g_k} \in H^{\otimes m-k}$ such that

$$(F_m, g_k \otimes h)_{H^{\otimes m}} = (F_m^{g_k}, h)_{H^{\otimes m-k}} \tag{1}$$

for all $h \in H^{\otimes m-k}$. If k=m, $F_m^{g_m} = (F_m, g_m)_{H^{\otimes m}}$.

Let $\mathcal{F}(B)$ be the completion with respect to P of the Borel σ-field $\mathcal{B}(B)$. If K is any separable, real Hilbert space, denote by $L^2(K)$ the Hilbert space of K-valued, $\mathcal{F}(B)$-measurable random variables on B. The inner product and norm in $L^2(K)$ will be denoted by

$$\langle \xi, \eta \rangle = E(\xi, \eta),\ \|\xi\| = (E|\xi|^2)^{1/2} .$$

The inner product (,) and norm $|\cdot|$ refer to K for various choices of K. For $K = \mathbb{R}^1$, we write L^2 for $L^2(\mathbb{R}^1)$ and then $\langle \xi, \eta \rangle = E(\xi\eta)$. Every $\mu \in L^2$ has the well known chaos (or Wiener chaos) expansion

$$u = \sum_{m=0}^{\infty} u_m = \sum_{m=0}^{\infty} J_m(F_m) \tag{2}$$

where $F_m \in H^{\otimes m}$ and $u_m = J_m(F_m)$ is the mth order chaos expansion in terms of Fourier-Hermite functionals (see [5]). (2) may also be expressed as

$$L^2 = \sum_{m=0}^{\infty} \oplus \mathcal{H}_m \tag{2'}$$

the RHS being an orthogonal decomposition and $\mathcal{H}_m$, the mth order homogeneous chaos. The relation $\mathcal{H}_m \longrightarrow H^{\otimes m}$ is an isometry given by $u_m \longrightarrow \tilde{F}_m$ where $\tilde{F}_m = \mathcal{P}_m F_m$ and $\mathcal{P}_m$ is the orthogonal projection from $H^{\otimes m}$ onto $H^{\hat{\otimes} m}$. Furthermore we have

$$\langle J_m(F_m), J_n(F_n) \rangle = m!\, |\tilde{F}_m|^2 \ \text{ if } m=n, \text{ and } =0 \text{ if } m \neq n. \tag{3}$$

Definition. Let $u \in L^2$ be given by (2). For $g_k \in H^{\otimes k}$ define

$$D_{g_k} u := \sum_{m=k}^{\infty} \frac{m!}{(m-k)!} J_{m-k}(F_m^{g_k}) \tag{4}$$

provided the series on the R.H.S. converges in L^2.

Definition. The kth order Sobolev derivative of $u \in L^2$ is defined as

$$\nabla^k u := \sum_{j=1}^{\infty} (D_{e_j} u) e_j \tag{5}$$

for any CONS (e_j) in $H^{\otimes k}$ provided the RHS series converges in $L^2(H^{\otimes k})$ and the limit does not depend on the choice of (e_j). The domain of ∇^k is given by

$$\text{Dom } \nabla^k := \{u \in L^2: \sum_{m=k}^{\infty} \frac{(m!)^2}{(m-k)!} |F_m|^2 < \infty\}. \tag{6}$$

It is easily seen that Dom ∇^k is dense in L^2. For k=1, writing ∇ for ∇^1 from (5) $\nabla u = \sum_{j=1}^{\infty} (D_{e_j} u) e_j$ and hence, if $h \in H$, it is easy to see that

$$(\nabla u, h) = D_h u.$$

We now state two key results that are used in proving Theorems 1 and 2 below. The first result appears to be known while the second is an extension to abstract Wiener spaces of Itô's decomposition formula [2].

Definition. The divergence operator $\delta^k : H^{\otimes k} \longrightarrow L^2$ is defined by

$$\langle \nabla^k u, f_k \rangle = \langle u, \delta^k f_k \rangle$$

for $u \in \text{Dom}\nabla^k$ and $f_k \in H^{\otimes k}$. Let us denote by $J_k(f_k)$, the homogeneous chaos expansion in $\mathcal{H}_k$.

Proposition 4.1. For $f_k \in H^{\otimes k}$,

$$\delta^k f_k = J_k(f_k).$$

The proof follows from a simple verification that $\langle \nabla^k u, f_k \rangle$ where

$u = \sum_{m=0}^{\infty} J_m(f_m)$, equals $\langle u, J_k(f_k)\rangle$.

Proposition 4.2. (Itô decomposition formula for abstract Wiener spaces).

Let (φ_i) be an arbitrary CONS in H and $p\geq 2$. Then

$$\delta^p(\varphi_{i_1}\otimes\ldots\otimes\varphi_{i_p}) = \delta^{p-1}(\varphi_{i_1}\otimes\ldots\otimes\varphi_{i_p})\delta(\varphi_{i_p})$$

$$- \sum_{\ell=1}^{p-1} \delta^{p-2}(\varphi_{i_1}\otimes\ldots\hat{\otimes}\ldots\otimes\varphi_{i_{p-1}}(\varphi_{i_\ell},\varphi_{i_p})$$

where $\hat{\otimes}_{i_\ell}$ means that φ_{i_ℓ} is to be omitted.

Note that $\delta(h)$ is written for $\delta^1(h)$ and is the same as $L(h)$.

We are now in a position to state our main results. We assume, as above, that $f_p \in H^{\hat{\otimes}p}$.

Theorem 1. $R\psi_p(f_p)$ exists if and only if $\overrightarrow{Tr}^k f_p$ exists for $0\leq k\leq[p/2]$. When $R\psi_p(f_p)$ exists it is given by

$$R\psi_p(f_p) = \sum_{k=0}^{[p/2]} \frac{p!}{(p-2k)!2^k k!}\,\delta^{p-2k}(\overrightarrow{Tr}^k f_p). \tag{8}$$

Theorem 2. Let $\psi_p(f_p)$ be the p-linear form on H associated with f_p. Suppose that f_p has all its first and second order traces and that they are consistent. Then

$$\delta^p f_p = \sum_{k=0}^{[p/2]} (-1)^k \frac{p!}{(p-2k)!2^k k!} R[\psi_{p-2k}(\overrightarrow{Tr}^k f_p)] \tag{9}$$

where $R[\psi_{p-2k}(\overrightarrow{Tr}^k f_p)]$ is the $\mathcal{L}^2$-lifting of the $(p-2k)$-form ψ_{p-2k} associated with $\overrightarrow{Tr}^k f_p$ in $H^{\hat{\otimes}p-2k}$.

We give a brief indication of the proof of Theorem 1. Let $\{\Pi_N\}\in\mathcal{P}$ such that $\Pi_N \uparrow I$ strongly in H and $\dim(\Pi_N H) = d_N < \infty$. We can then obtain $\{\varphi_i\}$, a CONS in H such that $\{\varphi_i\}$ $(i=1,\ldots,d_N)$ is a CONS in $\Pi_N H$. For convenience, we set $d_N = N$. Then f_p has the expansion

$$f_p = \sum_{i_1 \cdots i_p} a_{i_1 \cdots i_p} \varphi_{i_1} \otimes \cdots \otimes \varphi_{i_p}$$

and

$$\psi_p(m) = (f_p, m^{\otimes p}) = \sum_{i_1 \cdots i_p} a_{i_1 \cdots i_p} (\varphi_{i_1}, m) \cdots (\varphi_{i_p}, m), \quad m \in H. \quad (10)$$

$$\psi_p \circ \Pi_N(m) = \psi_p(\Pi_N m) = \sum_{i_1, \ldots, i_p} a_{i_1, \ldots, i_p} (\varphi_{i_1}, \Pi_N m) \cdots (\varphi_{i_p}, \Pi_N m)$$

$$= \sum_{i_1, \ldots, i_p} (\Pi_N \varphi_{i_1}, m) \cdots (\Pi_N \varphi_{i_p}, m)$$

$$= \sum_{i_1, \ldots, i_p = 1}^{N} a_{i_1 \cdots i_p} (\varphi_{i_1}, m) \cdots (\varphi_{i_p}, m)$$

$$= (f_p^N, m^{\otimes p})$$

where f_p^N has the finite expansion $\sum_{i_1, \ldots, i_p = 1}^{N} a_{i_1, \ldots, i_p} \varphi_{i_1} \otimes \cdots \otimes \varphi_{i_p}$.

Recall that the abstract Wiener space $(B, \mathcal{F}(B), P)$ is our representation space with the corresponding lifting $R(\varphi_i) = L(\varphi_i)$. Since the lifting map R is a homomorphism (see [6]) it follows from (10) that $R(\psi_p \circ \Pi_N)$ exists and

$$R(\psi_p \circ \Pi_N) = \sum_{i_1, \ldots, i_p = 1}^{N} a_{i_1 \cdots i_p} L(\varphi_{i_1}) \ldots L(\varphi_{i_p}). \quad (11)$$

The right hand side of (11) can be reduced to the expression

$$\sum_{0 \leq k \leq [p/2]} \frac{p!}{(p-2k)! 2^k k!} J_{p-2k} (\overrightarrow{Tr}^k f_p^N) \quad (12)$$

by an argument which makes crucial use of Proposition 2 and induction on p. (See [4]).

First suppose that $\overrightarrow{Tr}^k f_p$ exists for all k. It is then enough to show that $\|\overrightarrow{Tr}^k f_p^N - \overrightarrow{Tr}^k f_p\|_{H^{\hat{\otimes} p-2k}} \to 0$ as $N \to \infty$. We omit the details.

Hence $E_P[R(\psi_p \circ \Pi_N) - \eta]^2 \to 0$ where η is the random variable defined by the right hand side of (8). Furthermore, it follows from the definition of $\overrightarrow{Tr}^k f_p$ that η does not depend on the choice of $\{\Pi_N\}$. It immediately follows that $R\psi_p$ exists and $\in \mathcal{L}^2(H,\mathcal{C},\mu)$ and is given by the formula (8).

Next, assume that $R\psi_p(f_p)$ exists. Then

$$E_P[R(\psi_p \circ \Pi_N) - R(\psi_p \circ \Pi_{N'})]^2 \to 0 \quad \text{as } N, N' \to \infty. \tag{13}$$

Using the formula (12) for $R(\psi_p \circ \Pi_N)$ and a similar one for $R(\psi_p \circ \Pi_{N'})$, (13) becomes

$$\sum_{0 \le k \le [p/2]} \left[\frac{p!}{(p-2k)!2^k k!} \right]^2 (p-2k)! \, \| \overrightarrow{\mathrm{Tr}}^{*k} f_p^N - \overrightarrow{\mathrm{Tr}}^{*k} f_p^{N'} \|^2 \to 0. \tag{14}$$

Hence $\| \overrightarrow{\mathrm{Tr}}^{*k} f_p^N - \overrightarrow{\mathrm{Tr}}^{*k} f_p^{N'} \| \to 0$ for $k = 0,1,\dots,[\frac{p}{2}]$. Let $h_p^k \in H^{\hat{\otimes} p-2k}$ be such that $\| \overrightarrow{\mathrm{Tr}}^{*k} f_p^N - h_p^k \| \to 0$. Hence $h_p^k = \overrightarrow{\mathrm{Tr}}^{*k} f_p$. (8) follows immediately.

Remarks.

(1) In view of Proposition 1 $\delta^p f_p$ can be identified with $J_p(f_p)$, the pth homogeneous chaos expansion of f_p. In the case of standard Wiener space, i.e. when $B = C_o(\mathbb{R}_+)$, the space of real continuous functions on $\mathbb{R}_+$ vanishing at 0 and P= standard Wiener measure $J_p(f_p)$ is the pth order multiple Wiener-Itô integral.

(2) In a recent paper of Sugita [8] the RHS of (8) is referred to as the pth order multiple Stratonovich integral and denoted by $\delta^{\circ p} f_p$. (9) is also given there with the lifting on the RHS of (9) replaced by $\delta^{\circ p-2k}(\mathrm{Trace}^k f_p)$. However, Sugita's traces are iterated k-traces whereas ours are limiting k-traces.

(3) Versions of Theorems 1 and 2 for the special case of standard Wiener space together with a discussion of traces of symmetric f_p belonging to $L^2(\mathbb{R}_+^p)$ have been given in [3]. Unfortunately, due to circumstances beyond the control of the authors, [3] contains several errors. The k-trace mentioned in Theorem 1 of that paper should be the limiting trace $\overrightarrow{\mathrm{Tr}}^{*k} f_p$. The assumptions of Theorem 2 of [3] should be the same as those made in Theorem 2 of the present article.

(4) One can derive analogous results by taking the representation space to be $(\mathcal{S}^*, P)$ where $S^* = S^*(\mathbb{R}^d)$ is the space of Schwartz distributions, P is Gaussian white noise measure and L is given by expression (4.54) of [6]. Such possibilities will be considered in a

later paper.

5. Acknowledgement

Research supported by the Air Force Office of Scientific Research Contract No. F49620 85C 0144.

6. References

1. Y.Z. Hu and P.A. Meyer, Chaos de Wiener et integrale de Feynman, Séminaire de Probabilités XXII, Université de Strasbourg, (1987) 51-71, *Lecture Notes in Math.*, **1321** (Springer-Verlag, Berlin, 1988).

2. K. Itô, Multiple Wiener integrals, *J. Math. Soc. Japan* **3** (1951) 157-169.

3. G.W. Johnson and G. Kallianpur, Some remarks on Hu and Meyer's paper and infinite dimensional calculus on finitely additive canonical Hilbert space, preprint, *Theory of Probab. Appl.* XXXIV (1989) 742-752.

4. G.W. Johnson and G. Kallianpur, Homogeneous chaos, p-forms, scaling and the Feynman integral, University of North Carolina Center for Stochastic Processes Technical Report No. 274, (Feb. 1990).

5. G. Kallianpur, Stochastic Filtering Theory, *Applications of Mathematics* **13** (1980) Springer-Verlag, Berlin.

6. G. Kallianpur and R.L. Karandikar, in *White Noise Theory of Prediction, Filtering and Smoothing*, Stochastic Monographs 3 (Gordon and Breach, New York, 1988).

7. G. Kallianpur and A.S. Ustunel, Distributions on abstract Wiener space and applications to the Feynman integral, to appear (1990).

8. H. Sugita, Hu-Meyer's multiple Stratonovich integral and essential continuity of multiple Wiener integral, Reprint.

WHITE - NOISE ANALYSIS AS A TOOL IN COMPUTING FEYNMAN INTEGRALS

DINKAR C. KHANDEKAR
Theoretical Physics Division
Bhabha Atomic Research Centre
Bombay, 400 085 India

ABSTRACT

The paper discusses the use of white-noise calculus in explicit evaluation of Feynman Propagators. First, a brief account of Streit and Hida's definition of Feynman integral is given. Feynman propagators corresponding to both local and non-local quadratic action functionals are explicitely evaluated. A new expansion for the Feynman propagators is derived which is different from the conventional Dyson and perturbation expansions.

1. Introduction to White Noise Calculus

The basic probability space used in white noise calculus[1] is (S^*,β,μ) where S^* is the dual of Schwartz $S(\mathbb{R})$, β is the σ algebra generated by cylinder sets

$$A_{\xi_1,\ \xi_2,\ \ldots,\ \xi_n;\ B} = \left\{x \in S^* : \left(\langle x, \xi_1\rangle, \langle x, \xi_2\rangle, \ldots, \langle x, \xi_n\rangle\right) \in B\right\}$$

where $(\langle x,\xi_i\rangle$ represents the value of the distribution x on the test function $\xi_i \in S(\mathbb{R})$ and B is a Borel set in $\mathbb{R}^n$. The probability measure (white noise measure) μ on B is defined through Bochner- Milnos [2] theorem by the requirement

$$\int_{S^*} \exp\left\{i\langle x, \xi\rangle\right\} d\mu(x) = \exp\left(-1/2\ \|\xi\|^2\right) \equiv C(\xi) \tag{2}$$

where, $\|\xi\|$ is the L^2 $(\mathbb{R})$ norm of ξ.

Next one defines the set of random variables $\{b_\xi : \xi \in S\}$ defined through the map

$$b_\xi : x \in S^* \longrightarrow b_\xi(x) = \langle x, \xi \rangle \quad (3)$$

If $\xi_1, \xi_2, \ldots, \xi_n$ is a sequence of elements of S converging in $L^2(\mathbb{R})$ to some f, the sequence of random variables $\langle x, \xi_n \rangle$ converges in L^2 (S^*, μ) to a random variable φ . This random variable is called a functional of white noise or Brownian Motion.

An integral representation for φ is obtained via *F* transform of φ defined by its action on a test function $\xi \in S$:

$$(f\ \varphi)\ (\xi) = \int_{S^*} \varphi(x)\ e^{i\langle x, \xi\rangle}\ d\mu(x) \quad (4)$$

Further, it can be shown that $F\ \varphi$ admits a unique expansion

$$(F\ \varphi)\ (\xi) = C(\xi) \sum_{n=0}^{\infty} i^n \int d^n \zeta\ F_n(\zeta_1, \ldots, \zeta_n)\ \xi(\zeta_1)\ \xi(\zeta_2) \ldots \xi(\zeta_n) \quad (5)$$

where F_n is a symmetric function of its arguments in $L^2(\mathbb{R}^n)$ with :

$$\|\varphi\|^2_{L^2(S^*(\mu)} = \sum_{n=0}^{\infty} n!\ \|F_n\|^2_{L^2(\mathbb{R}^n)} \quad (6)$$

with the above identification $(L^2) = L^2(S^*, \mu)$ is just the Bosonic-Fock space (n particle space). Thus φ is completely characterised by the set $\{F_n\}$. This characterisation serves as a very useful tool in practical computations . Next, let φ and ψ be two real elements of correspond respectively to kernals $\{F_n\}$, $\{G_n\}$ of S^*, we can easily verify

$$E(\varphi, \psi) = \sum_n n!\ \langle F_n, G_n \rangle_{L^2(\mathbb{R}^n)} \quad (7)$$

where, $E(\cdot)$ is the integral of $(\cdot)$ with respect to the white noise measure defined via (2). Moreover the right hand side of Eq.(7) may be meaningful even if F_n is highly singular s. t. (6) does not hold provided G_n compensates for this singularity. More generally $\langle \cdot \; \cdot \rangle_n$ can be interpreted as the canonical pairing between the extreme members of a Gel'fand triplet $(L^2)^+ \subset (L^2) \subset (L^2)^-$ in which the space of symmetric functions is imbedded. Thus the right hand side of Eq.(7) defines in fact the canonical pairing between the generalized function $\varphi \in (L^2)^-$ and the test function $\psi \in (L^2)^+$, for which with the usual abuse of notations $E(\varphi,\psi)$ is retained. This generalization is very significant as very soon we shall see that the Feynman propagator can be interpreted as a generalized functional of Brownian Motion.

It might be also interesting to note that expansion(5) is essentially a decomposition of φ in components each belonging to a given "n-particle space". Such a decomposition is routinely used by particle physicists to project the total wave-function on to a "n-particle space".

2. Feynman Integral

In 1948, Feynman[3] introduced the quantum mechanical propagator or the kernel of Schrödinger equation K(q", t"; q', t') as an integral over the space of functions taking two fixed values namely q(t")=q" and q(t')=q'. Physically K represents the transition amplitude for the system to move to a position $q'' \in \mathbb{R}^n$ at time $t'' \in \mathbb{R}$ if it was at $q' \in \mathbb{R}^n$ at an earlier time $t' < t''$. Formally, using intuitive arguments Feynman wrote

$$K(q'', t''; q', t') = \int \exp\left[i/h\, S\right] D\left[x(t)\right] \qquad t' < t < t'' \tag{8}$$

where, S is known as the action functional along trajectory q(t) and is given by the integral

$$S = \int_{t'}^{t''} L\,(q, \dot{q}\,, t)\; dt \tag{9}$$

L being the Lagrangian of the system and the integral is thought of as being over the paths q(t')=q', q(t")=q".

The representation in Eq.(8) has been very useful in several investigations[4]. Mathematically, however, it has not yet been possiable to give a proper definition for computing K. This is largely due to the fact that there is no flat measure in infinite dimensional space. There have been attempts to interpret Feynman Integral as a generalization of Fresnel Integral[5]. C. DeWitt Morette[6] has used probabilistic methods to give a rigorous definition of Feynman Integral which has been sbsequently used by Khandekar and Lawande[7] amongst several others to compute Feynman propagators associated with some dynamical systems.

In 1980, Streit[8] proposed a precise mathematical meaning to Feynman's intuitive arguments. Later in 1983, Streit and Hida[9] asserted that the integral in Eq.(8) can be interpreted as an expectation with respect to white noise measure, if the trajectroies q(t) are taken as the sample paths of Brownian motion superimposed on a well determined fixed trajectory taking the value q' at t=t'. More precisely these authors interpret as

$$q(t) = y(t) + \sqrt{h/m}\quad B(t) \tag{10}$$

where m is the mass of the particle, whose quantum dynamics is being studied and B(t) is Brownian motion. y(t) is a fixed trajectory s. t. y(t')=q'. For simplicity henceforth we shall take y(t) = q' and set h=m=1 by proper choice of units. Further, Streit and Hida assert that the Feynman propagator can be viewed as :

$$K(q'', q'\ T) = E\left(N \exp\left[i/2 \int_0^T \dot{q}^2(\zeta)\, d\zeta + 1/2 \int_0^T \dot{B}^2(\zeta)\, d\zeta\right] \delta(q(T) - q'') \exp\left(-i \int_0^T V(\zeta)\, d\zeta\right)\right)$$

$$\equiv E\left(I \exp\left(-i \int_0^T V(\zeta)\, d\zeta\right)\right) \tag{11}$$

$E(\cdot)$ being the expectation with respect to white noise measure. The first term in the exponent inside the expectation corresponds to the action functional associated with a free particle. The second term has been introduced to compensate for the formal density $\exp\left[-1/2 \int B^2(\zeta)\, d\zeta\right]$ of white noise measure μ with respect to the (ill defined) translationally invariant measure in path space considerd by Feynman using intuitive arguments. The δ function serves to pinch tranjectories at $q(T)=q''$ while the choice $y(t)=q'$ ensures the intial condition. The first term in the exponent and the argument of last exponential together make up the action functional for a dynamical system, V being the potential.

The main achievement of Streit and Hida[9] is to give precise matematical meaning to I (q'', q', T) as a generalized functional of Brownian motion. We will not go into any mathematical details about these generalized functionals, more about them can be found in a recent review by H. H: Kuo[10] and Hida[11] et al. With these remarks we proceed to compute the free particle propagator which would also provide us with an opportunity to gain insight into the defination of I.

2.1 Free Particle

We set V=0 in Eq.(8) and recall that for quadratic functionals of Brownian motion of the form $\langle x, K x\rangle \in (L^2)$, it is known[9] that

$$\left(F \exp\left[-1/2 \left\langle x, K x\right\rangle\right)\right) (\xi) = C(\xi) \left(\det(1+K)^{-1/2}\right)$$
$$\exp\left[1/2 \left\langle \xi, K(K+1)^{-1} \xi \right\rangle\right] \quad (12)$$

In case of a free particle, (V=0), it can be easily seen that

$$K(\zeta_1, \zeta_2) = -(1+i)\,\delta(\zeta_1 - \zeta_2) \qquad 0 < \zeta_1, \zeta_2 < T$$
$$= 0 \text{ otherwise} \quad (13)$$

However, K of Eq.(13) is not in trace class and hence for this kernel a multiplicative renormalization is needed, corresponding to the definition

$$F\left(N \exp\left(-1/2 \left\langle x, K x\right\rangle\right)\right)(\xi) = C(\xi) \exp\left[1/2 \left\langle \xi, K \ (K+1)^{-1} \xi\right\rangle\right] \quad (14)$$

Once this is done δ function appearing in I can be interpreted in the distributional sence by a fourier integral and we can write

$$E(I \cdot 1) = 1/2\pi \int_{\mathbb{R}} dp \exp(i(q''-q')p) \ (F(N \exp(-1/2 \langle x, Kx\rangle)(\xi) \quad (15a)$$

$$\text{with } \xi(\zeta) = p \qquad 0<\zeta<T$$
$$= 0 \text{ otherwise} \quad (15b)$$

which can be easily evaluated using (14) and leads to the familier expression for the free particle propagator. Thus we see that the unit functional does indeed serve as a test functional to the generalized functional I.

2·2 Quantum systems under the influence of potentials.

If V is not equal to zero , one has to indeed exmine as to when $\exp(-i\int V(\zeta)\, d\zeta)$ is an admissible test functional. However, if V is at best quadratic in q, Streit and Hida show that $\exp(-i\int V(\zeta)d\zeta)$ is an admissible test functional. The derivation of the quantum propagator is essentially along the similar steps as outlined for obtaining free particle propagator except one uses the result

$$\left(\mathcal{F}(\mathcal{N}\ \exp(-1/2 \langle x, K x\rangle) \exp(-1/2 \langle x, L x\rangle)\right)(\xi)$$
$$= \det (1+L (1+K)^{-1})^{-1/2} \exp(-1/2\ \xi, (1+K+L)^{-1}\ \xi) \qquad (16)$$

where K is as given in (13) and for the particular case of $V=1/2\ \omega^2\ q^2$, L is given by

$$L = \omega^2 (T - \zeta_1 \vee \zeta_2) \qquad (17)$$

The present author in collaboration with Diago de Falco[12] has extended the class of test functionals further to include non-local interactions of the form

$$\exp\left(-i \int_0^T V(\zeta)\, d\zeta\right) \equiv \exp\left(-i \int_0^T dt \int_0^T ds\ M(t,s)\ (q(t)-q(s))^2\right) \qquad (18)$$

The interactions of the type (18) very frequently occur while studying the quantum dynamics of a particle in random enviournment. For these interactions, it is very easy to see that

$$L(\zeta_1, \zeta_2) = 4i\left[\int_{\zeta_1 \vee \zeta_2}^{T} W^2(t)\, dt - \int_{\zeta_1}^{T} dt \int_{\zeta_2}^{T} ds\ M(t,s)\right] \qquad (19a)$$

with

$$W^2(t) = \int_0^T M(t,s)\, ds.$$

and the quantum propagator is given by E (I. exp (-iV)) which can be evaluated easily and reads as :

$$E\left(I. \exp\left(-i \int_0^T dt \int_0^T ds\ M(t,s)\right)\right) = \int dp\ \left(\mathcal{F}(\mathcal{N}\ \exp(-1/2 \langle x, K x\rangle)\right.$$
$$\left.\exp(-1/2\langle x, L x\rangle)\right)(\xi) \qquad (20a)$$

$$\text{with} \quad \xi = p \qquad o < \zeta < T$$
$$= 0 \qquad \text{otherwise} \qquad (20b)$$

which when computed explicitly reads as :

$$K(q'', q', T) = \left\{2\pi \det (1+L+K)^{-1}\right\}^{-1/2} B^{-1/2} \exp\left(-\frac{(q''-q')}{2B}\right) \qquad (21a)$$

where, the constant B is given by

$$B = \int_0^T \int_0^T A(\zeta_1, \zeta_2)\, d\zeta_1\, d\zeta_2 \tag{21b}$$

and the corresponding integral $A(\zeta_1, \zeta_2)$ is given by :

$$A\,(\zeta_1, \zeta_2) = (1+L+K)^{-1}\,(\zeta_1, \zeta_2) \tag{21c}$$

2·3 The Velocity Dependent potentials

In problems involving magnetic fields, the potential is usually expressed as $V=\dot{q} \cdot A(q)$ where **A** is a given function of $q \in \mathbb{R}^3$. For the special case when $A = H/2\,(-q_2, q_1, 0)$ de Falco and Khandekar[12] have shown that the quantum propagator can be related to Levy's Stochastic Area[13]. This fact has been further exploited by the present author, to study the polymer entanglements of uniformly distributed clusters.

3. A New Expansion For The propagator

The assertion

$$K\,(q'', q', T) = E\,(I \exp(-i \int_0^T V(\zeta) d\zeta))$$

by Streit and Hida can be exploited to obtain a very interesting expansion for the Feynman propagator which is quite different from the conventional Dyson and the perturbation expansions. In this paper we shall not give the full details of derivation but merely outline the details. First note that

$$(F(I))\,(\xi) = K\,(q'', q', T)\, C(\xi) * \exp\left[1-i/2\, \|\xi\|^2 + i/2T \left\{ \left(\int_0^T \xi(\zeta) d\zeta\right)^2 + 2\,(q'', q') \int_0^T \xi(\zeta) d\zeta \right\}\right] \tag{22a}$$

where, K_0 is free particle propagator

$$K_0\ (q'', q', T) = (1/\sqrt{2\pi i T}\ \exp\ (i/2T\ (q'', q')^2) \tag{22b}$$

Since $F(I)$ is explicitly known one can immediately expand it and write

$$F(I)\ (\xi) = C\ (\xi) \sum_{n=0}^{\infty} i^n \int_{\infty}^{\infty} d^n \zeta\ F_n\ (\zeta_1, \ldots, \zeta_n)\ \xi\ (\zeta_1), \ldots, \xi\ (\zeta_n) \tag{23a}$$

where, F_n apart from some multiplicative constant has the form

$$i^n F_n = \sum_{p=0}^{[n/2]} \left(\left(1/\sqrt{2iT}\right)^{n-2p} \left(1-i/2\right)^p / p!\ (n-2p)! \right) H_{n-2p}\left(-2(q''-q')/\sqrt{2iT}\right)$$

$$\delta(\zeta_1 - \zeta_2)\ \delta(\zeta_3 - \zeta_4), \ldots, \delta(\zeta_{2n-1} - \zeta_{2n}) \prod_{l=1}^{n-2p} \chi\ (\zeta_{2n+l}) \tag{23b}$$

where χ is the characteristic function namely

$$\begin{aligned} \chi\ (\zeta) &= 1 \qquad 0 < \zeta < T \\ &= 0 \qquad \text{otherwise} \end{aligned} \tag{23c}$$

Next let us consider

$$\left(F\left(\exp\left[-i\int_0^T V\ (q(\zeta)d\zeta)\right]\right)\right)(\xi) = \left(F \exp\left[-i\int_0^T V\ (q'+B(\zeta))d\zeta\right]\right)(\xi) \tag{24}$$

The F transform in Eq.(24) must admit an expansion

$$\left(F \exp\left[-i\int (q'+B(\zeta))\ d\zeta\right)(\zeta) = \sum_{n=0}^{\infty} i^n \int V_n\ (q', T, \zeta_1, \zeta_2, \ldots, \zeta_n)\ \xi\ (\zeta_1)$$

$$\xi\ (\zeta_2), \ldots, \xi\ (\zeta_n)\ d^n\zeta \tag{25}$$

And hence Eq.(7) implies

$$K\ (q'', q', T) = \sum_{n=0}^{\infty} n! \int d^n\zeta\ \ F_n\ (\zeta_1, \ldots, \zeta_n)\ V_n\left(q', T, \zeta_1, \zeta_2, \ldots, \zeta_3\right) \tag{26}$$

A very interesting feature of the expansion is the occuerence of Hermite polynomials singnifying a natural decompiosition in Fock-space. More over the arguments of Hermite polymenomials contain factors proportional to $|q''-q'|/\sqrt{T}$, an immediate signature of Brownian motion. One can further do some more explict calculations for some specific choices of V. For the choice where V is at most quadratic (oscillator type) the expansion leads to familier expression for the propagator.

Acknowledgements :

The author is extremely greatful to Profs. T. Hida and L. Streit for very inspiring discussions. Most of the work reported in this paper was done in collaboration with Diago de Falco during our stay at Research Centre BiBoS.

4. References

1. T. Hida, Brownian Motion in Applications of Mathematics 11, (Springer Verlag, Berlin, 1980)
2. Ref 1. Page 116
3. R. P. Feynman, Rev. Mod. Phys. 20, 367 (1948)
4. D. C. Khandekar and S. V. Lawande Phys. Rep. 137, 115 (1986)
5. S. A. Albeverio, R. Hoegh-Krohn, Mathematical Theory of Feynman Integrals. Lect. in Maths 523, Springer Verlag 1976.
6. C. Dewitt Morette, Com. Math. Phys. 28, 47 (1972). Also see C. Dewitt Morette, A. Maheshwary and B. Nelson Phys. Rep. 50, 255 (1979)

7. D. C. Khandekar, Sunil Datta and S. V. Lawande Phys. Rev. 31A, 3574 (1985)

8. L. Streit in Functional Inteegration: Theory and Application p.b. 43-52 eds. J. P. Antoine and E. Tirapegui Plenum Press, New York (1980)

9. L. Streit and T. Hida Stochastic Processes and their Applications 16, 55 (1983)

10. H. H. Kuo, BiBoS, Universität Bielefeld preprint No 263/87

11. T. Hida, J. Potthoff and L. Streit BiBoS, Universität Bielefeld FRG preprint No 314/88

12. Diago de Falco and Dinkar C. Khandekar Stochastic Processes and Their Applications 29, 257 (1988)

13. P. Levy, Processes Stochastiques et Mouvement Brownien (Gauther Villars, Paris 1965)

14. D. C. Khandekar and F. W. Wiegel Polymer Entanglement: A Path Integral Approach (to appear)

DISTRIBUTIONS, GIRSANOV AND DEGREE THEOREMS ON WIENER SPACES

H. KOREZLIOGLU and A.S.USTUNEL
Ecole Nationale Supérieure des Télécommunications
46, rue Barrault 75634 Paris, Cedex 13, France.

ABSTRACT

Test functionals and distributions on abstract Wiener spaces are defined via the second quantizations of the powers of a self-adjoint positive operator. The probability densities and the degree theorem are expressed in terms of distributions corresponding to right translations.

INTRODUCTION

The construction of test functions and distributions on Wiener spaces by means of second quantizations has been considered in [2], [3], [4], [7], [8], [9],[13]. We develop here the approach we presented in [2] and [3], based on Watanabe's construction of Sobolev spaces $D_{p,k}$ as in [12]. Let (W, H, μ) be an abstract Wiener space where H is the Cameron-Martin space, A a positive self-adjoint operator having an inverse with norm less than one and $\Gamma(A^{\alpha/2})$ the second quantization of $A^{\alpha/2}$, $\alpha \in \mathbb{R}$. We construct the Sobolev space $D^{\alpha}_{p,k}$ by replacing the polynomials ψ by $\Gamma(A^{\alpha/2})\psi$, in the definition of the norm of $D_{p,k}$. We then define the space Φ of test functionals as the intersection of these Sobolev spaces $D^{\alpha}_{p,k}$, equipped with the projective limit topology and the space of distributions as the dual Φ' of Φ. We denote by H_∞ the intersection of $\mathrm{dom}(A^{\alpha})$, $\alpha \in \mathbb{R}$ and by H_α the completion of H_∞ under the Hilbertian norm defined by $|h|^2_\alpha = (A^{\alpha}h, h)_H$. We construct the Sobolev space $D^{\alpha}_{p,k}(H_\alpha)$ of H_α -valued random variables in the same way as $D^{\alpha}_{p,k}$ and define also the space of vector valued test functionals $\Phi(H_\infty)$ as the intersection of all $D^{\alpha}_{p,k}(H_\alpha)$, also equipped with the projective limit topology. In the case of $H = H_1(\mathbf{R}^d)$, the elements of $\Phi(H_\infty)$ can be considered as smooth random fields , where the smoothness is to be understood as that in $\omega \in W$ and in the parameter $x \in \mathbf{R}^d$.

The space D of Watanabe's test functionals is the intersection of all $D^0_{p,k} = D_{p,k}$'s implying that Watanabe's distributions are elements of Φ'. Our construction provides other spaces of test functionals D^α, between Φ and D, obtained by the sole intersection of $D^\alpha_{p,k}$'s for fixed α.

We have shown in [3] that Φ and $\Phi(H_\infty)$ are equal to the intersection of $\{D^\alpha_{2,0}, \alpha \in \mathbb{R}\}$ and $\{D^\alpha_{2,0}(H_\alpha), \alpha \in \mathbb{R}\}$, respectively , as a consequence of the hypercontractivity property of the Ornstein-Uhlenbeck semigroup. Although in the definition of Φ the projective limit can be taken only with $p=2$, $k=0$, the use of the Sobolev spaces $D^\alpha_{p,k}$ has its own importance, for, the regularity, represented by k, can be balanced by the integrability, represented by p.

In case A is a Hilbert-Schmidt operator with norm less than one, it can be shown that Φ is a nuclear Fréchet space. Kubo and Yokoi's construction in [4], also considered in [7], corresponds to this case in a more abstract form, whereas the approaches of [8], [9] and [13] are based on a particular choice of a Hilbert-Schmidt operator on $H = L^2(\mathbb{R})$ with $H_\infty = \mathcal{S}(\mathbb{R})$. Many results obtained in this frame have their more general versions with our approach, some of which are presented in [3] .

We develop here our approach of [2] and [3] for the construction of Φ and $\Phi(H_\infty)$ and consider, as an application, the study of the right translation on an abstract Wiener space with subsequent probability density functional [5], [10] and the related degree theorem (cf.[1] and [5]), expressed in terms of distributions of Φ'.

1. SPACES OF TEST FUNCTIONS AND DISTRIBUTIONS

Let (W, H, μ) be an abstract Wiener space : H is a separable Hilbert space, W is a Banach space into which H is densely and continuously injected and μ is the canonical Gaussian cylindrical measure on H extended to W as a Radon measure. We denote by $-L$ the infinitesimal generator of the Ornstein-Uhlenbeck process with values in W ([6], [12]). In what follows X represents a given separable Hilbert space.

For $p \in (1, \infty)$ and $k \in Z$, $D_{p,k}(X)$ is the completion of the vector space of X-valued polynomials on W, with respect to the (Sobolev) norm

$$(1.1) \qquad \|\zeta\|_{D_{p,k}(X)} = \|(I + L)^{k/2}\zeta\|_{L^p(\mu\,;\,X)}$$

where $L^p(\mu\,;X)$ is the L^p space of X-valued random variables on (W, H, μ). The Sobolev derivative, denoted by ∇, is a continuous linear mapping of $D_{p,k}(X)$ into $D_{p,k-1}(X \otimes H)$ for all $p \in (1, \infty)$ and $k \in Z$. The divergence operator δ, defined as the adjoint of ∇, is continuous from $D_{p,k}(X \otimes H)$ into $D_{p,k-1}(X)$. The intersection D(X) of all these spaces $D_{p,k}(X)$, equipped with the projective limit topology, is the space of X-valued (Watanabe) test functions on (W, H,μ), and its topological dual is called the space of (Watanabe) distributions. We recall the fact that, according to Meyer's inequalities [6], the norm (1.1) is equivalent to

$$(1.2) \qquad \sum_{i=0}^{k} \|\nabla^i \zeta\|_{L^p(\mu\,;\,X \otimes H^{\otimes i})}$$

In order to simplify the notations in the case of $X = \mathbb{R}$, we shall omit to write the inside of the parantheses in $D_{p,k}(X)$, D(X) etc.., and note simply $D_{p,k}$, D etc..

Let A be a self-adjoint, positive operator with domain in H and spectral support in $(1,\infty)$. Then $\|A^{-\alpha}\| < 1$ for all $\alpha > 0$,

$$H_\infty = \bigcap_{n>0} dom\,(A^n)$$

is dense in H and $\alpha \to (A^\alpha h, h)$ is increasing for all $h \in H_\infty$. For $\alpha \in \mathbb{R}$, H_α denotes the completion of H_∞ under the norm $|\,.\,|_\alpha$, defined by

$$(1.3) \qquad |h|_\alpha^2 = (A^\alpha h, h)_H$$

The dual of H_α is identified with $H_{-\alpha}$. The set $\{ H_\alpha \,;\, \alpha \in \mathbb{R} \}$ is a decreasing scale of Hilbert spaces such that, for $\beta < \alpha$, the injection of H_α into H_β is continuous and the set of norms $\{ |\,.\,|_\alpha , \alpha \in \mathbb{R} \}$ is compatible. H_∞, which coincides with the intersection of H_α 's, $\alpha \in \mathbb{R}$, is endowed with the projective limit topology.

Let ψ be a r.v. in $L^2(\mu)$ having the following Wiener chaos representation :

$$(1.4) \qquad \psi = E(\psi) + \sum_{n=1}^{\infty} I_n(\psi_n)$$

where ψ_n is an element of the nth symmetric tensor power of H, denoted here by $H^{\circ n}$ and I_n is the nth order multiple Wiener integral, defined as the nth order divergence $\delta^{(n)}$, the adjoint of ∇^n. For $\alpha \in \mathbb{R}$, the second quantization of A^α [11] is defined by

$$(1.5) \qquad \Gamma(A^\alpha)\,\psi = E(\psi) + \sum_{n=1}^{\infty} I_n((A^\alpha)^{\otimes n}\,\psi_n),$$

where $(A^\alpha)^{\otimes n}$ is the nth tensor product of A^α, for all ψ for which the series of the r.h.s. makes sense.

Let P be the space of polynomials : $\psi = p(\delta h_1, ..., \delta h_n)$, where p is any polynomial on $\mathbb{R}^n$, $n \geq 1$ and $h_i \in H_\infty$, $i = 1,..,n$. We denote by $P(X)$ the space of X-valued polynomials defined by

$$\psi = \sum_{i=1}^{m} \psi_i k_i$$

with $\psi_i \in P$ and $k_i \in X$.

DEFINITION 1.1: *For $p > 1$, $k \in Z$, $\alpha \in \mathbb{R}$, the (Sobolev) space $D^\alpha_{p,k}(X)$ is the space of X-valued random variables, obtained by the completion of P(X) with respect to the (Sobolev) norm*

$$(1.6) \qquad \|\psi\|_{D^\alpha_{p,k}(X)} = \|(I+L)^{k/2}\,\Gamma(A^{\alpha/2})\,\psi\|_{L^p(X;\mu)}$$

The differential operator ∇ is defined on $P(X)$ as usual [12]. It is continuously extended to $D^{\alpha}{}_{p,k}(X)$, according to the following lemma.

LEMMA 1.2:*(i) for all* $\psi \in P(X)$ *and* $\alpha \in \mathbb{R}$, *we have*

$$\nabla^i \Gamma(A^\alpha)\psi = (A^\alpha)^{\otimes i}\Gamma(A^\alpha)\nabla^i \psi \tag{1.7}$$

and the norm $\|\psi\|_{D^{\alpha}{}_{p,k}(X)}$ *is equivalent to*

$$\sum_{i=0}^{k} \|(A^{\alpha/2})^{\otimes i}\Gamma(A^{\alpha/2})\nabla^i \psi\|_{L^p(\mu;X\otimes H^{\otimes i})} \tag{1.8}$$

(ii) ∇ *extends to a continuous operator from* $D^{\alpha}{}_{p,k}(X)$ *into* $D^{\alpha}{}_{p,k-1}(X \otimes H_\alpha)$.

Proof : For $X = \mathbb{R}$, (1.7) is deduced from (1.4) for a finite sum and the continuity of ∇ is a consequence of its continuity on $D_{p,k}$, as shown below :

$$\|\nabla\psi\|_{D^{\alpha}_{p,k-1}(H_\alpha)} = \|\Gamma(A^{\frac{\alpha}{2}})\nabla\psi\|_{D_{p,k-1}(H_\alpha)} = \|A^{\frac{\alpha}{2}}\Gamma(A^{\frac{\alpha}{2}})\nabla\psi\|_{D_{p,k-1}(H)}$$

$$= \|\nabla(\Gamma(A^{\frac{\alpha}{2}})\psi)\|_{D_{p,k-1}(H)} \le C\|\Gamma(A^{\frac{\alpha}{2}})\psi)\|_{D_{p,k}} = C\|\psi\|_{D^{\alpha}_{p,k}},$$

for $\psi \in P$, where C is a positive constant. The proof for X-valued ψ is similar and (1.8) is a direct consequence of (1.2). □

LEMMA 1.3: *For* $\beta < \alpha, p < p', k < k'$ *we have*

$$\|\psi\|_{D^{\beta}_{p,k}(H_\beta \otimes X)} \le \|\psi\|_{D^{\alpha}_{p,k}(H_\alpha \otimes X)} \le \|\psi\|_{D^{\alpha}_{p',k'}(H_\alpha \otimes X)} \tag{1.9}$$

and the set of norms

$$\{\|\ \ \|_{D^{\alpha}_{p,k}(H_\alpha \otimes X)}; \alpha \in \mathbf{R}, p > 1, k \in Z\} \tag{1.10}$$

forms a compatible family. Therefore,

$$D^{\alpha}_{p',k'}(H_{\alpha}\otimes X)\subset D^{\alpha}_{p,k}(H_{\alpha}\otimes X)\subset D^{\beta}_{p,k}(H_{\beta}\otimes X)$$

and the corresponding injections are continuous.

Proof : For the sake of simplicity we give the proof for $X = \mathbb{R}$. Since for $\beta<\alpha$, $A^{(\beta-\alpha)/2}$ is a contraction, so is $\Gamma(A^{(\beta-\alpha)/2})$ on $L_p(\mu; H)$ ([11] , Theorem I.17). We have, for $\psi \in P$,

$$\|\psi\|^p_{D^{\beta}_{p,k}(H_{\beta})} = E\,|(I+L)^{\frac{k}{2}}\,\Gamma(A^{\frac{\beta}{2}})\,A^{\frac{\beta}{2}}\psi\,|^p_H < E\,|(I+L)^{\frac{k}{2}}\,\Gamma(A^{\frac{\beta-\alpha}{2}})\,\Gamma(A^{\frac{\alpha}{2}})\,A^{\frac{\alpha}{2}}\psi\,|^p_H$$

$$\leq E\,|(I+L)^{\frac{k}{2}}\,\Gamma(A^{\frac{\alpha}{2}})\,A^{\frac{\alpha}{2}}\phi\,|^p_H = \|\psi\|^p_{D^{\alpha}_{p,k}(H_{\alpha})}$$

The second inequality of (1.9) is already known [12]. The compatibility of the set of norms (1.10) is well known for fixed α. It is then enough to prove it for fixed (p,k). We consider again the case of $X = \mathbb{R}$. Let $\beta < \alpha$, let $\{\psi_n, n \in \mathbb{N}\} \subset P$ converge to 0 in $D^{\beta}_{p,k}(H_{\beta})$ and let it be a Cauchy sequence in $D^{\alpha}_{p,k}(H_{\alpha})$. Then

$$\{\Gamma(A^{\frac{\alpha}{2}})\,A^{\frac{\alpha}{2}}\psi_n = \Gamma(A^{\frac{\alpha-\beta}{2}})\,A^{\frac{\alpha-\beta}{2}}\,\Gamma(A^{\frac{\beta}{2}})\,A^{\frac{\beta}{2}}\psi_n\,;\, n \in \mathbb{N}\}$$

is a Cauchy sequence in $D_{p,k}(H)$. Hence

$$\{\Gamma(A^{\frac{\alpha-\beta}{2}})\,A^{\frac{\alpha-\beta}{2}}\,\psi_n\,;\, n \in \mathbb{N}\}$$

is a Cauchy sequence in $D^{\beta}_{p,k}(H_{\beta})$. Let ψ be its limit. Since, for $\gamma > 0$, $\Gamma(A^{-\gamma})A^{-\gamma}$ is continous in $D^{\beta}_{p,k}(H_{\beta})$, by putting $\gamma = (\alpha-\beta)/2$, we see that $\Gamma(A^{-(\alpha-\beta)/2})A^{-(\alpha-\beta)/2}\psi = 0$. Therefore, $\psi = 0$ and hence, $\|\psi_n\|_{D^{\alpha}_{p,k}} = \|\Gamma(A^{(\alpha-\beta)/2})A^{(\alpha-\beta)/2}\psi_n\|_{D^{\beta}_{p,k}} \to 0$. □

This lemma allows the definition of projective limit topologies on various intersections of the above spaces.

DEFINITION 1.4: *We put*

$$(1.11) \qquad D^{\alpha}(X) = \bigcap_{p,k} D^{\alpha}_{p,k}(X)$$

$$(1.12) \qquad \Phi(X) = \bigcap_{\alpha} D^{\alpha}(X)$$

$$(1.13) \qquad \Phi(H_{\infty}) = \bigcap_{\alpha,p,k} D^{\alpha}_{p,k}(H_{\alpha})$$

These spaces are spaces of test functions and their topological duals, denoted $D^{-\alpha}(X)$, $\Phi'(X)$ *and* $\Phi'(H_{-\infty})$, *respectively, are spaces of distributions. We note that the dual of* $D^{\alpha}{}_{p,k}(X)$ *is* $D^{-\alpha}{}_{q,-k}(X)$, *where* q *is the conjugate number of* p, *(we identify* X *with its dual), and we define the divergence operator* δ *as the adjont of* ∇ *on* $D^{\alpha}{}_{p,k}(X)$. *Then* δ *is a continuous operator from* $D^{\alpha}{}_{p,k}(X\otimes H_{\alpha})$ *into* $D^{\alpha}{}_{p,k-1}(X)$.

REMARK 1.5: For arbitrary reals α, α', β and β' the operator $T=\Gamma(A^{(\alpha-\beta)/2})(A^{(\alpha'-\beta')/2})^{\otimes i}$ on $P(X \otimes H^{\otimes i})$ can be extended to an isometry from $D^{\alpha}{}_{p,k}(X \otimes H_{\alpha'}{}^{\otimes i})$ into $D^{\beta}{}_{p,k}(X\otimes H_{\beta'}{}^{\otimes i})$, that we denote by the same symbol. For $\psi \in D^{\alpha}{}_{p,k}(X \otimes H_{\alpha'}{}^{\otimes i})$, we define $T\psi$ as the isometric image of ψ in $D^{\beta}{}_{p,k}(X\otimes H_{\beta'}{}^{\otimes i})$. We note that if $\alpha < \beta$ and $\alpha' < \beta'$, then T is continuous on $D^{\alpha}{}_{p,k}(X \otimes H_{\alpha'}{}^{\otimes i})$ and $T\psi$ is an element of this spacc.

According to the definition given here, for $\psi \in D^{\alpha}{}_{p,k}(X)$, the relation

$$(1.7') \qquad \nabla^{i}\,\Gamma(A^{\frac{\alpha-\beta}{2}})\,\psi = (A^{\frac{\alpha-\beta}{2}})^{\otimes i}\,\Gamma(A^{\frac{\alpha-\beta}{2}})\,\nabla^{i}\psi$$

is valid on $D^{\alpha}{}_{p,k-1}(X \otimes H_{\alpha}{}^{\otimes i})$.

The operators ∇ and δ are continously extended to the spaces of test functions and of distributions. We have in particular,

THEOREM 1.6: *∇ is continuous from $\Phi' = (\Phi'(\mathbf{R}))$ into $\Phi'(H_{-\infty})$ and δ is continuous from $\Phi(H_{-\infty})$ into Φ'. Moreover, (1.7) is valid on Φ' and we have, for all $\xi \in \Phi'(H_{-\infty})$,*

$$(1.14) \qquad \Gamma(A^a)\,\delta\,\xi = \delta(\Gamma(A^a)\,A^a\,\xi)$$

Proof : The continuity properties are immediate consequences of Lemma 1.1 together with the definition. The validity of (1.7) on Φ' is a consequence of Remark 1.5. We can write, for $\psi \in \Phi$,

$$\begin{aligned} &E[(\Gamma(A^a)\,\delta\,\xi).\,\psi] = E[(\delta\,\xi)(\Gamma(A^a)\;\psi] = \\ &= E[(\xi, \nabla(\Gamma(A^a)\psi))_H] = E[(\xi, A^a\Gamma(A^a)\,\nabla\,\psi)_H] \\ &= E[(A^a\,\xi, \Gamma(A^a)\,\nabla\,\psi)_H] = E[(\Gamma(A^a)\,A^a\,\xi, \nabla\,\psi)_H] \\ &= E[\delta(\Gamma(A^a)\,A^a\,\xi).\,\psi]\;. \qquad \square \end{aligned}$$

Let (W^a, H_a, μ_a) be an abstract Wiener space whose Cameron - Martin space is H_a. Let $L^{p,a}$ denote the space $L^p(W^a, \mu_a; X)$ and $D^{(a)}{}_{p,k}(X)$ the Sobolev space $D_{p,k}(X)$ constructed on (W^a, H_a, μ_a). Then we have the following

THEOREM 1.7: *There is an isometric isomorphism between $D^{(a)}{}_{p,k}(X)$ and $D^a{}_{p,k}(X)$.*

Proof : Let $\{h_n\,;\, n \in \mathbb{N}\} \subset H_\infty$ be an orthonormal basis in H_a and let δ^a be the divergence operator and $-L^a$ the Ornstein-Uhlenbeck operator on (W^a, H_a, μ_a). Then for any Borel function F on $\mathbf{R}^k$, the random variables $F(\delta^a h_{n_1},..,\delta^a h_{n_k})$ and $F(\delta h_{n_1},..,\delta h_{n_k})$ have the same distributions. Therefore,

$$\Big\| \sum_{n=0}^{K} I^a_n(\psi_n) \Big\|_{L^{p,a}} = \Big\| \sum_{n=0}^{K} I_n((A^{\frac{a}{2}})^{\otimes n}\psi_n) \Big\|_{L^p(\mu,X)} = \Big\| \Gamma(A^{\frac{a}{2}}) \sum_{n=0}^{K} I_n(\psi_n) \Big\|_{L^p(\mu,X)}$$

where $\psi_n \in H_\infty{}^{on}$ and $I^a{}_n$ is the nth order multiple Wiener integral on (W^a, H_a, μ_a). This defines an isometric isomorphism from $L^{p,a}$ onto $L^p(\mu,X)$ that we denote by $\Gamma'(A^{a/2})$. We finally have the following chain of isometric isomorphisms.

$$D^{(\alpha)}_{p,k}(X) \xrightarrow{(I+L^{\alpha})^{\frac{k}{2}}} L^{p,\alpha} \xrightarrow{\Gamma'(A^{\frac{\alpha}{2}})} L^{p}(\mu\,;X)$$

$$L^{p}(\mu\,;X) \xrightarrow{(I+L)^{\frac{-k}{2}}} D_{p,k}(X) \xrightarrow{\Gamma(A^{\frac{\alpha}{2}})} D^{\alpha}_{p,k}(X). \quad \square$$

In what follows we are going to show that, as for the Sobolev spaces on a finite dimensional Euclidean space, Φ can be written as $\cap_\alpha D^{\alpha}{}_{2,0}$. We get the result by means of a series of lemmas.

We first remind the hypercontractivity of the Ornstein-Uhlenbeck semigroup [12]. Let $M_t : H \to H$ be defined by : $M_t\, h = e^{t}h$, for $h \in H$ and $t \in \mathbb{R}$. Then, for $p > 1$, $t > 0$, $q(t) = e^{2t}(p-1)+1$ and $\psi \in L^{q(t)}(\mu)$, $\|\Gamma(M_{-t})\,\psi\|_{L^{q(t)}} \leq \|\psi\|_{L^p}$. In particular, for $q > 2$, there is a $t > 0$ such that $\|\Gamma(M_{-t})\,\psi\|_{L^q} \leq \|\psi\|_{L^2}$.

LEMMA 1.8 *: For any α and $t > 0$, there exists $\beta > \alpha$ such that $\Gamma(M_t) : D^{\beta}{}_{2,k} \to D^{\alpha}{}_{2,k}$ is a contraction.*

Proof : Since $\|A^{-1/2}\| < 1$, there is $n \in \mathbb{N}$ such that $\|M_t\, A^{-n/2}\| \leq e^{t}\|\, A^{-1/2}\|^{n} < 1$. Therefore, $\Gamma(M_t\, A^{-n/2})$ is a contraction on $L^p(\mu, H)$. Let $\beta = \alpha + n$. Then we can write

$$\|\Gamma(M_t)\psi\|_{D^{\alpha}_{2,k}} = \|\Gamma(A^{\frac{\alpha-\beta}{2}})\,\Gamma(M_t)\psi\|_{D^{\beta}_{2,k}} = \|\Gamma(M_t A^{-\frac{n}{2}})\psi\|_{D^{\beta}_{2,k}} \leq \|\psi\|_{D^{\beta}_{2,k}}. \quad \square$$

From this lemma and the hypercontractivity property we get the following.

COROLLARY 1.9: *For $p > 2$, $k \in Z$, $\alpha \in \mathbb{R}$ there is $\beta > \alpha$ such that $\|\psi\|_{D^{\alpha}_{p,k}} \leq \|\psi\|_{D^{\beta}_{2,k}}$, for $\psi \in \Phi$.*

A similar conclusion can be obtained for the Sobolev norms of X-valued random variables.

LEMMA 1.10: *For p >2, k ∈ Z, α ∈ ℝ, there exist some β >α and a positive constant K such that* $\| \xi \|_{p,\alpha} \le K \| \xi \|_{2,\beta}$, *for* $\xi \in \Phi(X)$, *where we denote* $\| \ \|_{D^\alpha_{p,k}(X)}$ *by* $\| \ \|_{p,\alpha}$.

Proof: Let $(e_i)_{i \in N}$ be a complete orthonormal sequence in X. Let $\xi \in \Phi(X)$. Then we have

$$\xi = \sum \xi_i e_i \quad and \quad \| \xi \|^p_{p,\alpha} = E[\sum ((I+L)^{\frac{k}{2}} \Gamma(A^{\frac{\alpha}{2}}) \xi_i)^2]^{\frac{p}{2}}$$

Let $(e_i)_{i \in N}$ be an independent sequence of random variables with values -1 and +1 and probability γ. The Khintchine inequalities [12] imply that there exist positive constants c_1 and c_2 such that

$$\| \xi \|^p_{p,\alpha} \le c_1 E_\gamma \| \sum \gamma_i \xi_i \|^p_{p,\alpha} \quad and \quad E_\gamma \| \sum \gamma_i \xi_i \|^2_{2,\beta} \le c_2 \| \xi \|^2_{2,\beta}.$$

Corollary 1.9 aplied to $\Sigma \gamma_i \xi_i$ gives the desired inequality. □

LEMMA 1.11: *For p >2, α ∈ ℝ, k ∈Z there exists some β' > α and a positive constant ρ such that*

$$\| \xi \|_{D^\alpha_{p,k}(X)} \le \rho \| \xi \|_{D^{\beta'}_{2,0}(X)}$$

for all $\xi \in \Phi(X)$. *In the case of X = ℝ, we have ρ=1.*

Proof: We consider the case of X=ℝ. The proof for arbitrary X can be done as in Lemma 1.9. Let β > α be as in Lemma 1.8. Then for ξ ∈ Φ and t > 0 such that $(1+n)^{k/2} \le e^{nt}$ for all n ∈ ℕ according to Lemma 1.8, there is β'>β such that

$$\|\xi\|_{D^\beta_{2,k}} = \|(I+L)^{\frac{k}{2}} \xi\|_{D^\beta_{2,0}} = \|(I+L)^{\frac{k}{2}} \Gamma(M_{-t}) \Gamma(M_t) \xi\|_{D^\beta_{2,0}} \le \|(I+L)^{\frac{k}{2}} \Gamma(M_{-t}) \xi\|_{D^{\beta'}_{2,0}} \le \|\xi\|_{D^{\beta'}_{2,0}}$$

where,for the last equality, we used the fact that $(I+L)^{k/2}\Gamma(M_{-t})$ is a contraction on $D^{\beta'}_{2,0}$. The desired inequality is then a consequence of Corollary 1.9. □

We conclude the above lemmas with the following result.

THEOREM 1.12: *We have*

$$\Phi(X) = \bigcap_{p,k,\alpha} D^{\alpha}_{p,k}(X) = \bigcap_{k,\alpha} D^{\alpha}_{2,k}(X) = \bigcap_{\alpha} D^{\alpha}_{2,0}(X)$$

and the corresponding projective limit topologies on $\Phi(X)$ are the same.

We note that the same conclusion holds for $\Phi(H_\infty)$, i.e.

THEOREM 1.13: *We have*

$$\Phi(H_\infty) = \bigcap_{p,k,\alpha} D^{\alpha}_{p,k}(H_\alpha) = \bigcap_{k,\alpha} D^{\alpha}_{2,k}(H_\alpha) = \bigcap_{\alpha} D^{\alpha}_{2,0}(H_\alpha),$$

and the corresponding projective limit topologies on $\Phi(H_\infty)$ coincide.

2. APPLICATIONS

Many interesting application are given in [3] and in a forthcoming paper by G. Kallianpur and A.S. Ustunel. We will only consider here the right translation on the Wiener space and present some of its properties. We first reproduce the following theorem, proved in [3].

THEOREM 2.1: *Suppose that $u \in D^0{}_{2,0}(H)$ satisfies the condition*

$$\sum_{n=0}^{\infty} \frac{1}{\sqrt{n!}} [E(|u|_H^{2n})]^{\frac{1}{2}} < \infty \qquad (2.1)$$

Then the mapping $\psi \to E[\psi(. + u(.))]$ of Φ into $\mathbb{R}$ defines a distribution $X(u) \in \Phi'$, characterized by the series

$$X(u) = \sum_{n=0}^{\infty} \frac{1}{n!} \delta^{(n)}(u^{\otimes n}), \qquad (2.2)$$

converging weakly uniformly in $D^{-\alpha}{}_{2,0}$ for any α for which $\| A^{-\alpha/2} \|^2 \leq 1/2$.

Right translations on the Wiener space have been considered by several authors, among which Ramer [10], Kusuoka [5]. Ramer showed that if the mapping $h \to u(\omega + h)$, $h \in H$, is Lipschitz-continuous on H for all $\omega \in W$ with Lipschitz constant less than 1/2, we have, for any bounded Borel function on W,

$$(2.3) \qquad \int_W f(\omega)\,\mu(d\omega) = \int_W f(\omega + u(\omega))\,|\,d_2(u)(\omega)\,|\,\mu(d\omega)$$

where,

$$(2.4) \qquad d_2(u) = det_2(I_H + \nabla u)\,exp\,(-\delta u - \frac{1}{2}\,|\,u\,|_H^2)$$

and $\det_2$ denotes the Carleman-Fredholm determinant.

As $(I + u)^{-1}$ is a map of the form I - v where v has the same properties as u (with Lipschitz constant not necessarily less than 1/2) (cf [10]), we have for any bounded cylindrical function on W

$$(2.5) \qquad E[F(\omega + u(\omega))] = < X(u), F > = E[F\,.\,|\,d_2(-v)\,|\,].$$

Hence $X(u) = |d_2(-v)|$ and symmetrically, $X(-v) = |d_2(u)|$.

Even if $I + u$ is not invertible, $d_2(u)$ may be well defined and in this case, from the degree theorem on the Wiener space (cf [1], [5]), for any nice cylindrical function f on W, we have

$$E\,|\,f((I + u)(\omega))\,d_2(u)(\omega)\,| = deg\,(I + u)\,E\,(f)$$

On the other hand, we deduce from (2.2)

$$E\,|\,fo\,(I + u)\,d_2(u)\,| = E\,|\,\sum_{n=0}^{\infty} \frac{(\nabla^n f, u^{\otimes n})}{n!}\,d_2(u)\,| =$$

$$= E\,|d_2(u)\,f + \sum_{n=1}^{\infty} \frac{(\nabla^n f, u^{\otimes n})}{n!}\,d_2(u)\,|$$

$$= E\,|d_2(u)\,f + \sum_{n=1}^{\infty} f\,\frac{\delta^n\,(u^{\otimes n}\,d_2(u))}{n!}\,]$$

for all $f \in \Phi$.

These relations together with (2.5) allow to conclude with the following theorem.

THEOREM 2.2: *Under, the above conditions on u, we have*

$$X(u) = | d_2(-v) | \quad and \quad X(-v) = | d_2(v) | ;$$

and moreover,

$$deg\,(I + u) = d_2(u) + \sum_{n=1}^{\infty} \frac{\delta^n (u^{\otimes n} d_2(u))}{n!} .$$

We would like to end this paper by pointing out the connection between our approach and I. Kubo and Y. Yokoi's in [4], where they consider a sequence of Hilbert spaces H_n (like H_α 's here) whose intersection H_∞ is a nuclear space. The construction of the space of test functionals is based on an isometry, between the Fock spaces on H_n 's and subspaces of random variables expressed by their chaos representations. The approaches in [8], [9] and [13] correspond to a particular choice of A, for instance $A = I + x^2/2 - 2(d^2/dx^2)$, acting on $H = L^2(\mathbb{R})$. In all these cases Φ and $\Phi(H_\infty)$ are nuclear Fréchet spaces. More generally, let A be an operator having the properties considered in Section 1 and suppose that, for some $\alpha_0 > 0$, $A^{-\alpha_0}$ is a Hilbert-Schmidt operator with Hilbert-Schmidt norm less than one. Let $\beta < \alpha$ and i denote the canonical injection of $D^\alpha_{2,0}$ into $D^\beta_{2,0}$. Then for $\psi \in D^\alpha_{2,0}$, we have

$$\| i(\psi) \|_{D^\beta_{2,0}} = \| \Gamma (A^{\frac{\beta-\alpha}{2}}) \psi \|_{D^\alpha_{2,0}} .$$

Using this relation, one can easily show that for $(\alpha - \beta)/2 > \alpha_0$, the injection i is a Hilbert-Schmidt operator. This implies that Φ is a nuclear Fréchet space and therefore, Φ' is a nuclear space under the strong topology. It can be similarly shown that the pair $(\Phi(H_\infty), \Phi'(H_{-\infty}))$ has the same nuclearity property. In this case, as an immediate application, it is proved in [3] (as it was also done by Kubo and Yokoi in [4]) that, for a.a. $\omega \in W$, the Dirac measure at ω is a distribution in Φ'.

REFERENCES

1. E. Getzler: Degree theory of Wiener maps. J. Funct. Anal. 68, p.388-403, 1986.

2. H. Korezlioglu and A.S. Ustunel:Construction de distributions sur l'espace de Wiener via la second quantification. C.R. Acad. Sci. Paris, t. 309, Série I, p.781-784, 1989.
3. H.Korezlioglu and A.S. Ustunel: A new class of distributions on Wiener spaces. To be published in "Stochastic Analysis and Related Topics II" Proceedings of the Second Silivri Workshop, Lect. Note in Math. , Springer-Verlag, 1990.
4. I. Kubo and Y. Yokoi: A remark on the space of testing random variables in the white noise calculus . Nagoya Math. J.,115, 1989.
5. S. Kusuoka: Degree Theorem in certain Wiener Riemanian manifolds. Stochastic Analysis, Lect. Notes in Math. , 1322, p. 93-108,Springer-Verlag, 1988.
6. P.A. Meyer : Transformation de Riesz pour les lois Gausiennes. Séminaire de Probabilités, XVIII, Lect. Notes in Math., 1059, p.179-193, Springer-Verlag, 1984.
7. P.A. Meyer and Yan, Jia-an: Distributions sur l'espace de Wiener (suite). Séminaire de Probabilités,XXVIII,Lect. Notes in Math. , 1372, p.382-392, Springer-Verlag, 1989.
8. J. Potthoff and L.Streit: A characterization of Hida distributions. Preprint, 1990.
9. J.Pothoff and Yan Jia-an: Some results on test and generalized functionals of white noise. Universität Bielefeld , BIBOS , Nr.368/89.
10. R.Ramer: On nonlinear transformations of Gaussian measures . J. Funct. Anal. 15, p.166-187, 1974.
11. B.Simon: The $P(\Phi)_2$ Euclidean (Quantum) Field Theory. Princeton University Press , 1974 .
12. S.Watanabe: Stochastic Differential Equations and Malliavin Calculus . Tata Institute of Fundamental Research, Bombay, 1984 .
13. Y.Yokoi: Positive generalised white noise functionals. Hiroshima Math. J. 20. 1990.

AN INTERPRETATION OF KOLMOGOROFF'S AXIOMS IN AGREEMENT WITH WHITE NOISE AND DISTRIBUTIONS METHODS

PAUL KREE
Département de Mathématiques et U.A. 213
Université de Paris VI
tour 46, 4, Place Jussieu, 75005 Paris

ABSTRACT

Considering that distributions methods and white noise methods are "not probabilistic", these methods are some times retranscripted in theories apparently more close from usual probability theory. We prove that the retranscriptions are not necessary since distributions and white noise methods are consistent with Kolmogoroff's axioms and also more effective.

I - Difficulties and ambiguities connected with some interpretations of Kolmogoroff's axioms.

I.1 Kolmogoroff's axioms.

Set theoretical integration theory has been developped (even also with some topological and physical aspects) in 1905-1923. In 1933, A.N. Kolmogoroff founds Probability Theory (P.T.) as a mathematical discipline ("as geometry, as algebra") using set theoretical integration theory. Without taking into account the scientific context of 1933, Kolmogoroff's axioms can be summarized in the following way.

I.1 Definition. An abstract probabilized space is a triple (X,F,m) where F is a σ-field of subsets of the set X and where m denotes a probability measure on F i.e. a positive measure of total mass one on F.

I.2 Kolmogoroff's axioms have been a great discovery because underlining the fact that the structure (X,F,m) is fundamental. All existing probability theories has been created on these foundations and more precisely on the interpretations of these foundations. Hence important mathematical results. Also, thanks to the deep reflexions of some probabilists on physics, important applications have been discovered as for example :

(1.1)
- a) Ito's stochastic calculus and related topics
- b) Varadhan large deviations theory
- c) Nelson's work on functional integrals in quantum physics
- ....

But since Kolmogoroff foundations give perhaps to much freedom, very large deviations are observed. Also, today P.T. is even really exploding in numerous theories which are not all consistent : this without showing for non specialists the main lines of forces of the theory and where the fundamental problems lie, and also without giving to other mathematicians and to physicists the methods they need.

I.3 Some important probabilistic problems of physics and mathematics.

Below physics in understood in a large sense i.e. as including not only what is usual called "physics" but also mechanics, economy...

In applied physics, probabilistic methods are more and more fundamental. For example engineers design communication systems, satellites, rockets... using very advanced probabilistic methods whose relation with Kolmogoroff axioms are not obvious for example involving advanced functional analysis.

Probabilistic problems are very important today in fundamental physics.

In the last century, the works of Boltzmann, Gibbs, Maxwell... showed the fundamental interest in physics of a new discipline : statistical mechanics. Noticing in 1900 that statistical mechanics combines P.T. with mechanics, that P.T. was not rigourous and that O.D.E. theory is too simple in order to build foundations of mechanics, D. Hilbert did formulate in the following way his sixth problem.

"Les recherches sur la géométrie nous conduisent au problème : traiter sur ce modèle, les branches de la physique où les mathématiques jouent un rôle prépondérant : avant tout le calcul des probabilités et la mécanique."

This problem has been solved concerning the "calcul des probabilités" by A.N. Kolmogoroff's Grund begriffe (1933)

Let us note that N. Wiener did in 1923 the theory of Brownian Motion (B.M.) i.e. the first mathematical theory concerning statistical mechanics ; and that the corresponding mathematical structure of

P.V.S. = Probabilized Vector Space

is implicit (but not explicit) in Kolmogoroff's theory of continuous processes (1934). Althought numerous mathematical works concerning statistical mechanics and dynamical systems (Ruelle, Sinai...) have been realized, many problems suggested by statistical physics remain.

Nelson's work and related works [SIM 74][SIM 79] also show the fundamental importance in quantum physics and in the study of functional integrals of statistical mechanics, and also some times of some "non commutative, integration theory" (I. Segal, D. Kastler, A. Connes...). All these problems are actively studied. Also a very challenging problem is Nelson's stochastic mechanics and related theories whose goal is to explain quantum dynamics.

Let us indicate several interpretations of Kolmogoroff's axioms.

I.4 The Doob interpretation. This interpretation has motivated Doob's work. Today's the classical theory of stochastic processes is constructed just in the line of this interpretation. The important concepts there are mainly set theoretical :

(1) (continuous processes, right processes, filtrations, Markov processes. Martingales. Semi-martingales

I.5 Theory of P.V.S. and Analysis on P.V.S. Some probabilistic problems related with analysis and physics have motivated the study of probabilized locally convex Hausdorff spaces (Wiener, Segal, Prohoroff, Sazonov, Gelfand, Minlos, X. Fernique, J. Dudley, Kolmogoroff, L. Schwartz...). Hence in relation with more recent works the following basic concepts :

(2) (General P.V.S., Gauss spaces, Poisson spaces, P.V.S.of continuous functions, of right functions, of distributions Markoff fields white noise spaces tiltered P.V.S.

The simplest examples of P.V.S. spaces like the filtered Wiener space, the filtered space of the Poisson process are implicit in the classical theory. Problems of statistical mechanics, of quantum fields theory also the mathematical study of functional derivates and of corresponding equations arising there have motivated the elaboration of a mathematical analysis on Gauss spaces and more recently on filtered P.V.S. [Krée 88][Krée 89]. In the analysis, vectorial structures, dual tensorial structures and functional analysis work in an essential way. Hence the following theories.

(3) | Study of the moments of a probability measure on a locally convex space. Nelson's Markoff field. White Noise methods. Sutdy of operator in rigged Fock spaces using formal power series, kernels and Wick symbol techniques. Chaotic transform. Differential calculus in the sense of cylindrical distribution and relatid functional spaces. Skohorod integral.

The general theories (2)+(3) are not recognized in the following set theoretical interpretation of Kolmogoroff's axioms, appeared in the sixties.

I.6 The literal interpretation of Kolmogoroff's axioms. P.T. is founded on the structure of abstract probabilized space. P.T. is constructed independantly of all other theories, mainly using methods of set theoretical integration theory. Constructed in this way, the theory can be applied in analysis and in physics.

Only supposing that the usual differential calculus on $V = \mathbb{R}^n$ is given (in order to write Ito's formulas) and also retranscripting in set theoretical terms Choquet's capacities theory, the theory of Souslin spaces, of Lusin spaces... the classical theory of stochastic processes and some applications can be presented in this way, in an extraordinary unified style of set theoretical integration theory. But living aside at its beginving the theories (2) of P.V.S., the literal interpretation lives also aside the theories (3) concerning the analysis on P.V.S. and the corresponding problematic connecteding strongly analysis, P.T. and physics. Denying the existence of this strong connection and the fact that the additional structures of abstract probabilized spaces work in an essential way in these theories, some isomorphisms preserving measurable structures but killing all other structures and even filtrations are used in order to reduce problems concerning the general case (2) to the simplest cases. In these conditions retranscriptions against nature of selectas of some theories (3) specialized to these simplest cases are established. In physics, according to P.A. Meyer (Sém. XX = L.N.M. (204)) this generates doubts on Kolmogoroff axioms... therefore other probability theories have been build up. Let us recall some facts in order to see how this situation has been generated.

II - Differential analysis and distribution methods on P.V.S. in 1977

II.1 Generalities. L. Schwartz gave a talk on the theory of cylindrical distributions at the 1977 Jerusalem conference on functional analysis but L. Schwartz did stop his interest for this subject in 1978 Mathematicians aware of these problems did participate at the 1977 Dublin conference, whose proceedings (Lect. Notes in Math n°644) show that the field was very active and involving numerous analysts (Aron A. Badrikian, P. Boland, S. Dineen, D. Ellworthy, V. Goodman, L. Gross, H.H. Kuo, B. Lascar, P. Raboin, W. Slowikowski, A. Truman, J. P.R. Christensen, T.A.W. Dwyer, P.J. Guerra, P. Lelong, J. Mujica) These proceedings show that the general methods of dimension free analysis was at that time the following :

- measure theory on locally convex spaces
- the Wiener-Segal theory of chaos
- the corresponding theory of I. Segal of the basic operators of $L^2_\xi(\Omega)$ for any Gauss linear process ξ associated with any separable Hilbert space

(4) $(N, a(u), a^*(u), H_0 ...)$

- the H'-differential calculus extensively developped by L. Gross and his group for function of defined on any abstract Wiener space $(H' \subset \Omega, P)$
- infinite dimensional holomorphy developped by P. Lelong, L. Nachbin, P. Boland, S. Dineen, T. Dwyer...

The Dublin proceedings show that numerous research group did developp after 1970 different concepts and methods of distributions. But that these various theories was each time related only with one kind of applications (C. de Witt, Meljbo and Poulsen, L. Nachbin, H.H. Kuo, Dubin, Fomin, Elson, P. Krée...). For example in his Dublin lecture H.H. Kuo developps the theory (see also Dudin, Fomin...) of absolute derivatives $\tilde{\nabla} m$, $\tilde{\nabla}^2 m$... of a given bounded measure m. In his Dublin lecture, L. Gross discuss his beautifull logarithmic Sobolev inequality concerning $L^p(\mathbb{R}^d, \gamma)$ and $1<p<\infty$ [GRP 75]. I did learn recently form E. Carlen that the particular case p = 2 i.e.

$$\int_{\mathbb{R}^2} |f|^2 \ell n\, |f|\, d\gamma \leq \| \nabla f \|^2 + \| f \|^2 \ell n \| f \|$$

is due to Stam (Inf. and Control 2 p.101-102 ; 1959).

L. Gross also notes writting $\| \nabla f \| = \langle Nf, f \rangle$, that this inequality kipps a meaning for $n = \infty$.

III.2 - The theory of cylindrical distributions.

III.2.1. The Dublin lectures of B. Lascar and P. Krée and references included there show that a general and dimension free theory of distributions extending some of the general methods of dim. free analysis (4) and also all particular distributions methods and corresponding application mentioned before did exist at that time. See the table. Let us discuss some particular topics only considering the Gauss space.

$$(H \supset U ... \Omega, P).$$

Here P denotes a Gauss measure on a quasi-complete locally convex Lusin space, U denotes a subspace of Ω' separating point of Ω, H denotes the covariance Hilbert space i.e. the completion of U for the covariance scalar product. Finally $(U_\alpha, \alpha \in \Lambda)$ denotes a filtering increasing family of finite subspaces of U and s_α denotes the canonical surjection of Ω on to $\Omega_\alpha = \Omega / U_\alpha^\perp$. The reproducing Hilbert space $H' \subset \Omega$ of P is the dual of H.

Theory of cylindrical distributions	
Methods used for the elaboration of these parts	Some parts of the theory of cylindrical distributions
Differential Calculus of Gross + methods of Schwartz (bitransposition, spaces of distributions, cylindrical measures...)	A. Definition of the distributions derivatives $\nabla, \delta, \tilde{\nabla}$ + methods of spaces of distributions + study of Gauss Sobolev spaces [KRE 73-75][KRE 74-77][LAS 74-76]
Segal's approach of Wiener chaos (1956) + Dwyer representation of Fock space using infinite dim holomorphy + Schwartz's theory of F.T.	B. Reformulation and extension of Segal's decomposition in chaos producing an analogous of Schwartz theory of F.T. [KRE 74-75]
Grothendick thesis + Schwartz's kernels theory + Berezin book (1965)	C. Methods of kernels and of Wick symbols [KRE-RAC 77][LAS 78]
	Some application published in 77-79.
Work of Albeverio and Hoegh-Krohn [ALB-HOEG 77]	D. Energetic Capacity Theory : thesis of Paclet [PAC 79]
Berezin's book (1965)	E. Methods of kernels and symbols for fermions : Thesis of M. Marias [MAR 78]
finite dimensional PDE and PDO theory + all previous works	F. Dimension free theory of P.D.E. and P.D.O. Thesis of B. Lascar [LAS 78]

II.2.2 Cylindrical test functions and triplets.

The framework of L. Gross differential calculus the main motivation is potential theory on Hilbert space. Hence the natural spaces of test functions are spaces of C^∞-Gross derivable functions on Ω, hence distributions are elements of dual spaces and the theory is applied to Δ : see Elson's theory of distributions. Our approach was different : a) Having in mind fields theory Segal's work on N, a(u)... and Nelson's work, we work with equivalence classes of functions as test functions and triplets centered on L^2 b) having in mind Schwartz's work [SCH 73], any bounded measure (or more generally any cylindrical distribution) in on Ω is characterized by all its direct images $m_\alpha = s_\alpha(m)$. In particular any $f \in L^p(\Omega)$ is characterized in [KRE 74-75] by the martingale $\alpha \to E[f \parallel s_\alpha]$; in the same way ∇m, δm, $\tilde{\nabla} m$ are characterized by consistent systems of usual Schwartz's distribution ∇m_α, δm_α, $\tilde{\nabla} m_\alpha$ on finite dimensional spaces. Therefore our test functions was cylindrical. In our Dublin lecture, various spaces of test functions are experimented, bu the space $P(\Omega)$ of cylindrical polynomials has been discovered and recognized as the good space of test functions only later one [KRE 79-81].

II.2.3 Chaotic transform I^{-1}. In the 70 probabilists did formulate systematically Wiener's bijective isometry I of decomposition in chaos (where Ω_W denotes the Wiener space).

(5) $\bigoplus L^2_{sym}(\mathbb{R}^k_+, dt) \overset{I=\Sigma I_k}{\rightarrow} L^2(\Omega_\omega)$

in terms of multiple stochastic integrals

$$f = \Sigma_o^\infty f_k \rightarrow \Sigma_o^\infty \int_{Q_k} f(t_1,\ldots t_k)\, dW_{t_1}\ldots dW_{t_k}.$$

Where $Q_k = (t = (t_1,\ldots,t_k)\ ;\ 0 < t_1 <\ldots< t_k)$. In view of the literal interpretation of Kolmogoroff's axioms, never they use Segal's work (1956) extending and clarifying Wiener's work, since giving for any separable Hilbert space H a bijective isometry

$$\mathrm{Fock}_A(H) := \bigoplus S_k(H) \rightarrow L^2_\xi(\Omega)$$

where $\xi : H \rightarrow L^2(\Omega)$ denotes any linear Gauss process based on H. In 1971, working with methods of dim. free holomorphy Dwyer did introduce his Dwyer picture Fock (H) of abstract Fock space $\mathrm{Fock}_A(H)$ where any element $f = \Sigma f_k$ is interpreted as a converging power series on H. Using this Dwyer's picture and also the progresses of measure theory after 1956, [KRE 74] gives for "any" Gauss space ($H \supset U \ldots \Omega, P$) the following explicit formula for I^{-1} of I :

(6) $L^2(\Omega) \ni g \overset{I^{-1}}{\rightarrow} \hat{g}(h) = \int_\Omega e^{h.w - |h|^2/2}\, dP(w).$

In [SEG 78], Segal explains that Dwyer's picture of abstract Fock spaces is more effective that the picture introduced by Segal in 1960 in his Boulder lecture, Refering to [KRE 74] I. Segal presents (6). The formula (6) is also fundamental in the theory of cylindrical distributions : first since cylindrical polynomials are invariant by I^{-1}, I^{-1} can be extended by duality to cylindrical distributions (Krée-Raczka 1978, M. Krée 1977) providing in this way an isomorphism of triplets :

$$\begin{array}{ccccccc} P(\Omega) & \subset & L^2(\Omega) & \simeq & L^2(\Omega)' & \subset & P(\Omega)^* \\ \simeq\downarrow & & & & \simeq\downarrow I^{-1} & & \simeq\downarrow \\ S(U) & \subset & \mathrm{Fock}\, H & \simeq & (\mathrm{Fock}\, H)' & \subset & \widehat{\mathrm{Pol}}(U) \end{array}$$

By another hand, extending I^{-1} to vectorial distributions and vectorial Fock spaces, I^{-1} associates to the gradient ∇ the derivation D of formal power series, hence by transposition the divergence to D^T.

II.2.4 Normal spaces of cylindrical distributions, Gauss Sobolev spaces $W^{p.k}(\Omega)$

a) A normal space of cylindrical distribution on Ω denotes any sequentially complete locally convexe space $\mathcal{N}$ such that

(7) $P(\Omega) \subset \mathcal{N} \subset P(\Omega)^*$

where the first injection his a dense range and where the second injection is continuous endowing $P(\Omega)^*$ with the weak topology. Transposing (7) we see that the strong dual of any normal space $\mathcal{N}$ normal space $\mathcal{N}'$ of cylindrical distribution. Also since the chaotic transform is an isomorphism, the space $I^{-1}\mathcal{N}$ of all chaotic transforms of all elements of $\mathcal{N}$ is a normal space of formal power series

(8) $$S(U) \subset I^{-1}(\mathcal{N}) \subset \widehat{Pol}(U).$$

As in L. Schwartz's distributions theory, normal spaces of cylindrical distributions can by defined and studied in this way in view of specific applications.
b) The Dublin proceedings give references to the following results of M.P. Krée and B. Lascar concerning the Gauss Sobolev spaces $W^{p,k}(\Omega)$ for $1<p<\infty$,
$k = \pm 1$... Normality, chain rule, dual space ; and for $p = 2$ only : equivalence of the norms

(9) $$\| f \|_{p,k} \sim \| (1+N)^{k/2} f \|_p$$

continuity of δ, partition of unity, support, characterisation of chaotic transforms, theorem of Paley-Wiener type, behaviour by diffeomorphisms, characterisations of traces on hyperplanes...

M. Visik did tell us in 1975 that N. Frolov [FRO 73] did define the spaces $W^{2,k}(\Omega)$ for $k > 0$ completing some space of C^∞-Gross derivable functions for the Sobolev norm, but without proving that the natural mapping $W^{2,k}(\Omega) \to L^2(\Omega)$ is injective and without establishing any property of these spaces.

II.2.5 Kernels (in the Schwartz-Grothendick sense) and Wick symbols. These works extend previous works of F.A. Berezin concerning annihilators and creators. Here the fundamental operators are no more a and a* but D and its transpose D^T.

III - White noise and distributions methods interacting with the stochastic calculus in 1977-1979.

III.1 Hida's White Noise. In 1978, T. Hida begins his investigation on Ito's calculus in terms of the white Noise W' i.e. of the distribution derivative W' of
W = BM(1) :

(10) $$(\mathcal{D} \subset H = H' \subset \mathcal{D}',P)$$

where $H = L^2(\mathbb{R}_+,dt)$, $\mathcal{D} = \mathcal{D}(]0,+\infty[)$. Using also the Dwyer picture of abstract Fock spaces, T. Hida works with the triplet

$$Hid(\mathcal{D}) \subset Fock(h) \simeq Fock(H)' \subset Hid(\mathcal{D})'$$

where

$$Hid(\mathcal{D}) = \bigoplus \mathcal{D}_{sym}(\mathbb{R}_+^k) \Rightarrow Hid(\mathcal{D})' = \Pi \mathcal{D}'_{sym}(\mathbb{R}_+^k)$$

Therefore $Hid(\mathcal{D})$ and $Hid(\mathcal{D})'$ are chaotic transforms fo some dual normal spaces $\mathcal{N}$ and $\mathcal{N}'$ of cylindrical distributions :

$$P(\Omega) \subset \mathcal{N} \subset L^2(\Omega) \simeq L^2(\Omega)' \subset \mathcal{N}' \subset P(\Omega)^*$$

T. Hida also use (6). In order to make more explicit the relation of Hida's work with other works, let us introduce the usual Wiener space (modeling W) appearing on the first line of the foolowing diagram

$$(H_W \supset U_W \quad ... \quad \Omega_W , P_W)$$
$$\simeq \uparrow \text{-d/dt} \qquad \simeq \downarrow \text{d/dt}$$
$$(U_{W'} \subset H=H' \subset \Omega_{W'}, P')$$

where $U_W ... \Omega_W$ denotes the Wiener space Ω_W in duality with U_W = Span (δ_s,s>0). the second line of this diagram gives an isomorphic Gauss space modeling W' : $\Omega_{W'}$ denotes the space of distributions derivatives of all w in Ω_W.

II.2 <u>The principle of Malliavin's stsrategy for the regularity of laws</u>. This strategy concerns the regularity of the law m = $f(P_W)$ of an equivalence class $f : \Omega_W \to \mathbb{R}^d$ such that <u>informally</u> the gradient $\nabla f : H_W' \to \mathbb{R}^d$ is allmost surely surjective. Malliavin him self presents his strategy in [MAL 76-78] part 1, as inspired by the theory of cylindrical distribution and by Hida's white noise since he first analyzes the problem substituing Q_W by a finite dimensional Gauss space, identified with the dual space ; then he writes using Schwartz theory his lifting formula for j=1...d and $\varphi \in \mathcal{D}(\mathbb{R}^d)$

(11) $$E\left[\frac{\partial}{\partial y_j} \varphi o f\right] = E[\varphi o f . T_j]$$

where $T_j = (Nf)\gamma_j - \nabla\gamma_j . \nabla f$ and finally Malliavin obtains aplying Holder

(12) $\quad$ Nf and ∇f in L^p, γ_j and $\nabla\gamma_j$ in $L^q \Rightarrow T_j \in L^1$

i.e. in view of (11) the distribution derivative $\partial m/\partial y_j$ is a bounded measure. But curiously Malliavin forgets distributions as he comes back to the infinite dimensional case and he proposes a definition of ∇f for f in L^p using an informal Ito's formula. We learned in 1984 that D. Stroock did notice soon in 1977 that this differential calculus on Ω_W is not fully correct. During a talk of Malliavin on [MAL 76-78] at our seminar (9 march 1977) we did indicate him that the theory of cylindrical distributions gives a natural proof of [MAL 76-78] part 1: just introducing the normal space $\mathcal{N}_0$ of cylindrical distributions suggested by (12) and by the finite dimensional approach. A young japanese probabilist [SHI 79-80] did publish 3 years later a similar proof but where $\mathcal{N}_0$ is defined directly by completion where a selecta of methods of cylindrical distributions are converted in a calculus. Note also that Malliavin's strategy is related with [ALB-HK 75-77].

III.3 <u>Mathematical formulation of all the Malliavin's strategy</u>. Receiving in 1979 the preprint of [SHI 789-80] and learning that the calculus there has been taught by Malliavin (see the lecture of D. Stroock at the Int. Congress of 1983 p.1110) D. Stroock did call this : "Malliavin Calculus". Combining this with his own work, D. Stroock did establish [STR 78-81] the main steps of a mathematical proof of all Malliavin's strategy. This procedure did generate a big confusion among probabilists (see for ex. the proceedings of

Durham's conference of 1981) for the two following reasons a) presented in this way Malliavin's strategy and the calculus taught by Malliavin did appear as surnatural.
b) soon after, Bismut did discover a completely different probabilistic proof using Girsanov's formula. In fact there was no mystery : Malliavin's strategy was inspired by distributions methods ; Bismut's proof was inspired by stochastic calculus ; hence this was the first interaction which has been observed between these two theories.

III.4 Professor T. Hida did began in 1978 his analysis of Ito's calculus combining white noise and distributions methods. Also D. Stroock discovered important results in this direction. Since [SHI 79-80] presents erronously ∇ and N as the fundamental differential operators and forgets δ this explain why probabilists have worked during some years only with ∇ and N.

IV - Further progresses. Conclusions.

IV.1 We refeer to the lectures of T. Hida, H.H. Kuo, Kubo, J. Pothoff at this conference for further progresses. Numerous important results have been first discovered using white noise and distributions methods, and then retranscripted in terms of some calculus. Let us mention shortly some examples : the work [KU-TA 80] showing for the first time that Ito's integration is induced by δ, the H.H. Kuo's result [KUO 83] introducing for the first time the pullback of cylindrical distributions, see also [WAT 83] ; M.P. Krée's result [KRE 83] showing for the first time the L^p continuity of δ and the equivalence (9) for $1<p<\infty$, the work [KUB 83] concerning the generalized Ito's formula, the result of J. Pothoff [POT 87] concerning positive cylindrical distributions, the solution [KRE 89] of the problem of representations of quantum martingales, the first result [DER 90] concerning chaotic martingales... Since distributions methods have been some times presented as non probabilistic and even as trivial (this reducing erronously the theory to the definition of distribution-derivatives), let us mention that the modern theory of P.D.E. and of P.D.O. is not born in 1934 but in the 50' and 60' using all the methods created by L. Schwartz in 1950-1951. Let us also recall therefore an interpretation of Kolmogoroff's foundations which is consistent with the theories (1)+(2)+(3) and also with physical probability theory in its present state.

IV.2 The natural interpretation of the Grund Bregriffe.

Clearly as Kolmogoroff did apply the definition of abstract prabilized spaces to $X = \mathbb{R}^n$, he did consider not only that set theoretical integration theory can be used there, but also all finite dimensional analysis as for ex. the usual differential calculus, F.T., and even kernels and symbols, Plancherel theorem, expansions in polynomials can be applied. Now we know that for P.V.S. the corresponding analysis needs distributions and white noise methods since all efforts in order to avoid these methods have failed. Therefore methods of cylindrical distributions and white noise methods can be used directly without retranscription. Note that this interpretation of Kolmogoroff's axioms is in agreement with the lecture of K. Ito at his conference.

IV 3. Acknowledgment. The author thanks T. Hida, H.H. Kuo and J. Pothoff for their invitation at this Bielefeld conference. Also the great idea of the organizers to give again life to dimension free analysis.

References

[ALB-HK 75-77] S. Albeverio and R. Hoegh Krohn.Dirichlet forms and Diffusion processes on rigged Hilbert Spaces. Z. für Wahrsc and Ver : Get. t 40 (1977) p.1-57. Manuscript diffused in 1975.

[DE 77-90] A. Dermoune. Distributions sur les espaces de Paul Levy à paraître aux Annales de l'I.H.P. (1990).

[DWY 71] T. Dwyer. Fisher Fock spaces Bull. A.M.S. 1971.

[FRO 73] N. Frolov. On a coercive inequality for an elliptic operator in infinite many independant variables. Mat. Sbornik Tom 90 (132)(1973) n°3. English Transl. in Math. USSR Sbornik vol.19 (1973) n°3 p.395-406.

[GRO 67] L. Gross. Potential Theory on Hilbert spaces J. of Funct. Analysis 1 (1967) kjp.123-181.

[GRO 73-75] L. Gross. Logarithmic Sobolev Inequalities Amer. Journal of kMath. 97 (1975) p.1061-1093. Manuscript received by A.M.J. in 1973;

[KRE 89] P. Krée. Dimension free stochastic calculus in the distribution sense in Pitman Research Notes in Math. n°200 K. Elworthy and J.C. Zambrini editors Stochastic Analysis, paths integration and dynamics p.94-129. Longman 1989.

[KRE 88] P. Krée. La théorie des distributions en dim. quelconque et l'intégration stochastique. in Stochastic Analysis and Related Topics (Proceed Silivri 1986) p.170-233. Lect. Notes in Math. n°1316 Springer Verlag 1988.

[KRE 71] P. Krée. Courants et courant cylindriques sur les variétés banachiques. Linear Operator and Approximation Oberwollfack 1971. Birhauser Basel (1972).

[KRE 73-75] P. Krée. Equations aux dérivées fonctionnelles. Deux exposés au Sém. P. Lelong 1973-1974 et 1974-1975. Lect. Notes in Math n°410 (1974) and n°474 (1975) Springer Verlag.

[KRE 74-77] M. Krée. Propriété de traces pour des espaces de Sobolev en dim. infinie. Comptes Rendus t.270 (1974) série A pp.157-160 et Bull. Soc. Math. France (1977) p.141-163.

[KRE-RAC 78] P. Krée and R. Raczka. Kernels and symbols in quantum fields theory Annales de l'Inst. H. Poincaré Section A vol. 28 n°1 (1978) kp.41-73.

[KRE 79-81] P. Krée. Linear Differential operators on Vector Spaces in Rio Conference (1979). Lect. Notes in Math. 843. Springer Verlag 1981.

[KRE 83] M.P. Krée. Comptes Rendus t.296 (6 juin 1983) série I pp.833-836.

[KU-TA 80] I. Kubo and S. Takenaka. Calculus on Gaussian White Noise Proceed Japan Academy I. 56.A (1980) p.376-380 II. 56.A (1980) p.411-416 III. 57.A (1981) p.433-436 IV. 58.A (1982) p.186-189.

[KUB 83] I. Kubo. Ito formula for Generalized Brownian functionals. Lect. Notes in Control and Inf. Sc. vol 49 (1983) kp.156-166. Springer Verlag.

[KUO 83] H.H. Kuo. Donsker delta function as a generalized Brownian functional and its application Lect. Notes in Control and Inf. Sc. vol.49 (1983) p.167-178. Springer Verlag.

[LAS 74-76] B. Lascar. Propriétés lcoales d'espaces de Sobolev en dim. infinie. Comm. in Part. Diffu. Equ. 1-6 pp.561-584 (1976).

[LAS 78] B. Lascar. Equations aux Dérivées Partielles en Dim. Infinie Thèse de l'Université de kParis VI (nov. 1978 - Malliavin referee). See Lect. Notes in Math. n°644 (Proceedings of Dublin conference 1977) for a Summary and further references concerning this thesis.

[MAL 76-78] kjP. Malliavin. C^k hypoellipticity with degeneracy Stochastic Analysis. Academic Press (1978) pp.199-214 + 327-340. Manuscript diffused in nov. 1976.

[MAR 78] M. Marias. Transformations canoniques dans les espaces de Hilbert. Thèse de l'Université de Paris VI. Publi. dans le sém. 1977-1978 d'E.D.P.∞ (Institut H. Poincaré. Secrétariat Mathématique).

[PAC 79] P. Paclet. Théorie de la capacité en dim. infinie. Sém. 1978-1979 sur les E.D.P.∞ (Institut H. Poincaré. Secrétariat Mathématique).

[POT 87] J. Pothoff. Positive Distributions. Journal of Functional analysis 1987.

[SCH 73] L. Schwartz. Radon measures on arbitrary topological spaces (1973).

[SEG 78] I. Segal. The Complex-Wave representation of the free Boson field Topics in Funct. Analysis Adv. in Math. Supplementary Studies vo. 3 Academic Press 1978.

[SHI 79-80] D. Shigekawa. manuscrit diffusé en janvier 1979 et publié dans J. Math. Kyoto Univ. n°2O (1980) jpp.263-289.

[STR 79-81] D. Stroock. The Malliavin Calculus and its application to second order parabolic differential equations. Manuscrits diffusés en 1979 et publiés dans Math. Système Th. 14 (1981) pp.25-65 et 1411-1471.

[WAT 83] S. Watanabe. in Lecture Notes in Control and Inf. Sc. vol. 49 (1983) Springer Verlag.

Anticipating Stochastic Integrals and Stochastic Differential Equations

H.–H. Kuo and J. Potthoff

Department of Mathematics
Louisiana State University
Baton Rouge, LA 70803
U.S.A

§1. Introduction and Background

In the recent years there has been an increasing interest in stochastic analysis involving anticipating (or "non-adapted") processes from both points of view : theory and application. Applications arise in nonlinear filtering, stochastic bifurcation theory, stochastic flows on manifolds and other domains. Theoretical interest has been stimulated by the observation that in Malliavin calculus [GT 82, NP 88, NZ 86, Oc 84] and in white noise analysis [AP 88, KT 80, Kuo 88, KR 88] Itô's stochastic integral for non-anticipating integrands has a very simple formulation, which generalizes to anticipating integrands in a more or less straightforward manner. However, this formulation of stochastic integration (with respect to Brownian motion) is also interesting because it offers a more functional-analytic approach.

Here we shall use Hida's white noise analysis as a framework [Hi 75, Hi 80, HKPS, HP 89, KT 80, Kuo 83]. The basic feature is the time derivative of Brownian motion, i.e. white noise, as the basic entity. It turns out that this is particularly suited for stochastic integrals and stochastic differential equations with respect to Brownian motion.

Let $(\mathcal{S}'(\mathbb{R}), \mathcal{B}, \mu)$ be the probability space of white noise, where $\mathcal{S}'(\mathbb{R})$ is the Schwartz space of tempered distributions over $\mathbb{R}$, $\mathcal{B}$ the σ-algebra generated by the cylinder sets and μ the Gaussian measure on $\mathcal{B}$ whose characteristic function is given by

$$\int_{\mathcal{S}'(\mathbb{R})} e^{i\langle x,\xi\rangle} d\mu(x) = e^{-\frac{1}{2}|\xi|_2^2}. \tag{1.1}$$

In (1.1) $\langle \cdot, \cdot \rangle$ denotes the duality of $\mathcal{S}'(\mathbb{R})$ and $\mathcal{S}(\mathbb{R})$, $|\cdot|_2$ is the norm of $L^2(\mathbb{R}, dt)$ and $\xi \in \mathcal{S}(\mathbb{R})$.

Let X_ξ denote the random variable $x \longmapsto \langle x, \xi \rangle$, $\xi \in \mathcal{S}(\mathbb{R})$. Note that the mapping $\xi \longmapsto X_\xi$, from $\mathcal{S}(\mathbb{R})$ into $L^2(\mathcal{S}'(\mathbb{R}), \mu)$ can be extended continuously to $L^2(\mathbb{R})$. In particular, we may choose for ξ the indicator $1_{[0,t)}$ of the interval $[0,t)$. In this case, we shall denote X_ξ by $B(t)$ and it follows from (1.1) that $B(t)$ is a version of Brownian motion. Writing informally $B(t)(x) = \langle x, 1_{[0,t)} \rangle$, $x \in \mathcal{S}'(\mathbb{R})$, we see that "$x(t)$ is a version of $B'(t)$". This observation provides the intuition behind our approach.

For brevity, (L^2) will denote the space $L^2(\mathcal{S}'(\mathbb{R}), \mu)$. Define a transformation S on (L^2) [KT 80] as follows : if $\varphi \in (L^2)$

$$S\varphi(\xi) := \int_{\mathcal{S}'(\mathbb{R})} \varphi(x + \xi) d\mu(x), \quad \xi \in \mathcal{S}(\mathbb{R}). \tag{1.2}$$

The S-transform (1.2) serves two purposes simultaneously: 1) it transforms a function on $\mathcal{S}'(\mathbb{R})$ into a function on $\mathcal{S}(\mathbb{R})$, which is often easier to deal with, and 2) it implements the unitary isomorphism between (L^2) and the symmetric Fock space $\bigoplus_{n=0}^{\infty} L^2(\widehat{\mathbb{R}^n})$. In particular, if $\varphi \in (L^2)$ has the homogeneous chaos expansion (cf. [HP 89] for details)

$$\varphi(x) = \sum_{n=0}^{\infty} \langle : x^{\otimes n} :, f^{(n)} \rangle, \quad x \in \mathcal{S}'(\mathbb{R}),$$

where $f^{(n)} \in L^2(\widehat{\mathbb{R}^n})$, then

$$S\varphi(\xi) = \sum_{n=0}^{\infty} \int_{\mathbb{R}^n} f^{(n)}(u) \xi^{\otimes n}(u) du$$

and $S\varphi$ as an element in Fock space can be identified with the sequence $(f^{(n)}, n \in \mathbb{N}_0)$.

In the present article we shall use the spaces $(\mathcal{S})$ and $(\mathcal{S})^*$ of test and gener alized functionals, respectively, of white noise $x \in \mathcal{S}'(\mathbb{R})$. These spaces, being a special choice in the class introduced by Kubo and Takenaka [KT 80], have been investigated in a number of papers [HPS 88, KY 87, MY 88, PY 89, Yo 87] (cf.

also the articles [HP 89, IKT 89, Lee 89, Yo 89] in these proceedings). $(\mathcal{S})$ is the projective limit of the Hilbert spaces $(\mathcal{S})_p$, $p \in \mathbb{N}_0$, and $(\mathcal{S})^*$ is the inductive limit of $(\mathcal{S})_p^*$, $p \in \mathbb{N}_0$ (cf. e.g. [HP 89]). $(\mathcal{S})^*$ and $(\mathcal{S})_p^*$ will be used in Section 3 to state an existence and uniqueness theorem of weak solutions of anticipating stochastic differential equations.

Note that due to the translation formula for Gaussian measures [Kuo 75]

$$S\varphi(\xi) = e^{-\frac{1}{2}|\xi|_2^2} \int_{\mathcal{S}'(\mathbb{R})} e^{\langle x,\xi\rangle} \varphi(x) d\mu(x).$$

On the other hand, $\exp(\langle\cdot,\xi\rangle) \in (\mathcal{S})$ for all $\xi \in \mathcal{S}(\mathbb{R})$ (e.g. [HPS 88]). Thus the S-transform may be extended to $(\mathcal{S})^*$ by: If $\Phi \in (\mathcal{S})^*$,

$$S\Phi(\xi) = e^{-\frac{1}{2}|\xi|_2^2} \ll \Phi, e^{\langle\cdot,\xi\rangle} \gg, \quad \xi \in \mathcal{S}(\mathbb{R}), \tag{1.3}$$

where $\ll \cdot,\cdot \gg$ is the duality of $(\mathcal{S})^*$ and $(\mathcal{S})$. We shall refer to $S\Phi$ as the *U-functional of Φ*.

Note that the product of two U-functionals is again a U-functional (e.g. [KT 80, MY 88]). Thus if $\Phi, \Psi \in (\mathcal{S})^*$, then there exists an element in $(\mathcal{S})^*$, denoted by $\Phi \diamond \Psi$, such that

$$U[\Phi \diamond \Psi](\xi) = U[\Phi](\xi) \cdot U[\Psi](\xi). \tag{1.4}$$

Following Hida [Hi 75] and Kubo and Takenaka [KT 80], we introduce for suitable $\varphi \in (L^2)$ the following operation

$$\partial_t \varphi = S^{-1} \frac{\delta}{\delta\xi(t)} S\varphi,$$

where $t \in \mathbb{R}$ and $\frac{\delta}{\delta\xi(t)}$ denotes the Fréchet functional derivative. We will call $\partial_t \varphi$ the *Hida derivative of φ*.

It is not hard to see that if $\varphi \in (\mathcal{S})$, then $\partial_t \varphi$ can be identified with the Gâteaux derivative of φ in the direction of δ_t, $t \in \mathbb{R}$ (cf. [HP 89, Kuo 83, PY 89]). Therefore, all standard rules of differentiation hold for ∂_t on appropriate domains.

For the purpose of the present article it is important to observe that the operation

$$\varphi \longmapsto (\partial_t^* + \partial_t)\varphi \tag{1.5}$$

implements the multiplication by white noise $B'(t)$, e. g. [Kuo 82]. It is this observation that leads to the starting point of stochastic integrals in Section 2. Note that ∂_t^* is the adjoint of ∂_t with respect to μ and observe that the U-functional of $\partial_t^*\varphi$ is given by

$$U[\partial_t^*\varphi](\xi) = \xi(t)U[\varphi](\xi). \tag{1.6}$$

Thus (1.5) has a rigorous meaning as a mapping from $(\mathcal{S})$ into $(\mathcal{S})^*$.

§2. Stochastic Integration

Since 1972 several extensions of Itô's integral and Itô's lemma to anticipating integrands have been obtained with various methods by many authors [AP 88, BM 82, GT 82, Hit 72, Itô 78, Ki 75, KR 88, NP 88, OP 88, Sk 75]. In this section we describe briefly the theory of stochastic integration from the viewpoint of white noise calculus.

Assume that $X : [0,1] \longrightarrow (L^2)$ is an appropriate stochastic process. Write informally $dB(t) = B'(t)dt$ for the differential of Brownian motion B at time $t \in [0,1]$. Thus it seems to be natural trying to define the stochastic integral of X with respect to B as follows

$$\int_0^1 B'(t)X_t\, dt. \tag{2.1}$$

Clearly, as B is a.s. nowhere differentiable, (2.1) is highly informal and in the following we shall describe how to give it a rigorous meaning. To this end, it is crucial to express the "multiplication by $B'(t)$" in terms of the white noise differentiation ∂_t and its adjoint ∂_t^* as in (1.5), i.e.

$$B'(t)\varphi = (\partial_t^* + \partial_t)\varphi. \tag{2.2}$$

For instance, for every $t \in \mathbb{R}$, $\partial_t^* + \partial_t$ maps $(\mathcal{S})$ continuously into $(\mathcal{S})^*$ (equipped with the weak or strong topology). However, considering (2.2) "pointwise with respect to t" leads to unnecessary restrictions. In fact, what we want is to integrate with respect to t, i.e.

$$\int_0^1 (\partial_t^* + \partial_t)X_t dt. \tag{2.3}$$

Consider first the following integral from (2.3)

$$\int_0^1 \partial_t^* X_t dt.$$

Put

$$\nabla\varphi(t,x) := \partial_t\varphi(x), \quad t \in \mathbb{R},\ x \in \mathcal{S}'(\mathbb{R}),$$

then we have the following facts (see e.g.[HP 89, PY 89]):

i) ∇ is a "true" gradient.

ii) ∇ is a continuous mapping from $\mathcal{D}(N^{1/2})$ into $L^2([0,1];(L^2))$. Here, N is the number operator.

iii) $\nabla^* X = \int_0^1 \partial_t^* X_t dt$, where ∇^* is the adjoint of ∇ with respect to $dt \otimes d\mu$ and ∇^* maps $L^2([0,1];\mathcal{D}(N^{1/2}))$ into (L^2). In fact, we have the following [KR 88, NZ 86]:

$$\mathbb{E}\left(\int_0^1 \partial_t^* X_t dt\right)^2 = \mathbb{E}\int_0^1 X_t^2 dt + \mathbb{E}\int_0^1\int_0^1 (\partial_s X_t)(\partial_t X_s)\,ds dt, \qquad (2.4)$$

and this implies that

$$\mathbb{E}\left(\int_0^1 \partial_t^* X_t dt\right)^2 \le \mathbb{E}\int_0^1 X_t^2 dt + \mathbb{E}\int_0^1 \left(N^{1/2} X_t\right)^2 dt.$$

In fact, it should be pointed out that the term

$$\nabla^* X = \int_0^1 \partial_t^* X_t dt \qquad (2.5)$$

is just the white noise version of the *divergence* studied in abstract Wiener space [Go 72, Kuo 75, Ra 74], Malliavin calculus (e.g. [GT 82, NP 88, NZ 86, Oc 84]), and calculus of infinite dimensional Sobolev spaces (e.g. [Kr 83]). (2.5) is now commonly referred to as the *(generalized) Skorokhod integral of* X (cf. in particular [NP 88]). It is well-known [GT 82, KT 80, KR 88, NP 88, NZ 86] that the Skorokhod integral (2.5) is an extension of Itô's integral to anticipating processes X.

We now turn to the second integral in (2.3), i.e.

$$\int_0^1 \partial_t X_t dt. \qquad (2.6)$$

As long as X is a continuous mapping from $[0,1]$ into $(\mathcal{S})$, its definition presents no problem. However, such processes are very limited for stochastic integration : for example $B(t) \notin (\mathcal{S})$.

Assume that X belongs to $L^2([0,1];\mathcal{D}(N^{1/2}))$ so that its Skorokhod integral $\nabla^* X$ is in (L^2). The fact ii) above implies that $\nabla X \in L^2([0,1]\times[0,1];(L^2))$. Therefore, in (2.6) we attempt to integrate a function which is measurable on $[0,1]\times[0,1]$ over the diagonal. Obviously, this is not well-defined in general and informal calculations with (2.6) lead to a paradox: for example, both $B_t = \langle\cdot, 1_{[0,t)}\rangle$ and $\widetilde{B}_t = \langle\cdot, 1_{[0,t]}\rangle$ are versions of Brownian motion, but "$\partial_t B_t = 0$" and "$\partial_t \widetilde{B}_t = 1$".

The solution of the preceding problem has been given in [KR 88] (cf. also [AP 88]). Suppose that ∇X has a version such that its restriction to the open set which is "right" or "left" to the diagonal has an integrable boundary value on the diagonal, i.e. *a trace from the right* or *from the left*. We shall denote the trace from the right by $\int_0^1 \partial_{t+} X_t dt$, and the one from the left by $\int_0^1 \partial_{t-} X_t dt$.

A convenient and not very restrictive criterion which implies the existence of a trace from the right or from the left of ∇X in (L^2) has been developed in [KR 88]. Let

$$D_+^n := \{(t,u) \in (0,1)\times \mathbb{R}^n;\ u=(u_1,\ldots,u_n),\ u_1 > t\}\,,\ n \in \mathbb{N}$$

and

$$D_-^n := \{(t,u) \in (0,1)\times \mathbb{R}^n;\ u=(u_1,\ldots,u_n),\ u_1 < t\}\,,\ n \in \mathbb{N}.$$

Note that the diagonal of $[0,1]\times[0,1]\times\mathbb{R}^{n-1}$ is a subset of the boundaries of $D_\pm^n$. For $\alpha \geq 0$, let $H_\alpha(D_\pm^n)$ denote the Sobolev space of order α over $D_\pm^n$. If $f^{(n)} \in H_\alpha(D_+^n)$, $\alpha > 1/2$, the trace theorem (cf. e.g. [Ad 75]) implies that $f^{(n)}$ has a version (again denoted by $f^{(n)}$) so that

$$t \longmapsto f^{(n)}(t;t+,\cdot)$$

belongs to $L^2([0,1];\ L^2(\mathbb{R}^{n-1}))$. Now if X_t has the homogeneous chaos expansion

$$X_t(x) = \sum_{n=0}^{\infty} X_t^{(n)}(x)$$

with

$$X_t^{(n)}(x) = \langle : x^{\otimes n} :, f^{(n)}(t;\cdot)\rangle,$$

then $\nabla X(t,s) = \partial_s X_t$ is given by

$$\sum_{n=1}^{\infty} n\langle : x^{\otimes(n-1)} :, f^{(n)}(t;s,\cdot)\rangle.$$

Therefore, if we assume that for every n, $f^{(n)} \in H_\alpha(D_+^n)$, $\alpha > 1/2$, and some technical summability condition (see [KR 88]), then ∇X has a trace from the right. *Mutatis mutandis*, we have a trace from the left.

From now on suppose that X satisfies the above conditions. The resulting stochastic integrals will be denoted as follows

$$\int_0^1 X_t dB(t+) := \int_0^1 (\partial_t^* + \partial_{t+}) X_t dt$$

$$\int_0^1 X_t dB(t-) := \int_0^1 (\partial_t^* + \partial_{t-}) X_t dt$$

and we call them *forward* ("+") and *backward* ("−") *integrals of* X, respectively.

It turns out that if X is non-anticipating then $\partial_{t+} X_t = 0$ ($dt \otimes d\mu$-a.e.) and therefore in this case the forward integral coincides with the Itô and the Skorokhod integrals of X [KR 88]. For general X, however, the forward and Skorokhod integrals are *different* extensions of Itô's integral.

Let us also mention that $\frac{1}{2}\left(\int_0^1 X_t dB(t+) + \int_0^1 X_t dB(t-)\right)$ provides an extension of Stratonovich's integral to anticipating integrands [KR 88], which coincides with the generalized Stratonovich integral of Nualart and Pardoux [NP 88].

As opposed to the Skorokhod integral the forward and backward integrals have the following useful properties [AP 88, KR 88]:

i) If φ is a random variable and X is a process as above such that the forward (resp. backward) integral of $\varphi \cdot X$ exists in (L^2), then

$$\int_0^1 \varphi X_t dB(t\pm) = \varphi \int_0^1 X_t dB(t\pm).$$

ii) If $\{\Delta_k = [t_k, t_{k+1});\ k = 1, \ldots, N\}$ is a partition of $[0,1)$ and X is as above, then

$$\sum_{k=1}^N X_{t_k} \cdot \left(B_{t_{k+1}} - B_{t_k}\right) \longrightarrow \int_0^1 X_t dB(t+)$$

and

$$\sum_{k=1}^N X_{t_{k+1}} \cdot \left(B_{t_{k+1}} - B_{t_k}\right) \longrightarrow \int_0^1 X_t dB(t-)$$

in (L^2) as the mesh of the partition tends to zero.

These two properties and their generalization form the basis of the following result [AP 88]: *Itô's lemma holds for the forward integral in its usual form.*

§3. Stochastic Differential Equations

In this section, we consider anticipating stochastic differential equations. Although stochastic integrals of anticipating integrands have been studied since 1972, only very recently some results about anticipating stochastic differential equations of special types have appeared [Bu 87, Bu 89, OP 88, PP 88, Sht 86] (cf. also the article of Ocone [Oc 89] in these proceedings and the references therein). The general case of such equations seems to be very hard to handle. Consider, for example, the following stochastic differential equation

$$dX_t = \partial_t^* f(t, X_t, \cdot)dt, \quad X_0 = x$$

or the equivalent stochastic integral equation

$$X_t = x + \int_0^t \partial_s^* f(s, X_s, \cdot)ds, \tag{3.1}$$

where f may be anticipating. If one tries the classical L^2-theory to solve this equation, then in view of formula (2.4), the unbounded white noise differentiation ∂_s appears so that neither the contraction principle nor the iteration method will work. However, the U-functional in white noise calculus provides an approach to avoid this difficulty and can be used to show the existence and uniqueness of a solution for a fairly general equation. It should be pointed out that in the U-functional method Picard's iteration works rather well, while the contraction principle fails. Moreover, the U-functional method gives a direct way to calculate solutions of certain stochastic differential equations. In fact, it was first introduced by Hida [Hi 83] to solve a linear stochastic partial differential equation.

In order to explain the U-functional method clearly, we will state a theorem and sketch its proof only for a special stochastic integral equation, i.e. the equation in (3.1). The details of the proof and some generalizations will appear elsewhere.

Recall that $U[\partial_s^*\varphi](\xi) = \xi(s)U[\varphi](\xi)$ from (1.6). Thus if we take the U-functionals on both sides of (3.1), then

$$U[X_t](\xi) = x + \int_0^1 \xi(s)U[f(s,X_s,\cdot)](\xi)ds, \quad \xi \in \mathcal{S}(\mathbb{R}).$$

Thus the idea is to put conditions on $U[f(s,X_s)]$ so that the iteration procedure can be applied.

Theorem. *Suppose $f : [0,1] \times (\mathcal{S})^* \longrightarrow (\mathcal{S})^*$ satisfies the following conditions:*
1) For any $t \in [0,1]$, $x \in \mathbb{R}$, and $\xi \in \mathcal{S}(\mathbb{R})$,

$$\int_0^1 |U[f(t,x)](\xi)|dt < \infty.$$

2) For each $\Phi \in (\mathcal{S})^*$, there exists $p \geq 1$ such that $f(t,\Phi) \in (\mathcal{S})_p^*$ for all $t \in [0,1]$.

3) There exists a constant $c > 0$ such that for any Φ, $\Psi \in (\mathcal{S})^*$, $\xi \in \mathcal{S}(\mathbb{R})$, and $t \in [0,1]$,

$$|U[f(t,\Phi)](\xi) - U[f(t,\Psi)](\xi)| \leq c|U[\Phi](\xi) - U[\Psi](\xi)|.$$

Then the stochastic integral equation

$$X_t = x + \int_0^t \partial_s^* f(s,X_s)ds$$

has a unique weak solution X_t in $(\mathcal{S})^*$.

Remark. It is possible to state a sufficient Lipschitz condition without referring to the U-functional. For instance, the following condition implies 3) above: There exists a constant $c > 0$ such that for any $\Phi, \Psi \in (\mathcal{S})^*$, $\varphi \in (\mathcal{S})$, and $t \in [0,1]$,

$$| \ll f(t,\Phi) - f(t,\Psi), \varphi \gg | \leq c \mid \ll \Phi - \Psi, \varphi \gg |.$$

Proof. The uniqueness of a solution is easy to show. Suppose Y_t is another solution. Let F_t and G_t be the U-functionals of X_t and Y_t, respectively. Then, by the Lipschitz condition 3), we have

$$|F_t(\xi) - G_t(\xi)| \leq c|\xi|_\infty \int_0^t |F_s(\xi) - G_s(\xi)|ds,$$

where $|\xi|_\infty = \sup\{|\xi(t)|;\ 0 \le t \le 1\}$. It follows from the Gronwall inequality that $F_t(\xi) - G_t(\xi) = 0$ for all $t \in [0,1]$ and $\xi \in \mathcal{S}(\mathbb{R})$. Thus $X_t = Y_t$.

Now, we sketch a proof for the existence of a solution. Define

$$\begin{aligned} X_t^{(0)} &= x \\ X_t^{(n+1)} &= x + \int_0^t \partial_s^* f(s, X_s^{(n)})ds, \quad n \ge 0. \end{aligned}$$

Let $F_t^{(n)}$ be the U-functional of $X_t^{(n)}$. Then

$$\begin{aligned} F_t^{(0)} &= x \\ F_t^{(n+1)}(\xi) &= x + \int_0^t \xi(s) U[f(s, X_s^{(n)})](\xi)ds. \end{aligned} \tag{3.2}$$

By using the Lipschitz condition 3), we get

$$|F_t^{(n+1)}(\xi) - F_t^{(n)}(\xi)| \le c|\xi|_\infty \int_0^t |F_s^{(n)}(\xi) - F_s^{(n-1)}(\xi)|ds.$$

Therefore, for $t \in [0,1]$ and $\xi \in \mathcal{S}(\mathbb{R})$,

$$|F_t^{(n+1)}(\xi) - F_t^{(n)}(\xi)| \le (n!)^{-1} c^n t^n |\xi|_\infty^{n+1} \int_0^1 |U[f(s,x)](\xi)|ds. \tag{3.3}$$

Hence

$$|F^{(n+1)}(\xi) - F^{(n)}(\xi)|_\infty \le (n!)^{-1} c^n |\xi|_\infty^{n+1} \int_0^1 |U[f(s,x)](\xi)|ds,$$

where $|\cdot|_\infty$ is the sup-norm in the t-variable. Thus for each $\xi \in \mathcal{S}(\mathbb{R})$, $\{F^{(n)}(\xi); n \ge 0\}$ converges uniformly on $[0,1]$. Let

$$F_t(\xi) = \lim_{n\to\infty} F_t^{(n)}(\xi).$$

Note that from (1.3)

$$F_t^{(n)}(\xi) = e^{-|\xi|_2^2/2} \ll X_t^{(n)}, e^{\langle\cdot,\xi\rangle} \gg .$$

Therefore,

$$\lim_{n\to\infty} \ll X_t^{(n)}, e^{\langle\cdot,\xi\rangle} \gg = F_t(\xi) e^{|\xi|_2^2/2}. \tag{3.4}$$

Note that the collection $\{e^{\langle\cdot,\xi\rangle};\ \xi \in \mathcal{S}(\mathbb{R})\}$ spans a dense subspace of $(\mathcal{S})$. Hence, the sequence $\{X_t^{(n)}; n \geq 0\}$ converges on a dense subset of $(\mathcal{S})$. On the other hand, it follows from (3.3) and condition 2) that

$$|F_t^{(n)}(\xi)| \leq c_1 \exp(c_2 |\xi|_{2,p}), \tag{3.5}$$

where $|\cdot|_{2,p}$ is the norm of $\mathcal{S}_p(\mathbb{R})$, i.e. $|\xi|_{2,p} = |A^p \xi|_2$, $A = -d^2/dx^2 + x^2 + 1$. The inequality (3.5) implies that there exist positive constants q and L such that for all t and n,

$$\|X_t^{(n)}\|_{2,-q} \leq L,$$

where $\|\cdot\|_{2,-q}$ is the norm of $(\mathcal{S})_q^*$. Therefore, $\{X_t^{(n)};\ n \geq 0\}$ converges weakly in $(\mathcal{S})^*$ to a limit, say X_t, i.e.

$$\lim_{n\to\infty} \ll X_t^{(n)}, \varphi \gg = \ll X_t, \varphi \gg \quad \text{for all} \quad \varphi \in (\mathcal{S}).$$

Thus, by (3.4), we have

$$\ll X_t, e^{\langle\cdot,\xi\rangle} \gg = F_t(\xi) e^{|\xi|_2^2/2},$$

or

$$F_t(\xi) = e^{-|\xi|_2^2/2} \ll X_t, e^{\langle\cdot,\xi\rangle} \gg .$$

This says that F_t is the U-functional of X_t. Finally, letting $n \to \infty$ in (3.2), we obtain

$$F_t(\xi) = x + \int_0^t \xi(s) U[f(s, X_s)](\xi) ds.$$

Therefore, X_t satisfies the following equation in the weak sense

$$X_t = x + \int_0^t \partial_s^* f(s, X_s) ds.$$

This completes the proof.

Example 1.

$$\begin{aligned} X_t &= x + \int_0^t \partial_s^* X_s ds \\ &= x + \int_0^t X_s dB(s) \qquad \text{(in Itô's sense).} \end{aligned}$$

It is well-known that the solution is $X_t = x\exp[B(t) - \frac{1}{2}t]$, which can be checked by using Itô's lemma. However, we can derive this solution directly by the U-functional method. Let F_t be the U-functional of X_t. Then F_t satisfies the following equation for each ξ in $\mathcal{S}(\mathbb{R})$:

$$\frac{d}{dt}F_t(\xi) = \xi(t)F_t(\xi), \quad F_0(\xi) = x.$$

Thus $F_t(\xi)$ is given by

$$\begin{aligned} F_t(\xi) &= x\exp\left[\int_0^t \xi(s)ds\right] \\ &= x\sum_{n=0}^{\infty}\frac{1}{n!}\left(\int_0^t \xi(s)ds\right)^n. \end{aligned}$$

Note that $U[H_n(B(t)/\sqrt{2t})](\xi) = (2/t)^{n/2}\left(\int_0^t \xi(s)ds\right)^n$, (see e.g. [Kuo 83]). Here H_n is the Hermite polynomial of degree n, i.e. $H_n(x) = (-1)^n e^{x^2}(d/dx)^n e^{-x^2}$. Therefore, X_t is given by

$$X_t = x\sum_{n=0}^{\infty}\frac{1}{n!}\left(\frac{t}{2}\right)^{n/2} H_n\left(B(t)/\sqrt{2t}\right).$$

But, by using the generating function of the Hermite polynomials, we see easily that

$$X_t = x\exp\left[B(t) - \frac{1}{2}t\right].$$

Example 2.

$$X_t = x + \int_0^t \partial_s^*\left(B(1)\diamond X_s\right)ds,$$

where $\diamond$ denotes the Wick multiplication in (1.4).

Let F_t be the U-functional of X_t. Then F_t satisfies the following equation for each ξ in $\mathcal{S}(\mathbb{R})$:

$$\frac{d}{dt}F_t(\xi) = \xi(t)\left(\int_0^1 \xi(s)ds\right)F_t(\xi), \quad F_0(\xi) = x.$$

The solution is given by

$$\begin{aligned} F_t(\xi) &= x\exp\left[\int_0^1 \xi(s)ds\int_0^t \xi(u)du\right] \\ &= x\exp\left[-\frac{1}{2}(\xi,\ L\xi)\right], \end{aligned}$$

where $(\cdot\ ,\ \cdot)$ is the $L^2(\mathbb{R})$ inner product and L is the following operator

$$Lf(s) = -2 \cdot 1_{(0,1]}(s) \int_0^t f(u)du, \quad f \in L^2(\mathbb{R}).$$

It can be checked [Kuo 75, KP 89, SH 83] that F_t is the U-functional of

$$X_t(z) = x \left\{\det[I + L(I-L)^{-1}]\right\}^{\frac{1}{2}} \exp\left[-\frac{1}{2}(z,\ L(I-L)^{-1}z)\right], \quad z \in \mathcal{S}'(\mathbb{R}).$$

Moreover, a direct calculation shows that

$$\det\left[I + L(I-L)^{-1}\right] = (1+2t)^{-1}$$

and

$$\begin{aligned}\exp\left[-\frac{1}{2}(z,\ L(I-L)^{-1}z)\right] &= \exp\left[(1+2t)^{-1}\langle z,\ 1_{(0,1]}\rangle\langle z,\ 1_{(0,t]}\rangle\right] \\ &= \exp\left[(1+2t)^{-1}B(1)B(t)\right].\end{aligned}$$

Here we have substituted z by the white noise B' so that $\langle z,\ 1_{(0,1]}\rangle = \int_0^1 B'(u)du = B(1)$ and $\langle z,\ 1_{(0,t]}\rangle = B(t)$. Therefore, the solution X_t is given by

$$X_t = x(1+2t)^{-\frac{1}{2}} \exp\left[(1+2t)^{-1}B(1)B(t)\right].$$

Example 3.

$$X_t = x + \int_0^t \partial_s^* \left(B(s) \diamond X_s\right) ds.$$

The U-functional F_t of X_t satisfies following the equation for each $\xi \in \mathcal{S}(\mathbb{R})$:

$$\frac{d}{dt} F_t(\xi) = \xi(t) \left(\int_0^t \xi(s)ds\right) F_t(\xi), \quad F_0(\xi) = x.$$

Thus $F_t(\xi)$ is given by

$$F_t(\xi) = x \exp\left[\frac{1}{2}\left(\int_0^t \xi(s)ds\right)^2\right]$$

It can be checked [KP 89] that the solution X_t is then given by

$$X_t = x(1+t)^{-\frac{1}{2}} \exp\left[(1+t)^{-1}B(t)^2\right].$$

Example 4.

$$X_t = x + \int_0^t \partial_s^*(B'(s) \diamond X_s)ds.$$

Since $U[B'(t)](\xi) = \xi(t)$, the U-functional F_t of X_t satisfies the equation for each $\xi \in \mathcal{S}(\mathbb{R})$:

$$\frac{d}{dt}F_t(\xi) = \xi(t)^2 F_t(\xi), \quad F_0(\xi) = x.$$

Hence $F_t(\xi)$ is given by

$$F_t(\xi) = x \exp\left[\int_0^t \xi(s)^2 ds\right]$$

It has been shown in [Kuo 83, Kuo 88, SH 83] that

$$U\left[\mathcal{N}\exp\left(c\int_0^t B'(s)^2 ds\right)\right](\xi) = \exp\left(\frac{c}{1-2c}\int_0^t \xi(s)^2 ds\right), \quad c \neq \frac{1}{2},$$

where $\mathcal{N}\exp$ denotes the renormalized exponential functional. Therefore, the solution X_t is given by

$$X_t = x\mathcal{N}\exp\left[\frac{1}{3}\int_0^t B'(s)^2 ds\right].$$

§4. Concluding Remarks

In this section, we make some remarks about possible extensions of the theorem in §3 and several related problems. The details will appear elsewhere.

1) A trivial extension is to add a drift term, i.e.

$$X_t = x + \int_0^t \partial_s^* f(s, X_s)ds + \int_0^t g(s, X_s)ds. \tag{4.1}$$

If $g : [0,1] \times (\mathcal{S})^* \to (\mathcal{S})^*$ satisfies the same conditions as f in the theorem, then the equation in (4.1) has a unique weak solution in $(\mathcal{S})^*$.

2) If we observe that $\partial_s^* f(s, X_s) = B'(s) \diamond f(s, X_s)$, then it becomes clear that we can generalize the theorem to the following equation

$$X_t = x + \int_0^t \theta(s) \diamond f(s, X_s)ds + \int_0^t \varrho(s) \diamond g(s, X_s)ds, \tag{4.2}$$

where f and g satisfy the same conditions as in the theorem, θ and $\varrho : [0,1] \to (\mathcal{S})^*$ satisfy the following conditions ($\lambda = \theta, \varrho$) :

a) $\sup_{0\le s\le 1} |U[\lambda(s)](\xi)| < \infty, \quad \xi \in \mathcal{S}(\mathbb{R})$.
b) There exists $q \ge 1$ such that $\lambda(s) \in (\mathcal{S})^*_q$ for all $s \in [0,1]$.

A special case of (4.2) is the famous Oosawa equation [Oo 89]:

$$X_t = x + \int_0^t [-k_+ E_s(-f)X_s + k_- E_s(f)(1 - X_s)]ds,$$

where $k_\pm > 0$ and $E_t(f) = \exp\left[\int_0^t f(t-u)\partial_u^* du\right]$.

3) The U-functional method can be applied to equations involving not only ∂_t^*, but also the one-sided white-noise differentiation ∂_{t+} and ∂_{t-} [KR 88], e.g.

$$X_t = x + \int_0^t \partial_s^* f(s, X_s)ds + \int_0^t \partial_{s+} g(s, X_s)ds + \int_0^t \partial_{s-} h(s, X_s)ds,$$

where f satisfies the same conditions as in the theorem, while g and h satisfy certain conditions. However, in order to get the uniqueness of a solution, we must impose conditions relating g and h to f. For example, consider the equation

$$X_t = x + \int_0^t \partial_{s-} X_s ds. \tag{4.3}$$

It can be checked that for any C^1-function θ on $\mathbb{R}$ with $\theta(0) = 0$ and any $L^2(0,1)$ function u with left hand limit $u(t-)$ for all $0 \le t \le 1$, the following Brownian functional is a solution of the equation in (4.3)

$$X_t = x + \theta\left(\int_0^t u(s)ds + \int_0^t u(s)dB(s)\right).$$

4) The use of Wick multiplication $\diamond$ in the examples in Section 3 makes the equations easy to solve. On the other hand, instead of the equation in Example 2, we can consider the equations $dX_t = \partial_t^*(B(1)X_t)dt$ [Bu 87] and $dX_t = B(1)X_t \circ dB(t)$, where $\circ$ denotes the Stratonovich integral. These two equations cannot be handled by our theorem. Nevertheless, they can be solved in terms of series by the U-functional method.

5) The U-functional method transfers a problem about generalized Brownian functionals in $(\mathcal{S})^*$ into another problem about functions defined on $\mathcal{S}(\mathbb{R})$. The analysis of functions on $\mathcal{S}(\mathbb{R})$ was initiated by Lévy [Lé 51]. For further development, see the book [Pol 88] and references therein. On the other hand, for the analysis of generalized Brownian functionals, there is an important question as when the solution X_t is in the space (L^2) of μ-square integrable functions on $\mathcal{S}'(\mathbb{R})$, or when X_t is in $(\mathcal{S})_p^*$ for some p. This type of regularity problem will be addressed elsewhere.

Acknowledgement:

We would like to thank many friends for their interest in our work, their criticism and the discussions we have had with them : S. Albeverio, J. Asch, P. L. Chow, L. Gross, G. Kallianpur, M. Röckner, T. C. Sun, and J. A. Yan.

Most of all we feel deeply grateful to T. Hida and L. Streit. Their constant encouragement and neverending curiosity in white noise analysis and its applications have inspired us in the past through so many edifying experiences.

References

[Ad 75] Adams, R.A.: *Sobolev Spaces.* New York, London: Academic Press (1975)

[AP 88] Asch, J. and Potthoff, J.: Itô's lemma without non-anticipatory conditions; *BiBoS Publications* **316** (1988), to appear in *Probab. Th. Rel. Fields*

[BM 82] Berger, M.A. and Mizel, V.J.: An extension of the stochastic integral; *Ann. Probab.* **10** (1982) 435–450

[Bu 87] Buckdahn, R.: Skorohod's integral and linear stochastic differential equations; *Preprint* (1987)

[Bu 89] Buckdahn, R.: Anticipating linear stochastic differential equations, *Lecture Notes in Control and Information Sciences* **136** (1989) 18–23

[GT 82] Gaveau, B. and Trauber, P.: L'intégrale stochastique comme opérateur de divergence dans l'espace fonctionnel; *J. Func. Anal.* **46** (1982) 230–238

[Go 72] Goodman, V.: A divergence theorem for Hilbert space; *Trans. Amer. Math. Soc.* **164** (1972) 411–426

[Hi 75] Hida, T.: *Analysis of Brownian Functionals.* Carleton Mathematical Lecture Notes, no. **13** (1975)

[Hi 80] Hida, T.: *Brownian Motion.* Berlin, Heidelberg, New York: Springer-Verlag (1980)

[Hi 83] Hida, T.: Causal analysis and an application to prediction theory; in: *Prediction Theory and Harmonic Analysis*; V. Mandrekar and H. Salehi (eds.) (1983)

[HKPS] Hida, T., Kuo, H.-H., Potthoff, J. and Streit, L.: *White Noise: An Infinite Dimensional Calculus.* Monograph in preparation

[HP 89] Hida, T. and Potthoff, J.: White noise analysis – an overview; these proceedings

[HPS 88] Hida, T., Potthoff, J. and Streit, L.: Dirichlet forms and white noise analysis; *Commun. Math. Phys.* **116** (1988) 235–245

[Hit 72] Hitsuda, M.: Formula for Brownian partial derivatives; in: *Proc. 2nd Japan-USSR Symp. Probab. Th.* **2** (1972) 111–114

[Itô 78] Itô, K.: Extension of stochastic integrals; in: *Proc. Intern. Symp. Stochastic Differential Equations*; K. Itô (ed.) (1978) 95–109. Tokyo: Kinokuniya, New York: Wiley

[IKT 89] Ito, Y., Kubo, I. and Takenaka, S.: Calculus on Gaussian white noise and Kuo's Fourier transformation; these proceedings

[Ki 75] King, R.: Stochastic integrals and metadistributions. Applications to stochastic partial differential equations and quantum field theory. Ph. D. dissertation (1975), Cornell Univ.

[Kr 83] Krée, M., Krée, P.: Continuité de la divergence dans les espaces de Sobolev relatifs à l'espace de Wiener; *C. R. Acad. Sci. Paris* **296** (1983) 833–836

[KT 80a] Kubo, I. and Takenaka, S.: Calculus on Gaussian white noise I; *Proc. Japan Acad.* **56A** (1980) 376–380

[KT 80b] Kubo, I. and Takenaka, S.: Calculus on Gaussian white noise II; *Proc. Japan Acad.* **56A** (1980) 411–416

[KY 87] Kubo, I. and Yokoi, Y.: A remark on the space of testing random variables in the white noise calculus; to appear in *Nagoya Math. J.*

[Kuo 75] Kuo, H.-H.: *Gaussian Measures in Banach Spaces.* Lecture Notes in Math., Vol. 463. Berlin, Heidelberg, New York: Springer-Verlag (1975)

[Kuo 82] Kuo, H.-H.: On Fourier transform of generalized Brownian functionals; *J. Multivariate Anal.* **12** (1982) 415–431

[Kuo 83] Kuo, H.-H.: Brownian functionals and applications; *Acta Appl. Math.* **1** (1983) 175–188

[Kuo 88] Kuo, H.-H.: White noise calculus; in: *Algebra, Analysis and Geometry;* M.C. Kang and K.W. Lih (eds.) (1988) 77–119, World Scientific

[KP 89] Kuo, H.-H. and Potthoff, J.: Anticipating stochastic differential equations; to appear in: *Proc. 5th Intern. Vilnius Conf. on Probab. Th. and Math. Stat.*, Vilnius (1989)

[KR 88] Kuo, H.-H. and Russek, A.: White noise approach to stochastic integration; *J. Multivariate Analysis* **24** (1988) 218–236

[Lee 89] Lee, Y.-J.: A reformulation of white noise calculus; *Preprint* (1989)

[Lé 51] Lévy, P.: *Problèmes Concrets d'Analyse Fonctionnelle.* Paris: Gauthier-Villars (1951)

[MY 88] Meyer, P.A. and Yan, J.A.: Distributions sur l'espace de Wiener (suite); to appear in *Séminaire de Probabilités*

[NP 88] Nualart, D. and Pardoux, E.: Stochastic calculus with anticipating integrands; *Probab. Th. Rel. Fields* **78** (1988) 535–581

[NZ 86] Nualart, D. and Zakai, M.: Generalized stochastic integrals and the Malliavin calculus; *Probab. Th. Rel. Fields* **73** (1986) 255–280

[Oc 84] Ocone, D.: Malliavin's calculus and stochastic integral representation of functionals of diffusion processes; *Stochastics* **12** (1984) 161–185

[OP 88] Ocone, D. and Pardoux, E.: Non-adapted solutions of stochastic differential equations; *Preprint* (1988)

[Oo 89] Oosawa, F.: Hierarchy of noise production in living cells; these proceedings

[PP 88] Pardoux, E. and Protter, P.: Stochastic Volterra equations with anticipating coefficients; *Preprint* (1988)

[Pol 88] Polishchuk, E. M.: *Continual Means and Boundary Value Problems in Function Spaces.* Birkhäuser Verlag, Basel, Boston, Berlin (1988)

[PY 89] Potthoff, J. and Yan J.A.: Some results about test and generalized functionals of white noise; to appear in: *Proc. Singapore Probab. Conf.* (1989), L.Y. Chen (ed.)

[Ra 74] Ramer, R.: On nonlinear transformations of Gaussian measures; *J. Func. Anal.* **15** (1974) 166–187

[Sht 86] Shiota, Y.: A linear stochastic integral equation containing the extended Itô integral; *Math. Rep. Toyama Univ.* **9** (1986) 43–65

[Sk 75] Skorokhod, A. V.: On a generalization of a stochastic integral; *Th. Probab. Appl.* **20** (1975) 219–233

[SH 83] Streit, L. and Hida T.: Generalized Brownian functionals and the Feynman integral; *Stoch. Processes Appl.* **16** (1983) 55–69

[Yo 87] Yokoi, Y.: Positive generalized functionals; to appear in *Hiroshima Math. J.*

[Yo 89] Yokoi, Y.: Positive generalized Brownian functionals; *Preprint* (1989)

A Reformulation of White Noise Calculus

Yuh–Jia Lee
Department of Mathematics
National Cheng–Kung University
Tainan, Taiwan, ROC.

Abstract

Taking the space ($\mathscr{S}$) as test functionals, the white noise calculus initiated by T. Hida is reformulated in such a way that the generalized white noise functionals are defined and studied through their functional representations instead of employing their U–functionals.

1. Introduction

In our previous paper [13], the white noise calculus was established on an arbitrary abstract Wiener space using a class of analytic functions of exponential growth as the test functionals. There the generalized white noise functionals are defined and studied through their linear functional representations instead of employing their U–functionals which are essential in Hida's approach [13]. The U–functional of an ordinary functional f is nothing but the convolution $\mu * f$ of the centered Gaussian measure μ with f. Hida's approach basically enlarges the U–functionals of ordinary functionals to obtain a large class of U–functionals of certain class of "functions", known as the generalized white noise functionals, through which the theory of white noise calculus are developed. In our approach, we avoid using the extended U–functionals to define and study the generalized white noise functional but use only the U–functionals of ordinary functionals (in fact, the test functionals) to represent the generalized white functionals.

In the recent development of white noise calculus, the space $(\mathscr{S})$, which was first introduced by Kubo, Takenaka and Yokoi[5,6], has become an increasingly important one among those existent spaces of test functionals (see [3,4,5,6,13,16,17] and the papers cited there). As the connection of Hida's approach and our approach is concerned, it is desirable to reformulate the white noise calculus by our approach with the space $(\mathscr{S})$ being taken as the test functionals. In this note, we show that this can be done by applying the results of [15] in which it is shown that $(\mathscr{S})$ has an analytic version.

Notations and Definitions

$\mathscr{L}^n[X]$: the space of all n–linear operators on a normed linear space X.

$\mathscr{L}_2^n[H]$: the space of all n–linear Hilbert–Schmidt operators on a Hilbert space H.

$\mathscr{C}X$: the complexification of X, with Euclidean norm induced by the X–norm.

$Tx^n := T(x,\ldots,x)$ (n copies), where $T \in \mathscr{L}^n[X]$.

$Tx_1\ldots x_n := T(x_1,\ldots,x_n)$.

$Df(x)$: For notational convenience, this notation denotes both the Fréchet derivative and L^2–derivative of f at x. (See [2]).

$$\sigma f(x) = \int_{\mathscr{S}^*} f(x+iy)\mu(dy). \quad \text{(See [11]).}$$

$$\lambda^n T(x) = \int_{\mathscr{S}^*} T(x+iy)^n \mu(dy), \quad \text{for } T \in \mathscr{L}^n[\mathscr{S}_{-p}].$$

$$\lambda^n T(x) = L^2(\mu)\text{–}\lim_{k\to\infty} \lambda^n T_k(x), \quad \text{where } T \in \mathscr{L}_2^n[L^2] \text{ and } \{T_k\} \text{ a sequence}$$

in $\mathscr{L}^n[\mathscr{S}_{-p}]$ such that $T_k \longrightarrow T$ in $\mathscr{L}_2^n[L^2]$. (See [12]).

2. The test functionals

Let A denote the operator $Au = -u''+1+u^2$ with domain $\mathscr{D}(A) \subset L^2 = L^2(\mathbb{R}^1)$ and $\{e_n : n \in \mathbb{N}_0\}$ the CONS of L^2 consisting of eigenfunctions of A with corresponding eigenvalues $\{2n+2 : n \in \mathbb{N}_0\}$. e_n's are known as Hermite functions.

Let $\mathscr{S}$ and $\mathscr{S}^*$ be respectively the Schwartz space and tempered distributions on $\mathbb{R}^1$. For $p \in \mathbb{Z}$, let $\mathscr{S}_p$ denote the space of functions $f \in \mathscr{S}^*$ satisfying the condition :

$$\|f\|_p^2 := \sum_{j=0}^{\infty} (2j+2)^{2p}(f,e_j)^2 < \infty$$

where $(\cdot,\cdot)$ denoted the $\mathscr{S}^*-\mathscr{S}$ pairing. Then $\mathscr{S}_p$ forms a real separable Hilbert space and, for $q > p \geq 0$, we have

$$\mathscr{S} \subset \mathscr{S}_q \subset \mathscr{S}_p \subset L^2 = \mathscr{S}_0 \subset \mathscr{S}_{-p} \subset \mathscr{S}_{-q} \subset \mathscr{S}^*,$$

where $\mathscr{S} = \bigcap_{p\geq 0} \mathscr{S}_p$ and $\mathscr{S}^* = \bigcup_{p\geq 0} \mathscr{S}_{-p}$. $\mathscr{S}$ is provided with the projective limit topology.

According to [15], one may define the space of test functionals $(\mathscr{S})$ as follows :

Let μ be the centered Gaussian measure (with variance parameter 1) on $\mathscr{S}^*$ and let $(L^2) = L^2(\mathscr{S}^*,\mu)$ be the square integrable Borel functions on $\mathscr{S}^*$ with respect to μ.

2.1. Definition. For $p \in \mathbb{N}$, let $(\mathscr{S}_p)$ be the collection of all functions $f \in (L^2)$ such that the number

$$\|f\|_{(\mathscr{S}_p)} := \sum_{n=0}^{\infty} \frac{1}{n!}\|D^n \mu f(0)\|^2_{\mathscr{L}_2^n[\mathscr{S}_{-p}]}$$

is finite, where $\mu f(x) = \mu * f(x)$.

Denote the dual of $(\mathcal{S}_p)$ by $(\mathcal{S}_{-p})$. Then we have

$$(\mathcal{S}_q) \subset (\mathcal{S}_p) \subset (L^2) \subset (\mathcal{S}_{-p}) \subset (\mathcal{S}_{-q})$$

for $q > p$. Set $(\mathcal{S}) = \bigcap_{p \geq 0} (\mathcal{S}_p)$ and endow $(\mathcal{S})$ with the projective limit topology, we obtain the Gelfand triplets

$$(\mathcal{S}) \subset (L^2) \subset (\mathcal{S})^*.$$

$(\mathcal{S})$ has an analytic version $\mathcal{A}_\infty$ given as follows :

For $p \in \mathbb{N}_0$, denote by $\mathcal{A}_p$ the space of functions f defined on $\mathcal{C}\mathcal{S}^*$ satisfying the following conditions :

(i) f is continuous on $\mathcal{C}\mathcal{S}^*$.

(ii) f, regarded as a functions defined on $\mathcal{C}\mathcal{S}_{-p}$, is analytic.

(iii) There exists a constant C_f, depending only on f, such that

$$|f(z)| \leq C_f \exp(\|z\|_{-p}^2/2)$$

for all $z \in \mathcal{C}\mathcal{S}_{-p}$.

For $f \in \mathcal{A}_p$, define

$$\|f\|_{\mathcal{A}_p} = \sup_{z \in \mathcal{C}\mathcal{S}_{-p}} \{|f(z)| \exp(-\|z\|_{-p}^2/2)\}.$$

Then $\{\mathcal{A}_p, \|\cdot\|_{\mathcal{A}_p}) : p \in \mathbb{N}_0\}$ is a decreasing sequence of Banach spaces.

Define $\mathcal{A}_\infty = \bigcap_{p \in \mathbb{N}_0} \mathcal{A}_p$, and endow $\mathcal{A}_\infty$ with the projective limit topology, then $\mathcal{A}_\infty$ becomes a complete topological linear space.

2.2. Proposition [15]. Let $f \in (\mathcal{S}_p)$, $p \in \mathbb{N}$.

(a) The function $V_p f(z) = \sum_{n=0}^{\infty} \frac{1}{n!} \lambda^n D^n \mu f(0)(z)$ $(z \in \mathcal{C}\mathcal{S}_{-p})$ is analytic in $\mathcal{C}\mathcal{S}_{-p}$.

(b) There exists a unique continuous function $\tilde{f}$, defined on $\mathscr{C}\mathscr{S}^*$ and independent of p, such that $\tilde{f} = V_p f$ on $\mathscr{C}\mathscr{S}_{-p}$.

2.3. Theorem [15]. Let $\tilde{f}$ be the function define in Proposition 2.2 for $f \in (\mathscr{S}_p)$.

(a) If $f \in (\mathscr{S})$, then $\tilde{f} \in \mathscr{A}_\infty$. Moreover,

$$\|\tilde{f}\|_{\mathscr{A}_{p-1}} \leq C_p \|f\|_{(\mathscr{S}_p)},$$

where C_p is a constant.

(b) $\mathscr{A}_\infty \subset (\mathscr{S})$. In fact, for $p \geq 3$, there exists a constant C_p' such that

$$\|f\|_{(\mathscr{S}_{p-3})} \leq C_p' \|f\|_{\mathscr{A}_p}$$

for $f \in \mathscr{A}_p$.

(c) $\mathscr{A}_\infty^* = \bigcup_{p \in \mathbb{N}} \mathscr{A}_p^*$, where $\mathscr{A}_\infty^*$ and $\mathscr{A}_p^*$ are respectively the dual spaces of $\mathscr{A}_\infty$ and $\mathscr{A}_p$.

The above Theorem shows that $\mathscr{A}_\infty$ is an analytic version of $(\mathscr{S})$. More properties of $\mathscr{A}_\infty$ are given in the following

2.4. Proposition.

(a) If $f \in \mathscr{A}_\infty$, then, for $h_1,\dots,h_n \in \mathscr{S}$ and for $p \in \mathbb{N}$,

$$|D^n f(z)h_1 \dots h_n| \leq \|f\|_{(\mathscr{S}_p)} \exp[\|z\|_{-p}^2 + (\sum_{j=1}^{\infty} \|h_j\|_{-p})^2].$$

(b) $\sum_{n=0}^{\infty} \frac{1}{n!} D^n f(0) z^n$ converges to f in $\mathscr{A}_\infty$ for any $f \in \mathscr{A}_\infty$.

(c) $\mathscr{A}_\infty$ is an algebra.

(d) The Wiener–Ito decomposition of $f \in \mathscr{A}_\infty$ converges to f in $\mathscr{A}_\infty$.

(e) For $f \in \mathscr{A}_\infty$, define $\mathscr{F}_{\alpha,\beta} f(y) = \int_{\mathscr{S}^*} f(\alpha x + \beta y)\mu(dx)$ for $\alpha,\beta \in \mathbb{C}$. Then $\mathscr{F}_{\alpha,\beta}(\mathscr{A}_\infty) \subset \mathscr{A}_\infty$ and $\mathscr{F}_{\alpha,\beta}$ is continuous on $\mathscr{A}_\infty$.

(f) Let $f \in \mathscr{A}_\infty$ and S a bounded linear operator on L^2. Let $P_n(x) = \sum_{j=0}^{n} (x,e_n)e_n$ for $x \in \mathscr{S}^*$. Then $f(SP_nx)$ converges to a function $f_S(x)$ in measure. Moreover, $f_S(x) \in \bigcap_{p\geq 1} L^P(\mathscr{S}^*,\mu)$.

Proof. (a), (b), (c), (d) and (e) has been verified in [15]. We need only to proved (f).

First we observe that, for any $p \in \mathbb{N}$ and for any $m,n \in \mathbb{N}$, we have

$$|f(SP_nx)-f(SP_mx)|$$
$$\leq (\exp[2(\|SP_nx\|_{-p}^2+\|SP_mx\|_{-p}^2])\|SP_nx-SP_mx\|_{-p}.$$

Since $\|SP_nx\|_{-p}$ converges in measure to a measurable function $\|Sx\|_{-p}^{\sim}$ according to Lemma 4.1 from [7; page 59] and Lemma 4.3 from [7; page 62], $\{f(SP_nx)\}$ is a Cauchy sequence in measure and hence there exists a measurable function, say $f_s(x)$, such that $f(SP_nx) \longrightarrow f_s(x)$ in measure.

Furthermore, for $p \in \mathbb{N}$, we have

$$|f(SP_nx)| \leq [\exp(\tfrac{1}{2}\|SP_nx\|_{-p}^2)]\|f\|_{\mathscr{A}_p}$$

so that

$$|f_s(x)| \leq \|f\|_{\mathscr{A}_p}\exp[\tfrac{1}{2}(\|Sx\|_{-p}^{\sim})^2]. \tag{1}$$

For any $\beta \geq 1$, choose p so large such that $\beta\|S\|4^{-P} < 1$. Then we have

$$\int_{\mathscr{S}^*} \exp[\tfrac{\beta}{2}(\|Sx\|_{-p}^{\sim})^2]\mu(dx)$$
$$= \int_0^\infty \beta u e^{\frac{\beta}{2}u^2} \mu\{\|Sx\|_{-p}^{\sim} > u\}du + 1$$

$$\leq \int_0^\infty \beta u e^{\frac{\beta}{2}u^2} \mu\{\|x\|_{-p} > u\|S\|^{-1}\}du + 1$$

(by Lemma 4.3 in [7, page 62])

$$= \int_{\mathscr{S}^*} \exp[\tfrac{1}{2}\beta\|S\|^2\|x\|_{-p}^2]\mu(dx) < \infty.$$

This implies that $f_s \in L^\beta(\mathscr{S}^*,\mu)$ for all $\beta \geq 1$. ///

2.5. Proposition. Let $P_n(x) = \sum_{j=0}^{\infty} (x,e_j)e_j$ and $\varphi \in (\mathscr{S})$. Then $\tilde{\varphi}(P_n x)$ converges to $\tilde{\varphi}(x)$ in $\mathscr{A}_\infty$.

Proof. Observe that, for any $p \in \mathbb{N}$, $P_n(x) = \sum_{j=0}^{n} <x,(2j+2)^p e_j>_{-p}(2j+2)^p e_j$ and $P_n(x) \longrightarrow x$ in $\mathscr{S}_{-p}$. Since P_n is bounded on $\mathscr{S}_{-p}$, $\tilde{\varphi}(P_n x) \in (\mathscr{S})$. Moreover,

$$|\tilde{\varphi}(P_n x) - \tilde{\varphi}(x)| \leq \exp(4\|x\|_{-(p+1)}^2)\|P_n x - x\|_{-(p+1)}.$$

$$\leq \exp(\|x\|_{-p}^2)\|x\|_{-p}\Big(\sum_{k=n+1}^{\infty} (2k+2)^{-2}\Big)^{\frac{1}{2}}$$

$$\leq \exp(\tfrac{1}{2}\|x\|_{-(p-1)}^2)\Big(\sum_{k=n+1}^{\infty} (2k+2)^{-2}\Big)^{\frac{1}{2}}$$

for all $x \in \mathscr{S}_{-(p-1)}$ so that $\|\tilde{\varphi}\circ P_n - \tilde{\varphi}\|_{(\mathscr{S}_{-(p-1)})} \longrightarrow 0$ for all $p \geq 2$. Hence $\tilde{\varphi}\circ P_n \longrightarrow \tilde{\varphi}$ in $(\mathscr{S})$. ///

3. Generalized white noise functionals

Let $\langle\langle\cdot,\cdot\rangle\rangle_s$ ($\langle\langle\cdot,\cdot\rangle\rangle_a$) be the dual pairing of $(\mathscr{S})^*$–$(\mathscr{S})$ ($\mathscr{A}_\infty^*$–$\mathscr{A}_\infty$). Then any $F \in (\mathscr{S})^*$ can be regarded as a mumber of $\mathscr{A}_\infty^*$ through the relation $\langle\langle F,\varphi\rangle\rangle_a := \langle\langle F,\varphi\rangle\rangle_s$ for all $\varphi \in \mathscr{A}_\infty \subset (\mathscr{S})$. Conversely, if $F \in \mathscr{A}_\infty^*$ then F can be identified as an element $\tilde{F} \in (\mathscr{S})^*$ by the relation $\langle\langle \tilde{F},\varphi\rangle\rangle_s := \langle\langle F,\tilde{\varphi}\rangle\rangle_a$.

Members of $(\mathscr{S})^*$ or $\mathscr{A}_\infty^*$ shall be called generalized white noise functionals.

3.1. Example. For $t_1,t_2,\ldots,t_n \in \mathbb{R}^1$, define

$$\langle\langle :\dot{B}(t_1)\cdots\dot{B}(t_n):,\varphi\rangle\rangle_s := D^n\mu\tilde{\varphi}(0)\delta_{t_1}\ldots\delta_{t_n}$$

for $\varphi \in (\mathscr{S})$. Then, by Proposition 2.4(a), $:\dot{B}(t_1)\cdots\dot{B}(t_n): \in (\mathscr{S})^*$. In particular, $\langle\langle :\dot{B}(t),\varphi\rangle\rangle_s = D\mu\tilde{\varphi}(0)\delta_t$.

3.2. Example (Generalized Multiple Wiener Integral). If $f \in \mathscr{S}^*(\mathbb{R}^n)$, then the generalized multiple Wiener integral of f is defined by

$$\langle\langle \int_{\mathbb{R}^n} f(t_1,\ldots,t_n)dB(t_1)\cdots dB(t_n),\varphi\rangle\rangle_s$$
$$:= \int_{\mathbb{R}^n} f(t_1,\ldots,t_n)D^n\mu\tilde{\varphi}(0)\delta_{t_1}\cdots\delta_{t_n}dt_1\cdots dt_n,$$

where the integral on the RHS is interpreted as the $\mathscr{S}^*(\mathbb{R}^n)$–$\mathscr{S}(\mathbb{R}^n)$ pairing.

3.3. Example. For $\alpha \in \mathbb{C}$, define $\langle\langle :\exp(\alpha\dot{B}(t)):,\varphi\rangle\rangle_s := \mu\tilde{\varphi}(\alpha\delta_t)$ for $\varphi \in (\mathscr{S})$. Then it follows from Proposition 2.4(e) that $:\exp(\alpha\dot{B}(t)): \in (\mathscr{S})^*$.

3.4. Example. Let $E \subset \mathbb{R}^1$ be a Borel set and define the operator $T_E : L^2 \longrightarrow L^2$ by $T_E\xi(s) = 1_E(s)\xi(s)$. Then T_E is bounded linear operator on L^2 and $S_\alpha(T_E) = (I-\alpha T_E)^{-\frac{1}{2}}$ exists and bounded for $\alpha \in \mathbb{C}\backslash\{1\}$. Define

$$\langle\langle :\exp[\tfrac{\alpha}{2}\int_E \dot{B}(s)^2ds]:,\varphi\rangle\rangle_s := \int_{\mathscr{S}^*} \tilde{\varphi}_{S_\alpha(T_E)}(x)\mu(dx). \tag{2}$$

By the estimation (1) in the proof of Proposition 2.4, we see that $:\exp[\frac{\alpha}{2}\int_E \dot{B}(s)^2ds]: \in (\mathscr{S})^*$.

In particular if $E = \mathbb{R}^1$, $T_E = I$, the definition (2) becomes

$$《:\exp[\frac{\alpha}{2}\int_{-\infty}^{\infty}\dot{B}(s)^2ds]》_s := \int_{\mathscr{S}^*}\tilde{\varphi}((1-\alpha)^{-\frac{1}{2}}x)\mu(dx). \tag{3}$$

We remark here that the definition (2) corrects the wrong definition given in example 5.4 in [13].

3.5. Example. Let f be a measurable function define on $\mathscr{S}^*$ satisfy the condition that

$$\int_{\mathscr{S}_{-p}}|f(x)|e^{\|x\|_{-p}^2}\mu(dx) < \infty$$

for some $p \geq 1$. Then f generates a generalized white noise functional, still denoted by f, defined by

$$《f,\varphi》_s := \int_{\mathscr{S}_{-p}} f(x)\tilde{\varphi}(x)\mu(dx),$$

for all $\varphi \in (\mathscr{S})$. In fact $f \in (\mathscr{S}_{-p})$. In particular $L^{\alpha}(\mathscr{S}^*,\mu) \subset (\mathscr{S})^*$ for $\alpha < 1$.

3.6. Example. Let η be a measure defined on $\mathscr{S}^*$ so that, for some $p \geq 1$,

$$\int_{\mathscr{S}^*} e^{\|x\|_{-p}^2}\eta(dx) < \infty.$$

Then η generates a generalized white noise functional, still denoted by η, given by

$$《\lambda\eta,\varphi》_s := \int_{\mathscr{S}^*}\tilde{\varphi}(x)\eta(dx),$$

for all $\varphi \in (\mathscr{S})$. In fact, $\eta \in (\mathscr{S}_{-p})$ and the measurable support supp(η) of η lies in $\mathscr{S}_{-p}$.

3.7. Example. (Donsker's delta function).

The Donsker's delta function $\delta_a(B(t))$ $(t > 0)$ may be defined by

$$\langle\!\langle \delta_a(B(t)),\varphi \rangle\!\rangle_s := \frac{1}{2\pi}\int_{-\infty}^{\infty} e^{-ias-2^{-1}s^2t}\, \mathcal{T}_{1,i}\tilde{\varphi}(sh_t)ds, \tag{4}$$

for $\varphi \in (\mathscr{S})$, where $h_t = 1_{(0,t]}$, for $t > 0$.

It can be shown that $\delta_a(B(t)) \in \mathscr{A}_1^* \subset (\mathscr{S}_{-2})$. After simplification, (4) can be redefined by

$$\langle\!\langle \delta_a(B(t)),\varphi \rangle\!\rangle_s := \frac{1}{\sqrt{2\pi t}} e^{-a^2/2t}\int_{\mathscr{S}^*} \tilde{\varphi}(y-\tfrac{1}{t}(<y,h_t>-a)h_t)\mu(dy). \tag{5}$$

From (5), one see that $\delta_a(B(t))$ is a positive white noise functional. (See also [9]).

3.8. Example. Let T_E and $S_\alpha(T_E)$ be as given in Example 3.4. Then we define

$$\langle\!\langle :\dot{B}(t)^n \exp[\tfrac{\alpha}{2}\int_E \dot{B}(s)^2 ds]:, \varphi \rangle\!\rangle_s := \int_{\mathscr{S}^*} \varphi^{t,n}_{S_\alpha}(x)\mu(dx),$$

where $\varphi^{t,n}(x) = D^n\varphi(x)\delta_t^n$. In particular,

$$\langle\!\langle :\dot{B}(t)^n \exp[\tfrac{\alpha}{2}\int_{-\infty}^{\infty} \dot{B}(s)^2 ds]:, \varphi \rangle\!\rangle_s = \int_{\mathscr{S}^*} D^n\varphi(\frac{1}{\sqrt{1-\alpha}}x)\delta_t^n\mu(dx).$$

4. Calculus of generalized white noise functionals

4.1. Multiplication by a test functional.

For $F \in (\mathscr{S})^*$, $\eta \in (\mathscr{S})$, define $\langle\!\langle \eta F, \varphi \rangle\!\rangle_s := \langle\!\langle F, \eta\varphi \rangle\!\rangle_s := \langle\!\langle F, \tilde{\eta}\tilde{\varphi} \rangle\!\rangle_a$, for $\varphi \in (\mathscr{S})$.

4.2. Differentiation.

For $F \in (\mathscr{S})^*$, $h \in \mathscr{S}$, define $\langle\!\langle D_hF, \varphi \rangle\!\rangle_s := \langle\!\langle F, \tilde{h}\tilde{\varphi} \rangle\!\rangle_a - \langle\!\langle F, D_h\tilde{\varphi} \rangle\!\rangle_a$, for $\varphi \in (\mathscr{S})^*$, where $\tilde{h}(x) = (x,h)$.

The mapping $(h,\varphi) \longrightarrow \langle\!\langle D_n F, \varphi\rangle\!\rangle_s$ is bilinear and continuous from $\mathscr{S} \times (\mathscr{S})$ into $\mathbb{R}^1$. Denote the mapping by DF and denote the "kernel" function of DF symbolically by $\partial_t F(x)$, i.e. we interpret the formal integral $\int_{\mathscr{S}^*}\int_{\mathbb{R}^1} \partial_t F(x)\varphi(x)dt\mu(dx)$ by $\langle\!\langle D_h F, \varphi\rangle\!\rangle_s$. Symbolically we realize $\partial_t F$ as a $(\mathscr{S})^*$–valued generalized function in t–variable such that the formal integral $\int_{\mathbb{R}^1} \partial_t F h(t)dt := D_h F$ for all $h \in \mathscr{S}$. If $F \in (\mathscr{S}_p)$, $\partial_t F$ is nothing but $D_{\delta_t} F$ which makes sense for each $t \in \mathbb{R}^1$.

Next, define $D_h^* F$ by $\langle\!\langle D_h^* F, \varphi\rangle\!\rangle_s := \langle\!\langle F, D_h \varphi\rangle\!\rangle_s$ for all $\varphi \in (\mathscr{S})$ and for $h \in \mathscr{S}^*$. In particular, define $\partial_t^* = D_{\delta_t}^*$.

4.3. Translation.

For $z \in \mathscr{S}$, $F \in (\mathscr{S})^*$, define the translation $\tau_z F$ by

$$\langle\!\langle \tau_z F, \varphi\rangle\!\rangle_s := \langle\!\langle F, \tau_{-z}\tilde{\varphi}\rangle\!\rangle_a e^{-|z|_0^2/2},$$

where $\tau_y \varphi(x) = \varphi(x-y)$ for $x,y \in \mathscr{S}^*$.

4.4. The number operator.

For $F \in (\mathscr{S})^*$, define $\langle\!\langle NF, \varphi\rangle\!\rangle_s := \langle\!\langle F, N\tilde{\varphi}\rangle\!\rangle_a$, for $\varphi \in (\mathscr{S})$, where $N\tilde{\varphi}(x) = -\Delta\tilde{\varphi}(x)+(x, D\tilde{\varphi}(x))$.

4.5. The operator Δ^*.

For $F \in (\mathscr{S})^*$, define $\langle\!\langle \Delta^* F, \varphi\rangle\!\rangle_s := \langle\!\langle F, \Delta\tilde{\varphi}\rangle\!\rangle_a$.

4.6. U–functional.

For $F \in (\mathscr{S})^*$, define its U–functional by

$$U_F(\xi) = \langle\!\langle F, e^{(\cdot,\xi)}\rangle\!\rangle \exp(-|\xi|_0^2/2)$$

for $\xi \in \mathscr{S}$.

4.7. Proposition. Let $F \in (\mathscr{S})^*$. Then we have

(i) $$NF = \lim_{n\to\infty}\left[-\sum_{j=0}^{n} D_{e_j}^2 F + \sum_{j=0}^{n} \tilde{e}_j D_{e_j} F\right]$$

(ii) $$NF = \int_{-\infty}^{\infty} \partial_t^* \partial_t F \, dt$$

(iii) $$\Delta^* F = \int_{-\infty}^{\infty} \partial_t^* \partial_t^* F \, dt$$

Proof.

(i) For any $\varphi \in (\mathscr{S})$,

$$\langle\!\langle NF, \varphi\rangle\!\rangle_s = \langle\!\langle F, N\tilde{\varphi}\rangle\!\rangle_a$$
$$= \lim_{n\to\infty} \langle\!\langle F, N(\tilde{\varphi}\circ P_n)\rangle\!\rangle_a \quad \text{(by Proposition 2.5)}$$
$$= \lim_{n\to\infty} \sum_{j=0}^{n} \langle\!\langle F, [-D_{e_j}^2\tilde{\varphi} + \tilde{e}_j D_{e_j}\tilde{\varphi}]\rangle\!\rangle_a.$$

Note that, for $h \in \mathscr{S}$ and $\psi \in (\mathscr{S})$, we have

$$\langle\!\langle D_h F, \psi\rangle\!\rangle_a = \langle\!\langle F, \tilde{h}\psi\rangle\!\rangle_a - \langle\!\langle F, D_h\psi\rangle\!\rangle_a$$

and

$$\langle\!\langle D_h^2 F, \psi\rangle\!\rangle_a = \langle\!\langle D_h F, \tilde{h}\psi\rangle\!\rangle_a - \langle\!\langle D_h F, D_h\psi\rangle\!\rangle_a$$

so that

$$\langle\!\langle -D_h^2 F + \tilde{h} D_h F, \psi\rangle\!\rangle_a = \langle\!\langle D_h F, D_h\psi\rangle\!\rangle_a$$
$$= \langle\!\langle F, -D_h^2\psi + \tilde{h} D_h F\rangle\!\rangle_a. \tag{6}$$

(i) then follows from (6).

(ii) and (iii) follows by checking their U–functionals which uniquely determine their generalized white noise functionals. ///

4.8. Remark.

Recall that, for $F \in (\mathscr{S})^*$ and $\varphi \in (\mathscr{S})$,

$$\begin{aligned}\langle\!\langle D_h F, \psi\rangle\!\rangle_s &:= \langle\!\langle F, \tilde{h}\tilde{\psi}\rangle\!\rangle_a - \langle\!\langle F, D_h\tilde{\psi}\rangle\!\rangle_a \\ &= \langle\!\langle \tilde{h}F, \tilde{\psi}\rangle\!\rangle_a - \langle\!\langle D_h^* F, \tilde{\psi}\rangle\!\rangle_a \\ &= \langle\!\langle D_h F, \tilde{\psi}\rangle\!\rangle_a \quad (h \in \mathscr{S}).\end{aligned}$$

In other words,

$$\tilde{h}F = D_h F + D_n^* F. \tag{7}$$

(7) may be formally written in the form

$$\left(\int_{-\infty}^{\infty} x(t)h(t)dt\right)F = \int_{-\infty}^{\infty} \partial_t F dt + \int_{-\infty}^{\infty} \partial_t^* F dt.$$

This leads to the following definition

$$\dot{B}(t)F = \partial_t F + \partial_t^* F. \tag{8}$$

More generally, for $F^*, G^* \in L(\mathscr{S}, (\mathscr{S})^*)$, we write $F^*(t) = G^*(t)$ symbolically if $F^*(h) = G^*(h)$. ///

4.9. Example.

(a) $D_h\dot{B}(t) = h(t)$.

In fact, for $\varphi \in (\mathscr{S})$, we have

$$\begin{aligned}&\langle\!\langle D_h\dot{B}(t), \varphi\rangle\!\rangle_s \\ &= \langle\!\langle \dot{B}(t), \tilde{h}\tilde{\varphi}\rangle\!\rangle_a - \langle\!\langle \dot{B}(t), D_h\tilde{\varphi}\rangle\!\rangle_a \\ &= D\mu(\tilde{h}\tilde{\varphi})(0)\delta_t - D\mu(D_h\tilde{\varphi})(0)\delta_t \\ &= h(t)\langle\!\langle 1, \varphi\rangle\!\rangle_a + \int_{\mathscr{S}^*} (x,h)D\varphi(x)\delta_t\mu(dx) - \int_{\mathscr{S}^*} D^2\varphi(x)\delta_t h\mu(dx) \\ &= h(t)\langle\!\langle 1, \varphi\rangle\!\rangle_a.\end{aligned}$$

Symbolically, we write $\partial_s\dot{B}(t) = \delta_s(t)$.

(b) $D_h(\int_{\mathbb{R}^n} f(t_1,\dots,t_n)dB(t_1)\cdots dB(t_n))$

$$= \frac{1}{(n-1)!}\sum_\pi \int_{\mathbb{R}^{n-1}}\int_{-\infty}^{\infty} f(t_1,\dots,t_n)h(t_{\pi(1)})dt_{\pi(1)}dB(t_{\pi(2)})\cdots dB(t_{\pi(n)}),$$

where π rums through all permutations of $\{1,\dots,n\}$.

In fact, for $\varphi \in (\mathscr{S})$, we have

$$\langle\!\langle D_h(\int_{\mathbb{R}^n} f(t_1,\dots,t_n)dB(t_1)\cdots dB(t_n)), \varphi\rangle\!\rangle_s$$

$$= \int_{\mathbb{R}^n} f(t_1,\dots,t_n) \langle\!\langle :\dot{B}(t_1)\cdots\dot{B}(t_n):, \tilde{h\varphi}\rangle\!\rangle_a \, dt_1\cdots dt_n$$

$$- \int_{\mathbb{R}^n} f(t_1,\dots,t_n) \langle\!\langle :\dot{B}(t_1)\cdots\dot{B}(t_n):, D_h\tilde{\varphi}\rangle\!\rangle_a \, dt_1\cdots dt_n$$

$$= \int_{\mathbb{R}^n} f(t_1,\dots,t_n) \Big\{\frac{1}{(n-1)!}\sum_\pi h(t_{\pi(1)})\int_{\mathscr{S}^*} D^{n-1}\tilde{\varphi}(x)\delta_{t_{\pi(2)}}\cdots\delta_{t_{\pi(n)}}\mu(dx)$$

$$+ \int_{\mathscr{S}^*}(x,h)D^n\tilde{\varphi}(x)\delta_{t_1}\cdots\delta_{t_n}\mu(dx)$$

$$- \int_{\mathscr{S}^*} D^{n+1}\tilde{\varphi}(x)h\delta_{t_1}\cdots\delta_{t_n}\mu(dx)\Big\}\, dt_1\cdots dt_n$$

$$= \langle\!\langle \sum_\pi \int_{\mathbb{R}^{n-1}}\int_{-\infty}^{\infty} f(t_1,\dots,t_n)h(t_{\pi(1)})dt_{\pi(1)}dB(t_{\pi(2)})\cdots dB(t_{\pi(n)}), \varphi\rangle\!\rangle_a.$$

In particular, $f = \mathrm{sym}(g \otimes \delta_{t_1}^{n-1})$, we have

$$D_h(\int_{-\infty}^{\infty} g(t):\dot{B}^n(t):dt) - n\int_{-\infty}^{\infty} g(t)h(t):\dot{B}^{n-1}(t):dt.$$

Symbolically, we write

$$\partial_s(\int_{-\infty}^{\infty} g(t):\dot{B}^n(t):dt) = ng(s):\dot{B}^{n-1}(s):.$$

(c) $$N(\int_{\mathbb{R}^n} f(t_1,\dots,t_n)dB(t_1)\cdots dB(t_n)) = n\int_{\mathbb{R}^n} f(t_1,\dots,t_n)dB(t_1)\cdots dB(t_n) \qquad (9)$$

In fact,

$$\langle\langle N\left(\int_{\mathbb{R}^n} f(t_1,\dots,t_n)dB(t_1)\cdots dB(t_n)\right), \varphi\rangle\rangle_s$$

$$= \int_{\mathbb{R}^n} f(t_1,\dots,t_n)D^n\mu(N\tilde{\varphi})(0)\delta_{t_1}\cdots\delta_{t_n}\, dt_1\cdots dt_n. \tag{10}$$

The formula (9) now follows form the identify (10) and the following fact :

$$D^n\mu(N\tilde{\varphi})(0)h_1\cdots h_n$$

$$= \int_{\mathscr{S}^*} \sigma(\tilde{h}_1\cdots\tilde{h}_n)(x)N\tilde{\varphi}(x)\mu(dx)$$

$$= \int_{\mathscr{S}^*} N(\sigma(\tilde{h}_1\cdots\tilde{h}_n)(x)\tilde{\varphi}(x)\mu(dx)$$

$$= n\int_{\mathscr{S}^*} \sigma(\tilde{h}_1\cdots\tilde{h}_n)(x)\tilde{\varphi}(x)\mu(dx)$$

$$= nD^n\mu\tilde{\varphi}(0)\mu h_1\cdots h_n,$$

for $h_j \in \mathscr{S}$.

4.10. Fourier transform.

For $F \in (\mathscr{S})^*$, the Fourier transform $\mathscr{F}F$ is defined by

$$\langle\langle \mathscr{F}F, \varphi\rangle\rangle_s := \langle\langle F, \mathscr{F}_{1,-i}\tilde{\varphi}\rangle\rangle_a,$$

for $\varphi \in (\mathscr{S})$. (Cf. [8,10]). Then $\mathscr{F}F \in (\mathscr{S})^*$ by Proposition 2.4. (e).

The inverse transform of F is defined by

$$\langle\langle \mathscr{F}F, \varphi\rangle\rangle_s := \langle\langle F, \mathscr{F}_{1,i}\tilde{\varphi}\rangle\rangle_a.$$

4.11. Example.

(a) $\mathscr{F}(1) = \delta$.

In fact, $\langle\langle \mathscr{F}1, \varphi\rangle\rangle_s = \int_{\mathscr{S}^*}\int_{\mathscr{S}^*}\tilde{\varphi}(x-iy)\mu(dy)\mu(dx) = \tilde{\varphi}(0) = \langle\langle\delta,\varphi\rangle\rangle_a = \langle\langle\delta,\varphi\rangle\rangle_s$.

(b) $\mathscr{F}\delta_y = :\exp[-i\int_{-\infty}^{\infty} y(t)\dot{B}(t)dt]:$, $y \in \mathscr{S}^*$, where

$\langle\langle :\exp[\alpha\int_{-\infty}^{\infty} y(t)\dot{B}(t)dt]: ,\varphi\rangle\rangle_s := \mu\varphi(\alpha y)$ for $\alpha \in \mathbb{C}$.

(c) $\mathscr{F}(\int_{-\infty}^{\infty} y(t)dB(t)) = -iD_y^*\delta$, $y \in \mathscr{S}^*$.

In fact,

$$\langle\langle \mathscr{F}(\int_{-\infty}^{\infty} y(t)dB(t)), \varphi\rangle\rangle_s = (\int_{-\infty}^{\infty} y(t)D\mu(\mathscr{F}_{1,i}\tilde{\varphi})(0)\delta_t\, dt$$

$$= -i(\int_{-\infty}^{\infty} y(t)D\tilde{\varphi}(0)\delta_t\, dt = -iD\tilde{\varphi}(0)y$$

$$= -i\langle\langle \delta, D_y\tilde{\varphi}\rangle\rangle_a = -i\langle\langle D_y^*\delta, \varphi\rangle\rangle_s.$$

(d) $\mathscr{F}:\dot{B}(t_1)\cdots\dot{B}(t_n): = (-i)^n D^*_{\delta_{t_1}}\cdots D^*_{\delta_{t_n}}\delta.$

(e) $\mathscr{F}\left[:\exp(\frac{\alpha}{2}\int_{-\infty}^{\infty}\dot{B}(t)^2dt):\right] = :\exp\left[\frac{1}{2\alpha}\int_{-\infty}^{\infty}\dot{B}(t)^2dt)\right]:$ for $\alpha \in \mathbb{C}\backslash\{1\}$.

(f) $\mathscr{F}:\exp(\alpha\int_{-\infty}^{\infty} y(t)dB(t)): = \delta_{-i\alpha y}$.

4.12. Remark. The Proposition 5.15 remains true in the present case. For more properties of the Fourier transform, we refer the reader to [13,15]. ///

5. The composition of tempered distribution with Brownian motion

In [13] the composition $f(B(t))$ of a tempered distribution f with Brownian motion $B(t)$ was defined in the following way

$$\langle\langle f(B(t)), \varphi\rangle\rangle_p := (f, \hat{G}_{t,\varphi}), \tag{11}$$

where $G_{t,\varphi}(u) = (1/\sqrt{2\pi})\mathscr{F}_{1,i}\varphi(uh_t)e^{-(u^2|t|/2)}$, $t \neq 0$, and

$$h_t = \begin{cases} 1_{(0,t]}, & t \geq 0 \\ -1_{(t,0]}, & t < 0, \end{cases}$$

and the test function φ is taken from a class $\mathscr{E}_p$ of analytic functions on $\mathscr{S}_{-p}$.

Being motivated by (11), we shall define $f(B(t))$ by

$$\langle\langle f(B(t)), \varphi\rangle\rangle_s := (f, \hat{G}_{t,\tilde{\varphi}}) \tag{12}$$

for $\delta \in (\mathscr{S})$.

To check the continuity of $f(B(t))$, we needs to show that the mapping $\tilde{\varphi} \longrightarrow G_{t,\tilde{\varphi}}$ from $\mathscr{A}_\infty$ into $\mathscr{S}$ is continuous. This follows from the following

5.1. Lemma. For any $m,k \in \mathbb{N}$, we have

$$\sup_{u\in\mathbb{R}^1}\left[\left|u^m \frac{d^k}{du^k} G_{t,\tilde{\varphi}}(u)\right|\right] \leq \text{const.}\|\tilde{\varphi}\|_{\mathscr{A}_p}.$$

Proof. Apply [15; Proposition 3.2 (a)], we have

$$\|D^n\psi(z)\|_{\mathscr{L}^n(\mathscr{S}_{-p})} \leq \left(\frac{\sqrt{n}}{e}\right)^n e^{ne^2} e^{\|z\|_{-p}^2}\|\psi\|_{\mathscr{A}_p} \tag{13}$$

for $\psi \in \mathscr{A}_\infty$ and for $p \in \mathbb{N}$.

It follows that

$$\begin{aligned}
&\left|\frac{d^n}{du^n}\mathscr{F}_{1,i}\psi(uh_t)\right| \\
&\leq \left\{\int_{\mathscr{S}^*}\|D^n\psi(x+iuh_t)\|_{\mathscr{L}^n(\mathscr{S}_{-p})}\mu(dx)\right\}\|h_t\|_{-p}^n \\
&\leq \left(\frac{\sqrt{n}}{e}\right)^n e^{ne^2}\|h_t\|_{-p}^n e^{u^2\|h_t\|_{-p}^2}\left(\int_{\mathscr{S}^*} e^{\|x\|_{-p}^2} u(dx)\right)\|\psi\|_{\mathscr{A}_p} \\
&\leq \left(\frac{\sqrt{n}}{e}\right)^n e^{ne^2}\left(\frac{5}{12}|t|\right)^{\frac{n}{2}} e^{\frac{5}{12}u^2|t|}\left(\int_{\mathscr{S}^*} e^{\|x\|_{-1}^2} u(dx)\right)\|\psi\|_{\mathscr{A}_p} \\
&\leq c(t,n)\|\psi\|_{\mathscr{A}_p} e^{\frac{5}{12}u^2|t|},
\end{aligned}$$

so that

$$\sup_{u\in\mathbb{R}^1}\left[|u|^m\left|\frac{d^n}{du^n}\mathscr{F}_{1,i}\psi(uh_t)\right|\right]e^{-(u^2|t|/2)}$$

$$\leq \sup_{u\in\mathbb{R}^1}\left[|u|^m e^{-(u^2|t|/2)}\right]\cdot c(t,n)\|\psi\|_{\mathscr{A}_p}. \tag{14}$$

The lemma clearly follows from (14). ///

Differentiating $\langle\langle f(B(t)), \varphi\rangle\rangle_s$, one obtain

$$\frac{d}{dt}\langle\langle f(B(t)), \varphi\rangle\rangle_s = \langle\langle \partial_t^* f'B(t)), \varphi\rangle\rangle_s + \frac{1}{2}\langle\langle f''B(t)), \varphi\rangle\rangle_s$$

which, in turn, implies Ito's formula given as follows.

5.2. Theorem. For $f \in \mathscr{S}^*$, we have

$$f(B(b)) = f(B(a)) + \int_a^b \partial_t^* f'(B(t))dt + \frac{1}{2}\int_a^b f''(B(t))dt,$$

where $b > a > 0$.

5.3. Theorem. Suppose f is a positive tempered distribution. Then $f(B(t))$ is also positive as a linear functional on $(\mathscr{S})$.

Proof. From the definition of $f(B(t))$, we have

$$\langle\langle f(B(t)), \varphi\rangle\rangle_s = (f_{[\xi]}, \langle\langle \partial_\xi(B(t)), \varphi\rangle\rangle_s)$$

where $f_{[\xi]}$ indicates that f applies the test functions with respect to the variable ξ.

Now $\varphi \geq 0$ implies that $g(\xi) = \langle\langle \partial_\xi(B(t)), \varphi\rangle\rangle_s$ is positive (see Example 3.7). Hence $f(B(t))$ is positive. ///

Acknowledgment

The author wishes to express his sincere gratitude to Professor T. Hida, Professor L. Streit, Professor H.–H. Kuo and Professor J. Potthoff for their kind invitation. The author also likes to thank ZIF, University of Bieleteld for offering accommodation and support during the conference.

References

1. L. Gross, Abstract Wiener space, in "Proceedings 5th Berkeley symp. Math. Stat. Prob., 1965", Vol. 2, pp.31–42.
2. L. Gross, Potential theory on Hilbert space, J. Funct. Anal. 1(1967), 123–181.
3. T. Hida, Analysis of Brownian functionals, Carleton Mathematical Lecture Notes No. 13(1975).
4. T. Hida, J. Potthoff and L. Streit, Dirichlet forms and white noise analysis, Commun. Math. Phys. 116(1988), 235–245.
5. I. Kubo and S. Takenaka, Calculus on Gaussian white noise, I, Proc. Japan Acad. Ser. A Math. Sci. 56(1980), 376–380; Calculus on Gaussian white noise, II, Proc. Japan Acad. Ser. A Math. Sci. 56(1980), 411–416; Calculus on Gaussian white noise, III, Proc. Japan Acad. Ser. A Math. Sci. 57(1981), 433–436; Calculus on Gaussian white noise, IV, Proc. Japan Acad. Ser. A Math. Sci. 58(1982), 186–189.
6. I. Kubo and Y. Yokoi, a remark on the space of testing random variables in the white noise Calculus, Preprint, to appear in Nagoya Math. J.
7. H.–H. Kuo, Gaussian Measure in Banach spaces, Lecture notes in Math. Vol. 463, Springer–Verlag, Berlin/New York, 1975.
8. H.–H. Kuo, On Fourier transform of generalized Brownian functionals, J. Multivariate Anal. 12(1982), 415–431.
9. H.–H. Kuo, Donsker's delta function as a generalized Brownian functional and its application, in "Lecture Notes in Control and information Sciences Vol. 49", pp.167–178, Springer–Verlag, Berlin/New York, 1983.
10. H.–H. Kuo, The Fourier transform in white noise calculus, Preprint.
11. Y.–J. Lee, Integral transform of analytic functions on abstract Wiener space, J. Funct. Anal. 47(1982), 153–164.
12. Y.–J. Lee, Sharp inequalities and regularity of heat semigroup on infinite dimensional space, J. Funct. Anal. 71(1987), 69–87.
13. Y.–J. Lee, Generalized functions on infinite dimensional spaces and its application to white noise calculus, J. Funct. Anal. 82(1989), 429–464.

14. Y.–J. Lee, On the convergence of Wiener–Ito decomposition, to appear in Bull. Acad. Sinica.

15. Y.–J. Lee, Analytic version of test functionals, Fourier transform and a characterization of measures in white noise calculus, Preprint.

16. J. Potthoff and Yan Jia–an, Some results about test and generalized functions of white noise, preprint.

17. Y. Yokoi, Positive generalized Brownian function, preprint.

WHITE NOISE FROM A HILBERT SPACE POINT OF VIEW

by P.A. MEYER
Université Louis–Pasteur, Strasbourg

1. Continuous bases in Hilbert spaces. In classical white noise analysis, as described for instance in Hida [3], there is a mixture of two different mathematical ideas, one of which may be called "linear", and the other one "quadratic". The linear aspect is linked with distribution theory, and thus with differentiability, while the quadratic aspect is purely hilbertian. As such, it is also more general.

Let us start with the most elementary case. In the Hilbert space $\mathcal{H} = L^2(\mathbb{R}_+)$ (we use the half–line rather than the line, to conform to probabilistic use), the curve $t \longmapsto X_t(x) = I_{]0,t]}(x)$ has orthogonal increments and satisfies the property that $\| X_t - X_s \|^2 = |t-s|$. Thus we may define for $f \in \mathcal{H}$ the elementary Wiener integral $\int f(s) dX_s(\cdot)$, which turns out to be equal to f. The distribution derivative $\dot{X}_t = \dfrac{dX(t)}{dt}$ is the measure ε_t (the unit mass at t), and we may as well write the integral representation of f as $\int f(s)\dot{X}_s(\cdot)ds = f(\cdot)$. On the other hand, what Dirac really wished to define was a system of "continuous basis vectors" $h_t = \dfrac{dX(t)}{\sqrt{dt}} = \varepsilon_t \sqrt{dt}$, unit vectors in $L^2(\mathbb{R})$ with the property that $< h_s, h_t > = \delta_{st}$. Then for $f \in L^2(\mathbb{R})$ we expect $< h_t, f >$ to have the value $f(t)\sqrt{dt}$ so that we have

$$< f, g > = \sum\nolimits_t < f, h_t >< h_t, g > = \int \overline{f(t)} g(t) dt \, .$$

This hasn't become a well defined object through distribution theory but, just as the unit mass $\varepsilon_t(dx) = \delta(t-x)\, dx$ is a rigorous mathematical substitute for the "delta function" (the formal density of the unit mass with respect to Lebesgue measure), the differential element $dX(t) = h_t\sqrt{dt}$ of the elementary Wiener integral is the mathematical substitute for the continuous basis h_t, and the symbolic relation $< dX(s), dX(t) > = \delta_{st}\, dt$ has a clearer meaning than $< \dot{X}(s), \dot{X}(t) > = \delta(s-t)$. Except for the stochastic integral notation, this is the interpretation von Neumann gave of Dirac's formalism, and there is nothing new about it.

We define an (ordered) continuous basis for a separable, infinite dimensional Hilbert space $\mathcal{H}$ as a curve $X(t)\, (t \geq 0)$ in $\mathcal{H}$, with $X(0) = 0$, orthogonal increments, $\| X(t) - X(s) \|^2 = |t - s|$, and such that the space of integrals $\int f(s)dX(s)$, $f \in L^2(\mathbb{R}_+)$, is the whole of $\mathcal{H}$. This means that we choose explicitly an isomorphism between $\mathcal{H}$ and $L^2(\mathbb{R}_+)$. Much thought has been given to unordered continuous bases, the more general spectral measures $X(A) \in \mathcal{H}$ indexed by a subset A in some measure space $(E, \mathcal{E}, \mu)$, such that $\| X(A) \|^2 = \mu(A)$. Since every reasonable non–atomic measure space is isomorphic to an interval of the line under Lebesgue's measure, the gain of generality is only apparent : to do exterior algebra it is anyhow essential to work with an ordered basis. Results in which the order structure of the line doesn't appear are thus immediately transferable to such general spaces. Using an ordered basis gives a general meaning to "causal" calculus.

On the other hand, P. Krée has rightly insisted that the *linear* aspects of white noise theory depend essentially on the additional structure given to the state space E by suitable spaces of "test–functions" and families of differential operators acting on them.

If $\mathcal{H}$ is given a continuous basis $X(t)$, the Hilbert sum of a finite number ν of copies of $\mathcal{H}$ has a *continuous basis* $X^\alpha(t)$ *of multiplicity* ν, whose formal properties are those of ν–dimensional white noise.

Fock space. Given a Hilbert space $\mathcal{H}$ with a continuous basis, the elements of the symmetric (or antisymmetric) Fock space $\Gamma(\mathcal{H})$ over $\mathcal{H}$ can be represented as sums of multiple stochastic integrals

$$\int_{s_1 < \ldots < s_n} f(s_1, \ldots, s_n) dX(s_1) \ldots dX(s_n)$$

(a symbol $\odot$ or $\wedge$ might be inserted between consecutive $dX(s_i)$, but generally it isn't). The squared norm of such a stochastic integral is

$$\int_{s_1 < \ldots < s_n} |f(s_1, \ldots, s_n)|^2 \, ds_1 \ldots ds_n \, .$$

This representation is the continuous analogue of the well known expansion of Fock space elements using a discrete basis, but the pleasant feature here is the closer relationship between the symmetric and antisymmetric cases. For simplicity we consider only the case of multiplicity $\nu = 1$, but for a multiple continuous basis we would have multiple integrals of different types involving differential elements $dX^{\alpha_1}(s_1) \ldots dX^{\alpha_n}(s_n)$, all types being orthogonal. The notation recalling distribution theory

$$\int_{s_1 < \ldots < s_n} f(s_1, \ldots, s_n) \, \dot{X}(s_1) \ldots \dot{X}(s_n) \, ds_1 \ldots ds_n$$

could be used here, but it would clearly be a mere notation, unless more structure is given.

Assume now that $\mathcal{H}$ is a subspace of $L^2(\Omega)$, where Ω is a probability space with a filtration $(\mathcal{F}_t)$, and that the curve $X(t)$ is a martingale with $X(0) = 0$, and such that $X(t)^2 - t$ is a martingale. In the multiple case each $X^\alpha(t)$ should be a martingale, as well as $X^\alpha(t)X^\beta(t) - \delta^{\alpha\beta}t$. Then the multiple integrals have a meaning as square integrable random variables, and the Fock space $\Gamma(\mathcal{H})$ is contained in $L^2(\Omega)$. Then the problem of *chaotic representation* arises naturally : when are these two spaces identical? The answer is positive for the Wiener space generated by brownian motion, and for the Poisson space generated by compensated Poisson processes (our normalization hypothesis on the generating martingale setting a relation between the size of the jumps and the intensity of the Poisson process). Thus Brownian motion and Poisson processes are isomorphic from the "quadratic" white noise point of view, while from the "linear" point of view their sample functions and distributional derivatives are completely different.

A way to understand the difference between Wiener and Poisson noises is the following : the Hilbert space structure is the same in both cases, the curve $X(t)$ is the same, but the *multiplicative* structure isn't the same : there are two different identifications between the Fock space $\Gamma(\mathcal{H})$ and a probability space $L^2(\Omega)$, and if we carry back on $\Gamma(\mathcal{H})$ the multiplication of random variables, each vector $X(t)$ "becomes" a self-adjoint operator on $\Gamma(\mathcal{H})$, in two different ways. From this point of view, the only multiplication which is intrinsic to the structure is the renormalized product from white noise theory, or Wick product, and instead of defining renormalization as in [3], the problem gets inverted : define the (Wiener or Poisson) multiplication from Wick multiplication.

Once this is understood, there is no need to consider only commutative multiplications : symmetric and antisymmetric Fock space have the same Hilbert space structure, and fermion theory gets included into white noise.

New probabilistic interpretations. This change of point of view has led to interesting problems in classical probability : to find as many martingales as possible which possess the chaotic representation property. A few years ago, only the Wiener and Poisson cases were known. This search has led to unexpected results : first, the discovery by Emery that a concrete example due to Azéma, linked to the zero set of brownian motion, indeed has the chaotic representation property, and the extension of this to a whole family of martingales indexed by a continuous parameter (Emery [2]). Then, even more striking because the result is so simple, the discovery (due to Biane [1]) that given a continuous time Markov chains over a finite state space, whose Q–matrix satisfies a simple algebraic property, it is possible to find a finite system of martingales (explicit additive functionals of the chain), which satisfy the orthogonality assumptions and whose multiple integrals generate the whole Hilbert space of the chain. The research on chaotic representations for more general Markov processes is in progress (Mohari–Sinha [5]), but it will require the use of Fock spaces of infinite multiplicity. All this now comes within the theory of "white noise", but since even the independent increment property is lost in these examples, some of the "whiteness" intuition gets lost.

REFERENCES

[1] BIANE (Ph.). Paper to appear in *Stochastics,* 1990. See also a report by P.A. MEYER to appear in *Sém. Prob. XXIV.*

[2] EMERY (M.). On the Azéma martingales. *Sém. Prob. XXIII,* LN 1372, 1989 p. 66–87.

[3] HIDA (T.). *Brownian Motion.* Springer–Verlag, 1980.

[4] MEYER (P.A.). Eléments de Probabilités Quantiques. *Sém. Prob. XX, XXI, XXII* LN in M. 1204 (1986), 1247 (1987), 1321 (1988)

[5] MOHARI (A.). and SINHA (K.B.). Quantum stochastic flows with infinite degrees of freedom and countable state Markov chains. *Preprint,* Indian Statistical Institute, Delhi, 1989.

Institut de Recherche Mathématique Avancée
(Laboratoire associé au CNRS)
7 rue René Descartes
F–67084 Strasbourg–Cedex

ANTICIPATING STOCHASTIC CALCULUS AND APPLICATIONS

Daniel Ocone
Mathematics Department, Rutgers University
New Brunswick, NJ 08903, USA

ABSTRACT

This paper presents an introduction to the theory of stochastic integrals for anticipating process defined on Wiener space. We pay special attention to stochastic integrals defined by a localization procedure of Nualart and Pardoux and its relationship to distribution-valued stochastic integrals. We illustrate with an example of a stochastic Fredholm equation with a unique distribution valued solution, but an infinite number of locally integrable solutions. We also briefly survey the applications of anticipating stochastic calculus to stochastic differential equations.

1. Introduction.

Let $\{W(t)\}$ be a Brownian motion. The classical Itô theory of stochastic integration requires that in defining the integral $\int_0^1 u(t)\,dW(t)$ the integrand be adapted to the Brownian motion. Recently, progress has been made from several points of view in extending stochastic integration to integrands which are not adapted and in applying these integrals to stochastic equations. In this paper, we wish to discuss the approach initiated by Nualart and Pardoux[7], which develops develops a stochastic calculus for the Skorohod and nonadapted Stratonovich integrals by applying concepts from the stochastic calculus of variations on Wiener space. Pardoux[12] provides good survey of the area. Our focus shall be somewhat different; we shall give an introduction to the Nualart-Pardoux theory, but we shall also introduce anticipating stochastic integrals with values in a space of distributions and compare them to stochastic integrals defined by a process of localization. We want to emphasize the importance and the special features of the localization definition and how it differs from the distributional definition. In section 3 we shall discuss how these theories are applied to stochastic evolution equations. Rather than attempt to give an exhaustive survey, we shall try to indicate which types of methods work and what are the present limits of the theory. Then we shall discuss an example of a Fredholm stochastic integral equation to illustrate the different results possible using distribution-valued as opposed to locally defined Skorohod integrals. This example possesses a unique distribution-valued solution, but an infinite number of locally integrable solutions.

The main theoretical tool in the Nualart-Pardoux theory is the stochastic calculus on Wiener space developed, for example, in Watanbe[22] for use in the Malliavin calculus. Many of the definitions and formulae can be, and have been, restated in the context of Professor Hida's white noise calculus. The reader familiar with the Hida calculus should be able to translate many of our statements into the Hida theory, but we shall not comment further on this point.

2. Nonadapted stochastic integration and calculus

The setting for the theory is the standard Wiener space $(\Omega, \mathcal{F}, P)$, where Ω is the space of continuous $\mathbb{R}^d$-valued paths on $[0,1]$ starting at 0, P is Wiener measure, and $\mathcal{F}$ is the Borel sigma-algebra of the uniform convergence topology on Ω. The canonical process itself shall be denoted by $W(t) = (W_1(t), \dots, W_d(t))$. In general, for ease of notation we shall restrict attention to the case $d = 1$. The easiest nonadapted integral to define is the *generalized Stratonovich or symmetric integral.* Let $u = \{u(t,\omega)\,;\,(t,\omega) \in [0,1] \times \Omega\}$ be a measurable and a.s. square integrable $\mathbb{R}^d$-valued process. For a sequence of finer and finer partitions $\{0 = t_1^n < \cdots < t_n^n < 1\}$, let

$$\bar{u}_k^n = (t_{k+1}^n - t_k^n)^{-1} \int_{[t_k^n, t_{k+1}^n)} u(s)\, ds \quad \text{and} \quad u^n(t) := \sum_k \bar{u}_k^n 1_{[t_k^n, t_{k+1}^n)}(t). \tag{1}$$

Then we define

$$\int_0^1 u(t) \circ dW(t) = \text{limit in prob.} \sum_1^n (\bar{u}_k^n)^* [W(t_{k+1}^n) - W(t_k^n)], \tag{2}$$

if the limit exists and is independent of the sequence of partitions. Certainly, no theory is needed to define this integral. However, we need to show it is defined for a large and useful class of nonadapted processes. This will be a consequence of further developments. The generalized symmetric integral is related to that of Ogawa[11] defined by orthogonal series expansion.

Next we define the *Skorohod* integral. For this it is necessary to recall the Itô-Wiener chaos decomposition of $L^2(\Omega)$. For convenience of notation we shall work only in the case $d = 1$ of a one dimensional Brownian motion W. Let $\hat{L}^2([0,1]^k)$ denote the symmetric square integrable functions on $[0,1]^k$. For any $f \in \hat{L}^2([0,1]^k)$ we define the k-fold multiple stochastic integral,

$$\begin{aligned} f \circ W^k &= \int_0^1 \cdots \int_0^1 f(t_1, \dots, t_n)\, dW(t_n) \dots dW(t_1) \\ &= k! \int_0^1 \int_0^{t_1} \cdots \int_0^{t_{n-1}} f(t_1, \dots, t_n)\, dW(t_n) \dots dW(t_1). \end{aligned}$$

The last integral is rigorously defined by iteration of adapted integrals. Multiple integrals of different order are always orthogonal and $E[(f \circ W^k)^2] = k!||f||^2$, where $||f||$ denotes the L^2 norm of the kernel. The Itô-Wiener theorem says that for each $F \in L^2(\Omega)$ there is a sequence of kernels $\{f_k\}$ such that

$$F = \sum_0^\infty f_k \circ W^k \tag{3}$$

converges in $L^2(\Omega)$. Now let $u = \{u(t,\omega),\ 0 \le t \le 1\}$ denote a process in $L^2([0,1] \times \Omega)$. We can thus represent u in the form

$$u(t) = \sum_0^\infty f_k(t,\ldots) \circ W^k, \tag{4}$$

where each f_k is now a function of $k+1$ variables such that $f_k(t,\ldots)$ is symmetric in the last k variables for a.e. t and $f_k \in L^2([0,1]^{k+1})$. Thus, for each k we may define $\hat{f}_k \circ W^{k+1}$, where $\hat{f}_k$ denotes the symmetrization of f_k. Then we define the Skorohod integral of u as

$$\int_0^1 u(s)\,dW(s) = \sum_0^\infty \hat{f}_k \circ W^{k+1}, \tag{5}$$

if the right hand side of this equation converges in mean square; in this case we say that u is *Skorohod integrable*. Skorohod[18] is the original reference; Ramer[14] defined an equivalent integral by a different method. When restricted to multiple integrals of a fixed order k, the Skorohod integral is a very natural integration concept. Morevover, if u is adapted and $u \in L^2([0,1] \times \Omega)$ then u is Skorohod integrable, and the Skorohod and Itô integrals coincide; for a proof see Ocone[23], Prop. 22.3.

To develop a stochastic calculus for Skorohod's integral we must be able to discuss integrals of the form

$$\int_0^t u(s)\,dW(s) := \int_0^1 1_{[0,t)}(s)u(s)\,dW(s)$$

as processes in the upper limit of integration, and we need to be able to localize the integral in order to extend its definition beyond the L^2 setting. Neither of these tasks is possible for general Skorohod integrable processes, especially given its unwieldy definition by orthogonal expansions. In fact, there are examples of processes u such that u is Skorohod integrable, but $\{u(s)1_{[0,t)}(s)\}$ is not for each $t < 1$; see Pardoux[11]. Nor is it known whether the operation of Skorohod integration is local on the set of Skorohod integrable processes. The central contribution of Nualart and Pardoux[7] is to place further restrictions on the integrands to resolve these problems. These restrictions are expressed in terms of the Wiener space gradient operator, which is the adjoint of the Skorohod integral, and which we now review.

Let $F \in L^2(\Omega)$ have the Itô-Wiener expansion given in Eq. 1. We shall say that $F \in \mathbb{D}^{1,2}$ if the series

$$D_t F = \sum_1^\infty k f_k(t, \ldots) \circ W^{k-1} \tag{6}$$

converges in $L^2([0,1] \times \Omega)$. In this case we call $DF = \{D_t F, 0 \le t \le 1\}$ the gradient of F. In fact, it is a true gradient. If F is a *smooth, cylinder function*, that is, if $F(W) = \phi(W(t_1), \ldots, W(t_n))$ where $\phi \in C_b^\infty$, then

$$\frac{d}{d\epsilon} F(W + \epsilon \int_0^\cdot h(t)\,dt) \Big|_{\epsilon=0} = \int_0^1 h(t) D_t F\,dt. \tag{7}$$

Let $||\cdot||$ denote the $L^2[0,1])$ norm, and for a Brownian functional $F \in \mathbb{D}^{1,2}$ define the norm

$$||F||_{2,1}^2 := E\big[F^2 + ||DF||^2\big]. \tag{8}$$

Then, in fact, the smooth cylinder functions are dense in $\mathbb{D}^{1,2}$ in the norm $||\cdot||_{2,1}$, and so D is a generalized gradient in the sense of Sobolev. For proofs see Ocone[23]. As a gradient, D is local in $\mathbb{D}^{1,2}$; that is,, if $F \in \mathbb{D}^{1,2}$, then $1_{\{F=0\}} D_t F = 0$ a.e. on $[0,1] \times \Omega$; see Nualart and Pardoux[7], or Shigekawa[16]. Higher order Sobolev spaces may be defined in the same way. For example, we may define

$$D_s D_t F = \sum_2^\infty k(k-1) f_k(s, t, \ldots) \circ W^{k-2},$$

if this series converges in $L^2([0,1]^2 \times \Omega)$. We then let $\mathbb{D}^{2,2}$ be the space of functionals, such that

$$||F||_{2,2}^2 := E\big[F^2 + \int_0^1 \int_0^1 (D_s D_t F)^2\,ds dt\big] < \infty.$$

In a similar way one may define $\mathbb{D}^{k,2}$ for all integers $k \ge 1$.

The gradient is the adjoint of the Skorohod integral in the following sense.

Proposition 1. *(Gaveau and Trauber[3]) A process $u \in L^2([0,1] \times \Omega)$ is Skorohod integrable if and only if there is a constant c such that*

$$\Big| E\Big[\int_0^1 u(t) D_t F\,dt\Big] \Big| \le c E^{1/2}[F^2] \quad \text{for all } F \in \mathbb{D}^{1,2}. \tag{9}$$

and then

$$E\Big[\int_0^1 u(t) D_t F\,dt\Big] = E\big[F \int_0^1 u(s)\,dW(s)\big]. \tag{10}$$

Corollary 2. *If $F \in \mathbb{D}^{1,2}$ and if u and Fu are Skorohod integrable processes, then*

$$\int_0^1 Fu(t)\,dW(t) = F\int_0^1 u(t)\,dW(t) - \int_0^1 u(t)D_tF\,dt. \tag{11}$$

Corollary 2 is a simple consequence of Proposition 1 obtained by computing $E[G\int_0^1 Fu(t)\,dW(t)]$ for arbitrary smooth cylinder functions G using Eq. 10 and the product rule for D; see Nualart and Pardoux[7]. Eq. 11 is the fact of central importance for the theory of Skorohod integrals and is largely responsible for the special features of Skorohod integration.

Now we can define an important subspace of Skorohod integrable processes. Let $\mathbb{L}^{1,2}$ denote the set of processes u such that $u(t) \in \mathbb{D}^{1,2}$ for a.e. t, $\{Du(t)\}$ admits a measurable version, and

$$\int_0^1 (||u(t)||_{2,1})^2\,dt < \infty.$$

The following theorem collects the important facts about Skorohod integrals of processes in $\mathbb{L}^{1,2}$.

Theorem 3. *(Nualart and Pardoux[7])*
(a). Every u in $\mathbb{L}^{1,2}$ is Skorohod integrable.
(b). For every t, $\{u(s)1_{[0,t)}(s)\} \in \mathbb{L}^{1,2}$ and hence $\int_0^t u(s)\,dW(s)$ is well-defined for all $0 \le t \le 1$.
(c).

$$E\Big(\int_0^1 u(t)\,dW(t)\Big)^2 = E\Big\{\int_0^1 u^2(t)\,dt + \int_0^1\int_0^1 D_su(t)D_tu(s)\,dsdt\Big\} \tag{12}$$

(d). $\int_0^1 u^n(t)\,dW(t)$ converges in $L^2(\Omega)$ as $n \to \infty$.
(e). The Skorohod integral is local on $\mathbb{L}^{1,2}$; that is, for $u \in \mathbb{L}^{1,2}$, $\int_0^1 u(t)\,dW(t) = 0$ a.s. on $A = \{\omega\,;\, u(t,\omega) = 0$ for a.e. $t \le 1\}$.

An important formula in the proof of this theorem is

$$\int_0^1 u^n\,dW(t) = \sum_1^{n-1}(\bar{u}_k^n)[W(t_{k+1}^n) - W(t_k^n)] - \frac{1}{t_{k+1}^n - t_k^n}\int\int_{[t_k^n,t_{k+1}^n]^2} D_tu(s)\,dtds,$$

which follows immediately from Eq. 11. This leads ultimately to statements (c) and (d) of the theorem. Statement (e) follows from statement (d) and the locality of D.

Theorem 3 enables one to build a nonadapted stochastic calculus for integrals of processes in $\mathbb{L}^{1,2}$. From statements (b) and (c) we know that $\int_0^t u(t)\,dW(t)$ is well-defined, mean-square continuous, and hence measurable. Various, additional, regularity assumptions, for example, requiring that $D_tu(s)$ be in $L^p([0,1]\times\Omega)$ for some

$p > 2$, guarantee the existence of an a.s. continuous modification of $\int_0^t u(t)\,dW(t)$; see Nualart and Pardoux[7].

For a reasonable stochastic calculus it is important to relax the moment conditions on u. Statement (e) allows one to do this by localization. We shall define $\mathbb{L}^{1,2}_{loc}$ to be the set of processes u such that u admits a *localizing sequence in* $\mathbb{L}^{1,2}$; that is, there is a sequence $\{(\Omega_k, u_k)\}$ such that $\{\Omega_k\}$ is an increasing sequence of measurable sets with $\Omega_k \uparrow \Omega$ a.s., $\{u_k\} \subset \mathbb{L}^{1,2}$, and

$$u \Big|_{\Omega_k} = u_k \Big|_{\Omega_k} \qquad \text{a.s. for every } k.$$

Statement (e) of Theorem 3 implies that the equalities

$$\int_0^1 u(t)\,dW(t)\,\Big|_{\Omega_k} = \int_0^1 u_k(t)\,dW(t)\,\Big|_{\Omega_k}, \quad \text{for all } k. \tag{13}$$

consistently define a stochastic integral for processes $u \in \mathbb{L}^{1,2}_{loc}$. In order to discuss some examples it is convenient to introduce also a localization of $\mathbb{D}^{1,2}$. We shall say that $F \in \mathbb{D}^{1,2}_{loc}$ if there is an increasing sequence $\{\Omega_k\}$ of measurable sets and a sequence $\{F_k\} \subset \mathbb{D}^{1,2}$ such that $\Omega_k \uparrow \Omega$ a.s., and

$$F \Big|_{\Omega_k} = F_k \Big|_{\Omega_k} \qquad \text{a.s. for every } k.$$

For $F \in \mathbb{D}^{1,2}_{loc}$ we define $D^{loc}F$ by

$$D^{loc}F \Big|_{\Omega_k} = DF_k \Big|_{\Omega_k}, \tag{14}$$

which is a consistent definition because D is local on $\mathbb{D}^{1,2}$.

Now we shall give a very simple, but instructive class of locally Skorohod integrable processes. Let $h \in L^2([0,1])$ and take any $F \in \mathbb{D}^{1,2}_{loc}$ and consider the process $u(t) := Fh(t)$, $0 \le t \le 1$. Certainly, $u \in \mathbb{L}^{1,2}_{loc}$; if (Ω_k, F_k) is a localizing sequence for F, then $(\Omega_k, u_k(t) := F_k h(t))$ is a localizing sequence for u_k. Moreover, Corollary 2, Eq. 4, implies that

$$\int_0^1 F_k h(t)\,dW(t) = F_k \int_0^1 h(t)\,dW(t) - \int_0^1 h(t) D_t F_k\,dt.$$

It follows from Eq. 13 that

$$\int_0^1 Fh(t)\,dW(t) = F \int_0^1 h(t)\,dW(t) - \int_0^1 h(t) D_t^{loc} F\,dt. \tag{15}$$

We consider two simple examples. Let $F = \operatorname{sgn}(W(1))$. Let $\{\phi_k(\cdot)\}$ be a sequence of approximations to $\operatorname{sgn}(\cdot)$ defined so that

(i) $\{\phi_k\} \subset C^1(\mathbb{R})$,
(ii) for each k, $\phi_k(x) = -1$ for $x \le -1/k$, and $\phi_k(x) = 1$ for $x \ge 1/k$.
For each k, $\phi_k(W(1)) \in \mathbb{D}^{1,2}$ and

$$D_t\phi_k(W(1)) = \phi_k'(W(1))D_tW(1) = \phi_k'(W(1))1_{[0,1]}(t).$$

If we let $\Omega_k = \{W;\ |W(1)| \ge 1/k\}$ then we see that $(\Omega_k, \phi_k(W(1)))$ is a localizing sequence for $\mathrm{sgn}(W(1))$, and hence that $\mathrm{sgn}(W(1)) \in \mathbb{D}^{1,2}_{loc}$. When $|W(1)| \ge 1/k$, $D_t\phi_k(W(1)) = 0$ for all $0 \le t \le 1$. Thus, by Eq. 14,

$$D^{loc}\mathrm{sgn}(W(1)) = 0. \tag{16}$$

It then follows from Eq. 15 that, for example,

$$\int_0^1 \mathrm{sgn}(W(1))\, dW(t) = \mathrm{sgn}(W(1)) \int_0^1 dW(t) = |W(1)|. \tag{17}$$

A similar type of calculation shows that $G := |W(1)|^{-1} \in \mathbb{D}^{1,2}_{loc}$ and also that $D_t^{loc}G = -\mathrm{sgn}(W(1))|W(1)|^{-2}1_{[0,1]}(t)$, and hence

$$\int_0^1 |W(1)|^{-1}\, dW(t) = \mathrm{sgn}(W(1)) - \mathrm{sgn}(W(1))|W(1)|^{-2}. \tag{18}$$

There is another way to extend the domain of Skorohod integration by allowing for distribution valued integrals; see, Ustunel[20]. To explain this, we shall define some spaces of distributions on Wiener space, following Watanabe[22]. For $k \ge 1$, let $\mathbb{D}^{-k,2}$ denote the topological dual space of $\mathbb{D}^{k,2}$. These spaces may be concretely represented using Itô-Wiener expansions. For example, $\mathbb{D}^{-1,2}$ may be identified with the set of formal Itô-Wiener series

$$T = \sum_0^\infty f_k \circ W^k,$$

such that

$$||T||^2_{-1,2} := \sum_0^\infty (1+k)^{-1} k! ||f_k||^2. \tag{19}$$

With this identification, the duality pairing between a $T \in \mathbb{D}^{-1,2}$ and a $G \in \mathbb{D}^{1,2}$, $G = \sum g_k \circ W^k$, is given by

$$\langle T, G\rangle := \sum_0^\infty k!(f_k, g_k)_{L^2}. \tag{20}$$

Notice that with these definitions, $L^2(\Omega) \subset \mathbb{D}^{-1,2}$ and, if $F \in L^2(\Omega)$ then $\langle F, G\rangle = E[FG]$.

Using the spaces $\mathbb{D}^{-k,2}$, we can define distributional generalizations of $\mathbb{L}^{1,2}$. Namely, for each $k \geq 1$, $\mathbb{L}^{-k,2}$ is the space of processes u with values in $\mathbb{D}^{-k,2}$ such that

$$\int_0^t ||u(t)||^2_{-k,2}\, dt < \infty.$$

$\mathbb{L}^{-k,2}$ is then the dual of $\mathbb{L}^{k,2}$ in the obvious way.

These distribution spaces allow one to give meaning to the series in Eq. 5 defining the Skorohod integral and the series in Eq. 6 defining the gradient, even when these series do not converge in mean square. For example, for any $F \in L^2(\Omega)$ the series defining DF in Eq. 6 will always converge to an element of $\mathbb{L}^{-1,2}$; c.f. Watanabe[22]. Hence D extends to a continuous, linear map $D : L^2(\Omega) \mapsto \mathbb{L}^{-1,2}$. In a dual manner, if $u \in L^2([0,1] \times \Omega)$ without further restrictions, then the series in Eq. 5 always defines an element of $\mathbb{D}^{-1,2}$; thus the Skorohod integral extends to a continuous linear operator from $L^2([0,1] \times \Omega)$ to $\mathbb{D}^{-1,2}$. All of these considerations extend to higher order Sobolev spaces on the Wiener space and their duals; see Watanabe[22]. In particular, the duality between D and Skorohod integration holds quite generally; For example, if $F \in L^2(\Omega)$ and $u \in \mathbb{L}^{1,2}$,

$$\langle DF, u\rangle = E[F \int_0^1 u(t)\, dW(t)]. \tag{21}$$

Let us now return to the examples for which we computed local gradients and integrals. These simple examples already reveal an interesting feature of the localized Skorohod integrals and gradients. Namely, the operations of local differentiation and integration can give different results than the extensions defined by distribution valued Wiener chaos expansions and can treat cases with no expansion interpretation. For example, consider $\mathrm{sgn}(W(1))$; because this random variable is in $L^2(\Omega)$, $\{D_t\mathrm{sgn}(W(1)),\ 0 \leq t \leq 1\}$ is a well-defined $\mathbb{L}^{-1,2}$-valued process, which we shall now calculate. Let δ_0 denote the Dirac delta function at 0. As a consequence of Lemmas 1 and 2, section 1.5 of Watanabe[22] , one can define a distribution $\delta_0(W(1))$ over Wiener space such that $\delta_0(W(1)) \in \mathbb{D}^{-2,2}$ and, if $F = f(W(t_1), \ldots, W(t_n))$ where $f \in C_b^\infty$, then

$$\langle \delta_0(W(1)), F\rangle = E\big[F \mid W(1) = 0\big]. \tag{22}$$

The conditional expectation in Eq. 22 is that calculated using the conditional density of $(W(t_1), \ldots, W(t_n))$ given $W(1) = 0$ as defined in elementary probability theory, so that the fact that $P(W(1) = 0) = 0$ causes no problem. We shall show that

$$D_t\mathrm{sgn}(W(1)) = 2\delta_0(W(1))1_{[0,1]}(t). \tag{23}$$

(It follows that actually $\delta_0(W(1)) \in \mathbb{D}^{-1,2}$.) To check Eq. 23 we will show that

$$\langle D\text{sgn}(W(1)), u\rangle = 2\langle \delta_0(W(1)), \int_0^1 u(t)\,dt\rangle, \tag{24}$$

for all processes $u = \{u(t)\,,\, 0 \le t \le 1\}$ of the form

$$u(t) = f(W(t_1), \dots, W(t_n))h(t), \tag{25}$$

where $f \in C_b^\infty(\mathbb{R}^n)$ and $h \in L^2(0,1)$. Linear combinations of these processes are dense in $\mathbb{L}^{1,2}$, and hence checking Eq. 24 will suffice to establish Eq. 23. Let $\{\phi_k\}$ be the sequence of functions defined after Eq. 15 for the localization of sgn$(W(1))$. Let $F := f(W(t_1), \dots, W(t_n))$, and set $g(y) := E[F \mid W(1) = y]$. Note that $g \in C_b^1$. Then using Eq. 21 and Eq. 11 several times

$$\begin{aligned}
\langle D\text{sgn}(W(1)), u\rangle &= E\big[\text{sgn}(W(1)) \int_0^1 u(t)\,dW(t)\big] \\
&= \lim_{n\to\infty} E\big[\phi_n(W(1)) \int_0^1 u(t)\,dW(t)\big] \\
&= \lim_{n\to\infty} E\big[\phi_n'(W(1)) \int_0^1 E[F \mid W(1)]h(t)\,dt\big] \\
&= \lim_{n\to\infty} E\big[\phi_n(W(1)) \int_0^1 E[F \mid W(1)]h(t)\,dW(t)\big] \\
&= E\Big[\text{sgn}(W(1))\Big(g(W(1) \int_0^1 h(t)\,dW(t) - g'(W(1)) \int_0^1 h(t)\,dt\Big)\Big].
\end{aligned}$$

The last expectation is simply a Gaussian integral, and a computation shows that it equals

$$2g(0) \int_0^1 h(t)\,dt = \int_0^1 \langle 2\delta_0(W(1)), F\rangle h(t)\,dt,$$

because $g(0) = E[F \mid W(1) = 0] = \langle \delta_0(W(1)), F\rangle$. This confirms Eq. 24.

It now follows from Eq. 23 that if we interpret $\int_0^1 \text{sgn}(W(1))\,dW(t)$ in the distributional, rather than localized, sense,

$$\int_0^1 \text{sgn}(W(1))\,dW(t) = \text{sgn}(W(1))W(1) - 2\delta_0(W(1)), \tag{26}$$

leading to a different answer than Eq. 17. Therefore, there is an ambiguity in the definition of Skorohod integral for processes belonging to $\mathbb{L}_{loc}^{1,2} \cap L^2([0,1]\times\Omega)$, and we must be careful to distinguish between the distributional and local interpretations when either are possible. It would be best to discriminate between them in our notation, but we do not do this here; the right interpretation should be clear from the context. It is easy to see why an ambiguity occurs in this example. The support of the distribution $\delta_0(W(1))$ is the zero measure set $\{W(1) = 0\}$. However, the definition of D^{loc} requires the existence of just one increasing sequence Ω_k with $\Omega_k \uparrow \Omega$ a.s.; we may and do choose $\{\Omega_k\}$ to avoid $\{W(1) = 0\}$, and thus the distributional term $\delta_0(W(1))$ does not appear in the integral defined by localization. By way of contrast, for the case of $|W(1)|^{-1}$ there seems to be neither a Wiener chaos expansion of $|W(1)|^{-1}$ nor a satisfactory interpretation of it as a distribution, and hence no way of defining $\int_0^1 |W(1)|^{-1}\, dW(t)$ in a distributional sense.

The ambiguity in the definition of integral for processes which are both local and in $L^2([0,1] \times \Omega)$ might at first seem a drawback, and one might try adding further restrictions in the definition of locally integrable processes to avoid it. However, this would severely restrict the range and utility of the localization technique. The definitions of $\mathbb{L}^{1,2}_{loc}$ and $\mathbb{D}^{1,2}_{loc}$ allow one to handle consistently a wide range of random functions and processes, such as $W(1)^{-1}$ which do arise in applications. We only must be careful to distinguish the operations of local differentiation and integration from the distributional ones. This is important in the question of technique used to solve stochastic equations with anticipating integrals. Chaos expansion techniques may say nothing about the possibility of solutions in $\mathbb{L}^{1,2}_{loc}$ as we shall see by example in section 3.

Next we consider the connection between Skorohod and symmetric integrals. Let $\mathbb{L}^{1,2}_c$ denote the set of processes in $\mathbb{L}^{1,2}$ such that, in a neighborhood V of the diagonal of $[0,1]^2$, $\{D_s u(t)\}$ has a version such that $t \to Du(t) \in L^2([0,1])$ is continuous on $V \cap \{s \le t\}$ uniformly in s, there is another version of $\{D_s u(t)\}$ for which the same is true on $V \cap \{s \ge t\}$, and $\operatorname{ess\,sup}_V E[D_s^2 u(t)] < \infty$. For such u we can define a trace

$$(\nabla u)_t := \lim_{s \to t, s > t} D_t u(s) + \lim_{s \to t, s < t} D_t u(s).$$

Then Nualart and Pardoux[7] show that if $u \in \mathbb{L}^{1,2}_c$, u is Stratonovich integrable and

$$\int_0^1 u(t) \circ dW(t) = \int_0^1 u(t)\, dW(t) + \int_0^1 (1/2)(\nabla u)_t\, dt. \tag{27}$$

Finally, one can prove chain rules, that is, Itô formulas, for stochastic integrals restricted to $\mathbb{L}^{1,2}$ or $\mathbb{L}^{1,2}_c$. The rigorous statement of these rules requires additional, technical regularity assumptions which we will not give here; see Nualart and Pardoux[7]. Rather, let us just give the formulas for comparison to the adapted case. For the Skorohod integral, let $U(t) = \int_0^t u(t)\, dW(t)$ and let $\Phi \in C^2$. Then under appropriate conditions on u,

$$\begin{aligned} \Phi(U(t)) = \Phi(0) + \int_0^t \phi'(U(s))u(s)\, dW(t) + \int_0^t \Phi''(U(s))u^2(s)\, ds \\ \int_0^t \Phi''(U(s))\Big[\int_0^s D_s u(r)\, dW(r)\Big] u(s)\, ds. \end{aligned} \tag{28}$$

Note the appearance of a new correction term that does not enter into Itô's formula for adapted integrands. Again, Eq. 11 is the source of this term, because, in proving Eq. (28) by discretization, one must interchange a random variable with Skorohod integration. Eq. 28 and Eq. 27 imply, under appropriate technical conditions, that anticipating Stratonovich integrals obey the rules of ordinary calculus. Ocone and Pardoux[8] prove generalizations of the Itô-Wentzell formula for anticipating integrals.

Extensions of this theory to multi-parameter, Hilbert space-valued, and cylindrical Wiener processes may be found in Shevlyakov[15], Thieullen[19], and Grorud and Pardoux[4].

3. Nonadapted stochastic equations

We shall begin this section with an informal discussion of some of the technical issues and difficulties that arise in trying to apply anticipating calculus for the solution of stochastic equations and so give an idea of what types of results are possible. We shall then discuss an example of a stochastic Fredholm equation which admits an infinite number of solutions in $\mathbb{L}^{1,2}_{loc}$, but a unique distributional solution. This serves to further illustrate the distinction between integration in $\mathbb{L}^{1,2}_{loc}$ and distribution-valued integrals and the care needed to formulate good models with anticipation.

The first difficulty one encounters in trying to solve stochastic evolution equations with anticipation is that L^p estimates of the anticipating integrals are very hard to obtain, despite the existence of a stochastic calculus. The problem arises from the additional term involving the gradient in Eq. 12 for computing the second moment of a Skorohod integral. We shall illustrate by showing how the usual estimates for obtaining uniqueness of stochastic differential equations by the Gronwall-Bellman inequality and existence by Picard iteration fail in the non-adapted case. Suppose we have two solutions $X(\cdot)$ and $Y(\cdot)$ of

$$X(t) = Z + \int_0^t \sigma(s, X(s))\, dW(s), \tag{29}$$

where Z depends on the entire path of W, and we wish to show them to be equal by the usual method of computing and then bounding $E[|X(t) - Y(t)|^2]$. If we assume that σ is continuously differentiable with a bounded derivative, we obtain from Eq. 12 and some elementary computation, that,

$$\begin{aligned} E[|X(t) - Y(t)|^2] \le KE[\int_0^t E[|X(s) - Y(s)|^2]\, ds \\ + KE[\int_0^t ||D(\sigma(X(s)) - \sigma(Y(s)))||^2\, ds. \end{aligned} \tag{30}$$

Thus, unless we have an independent, apriori estimate on the gradients of the solutions, we can not proceed with the Gronwall-Bellman argument to work toward showing that $E[|X(t) - Y(t)|^2] = 0$. To try to estimate the term

$$E\left[||D(\sigma(X(s)) - \sigma(Y(s)))||^2\right] = E\left[||\sigma'(X(s))DX(s) - \sigma'(Y(s))DY(s)||^2\right]$$

appearing in Eq. 30, one might try differentiating Eq. 29 by D and using this new equation for DX. However the bounds for $E[||DX(s)||^2]$, etc. will now involve $D^2X(s)$, etc. The introduction of ever higher derivative terms in the bounding procedure does not end, so it is difficult to derive a final bound on the original $E[|X(t) - Y(t)|^2]$. The same remarks apply to the procedure of Picard iteration for obtaining solutions to stochastic differential equations. Despite these remarks, the paper of Kuo and Pothoff[5] in this conference reports on a way to do Picard iteration. They use transforms from the Hida calculus to transfer the equation to a transform space and to do iteration in the transform space. However, now the solution is defined only in a weak sense and it is not clear whether the transform solution correspond to a Skorohod integrable process in the sense defined here.

Because of the problems with calculating bounds, most of the nonadapted solutions to stochastic evolution equations that are known have been obtained by direct, problem-specific construction. Moreover, it turns out that in general it is possible to construct solutions easily only for equations using the generalized Stratonovich integrals. For example, consider the Stratonovich version of Eq. 29:

$$X(t) = Z + \int_0^t \sigma(s, X(s)) \circ dW(s), \tag{31}$$

where Z is anticipating. Let $\phi_t(x)$ denote the adapted flow defined by

$$\phi_t(x) = x + \int_0^t \sigma(s, \phi_s(x)) \circ dW(s).$$

Then it is natural to conjecture that $\phi_t(Z)$ should solve Eq. 31; see Bismut[1]. Ocone and Pardoux[8], generalize this idea to allow anticipating drift coefficients in Eq. 31 and show that it works, in the sense that the stochastic integrals are all well defined. Ocone and Pardoux[9] analyze in detail the particular case of equations with linear coefficients and Stratonovich integrals in which nonadaptedness is introduced by the imposition of linear boundary conditions.

It is much more difficult to construct solutions for stochastic differential differential equations with Skorohod integrals, such as

$$X(t) = Z + \int_0^t b(W, s, X(s))\, ds + \int_0^t \sigma(s, X(s))\, dW(s), \tag{32}$$

where $b(W, t, x)$, $\sigma(W, t, x)$, and Z might all anticipate W. The simple use of adapted flows as in the Stratonovich case no longer works. Moreover, one cannot transform

Eq. 32 to a Stratonovich equation of the type handled in Ocone and Pardoux[8], because if one rewrites Eq. 32 with Stratonovich integrals using the formula in Eq. 27, the new equation will contain the term $\int_0^t \sigma'(s, X(s))(\nabla X)_t \, ds$, which depends not only on X but on the gradient of X, as well. Little is known in general about this case within the Nualart-Pardoux framework; (however, see Kuo and Pothoff[5].) Only special examples have been analyzed successfully. For example, Buckdahn[2] studies

$$X(t) = Z + \int_0^t a(s, W)X(s) \, ds + \int_0^t g(s, W)X(s) \, dW(t),$$

where a and g are anticipating and X is a scalar-valued process. His idea is to write the solution in the form $X(t, W) = Y(t, \Psi_t(W))$ where for each t, $\Psi_t : \Omega \mapsto \Omega$ and $Y(t, W)$ has the exponential form one would expect for a solution of an adapted linear equation; namely

$$Z \exp\{\int_0^t g(s, W) \, dW(t) + \int_0^t h(s) \, ds\},$$

for an appropriate h. This technique seems to be restricted essentially to the scalar case, and one cannot prove a uniqueness of solutions in the class in which one finds existence. Related results are obtained by Shiota[17] and Ustunel[21].

Our experience therefore suggests that in posing differential equations that require nonadapted solutions, the Stratonovich integral is the correct integral to use. In the remander of this paper, we wish to develop this theme by studying an example of the stochastic Fredholm integral equation

$$X(t) = \int_0^1 K(t, s)X(s) \, dW(t) + g(t), \tag{33}$$

where K is a deterministic kernel and g is a stochastic or deterministic process. We shall show that our example admits a unique distribution-valued solution, but an infinite number of locally Skorohod integrable solutions. This demonstrates that the Skorohod integral is too mathematically ill-behaved for stochastic integral equations, and it further demonstrates how entirely different results are possible depending on whether distribution-valued integrals or local Skorohod integrals are used. By way of contrast, if one replaces the Skorohod integral in Eq. 33 with some form of a generalized Stratonovich integral, reasonable theories of stochastic Fredholm equations can be developed; see, Ogawa[10], Shevlyakov[15], and Matskyavichus[6]. Protter and Pardoux[13] are able to solve a stochastic *Volterra* equation with Skorohod integrals, but their equation has a very special structure.

In what follows, we shall analyze the equation,

$$X(t) = \int_0^1 g(t)h(s)X(s) \, dW(s) + f(t) \tag{34}$$

in which we assume that g, f, and h are non-random measurable functions on $[0,1]$. (If one replaces the Skorohod integral in Eq. 34 by the Stratonovich integral, the resulting eqution is easy to solve, as the reader can check.) We first analyze distributional solutions of Eq. 34. This requires the introduction of new spaces of distributions. For any $b > 0$, let $\mathbb{K}^{b,2}$ denote those random variables F in $L^2(\Omega)$ such that

$$|||F|||_{b,2}^2 := \sum_0^\infty (k!)^2 b^{2k} ||f_k||^2$$

where $\{f_k\}$ is the sequence of kernels in the Itô-Wiener chaos expansion of F. $\mathbb{K}^{b,2}$ is a Hilbert space much smaller than $\cap \mathbb{D}^{k,2}$. . For every $b > 0$, the dual, $\mathbb{K}^{-b,2}$, of $\mathbb{K}^{b,2}$ may be identified with set of Itô-Wiener chaos expansions $J = \sum j_k \circ W^k$ such that

$$|||J|||_{-b,2}^2 := \sum b^{-2k} ||j_k||^2 < \infty,$$

and such that the duality pairing is given by

$$\langle J, F \rangle = \sum k!(f_k, j_k).$$

Thus $\langle J, F \rangle = E[JF]$, if both F and J are elements of $L^2(\Omega)$, and, in particular $L^2(\Omega) \subset \mathbb{K}^{-b,2}$ for any $b > 0$. $\mathbb{K}^{-b,2}$ is very large; for instance, $\cup_k \mathbb{D}^{-k,2} \subset \mathbb{K}^{-b,2}$ for any $b > 0$. Finally, for any b, define $\mathbb{M}^{b,2}$ to be the space of $\mathbb{K}^{b,2}$-valued processes such that

$$|||u|||_{b,2}^M := E[\int_0^1 |||u(t)|||_{b,2}^2 \, dt] < \infty.$$

Let the Wiener space gradient D and the Skorohod integral be defined on the spaces $\mathbb{K}^{b,2}$ and $\mathbb{M}^{b,2}$, respectively, by the chaos expansion definitions of Eqs. 5 and 6. Elementary calculations show that for $b > 0$,

$$|||DF|||_{b,2}^M = b^{-2} |||F|||_{b,2},$$

so D is a bounded map from $\mathbb{K}^{b,2}$ to $\mathbb{M}^{b,2}$. By duality, if $b > 0$, the operation D^* of Skorohod integration is bounded from $\mathbb{M}^{-b,2}$ to $\mathbb{K}^{-b,2}$. We shall say that Eq. 34 admits a solution in $\mathbb{M}^{-b,2}$ if there is a process $X(\cdot)$ in $\mathbb{M}^{-b,2}$ such that $h(\cdot)X(\cdot)$ is also in $\mathbb{M}^{-b,2}$ and X satisfies Eq. 34.

Proposition 4. *Assume that $g \in L^2(0,1)$, $f \in L^2(0,1)$, and gh and $fh \in L^2(0,1)$. Then Eq. 34 admits a unique solution in $\mathbb{M}^{-b,2}$ for any $b > ||gh||_{L^2}$. This solution is given by*

$$X(t) = \sum_0^\infty u_k(t, \ldots) \circ W^k,$$

where

$$u_0(t) = f(t)$$
$$u_1(t,s) = g(t)h(s)f(s) \quad \text{and}$$
$$u_k(t,s_1,\ldots,s_k) = g(t)\frac{1}{k}\sum_{i=1}^{k} h(s_i)f(s_i) \prod_{j=1,j\neq i}^{k} g(s_j)h(s_j). \tag{35}$$

Moreover, if f, g, and h are uniformly positive, then X is truly a distribution-valued solution in that sense that there exists a $b_0 > 0$ such that $X \notin I\!M^{-b,2}$ if $b < b_0$.

Proof. Suppose that a solution $X \in I\!M^{-b,2}$ exists, and let u_k, $k \geq 0$, denote the kernels of its chaos expansion. Then, by substitution into Eq. 34, one finds that the kernels must satisfy

$$u_{k+1}(t,s_1,\ldots,s_{k+1}) = g(t)\widehat{h(s_1)u_k}(s_1,\ldots,s_{k+1}),$$

where $\hat{f}$ denotes the symmetrization of f. It is easily seen by induction that the kernels defined in Eq. 35 are the only solutions to these equations. Therefore, it remains to verify that Eq. 35 actually defines a process in $I\!M^{-b,2}$ for sufficiently large b. Let $g_k(s_1,\ldots,s_k) := f(s_1)\prod_{j=2}^{k} g(s_j)h(s_j)$. Then, from Eq. 35,

$$u_k(t,s_1,\ldots,s_k) = g(t)\widehat{g_k}(s_1,\ldots,s_k).$$

Therefore

$$||u_k||^2_{L^2} \leq ||g||^2\,||f||^2\,||gh||^{2(k-1)}$$

and so,

$$(|||X|||^M_{-b,2})^2 \leq ||g||^2\,||f||^2 \sum b^{-2k}||gh||^{2(k-1)} < \infty,$$

if $b > ||gh||$. Finally, if f, g, and h are bounded below by $\alpha > 0$, then

$$||u_k||^2_{L^{k+1}} \geq \alpha^{2k},$$

and hence X will fail to be in $I\!M^{-b,2}$ if $b \leq \alpha$.

Remark. Let $H_k(Z)$ denote the projection of a random variable $Z \in L^2(\Omega)$ onto the space of kth order multiple Wiener integrals. Then the solution $X(t)$ given in Proposition 4 to Eq. 34 equals

$$X(t) = f(t) + g(t)\sum_{1}^{\infty} H_k\Big(\int_0^1 f(s)h(s)\,dW(s)\Big[\int_0^1 h(s)g(s)\,dW(s)\Big]^{k-1}\Big).$$

Next we seek solutions to Eq. 34 which are locally Skorohod integrable in the sense defined in section 2. Thus, we assume that $X(t) = Fg(t) + f(t)$, where $F \in I\!D^{1,2}_{loc}$. By substitution of X into Eq. 34 and the use of Eq. 15, it follows that

$$F = F\int_0^1 h(s)g(s)\,dW(s) - \int_0^1 D^{loc}_s F h(s)g(s)\,ds + \int_0^1 h(s)f(s)\,dW(s).$$

This is essentially a first order p.d.e. for F on the Wiener space. The next result shows how to find an infinite number of soltuions by a reduction to a first order p.d.e. on $I\!R^2$.

Proposition 5. *Assume that* $hg \in L^2(0,1)$ *and that that* $hf \in L^2(0,1)$. *Let* $a = \int_0^1 h^2(s)g^2(s)\,ds$ *and* $b = \int_0^1 h(s)f(s)g(s)\,ds$. *Let* η *be any smooth function of one variable and define*

$$\psi(x,y) = e^{-(a/2b^2)y^2+(x-1)(y/b)}\eta(x-ay/b) + \int_0^y (z/b)e^{-(a/2b^2)(y-z)^2+(x-1)(y-z)/b}\,dz$$

and

$$F = \psi\Big(\int_0^1 h(t)g(t)\,dW(t), \int_0^1 h(s)f(s)\,dW(t)\Big). \tag{36}$$

Then $F \in \mathbb{D}^{1,2}_{loc}$ *and* $X(t) := Fg(t) + f(t) \in \mathbb{L}^{1,2}_{loc}$ *solves Eq. 34.*

Proof. ψ satisfies the equation

$$\psi(x,y) = x\psi(x,y) - a\partial_x\psi(x,y) - b\partial_y(x,y) + y. \tag{37}$$

Let $Z := (\int_0^1 h(t)g(t)\,dW(t), \int_0^1 h(t)f(t)\,dW(t))$, so that $F = \psi(Z)$. Clearly, $F \in \mathbb{D}^{1,2}_{loc}$, and $D_tF = \partial_x\psi(Z)h(t)g(t) + \partial_y(Z)h(t)g(t)$. Thus $X(t) := Fg(t) + f(t)$ is locally Skorohod integrable, and, using Eq. 15 and Eq. 36,

$$\begin{aligned} g(t)\int_0^1 h(s)X(s)\,dW(s) &= g(t)\big[\psi(Z)Z_1 + Z_2 - a\partial_x(Z) - b\partial_y(Z)\big] \\ &= g(t)F = X(t) - f(t). \end{aligned}$$

6. References

1. J. M. Bismut, *Mécanique Aléatoire*, *Lecture Notes in Mathematics* **866**, Springer-Verlag, 1981.
2. R. Buckdahn, Skorohod's integral and linear stochastic differential equations, preprint **1843**, Humboldt Universität, Berlin, 1988.
3. B. Gaveau, P. Trauber, L'Intégrale stochastique comme operateur de divergence dans l'espace fonctionnel, *J. Funct. Anal.* **46** 230-238.
4. A. Grorud, E. Pardoux, Intégrales hilbertiennes anticipantes par rapport à un Wiener cylindrique et calcul stochastiques associé, to appear in *Applied Math. and Optimization*.
5. H. Kuo, J. Pothoff, Anticipating stochastic integrals and stochastic differential equations, this conference.
6. V. Matskyavichus, Symmetric stochastic integral equations of Fredholm type, *Litovsk. Mat. Sb.* **24** (1984), 121-130.
7. D. Nualart, E. Pardoux, Stochastic Calculus with Anticipating Integrands, *Probab. Th. Rel. Fields* **78** (1988), 535-581.

8. D. Ocone, E. Pardoux, A generalized Itô-Wentsell formula: Application to a class of anticipating stochastic differential equations, *Ann. Instit. Henri Poincaré* **25** (1989) 379-71.
9. D. Ocone, E. Pardoux, Bilinear Stochastic Differential Equations with Boundary Conditions, *Probab. Th. Rel. Fields* **82**, 1989, 489-526.
10. S. Ogawa, On the stochastic integral equations of Fredholm type, *Patterns and Waves: Qualitative Analysis of Nonlinear Differential Equations, Stud. in Mathematics* **18**, (1986), 597-605.
11. S. Ogawa, Quelques propriétés de l'intégrale stochastique du type noncausal, *Japan J. Appl. Math* **1** (1984), 405-416.
12. E. Pardoux, Applications of Anticipating Stochastic Calculus to Stochastic Differential Equations, Proceedings of the 1988 Silivri Conference on Stochastic Differential Systems, to appear in *Lecture Notes of Mathematics*.
13. E. Pardoux, P. Protter, Stochastic Volterra equations with anticipating coefficents, *Ann. Prob.*, to appear.
14. R. Ramer, On nonliner transformations of Gaussian measures, *J. Funct. Anal.* **15** (1974), 166-187.
15. A. Yu. Shevlyakov, On stochastic integral equations, *Theor. Probability and Math. Statist.* **38**, 1989, 127-135.
16. I. Shigekawa, Derivatives of Wiener functionals and absolute continuity of induced measures, J. Math. Kyoto Univ. **20** (1980), 263-269.
17. Y. Shiota, A linear stochastic integral equation containing the extended Itô integral, *Math. Rep. Toyama Univ.* **9** (1986) 48-55.
18. A. V. Skorohod, On a generalization of a stochastic integral, *Theor. Prob. Appl.* **20** (1975), 219-233.
19. M. Thieullen, Thesis, Université de Provence, Marseille, 1989.
20. A. S. Ustunel, Extensions of Itô's calculus via Malliavin's calculus, preprint (1985).
21. A. S. Ustunel, Some comments on the filtering of diffusions and Malliavin calculus, in *Stochastic Analysis and related topics*, H. Korezlioglu & A. S. Ustunel eds, *Lecture Notes in Math.* **1316** (1988) 247-266.
22. S. Watanabe, *Stochastic Differential Equations and Malliavin Calculus*, Tata Institute of Fundamental Research, Springer-Verlag, 1984.
23. D. Ocone, A guide to the stochastic calculus of variations, *Stochastic Analysis and rel. topics*, Korezlioglu, Ustunel eds, *Lect. Notes in Math.***1316** (1988) 1-79

HIERARCHY OF NOISE PRODUCTION IN LIVING CELLS

Fumio Oosawa
Aichi Institute of Technology,
Toyota 470-03, JAPAN.

ABSTRACT

A mathematical method is given to describe the swimming behavior of single cell organisms such as paramecia in an inhomogeneous and varying environment. The cells are treated as particles having internal variables which represent the state of molecular machines in them. They have a fluctuation generator to make the connection from sensors to motors indirect and probabilistic. Large and sharp fluctuation of the electric potential is produced by a two step device. This device is characterized by an equation containing fluctuating quantities in exponential terms. This equation is generally useful for the analysis of stochastic processes in a fluctuating field.

I. INTRODUCTION

Single cell organisms such as bacteria and paramecia swim in water and their swimming patterns are similar to those of particles undergoing the Brownian movement. They swim straight and occasionally change the swimming direction discontinuously. The time interval of successive directional changes shows approximately an exponential distribution and the angle of direction change shows an uniform distributuion. The discontinuous change of swimming direction seems to occur as a result of stochastic processes in the cell. Thus, the swimming pattern can be specified by two quantities; the velocity v(mm/s) during straight swimming and the average frequency $f(s^{-1})$ or the average time interval of directional changes.

If the environment is not uniform, the cells look for and come to the place which is better to live. If a temperature gradient is applied to the environment, they come to the place of temperature which is most comfortable for them. For example, when paramecium cells were put in a vessel, of which the left hand side was kept at 20 °C and the right hand side at 25 °C, the cells accumulate in the right hand side within several minutes. After accumulation, they do not stop swimming but continue to swim actively. When they swim towards lower temperature, they change the swimming direction to return to the place of 25 °C. On the other hand, when they swim towards 25 °C, they do not often change the swimming direction. This is the way of accumulation at 25 °C [1,2].

II. MATHEMATICAL DESCRIPTION

Let us consider a one-dimensional vessel in which a large number of cells are swimming [1]. They are classified into two groups; those which are swimming in the $+x$ direction and those in the $-x$ direction. The number of cells in the first group between the coordinate x and $x + dx$ is denoted as $n_+(x)dx$ and the number of cells in the second group as $n_-(x)dx$. The total number of cells between x and $x + dx$ is $n(x)dx = (n_+(x) + n_-(x))dx$. The cells in the first group and those in the second group have different swimming velocities on the average. The velocities are denoted as $v_+(x)$ and $v_-(x)$, respectively, which depend on the coordinate x. The average frequencies of directional changes are also distinguished between two groups and denoted as $f_+(x)$ and $f_-(x)$, respectively.

A gradient is given to the environmental condition in the x direction. Then, the number densities n_+ and n_- change with time and finally a stationary distribution of the cells is established. The following equations of continuity must be satisfied;

$$-\frac{d(v_+n_+)}{dx} - f_+n_+ + f_-n_- = 0 \qquad (1)$$

$$+\frac{d(v_-n_-)}{dx}+f_+n_+-f_-n_- \quad = \quad 0 \tag{2}$$

From these equations

$$\frac{d(-v_+n_+ + v_-n_-)}{dx} \quad = \quad 0 \tag{3}$$

$$\frac{d(-v_+n_+ - v_-n_-)}{dx} - 2(f_+n_+ - f_-n_-) \quad = \quad 0 \tag{4}$$

From (3) and the boundary condition

$$-v_+n_+ + v_-n_- = 0 \tag{5}$$

Using this result, from (4)

$$\frac{d(\ln(v_+n_+))}{dx} + \frac{f_+}{v_+} - \frac{f_-}{v_-} = 0 \tag{6}$$

These equations (5) and (6) must be always satisfied in the stationary distribution.

Let us assume that the cells swimming at x in the $+x$ direction and those swimming at the same x in the $-x$ direction have the same average velocity and the same average frequency of directional changes; that is,

$$v_+(x) = v_-(x)(= v(x)) \text{ and } f_+(x) = f_-(x)\ (= f(x)) \tag{7}$$

This means that the velocity and the frequency of the cells at x are determined by the environmental condition there, independently of from where they came to x. Under this condition, from (5) and (6)

$$n_+(x) = n_-(x)\ (= \frac{1}{2}n(x))\ \propto \frac{1}{v(x)} \tag{8}$$

The number density of the cells in the stationary distribution becomes inversely proportional to the swimming velocity. For accumulation, the cells slow down or stop swimming.

As described previously, paramecium cells which came to the place of temperature they like do not stop swimming but continue to swim. The large number density is not realised by slow swimming. This means, they do not satisfy the condition (7). In fact, the experiment showed that

$$f_+(x) \neq f_-(x) \text{ (whereas } v_+(x) = v_-(x)) \tag{9}$$

In the vessel, where the right hand side has a higher temperature 25 °C, f_+ is found to be smaller than f_-. Consequently, as derived from (6), v_+n_+ ($= v_-n_-$) increases with increasing x or temperature [2].

III. CELLS AS PARTICLES WITH INTERNAL STATE VARIABLES

The swimming cells have sensors for the environmental condition. The sensor sends information about the environment to a controller of swimming apparatus. If the connection from the sensor to the controller is direct and deterministic in such a manner that the condition (7) is satisfied, the only way of accumulation is to slow down or stop swimming. However, this is sometimes inconvenient for living cells. When a better place appears elsewhere, the cells which stopped once cannot find it.

To understand the behavior of living cells, we like to introduce the following idea [1]. Each cell is regarded as a particle which has a number of variables p, q, ..., representing its internal state. The swimming velocity and the frequency of directional changes are determined by the internal state;

$$v = v(p, q, ...) \tag{10}$$

$$f = f(p, q, ...) \tag{11}$$

The internal state changes with time depending on the environmental condition and the present state;

$$\frac{dp}{dt} = \dot{p}(p, q, ...; x, y, ...) \tag{12}$$

$$\frac{dq}{dt} = \dot{q}(p, q, ...; x, y, ...) \tag{13}$$

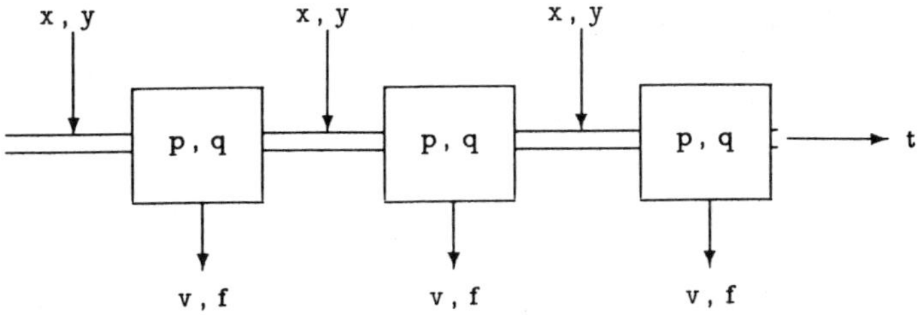

Figure 1: (x,y):the environment; (p,q):the internal state; (v,f):the swimming behavior.

where the spatial coordinate x, y, ... represent the environmental condition. These expressions (10) - (13) are physically reasonable and make the connection between the swimming behavior and the environmental condition indirect, as illustrated in Figure 1.

Such a two step mechanism produces a difference between f_+ and f_- and/or v_+ and v_-. Let us give a simple example. The frequency f is assumed to be proportional to the product of two internal state variables p and q;

$$f = f_o pq \tag{14}$$

and these variables change with time according to the following rate equations;

$$\frac{dp}{dt} = -\alpha_p p + \beta_p \tag{15}$$

$$\frac{dq}{dt} = -\alpha_q q + \beta_q \tag{16}$$

where α_p, β_q, α_q and α_q are functions of the environmental condition.

Let us suppose that the cells which have been kept at the temperature T_1 are transferred to T_2. Before transfer, the variable p took a value $p_\infty(T_1) = \beta_p(T_1)/\alpha_p(T_1)$. After transfer, p changes from $p_\infty(T_1)$ to $p_\infty(T_2) = \beta_p(T_2)/\alpha_p(T_2)$ with the rate $\alpha_p(T_2)$. To explain the behavior of paramecium cells between 20 °C and 25 °C, we assume that α_p is much larger than α_q, and $p_\infty(T)$ decreases and $q_\infty(T)$ increases with increasing temperature. Then, upon a temperature drop from 25 °C to 20 °C, the frequency f given by the product of p and q increases rapidly and then decreases slowly to a stationary value. Upon a temperature rise form 20 °C to 25 °C, the frequency decreases rapidly and then increases slowly. When the cell is swimming in a tempearture gradient, the variable p is always kept nearly equal to p_∞ , following immediately the change of the environmental temperature. On the other hand, the variable q cannot follow immediately but has some delay. It takes a value corresponding to the temperature in the past. Between 20 °C and 25 °C, the value of q of cells swimming at x in the $+x$ direction is smaller than that of cells swimming at the same x in the $-x$ direction. Thus, f_+ becomes smaller than f_-. As a result, the cells come towards 25 °C.

IV. TIME SEQUENCE OF RESPONSE

For swimming cells, the values of α_p, β_p, α_q, and β_q continue to change. The solution of the equation (15) is expressed in the form;

$$\begin{aligned} p(x(t)) &= p(x(t_o))exp(-\int_{t_o}^{t} \alpha_p(x(t'))dt') \\ &+ \int_{t_o}^{t} exp(-\int_{t'}^{t} \alpha_p(x(t''))dt'')\beta_p(x(t'))dt' \end{aligned} \quad (17)$$

where $x(t)$ means the trajectory of the cell. The difference between the value of p of the cell swimming at x and that of the cell hypothetically fixed at the same x is given by

$$p(x(t)) - p_\infty(x(t)) = \int^{t} exp(-\int_{t'}^{t} \alpha_p(x(t''))dt'')(dp_\infty(x(t'))/dt')dt' \quad (18)$$

Each cell at x has a different trajectory in the past and therefore, has a different value of p. Averaging the right hand side of (18) over all possible trajectories in the past of cells which are swimming at x in the $+x$ direction, we can have approximately [1]:

$$p(x) - p_\infty(x) = -(\frac{\bar{v}}{\bar{\alpha}_p + 2\bar{f}})(\frac{dp_\infty}{dx}) \quad (19)$$

where - means averaging over all cells swimming around x. For cells swimming at x in the $-x$ direction, the right hand side changes the sign from - to +.

If α_p is very much larger than f and if α_p is so large that in the distance of v/α_p the change in the environment is very small, the equation (19) states that the value of p of cells swimming at x is nearly equal to p_∞ at the same x. Then, cells swimming at x in the $+x$ direction and those swimming in the $-x$ direction have the same value of p, that is $p_\infty(x)$.

If the above condition is satisfied for all rate constants α_p, α_q, ... of variables p, q, ... which determine the frequency of directional changes and the velocity, then $f_+ = f_-$ and $v_+ = v_-$, corresponding to the case of direct connection from the sensor to the controller. Cells accumulate in the place they like by slowing down or stopping swimming.

The other extreme is the case where the rate constant α_p is very much smaller than the frequency f. If this condition is satisfied for all rate constants of variables, p, q, ..., then from (19) approximately,

$$(\frac{f_+}{v_+}) - (\frac{f_-}{v_+}) = (\frac{d(f/v)}{dx})(\frac{v}{f}) \tag{20}$$

Then, the equation (6) can be integrated to give

$$n(x) \propto \frac{f(x)}{v(x)^2} \; (= \frac{1}{D(x)}) \tag{21}$$

where D is the apparent diffusion constant of the cell. Thus, cells come to the place where the frequency of directional changes is high or the apparent diffusion constant is small. In this case, the behavior of cells can be described by a Fokker-Planck equation where the diffusion constant is a function of the spatial coordinate. The experiment showed that paramecium cells do not use this extreme either.

The above treatment indicates that the rate constants of internal state variables, α's, should be compared with the frequency of directional changes, f. If some of the variables have rate constants much larger than f and some others have rate constants much smaller than f, the number density of cells after accumulation is not simply related to v or f; neither (8) nor (21) is established.

Upon a temperature drop, paramecium cells begin frequent directional changes within a few seconds. This response ceases soon and the frequency decreases in one or two minutes. In the previous expression of f as a product of two variables p and q, we may put α_p to be larger than 1 s^{-1} and α_q to be of the order of 0.01 s^{-1}. On the other hand, the frequency f in the stationary state is of the order of 0.1 s^{-1}. Therefore, in paramecium cells, the following condition is satisfied;

$$\alpha_p > f > \alpha_q \tag{22}$$

V. STRATEGY OF THE CELL

As emphasized before, paramecium cells swim actively after coming to the place they like. They do not strongly condense there. They do not fall into a local best but continue to look for a better place. This is very important in native environment. The environment is heterogeneous and always changing.

Such behaviors are made possible by setting the time sequence of responses as expressed by (22). The rate constants of the fast response and the slow response, α_p and α_q, are determined by the sensor and possibly by an associated processor. On the other hand, the frequency f is determined by some stochastic process in the cell.

Let us suppose that the cell has a generator of fluctuation or noise. A fluctuating signal flows in the main route from the generator to a transducer which is connected to a switching apparatus of the motor. The signal from the transducer to the switching apparatus is probabilistically controlled by the signal from the generator. Another signal from the sensor flows in the secondary route connected to the main route and modifies the fluctuating signal from the generator. Figure 2 illustrates the whole system. The condition (22) means that the magnitude of fluctuation should be set up at an appropriate level in comparison with rate constants of the sensor and the processor.

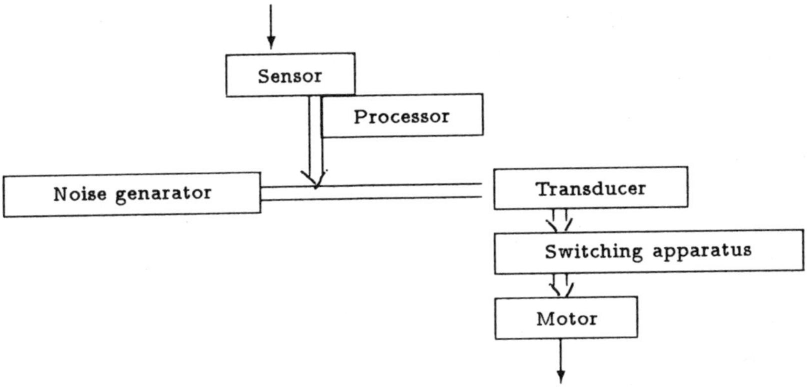

Figure 2: The organized system of molecular machines for regulation of the swimming behavior.

VI. ELECTRIC POTENTIAL FLUCTUATION

In paramecium cells, fluctuation of the electric potential inside the cell was found as an origin of spontaneous changes of the swimming direction [3]. A glass capillary electrode was inserted into a cell fixed on a glass plate and the electric potential difference between inside and outside the cell was recorded. The potential inside is about -30 mV on the average. It is not constant but fluctuates with the amplitude of one to a few mV. Without external stimuli, the potential occasionally shows a large spikelike fluctuation of the amplitude of several mV.

Paramecium cells swim by beating a large number of cilia on the cell surface. They change the swimming direction by reversing the direction of beating of cilia. The reversal of beating takes place transiently and locally on the cell surface. During such reversal, the cell changes its direction and with the recovery of normal beating, it begins to swim straight.

When the spikelike fluctuation of potential occurs in the fixed cell, we can observe under an optical microscope that the direction of beating is partially reversed. There is a good correlation between the reversal of beating of cilia and the spikelike fluctuation of potential. The spontaneous directional change of freely swimming cells is very likely initiated by the spikelike fluctuation.

The potential fluctuation consists of two components. One is the basic fluctuation and the other is the spikelike fluctuation. The basic fluctuation has a Gaussian distribution and its spectrum is of a Lorentzian type. According to the data obtained in various ionic conditions, the basic fluctuation is produced mainly by the open-close fluctuation of gates of channels in the cell membrane which are permeable to potassium ions. The rate constants of opening and closing are of the order of 10^3 s^{-1}. The number of those channels is estimated to be about 3 10^3.

The spikelike fluctuation occurs depending on the potential of basic fluctuation. The value of the potential at which a spike starts to rise was read in the record of fluctuating potential. At higher potentials of the basic fluctuation, the spikelike fluctuation starts more often. This suggests that the spike is produced by ion channels, of which the opening probability increases with a positive shift of the basic potential. The experimental data have shown that calcium ions flow through these channels into the cell and induce the reversal of beating of cilia.

VII. EFFECT OF FIELD FLUCTUATION

To describe the behavior of ion channels, which generate the spikelike fluctuation, we like to introduce the idea of "field fluctuation" [4,5]. Let us consider a channel fluctuating between two states, an open state and a closed state. The free energies of the channel in these states are assumed to depend on the electric field at the channel. This happens if, for example, the gate of the channnel has an electric dipole. If the direction of the dipole is different in two states, the free energy difference between two states becomes proportional to the field. The probability of the transition from the open state to the closed state and that of the reverse transition both depend on the field. If the field fluctuates, the probabilities fluctuate. The rate constants of two transitions, k_+ and k_-, are assumed to be given in the form;

$$k_+(t) \quad = \quad k_{+o} exp(+\gamma E(t))$$

$$k_-(t) = k_{-o}exp(-\gamma E(t)) \tag{23}$$

Here, E(t) is the fluctuating component of the electric field at the channel. Then, the probability of the open state, $P(t)$, changes with time according to the equation;

$$\frac{dP(t)}{dt} = -k_+(t)P(t) + k_-(t)(1 - P(t)) \tag{24}$$

In the above equation (24) with (23), fluctuating quantities are contained in the exponential term. The effect of fluctuation on the probability P is not additive but non-linear. This is because the field or the free energy is fluctuating.

The average values of two rate constants, $\langle k_+\rangle$ and $\langle k_-\rangle$, are both always larger than the rate constants without fluctuation, k_{+o} and k_{-o}. When the fluctuation of E(t) has a Gaussian distribution;

$$\begin{aligned}\langle k_+\rangle &= k_{+o}exp(+(\gamma^2/2)\langle E(t)^2\rangle)\\ \langle k_-\rangle &= k_{-o}exp(+(\gamma^2/2)\langle E(t)^2\rangle)\end{aligned} \tag{25}$$

Upon a change of the environmental condition, the probability $P(t)$ approaches to a new stationary value more quickly in a fluctuating field than in a constant field.

The probability $P(t)$, however, continues to fluctuate in a stationary state. The amplitude of fluctuation is given, if the field fluctuation is not very large, by

$$\begin{aligned}\langle (P(t) - \langle P\rangle)^2\rangle &= \frac{k_{+o}^2 k_{-o}^2}{(k_{+o} + k_{-o})^3}\\ &\times \int exp(-(\langle k_+ + \langle k_-\rangle)\tau)4sinh((\gamma^2/2)\langle E(\tau)E(0)\rangle)d\tau\end{aligned} \tag{26}$$

The equation (24) with (23) also indicates that the average of the probability, $\langle P(t)\rangle$, deviates from the probability in a constant field, P_o. The deviation is estimated approximately as

$$\begin{aligned}\langle P(t) - P_o\rangle &= \frac{k_{+o}k_{-o}(k_{+o} - k_{-o})}{(k_{+o} + k_{-o})^2}\\ &\times \int exp(-(\langle k_+\rangle + \langle k_-\rangle)\tau)2sinh((\gamma^2/2)\langle E(\tau)E(0)\rangle)d\tau\end{aligned} \tag{27}$$

This equation means that the average probability is always shifted towards 1/2 with increasing fluctuation of the field [5].

In paramecium cells, channels for potassium ions produce the fluctuating field $E(t)$. Channels for calcium ions behave as described by (24) with (23). Thus, the cell has a two step device composed of field-insensitive channels for the basic fluctuation and field-sensitive channels for the spikelike fluctuation. Such a device is very useful to amplify the fluctuation and generate spikes at random intervals.

To produce the basic fluctuation of a required amplitude, the cell prepares an appropriate number of field-insensitive channels. To generate spikes at a required average interval, the cell gives an appropriate sensitivity to the field-sensitive channels.

VIII. SPATIAL CORRELATION OF FLUCTUATION

When a mechanical stimulus is given to the front of a paramecium cell, the cell generates a large spike of the height of several tens mV. Then, beating of all cilia is reversed and the cell swims backward for a while. On the other hand, when the cell spontaneously changes the swimming direction, it reverses beating of cilia only in a limited area of the cell surface. The area is possibly regulated by the spatial correlation of the fluctuating field.

If a certain number of channels are in the same fluctuating field, the probability $P(t)$ in (24) gives the ratio of the number of channels in the open state, $n(t)$, to the total number of channels, N. In general, the field $E(t)$ in (23) is a function not only of the time t but also the coordinate x; the field and rate constants are given in the form; $E(x,t)$ and $k_{\pm}(x,t)$. Then, the probability of a channel or a group of channels at x in the open state is given by $P(x,t)$. Thus, we have a set of equations for $P(x,t)$ of channels at different x's. The extent of correlation of these channels is estimated by calculating the time average of the product of two probabilities at x_1 and x_2. From (24), we have approximately

$$\begin{aligned}\langle P(x_1,t)P(x_2,t)\rangle - \langle P(x_1,t)\rangle^2 &= \frac{k_{+o}^2 k_{-o}^2}{(k_{+o}+k_{-o})^3} \\ &\times \int exp(-(\langle k_+\rangle + \langle k_-\rangle)\tau)\, 4\, sinh((\gamma^2/2)\langle E(x_1,\tau)E(x_2,0)\rangle)d\tau \quad (28)\end{aligned}$$

Even when channels in different areas have no direct interaction, they seem to have interaction if the field fluctuation has correlation.

It is to be noted that the fluctuation of the probability $P(t)$ described by (24) with (23) has asymmetry with respect to time reversal. If a certain number of channels are in the same fluctuating field, the fluctuation of the number of open channels has such an asymmetry. On the other hand, if these channels are in independently fluctuating fields, the fluctuation of the number of open channels is symmetric with respect to time reversal.

In the above treatment of the field fluctuation, the effect of opened channels on the field was neglected. In fact, if one of the field- sensitive channels is opened, ions flow through this channel and the flow changes the field near the channel. The field $E(t)$ in (23) must contain, besides the basic fluctuation, an additional term which depends on the probability P itself. This term may enhance the cooperative transition of field-sensitive channels. In the case of paramecium cells, however, this effect is not very strong.

IX. SIGNAL FROM SENSORS

As described previously, upon a temperature drop from 25 °C to 20 °C, paramecium cells show very frequent changes of the swimming direction. As soon as the temperature is dropped, the electric potential inside the cell shifts towards zero. The basic fluctuation continues and the average of the potential is moved by several mV. The probability of opening of field-sensitive channels is increased and many spikes are generated to induce reversal of beating of cilia. Then, the potential returns slowly towards the original value and the frequency of spikes decreases.

Upon a temperature rise from 20 °C to 25 °C, the electric potential inside the cell shows a small shift towards a more negative value. The probability of opening of field-sensitive channels is decreased and spikes are not generated. The cell continues straight swimming. Then, the potential returns to the original value and the frequency of directional changes is restored.

Such responses of the potential are produced by "cold sensors" which are localized in the membrane in the anterior part of the cell [6]. They also form ion channels and the opening-closing of their gates depends on the temperature. The states of the sensors, channels with gates, may be described by two variables p and q, which were introduced in the previous section. The sensor may have two gates in sevies, the states of which are represented by p and q. The potential inside the cell changes with the change of p and q. Therefore, in (23), rate constants k_{+o} and k_{-o} and also $E(t)$ are functions of p and q. The probability P in (24) and consequently the frequency of spikes f become dependent on p and q. This should be the way of derivation of f as a function of p and q, corresponding to (14).

Paramecium cells have also "warm sensors" localized in the posterior part of the cell. Upon a temperature rise from 25 °C to a higher temperature, the sensors produce a positive shift of the potential which increases the frequency of spikes and reversal of beating of cilia. Upon a temperature drop to 25 °C, they produce a negative shift of the potential which inhibits generation of spikes to keep straight swimming. Thus, by using these two kinds of sensors, paramecium cells can come to the place of 25 °C, which is best for them, from both sides of lower and higher temperatures.

X. DIVERSITY OF INTERNAL STATES

In the previous treatment of cells in a temperature gradient, they were classified into two groups, those swimming in the $+x$ direction and those in the $-x$ direction. The average velocity and the average frequency of directional changes were compared between these two groups. In fact, however, the cells swimming at the same x in the same direction have various velocities and various frequencies since their trajectories in the past are not the same so that their internal states are not the same.

Let us consider, for example, a sensor, the state of which is represented by a variable p. The sensor assumes two states and p means the probability of one of these staes. The free energies of the sencer in these states depend on the environmental condition which changes with time following the trajectory of the cell. Therefore, the concept of field fluctuation is useful again. The behavior of the variable p may be described by an equation similar to (24) with (23). Here, instead of the fluctuating electric field $E(t)$ in (23), we have a fluctuating environment of the swimming cell. The magnitude

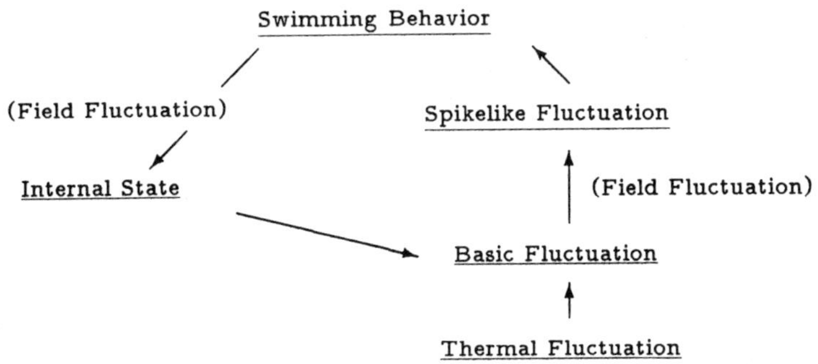

Figure 3: Hierarchy of noise production.

of fluctuation of internal states of the cell can be estimated by an equation similar to (26). The diversity of internal states results in the diversity of the electric potential fluctuation and of the swimming behavior.

In summary, Figure 3 illustrates the hierarchy of swimming cells to produce fluctuation and diversity which are indispensable to live in an inhomogeneous and varying environment. It is hoped to construct a more general theory to describe this hierarchy.

References

[1] F.Oosawa and Y.Nakaoka; *J. Theor. Biol.*,**66**, 747 (1977).

[2] Y.Nakaoka and F.Oosawa; *J. Protozool.*, **24**, 575 (1977).

[3] T.Majima; *Biophys. Chem.*,**11**, 101 (1980).

[4] F.Oosawa; *J. Theor. Biol.*,**52**, 175 (1975).

[5] F.Oosawa, T.Kubori and M.Tsuchiya; *Lecture Notes in Biomathematics*, **70**, 158 (1987).

[6] Y.Nakaoka, T.Kurotani, and H.Itoh; *J. exp. Biol.*, **127**, 95 (1987).

WEAK CONVERGENCE IN STOCHASTIC ANALYSIS

Philip Protter *
Mathematics and Statistics Departments
Purdue University
W. Lafayette, Indiana 47907 USA

In this article we discuss the weak convergence of stochastic integrals and of solutions of stochastic differential equations. The differentials are semimartingales, which have been shown by the Bichteler–Dellacherie theorem to be the most general reasonable stochastic differentials possible. See Protter [4] for all background on semimartingales, stochastic integration and differential equations, as well as for an exposition of the Bichteler–Dellacherie theorem. All results in this article were obtained jointly with Tom G. Kurtz, and proofs and details can be found in Kurtz–Protter [2].

Naively, one would like a theorem that if a sequence of semimartingales Y_n converges weakly to Y, and if predictable processes X_n converge weakly to X, then Y is a semimartingale and $\int X_n dY_n$ converges to $\int X dY$. However one immediately runs into the problem of interpreting the weak convergence of X_n to X, since one cannot use the reasonable path spaces C or $\mathbb{D}$ (continuous and right continuous, left limited functions, respectively; we call the latter cadlag henceforth, after its French acronym). One approach to this problem has been taken by Pages [3], but it is not useful for applications. Here we follow the approach of Jakubowski, Mémin, and Pages [1] and restrict our attention to integrands X_n which are adapted, cadlag processes. We then seek hypotheses such that if (X_n, Y_n) are defined on a sequence of spaces $(\Omega_n, \mathcal{F}_n, P_n)$ with X_n adapted, cadlag, and Y_n semimartingales, and such that if $(X_n, Y_n) \Rightarrow (X, Y)$, then $\int X_n(s-)dY_n(s) \Rightarrow \int X(s-)dY(s)$, where the notation "$\Rightarrow$" denotes weak convergence of the distribution measures on path space.

We begin with a technical definition. Fix an integer m and let $\mathbb{D}_{\mathbb{R}^m}[0,\infty)$ denote the space of cadlag functions mapping $[0,\infty)$ to $\mathbb{R}^m$. Let x denote a generic element of $\mathbb{D}_{\mathbb{R}^m}[0,\infty)$ and define

* Supported in part by NSF Grant No. DMS–8805595

$h_\delta : [0,\infty) \to [0,\infty)$ by $h_\delta(r) = (1-\delta/r)^+$. Define $J_\delta : D_{\mathbf{R}^m}[0,\infty) \to D_{\mathbf{R}^m}[0,\infty)$ by

$$J_\delta(x)(t) = \sum_{s\le t} h_\delta(|x(s) - x(s-)|)(x(s) - x(s-))$$

It is easy to check that $x \to J_\delta(x)$ and $x \to x - J_\delta(x)$ are continuous in the Skorohod topology, and further that if $(x_n, y_n) \to (x, y)$ in the Skorohod topology, then so also does

$$\int_o x_n(s-)dJ_\delta(y_n)(s) \to \int_0 x(s-)dJ_\delta(y)(s).$$

Let us next recall the classical definition of a semimartingale. Let $(\mathcal{F}_t)_{t\ge 0}$ be an increasing sequence of σ-algebras. A cadlag, $\{\mathcal{F}_t\}$-adapted process Y is a semimartingale if it can be decomposed as $Y = M + A$ where M is an $\{\mathcal{F}_t\}$-local martingale and the sample paths of A have finite variation on bounded time intervals, that is, there exists a sequence of $\{\mathcal{F}_t\}$-stopping times, τ_k, such that $\tau_k \to \infty$ a.s and for each k, $M^{\tau_k} \equiv M(\cdot \wedge \tau_k)$ is a uniformly integrable martingale, and for every $t > 0$, $T_t(A) = \sup \Sigma |A(t_{i+1}) - A(t_i)| < \infty$ a.s (where the supremum is over partitions of $[0,t]$).

An $\mathbf{R}^m$-valued process is an $\{\mathcal{F}_t\}$-semimartingale, if each component is a semimartingale. Let M^{km} denote the real-valued, $k \times m$ matrices. Throughout, $\int XdY$ will denote $\int X(s-)dY(s)$. Every semimartingale Y has associated to it an increasing, right continuous, adapted process $[Y]$, known as the quadratic variation process of Y. This process is often easy to calculate, or to estimate. Indeed one has the following facts concerning $[Y]$ (see [4] for proofs):

(1) let τ_n be a refining sequence of partitions of $[0,t]$ with $\lim_n \text{mesh}(\tau_n) = 0$. Then

$$\lim_{n\to\infty} \sum_{t_i\in\tau_n} (Y_{t_i} - Y_{t_{i-1}})^2 = [Y]_t$$

with convergence in probability;

(2) $Y_t^2 - 2\int_0^t Y(s-)dY(s) = [Y]_t$, with the convention $Y(0-) = 0$;

(3) if $Y = M$ is a local martingale, and if $E\{[M]_t\} < \infty$, then M is a martingale and $E\{M_t^2\} = E\{[M]_t\}$;

(4) if M is a local martingale, then $M_t^2 - [M]_t$ is also a local martingale;

(5) if Y is a continuous semimartingale with (unique) decomposition $Y = M + A$, with $A_0 = 0$, then $[Y] = [M]$;

(6) if $Z = \int X(s-)dY(s)$, then $[Z] = \int X(s-)^2 d[Y]_s$;

(7) if Z is the solution of

$$dZ = \sum_{i=1}^k f_i(Z(s-))dY^i(s),$$

for semimartingales Y_i, $1 \le i \le k$, then

$$[Z] = \sum_{i,j} \int f_i(Z(s-))f_j(Z(s-))d[Y^i, Y^j]_s,$$

where $[Y^i, Y^j] = \frac{1}{2}\{[Y^i + Y^j] - [Y^i] - [Y^j]\}$; in particular, if $(B^1, \ldots, B^k)$ is an $\mathbf{R}^k$–Brownian motion, and if

$$dZ = \sum_{i=1}^{k} f_i(Z(s-))dB^i(s),$$

then $[Z]_t = \sum_{i=1}^{k} \int_0^t f_i(Z(s-))^2 ds$, since $[B^i, B^j] = 0$ for $i \ne j$, and $[B^i]_t = t$.

In view of the preceding discussion, the next theorem is particularly useful for applications, since its hypotheses are usually easy to verify. (Another version of this theorem, with essentially equivalent hypotheses which are in a more obtuse form, can be found in [1]). For a finite variation process A let $T_t(A) = \int_0^t |dA_s|$ be the random variable denoting the total variation of the paths of A on $[0, t]$.

THEOREM 1. For each n, let (X_n, Y_n) be an $\{\mathcal{F}_t^n\}$–adapted process with sample paths in $D_{\mathbf{M}^{km}} \times \mathbf{R}^m[0, \infty)$, and let Y_n be an $\{\mathcal{F}_t^n\}$–semimartingale. Fix $\delta > 0$ (allowing $\delta = \infty$), and define $Y_n^\delta = Y_n - J_\delta(Y_n)$. (Note that Y_n^δ will also be a semimartingale.) Let $Y_n^\delta = M_n^\delta + A_n^\delta$ be a decomposition of Y_n^δ into an $\{\mathcal{F}_t^n\}$–local martingale and a process with finite variation. Suppose

(*) For each $\alpha > 0$, there exist stopping times $\{\tau_n^\alpha\}$ such that $P\{\tau_n^\alpha \le \alpha\} \le \frac{1}{\alpha}$ and $\sup_n E[[M_n^\delta]_{t\wedge\tau_n^\alpha} + T_{t\wedge\tau_n^\alpha}(A_n^\delta)] < \infty$.

If $(X_n, Y_n) \Rightarrow (X, Y)$ in the Skorohod topology on $D_{\mathbf{M}^{km}\times\mathbf{R}^m}[0, \infty)$, then Y is a semimartingale with respect to a filtration to which X and Y are adapted, and $(X_n, Y_n, \int X_n dY_n) \Rightarrow (X, Y, \int X dY)$ in the Skorohod topology on $D_{\mathbf{M}^{km}\times\mathbf{R}^m\times\mathbf{R}^k}[0, \infty)$. If $(X_n, Y_n) \to (X, Y)$ in probability, then the triple converges in probability.

<u>Remark</u> For $c > 0$, define $\tau_n^c = \inf\{t: |M_n^\delta(t)| \vee |M_n^\delta(t-)| \ge c \text{ or } T_t(A_n^\delta) \ge c\}$. Suppose the following conditions hold.

(i) $\{T_t(A_n^\delta)\}$ is stochastically bounded for each $t > 0$.

(ii) For each $c > 0$, $\sup_n E[M_n^\delta(t \wedge \tau_n^c)^2 + T_{t\wedge\tau_n^c}(A_n^\delta)] < \infty$.

Since $\sup_{t\le\alpha}|M_n^\delta(t)| = \sup_{t\le\alpha}|Y_n^\delta(t) - A_n^\delta(t)| \le \sup_{t\le\alpha}|Y_n(t)| + T_\alpha(A_n^\delta)$ is stochastically bounded in n for each α, there exists c_α so that $P\{\tau_n^{c_\alpha} \le \alpha\} \le \frac{1}{\alpha}$. In addition $E[[M_n^\delta]_{t\wedge\tau_n^{c_\alpha}}] = E[(M_n^\delta(t \wedge \tau_n^{c_\alpha})^2]$, and (*) is satisfied with $\tau_n^\alpha = \tau_n^{c_\alpha}$.

For $\delta < \infty$, (ii) above will usually be immediate since the discontinuities of Y_n^δ are bounded in magnitude by δ (making Y_n^δ a special semimartingale) and there will exist a decomposition with the discontinuities of each term bounded by 2δ.

We next turn our attention to stochastic differential equations. We wish to consider systems of equations of the form $(1 \leq i \leq k)$

$$X_t^i = U_t^i + \sum_{j=1}^{m} \int_0^t F_j^i(X, s-)dY_s^j$$

where U^i are cadlag, adapted processes, and F_j^i are coefficients which are functionals, allowing dependence on the past. Such equations arise in applications such as stochastic control theory. Natural hypotheses on a sequence F_n of coefficients (and a limiting coefficient F) might be:

(i) For each compact subset $\mathcal{X} \subset D_{\mathbb{R}^k}[0,\infty)$ and $t > 0$, $\sup_{x\in\mathcal{X}} \sup_{s\leq t} |F_n(x,s) - F(x,s)| \to 0$.

(**) (ii) For $\{x_n\}$ and x in $D_{\mathbb{R}^k}[0,\infty)$ and each $t > 0$, $\sup_{s\leq t}|x_n(s) - x(s)| \to 0$ implies $\sup_{s\leq t}|F(x_n,s) - F(x,s)| \to 0$.

(iii) For each compact subset $\mathcal{X} \subset D_{\mathbb{R}^k}[0,\infty)$ and $t > 0$, there exists a continuous function $\omega: [0,\infty) \to [0,\infty)$ with $\omega(0) = 0$ such that for all $\lambda \in \Lambda^1$, $\sup_{x\in\mathcal{X}} \sup_{s\leq t} |F(xo\lambda, s) - F(x, \lambda(s))| \leq \omega(\gamma(\lambda))$.

The above conditions imply that if (x_n, y_n) converges to (x, y) in the Skorohod topology, then so also $(x_n, y_n, F_n(x_n))$ converges to $(x, y, F(x))$. Nevertheless, for technical reasons, we need a slight strengthening of (**), involving properties of F_n and F under transformations of the time scale. Let $T_1[0,\infty)$ denote the collection of nondecreasing mappings λ of $[0,\infty)$ onto $[0,\infty)$ (in particular $\lambda(0) = 0$) such that $\lambda(t+h) - \lambda(t) \leq h$ for all $t,\ h \geq 0$. Let i denote the identity map $i(s) = s$. We will assume that there exist mappings $G_n,\ G: D_{\mathbb{R}^k}[0,\infty) \times T_1[0,\infty) \to D_{\mathbb{M}^{km}}[0,\infty)$ such that $F_n(x)o\lambda = G_n(xo\lambda, \lambda)$ and $F(x)o\lambda = G(xo\lambda, \lambda)$ for $(x,\lambda) \in D_{\mathbb{R}^k}[0,\infty) \times T_1[0,\infty)$. We further make the following assumptions on G_n and G:

(i) For each compact subset $\mathcal{X} \subset D_{\mathbb{R}^k}[0,\infty) \times T_1[0,\infty)$ and $t > 0$, $\sup_{(x,\lambda)\in\mathcal{X}} \sup_{s\leq t} |G_n(x,\lambda,s) - G(x,\lambda,s)| \to 0$.

(* * *)

(ii) For $\{(x_n,\lambda_n)\} \in D_{\mathbb{R}^k}[0,\infty) \times T_1[0,\infty)$, $\sup_{s\leq t}|x_n(s) - x(s)| \to 0$ and $\sup_{s\leq t}|\lambda_n(s) - \lambda(s)| \to 0$ for each $t > 0$ implies $\sup_{s\leq t}|G(x_n,\lambda_n,s) - G(x,\lambda,s)| \to 0$.

Note that (* * *) implies (**). Let M^{km} denote $k \times m$ real valued matrices.

<u>Examples</u> Let $g: \mathbf{R}^k \times [0,\infty) \to \mathbf{M}^{km}$ and $h: [0,\infty) \to [0,\infty)$ be continuous. The following functions satisfy (**) and have a representation in terms of a G satisfying (* * *):

a) $F(x,t) = g(x(t),t)$

b) $F(x,t) = \int_0^t h(t-s)g(x(s),s)ds$

For $k = m = 1$

c) $F(x,t) = \sup_{s \le t} h(t-s) g(x(s), s)$

d) $F(x,t) = \sup_{s \le t} h(t-s) g(x(s) - x(s-), s)$

Using vector and matrix notation, for $n = 1, 2, \ldots$ let $F_n\colon D_{\mathbb{R}^k}[0,\infty) \to D_{\mathbb{M}^{km}}[0,\infty)$, let U_n and Y_n be processes with sample paths in $D_{\mathbb{R}^k}[0,\infty)$ and $D_{\mathbb{R}^m}[0,\infty)$ respectively, adapted to a filtration $\{\mathcal{F}_t^n\}$. Suppose Y_n is a semimartingale and that F_n is nonanticipating in the sense that $F_n(x,t) = F_n(x^t, t)$ for all $t \ge 0$ and $x \in D_{\mathbb{R}_k}[0,\infty)$, where $x^t(\cdot) = x(\cdot \wedge t)$. Let X_n be adapted to $\{\mathcal{F}_t^n\}$ and satisfy

$$X_n(t) = U_n(t) + \int_0^t F_n(X_n, s-)\,dY_n(s), \tag{†}$$

and let X satisfy the limiting equation:

$$X(t) = U(t) + \int_0^t F(X, s-)\,dY(s). \tag{††}$$

We say that (X,τ) is a local solution of (††) above if there exists a filtration $\{\mathcal{F}_t\}$ to which X, U and Y are adapted, Y is an $\{\mathcal{F}_t\}$–semimartingale, τ is an $\{\mathcal{F}_t\}$–stopping time, and

$$X(t \wedge \tau) = U(t \wedge \tau) + \int_0^{t\wedge\tau} F(X, s-)\,dY(s).$$

We say that local uniqueness holds for (††) if any two local solutions (X_1, τ_1), (X_2, τ_2) satisfy $X_1(t) = X_2(t)$, $t \le \tau_1 \wedge \tau_2$, a.s. See Protter [4], Chapter V, for sufficient conditions for uniqueness.

The next theorem can be found, together with a proof, in Kurtz–Protter [2]. It improves upon results of Slominski [5].

THEOREM 2. Suppose that (U_n, X_n, Y_n) satisfies (†), $(U_n, Y_n) \Rightarrow (U, Y)$ in the Skorohod topology and that $\{Y_n\}$ satisfies $(*)$ for some $0 < \delta \le \infty$. Assume that $\{F_n\}$ and F have representations in terms of $\{G_n\}$ and G satisfying $(***)$. For $b > 0$, define $\eta_n^b = \inf\{t\colon |F_n(X_n,t)| \vee |F_n(X_n, t-)| \ge b\}$ and let X_n^b denote the solution of

$$X_n^b(t) = U_n(t) + \int_0^t 1_{[0,\eta_n^b)}(s-) F_n(X_n^b, s-)\,dY_n$$

that agrees with X_n on $[0, \eta_n^b)$. Then $\{(U_n, X_n^b, Y_n)\}$ is relatively compact and any limit point, (U, X^b, Y), gives a local solution (X^b, τ) of (††) with $\tau = \eta^c \equiv \inf\{t\colon |F(X^b,t)| \vee |F(X^b, t-)| \ge c\}$ for any $c < b$. If there exists a global solution X of (††) and local uniqueness holds, then $(U_n, X_n, Y_n) \Rightarrow (U, X, Y)$.

Applications of Theorems 1 and 2 to statistics, filtering theory, economics (finance theory), and M. Emery's martingale structure equation are given in Section 3 of Kurtz–Protter [2].

REFERENCES

1. Jakubowski, A., Mémin, J. and Pages, G. (1989). Convergence en loi des suites d'intégrales stochastiques sur l'espace D^1 de Skorokhod. *Probab. Th. Rel. Fields* **81**, 111–137.

2. Kurtz, T.G., Protter, P. (1989). Weak limit theorems for stochastic integrals and stochastic differential equations. To appear.

3. Pages, G. (1986). Un théorème de convergence fonctionnelle pour les intégrales stochastiques. Séminaire de Proba. XX, 572–611. Springer Lecture Notes in Math. 1204.

4. Protter, Philip (1990). *Stochastic Integration and Differential Equations: A New Approach.* Springer–Verlag, Berlin, Heidelberg, New York (in press: to appear February 1990).

5. Slominski, Leszek (1989). Stability of strong solutions of stochastic differential equations. *Stochastic Processes Appl.* **31**, 173-202.

A GENERALIZED ITO FORMULA FOR N–DIMENSIONAL TIME

Mylan Redfern
and
David Betounes

Department of Mathematics
University of Southern Mississippi
Hattiesburg, MS 39406–5045 USA

ABSTRACT
A differentiation rule for a certain class of generalized multiparameter stochastic processes is obtained. The integrals which arise in the corresponding Ito formula are studied and recognized as generalizations of well–known stochastic integrals.

§1. Introduction

Our purpose in this work is to derive an Ito formula for the multiparameter generalized Wiener functional F(W(t)) where F is a tempered distribution and W is a Wiener process on (L^2), the space of Wiener functionals. For 1–dimensional time this was done by Kubo .[5] In a previous work Redfern [6] derived a formula for the 2–parameter case. Both of these formulas use the operator ∂_t^*.

We will work in $(L^2)^-$ but the approach here will be somewhat different and applies the notions of differentiation and integration developed by the authors in reference 1. In our setting, for a process $\varphi(t)$ on $(L^2)^-$ we are able to interpret integrals such as $\int \frac{\partial^N \varphi(t)}{\partial t_1 \cdots \partial t_N} dt_1 ... dt_N$, $\int \varphi(t) \frac{\partial W}{\partial t_1}(t) dt_1$, and $\int \varphi(t) \frac{\partial W}{\partial t_1}(t) \hat{\otimes} \frac{\partial W}{\partial t_2}(t) dt_1 dt_2$ in a natural way. In particular, evaluation of the integral $\int \frac{\partial^N \varphi(t)}{\partial t_1 \cdots \partial t_N} dt_1 ... dt_N$ gives $\Delta_{[a,b]^n} \varphi$, the Stieltjes measure of $[a,b]^n$ with respect to φ. For instance

$$\int_{[a,b]^2} \frac{\partial^2 W(t)}{\partial t_1 \partial t_2} dt_1 dt_2 = W(b,b) - W(b,a) + W(a,a) - W(a,b).$$

Also, $\partial_t^* \varphi(t) = \varphi(t) \hat{\otimes} \frac{\partial^N W(t)}{\partial t_1 \cdots \partial t_N}$. For one – dimensional time the Ito integral is the only stochastic integral necessary for describing elements of (L^2) and $\int_a^b \partial_t^* \varphi(t)dt$ reduces to the Ito integral $\int_a^b \varphi(t)dB(t)$ when this latter integral exists. Within our framework this integral is formulated as $\int_a^b \varphi(t)\hat{\otimes}\frac{\partial B(t)}{\partial t}dt$. For higher dimensions, the Ito integral and other stochastic integrals are needed for the stochastic calculus. We show that these integrals occur naturally as a result of differentiation.

We will briefly review in section 2 the definitions and notation which we use. In section 3 we give our results. We will state and prove Theorem 1, our basic differentiation rule for F(W(t)). Repeated applications of this lead to our Ito formula, Theorem 2. Theorem 3 gives an alternate form for the stochastic integrals that arise in the integral version of the Ito formula. This reformulation allows us to recognize our formula as a generalization of Imkeller's .[4]

§2. Preliminaries

The probability space $(\mathscr{S}^*(\mathbb{R}^N), \mathscr{B}, \mu)$ determined by the characteristic function $C(\xi) = \exp(-\frac{1}{2}\|\xi\|^2)$ is called the N – dimensional white noise space. For $t \in \mathbb{R}_+^N$ and $[0,t)$ denoting the rectangle $\prod_{i=1}^{N}[0,t_i)$, there is the N–parameter Wiener process on $\mathscr{S}^*(\mathbb{R}^N)$ given by

$$W(t,x) = \langle x, 1_{[0,t)} \rangle.$$

Here $\langle \cdot,\cdot \rangle$ is the pairing between $\mathscr{S}^*$and $\mathscr{S}$. We also have the Wiener–Ito decomposition

$$L^2(\mathscr{S}^*(\mathbb{R}^N)) = \sum_{n=0}^{\infty} K_n$$

where $K_0 = \mathbb{R}$ and for $n > 1$, K_n is the space of n–tuple Wiener integrals based on the normal random measure $\{\langle \cdot, 1_A \rangle; A \subset \mathbb{R}^N$ and has finite Lebesgue measure$\}$. Let I_n: $\sqrt{n!}\,\hat{L}^2(\mathbb{R}^{Nn}) \to K_n$ be the isometry which sends each symmetric L^2 function $f(u^1,\dots,u^n)$ to its multiple Wiener integral $I_n(f)$. Just as in the case for $N = 1$ we have the following diagram where the inclusions are continuous and the arrows indicate isomorphisms, all of which we denote with I_n.

$$\begin{array}{ccccccc}
C_n & \hookrightarrow & K_n^+ & \hookrightarrow & K_n & \hookrightarrow & K_n^- \\
\updownarrow & & \updownarrow & & \updownarrow & & \updownarrow \\
\hat{C}_c^\infty(\mathbb{R}^{Nn}) & \hookrightarrow & \sqrt{n!}\,\hat{H}^{\frac{n(N+1)}{2}}(\mathbb{R}^{Nn}) & \hookrightarrow & \sqrt{n!}\,\hat{L}^2((\mathbb{R}^N)^n) & \hookrightarrow & \sqrt{n!}\,\hat{H}^{-\frac{n(N+1)}{2}}(\mathbb{R}^{Nn})
\end{array}$$

The space $\sum_{n=0}^{\infty} K_n^-$ is denoted as $(L^2)^-$ and is called the space of generalized Wiener functionals, see Hida.[3] Similarly, $(L^2)^+ = \sum_{n=0}^{\infty} K_n^+$ is called the space of test functionals. For our differentiation and integration we will also need $C = \sum_{\text{finite}} C_n$. If $\varphi = \sum_{n=0}^{\infty} \varphi_n$ in $(L^2)^-$ and $\psi = \sum_{n=0}^{\infty} \psi_n$ in $(L^2)^+$, then $\langle \varphi, \psi \rangle =: \sum_{n=0}^{\infty} \langle \varphi_n, \psi_n \rangle$.

If $\varphi(t)$ is in $(L^2)^-$, $\frac{\partial}{\partial t_i}\varphi(t)$ is an element of $(L^2)^-$ such that $\frac{\partial}{\partial t_i}\langle \varphi(t), \psi\rangle = \langle \frac{\partial}{\partial t_i}\varphi(t), \psi\rangle$ for every ψ in C, see reference 1. For example, one has the following:

(1) $\frac{\partial W}{\partial t_1}(t)$ is the element of K_1^- represented by T in $\hat{H}^{-\frac{(N+1)}{2}}(\mathbb{R}^N)$ where

$$T(\xi) = \int_0^{t_N} \cdots \int_0^{t_2} \xi(t_1, u_2, \ldots, u_N) du_2 \ldots du_N.$$

(2) $$\frac{\partial^N W}{\partial t_1 \partial t_2 \ldots \partial t_N} = I_1(\delta_t) = \text{the element of } (L^2)^- \text{ represented by } \delta_t$$

For $\varphi \in K_m^-$ and $\psi \in K_n^-$, the product $\varphi \hat{\otimes} \psi$ is defined via the above diagram and the tensor product of distributions. If φ and ψ are represented respectively by f $\in \hat{H}^{-\frac{m(N+1)}{2}}(\mathbb{R}^{Nm})$ and $g \in \hat{H}^{-\frac{n(N+1)}{2}}(\mathbb{R}^{Nn})$ then $\varphi \hat{\otimes} \psi$ is represented by $f \hat{\otimes} g$, the symmetric tensor product of the distributions f and g. Here $\langle f \otimes g, \xi \rangle = \langle f, \langle g, \xi \rangle \rangle$. One sees that $\varphi \hat{\otimes} \psi$ is in K_{m+n}^-. This product is just the extension of the usual Wick product.

The integral $\int \varphi(t)dt$ is similarly an element of $(L^2)^-$ such that $\langle \int \varphi(t)dt, \psi \rangle = \int \langle \varphi(t), \psi \rangle \, dt$ for every ψ in C.

<u>Example</u>

Let N = 2 and assume that $\varphi(t) = I_1(f(t))$ where $f \in \hat{L}^2([a,b]^2 \times [a,b]^2)$.

$$\int_{[a,b]^2} \varphi(t) \hat{\otimes} \frac{\partial W}{\partial t_1}(t) \hat{\otimes} \frac{\partial W}{\partial t_2}(t) dt = I_3(\hat{g})$$

for $g(p,q,u) = 1_{[a,b]}(p_1) 1_{[a,b]}(q_2) 1_{[0,q_2]}(p_2) 1_{[0,p_1]}(q_1) f((p_1,q_2),u)$.

§3. The Ito formula

We can now consider $\frac{\partial}{\partial t_j}F(W(t))$. For $F \in \mathscr{S}^*$, $\sigma = t_1 \cdots t_N$, and

$$H_n(x) = (-1)^n \exp(x^2) D^n \exp(-x^2),$$

let $$a_n(t) = \langle\, F, (\sqrt{n!2^n})^{-1} H_n(x/\sqrt{2\sigma}) \exp(-x^2/2\sigma)/\sqrt{2\pi\sigma}\,\rangle.$$

Russek [7] defined the generalized Wiener functional $F(W(t))$ as an element of $(L^2)^-$ by

$$F(W(t)) =: \sum_{n=0}^{\infty} a_n(t)\, (\sqrt{n!2^n})^{-1} H_n(\langle \cdot, 1_{[0,t)} \rangle/\sqrt{2\sigma})$$

Observing that

$$W^{\hat{\otimes} n}(t) = I_n(1_{[0,t)}{}^{\hat{\otimes} n}) = (\sqrt{\sigma/2})^n\, H_n(\langle \cdot, 1_{[0,t)} \rangle/\sqrt{2\sigma})$$

we can rewrite the series and get

$$F(W(t)) = \sum_{n=0}^{\infty} \frac{c_n(t)}{n!} W^{\hat{\otimes} n}(t)$$

where $c_n(t) = \langle\, F^{(n)}, \exp(-x^2/2\sigma)/\sqrt{2\pi\sigma}\,\rangle$. It is easy to show that the product rule for differentiation holds and that $\frac{\partial}{\partial t_j} W^{\hat{\otimes} n}(t) = n W^{\hat{\otimes} n-1}(t) \hat{\otimes} \frac{\partial}{\partial t_j} W(t)$. Differentiating the terms in the above series, then, becomes quite simple.

<u>Examples</u> We have that

$$W^2(t) = t_1 t_2 + W^{\hat{\otimes} 2}(t), \text{ therefore } \frac{\partial}{\partial t_1}(W^2(t)) = t_2 + 2W(t) \hat{\otimes} \frac{\partial}{\partial t_1} W(t).$$

Also,

$$W^3(t) = 3t_1 t_2 W(t) + W^{\hat{\otimes} 3}(t) \text{ and thus}$$

$$\frac{\partial}{\partial t_1}(W^2(t)) = 3t_2 W(t) + 3t_1 t_2 \frac{\partial}{\partial t_1} W(t) + 3W^{\hat{\otimes} 2}(t) \hat{\otimes} \frac{\partial}{\partial t_1} W(t).$$

These examples lead us to the following general rule.

<u>Theorem 1</u> For $F \in \mathscr{S}^*$

$$\frac{\partial}{\partial t_j} F(W(t)) = \tfrac{1}{2} \frac{\partial}{\partial t_j}(t_1 t_2 \cdots t_N) \hat{\otimes} F''(W(t)) + F'(W(t)) \hat{\otimes} \frac{\partial}{\partial t_j} W(t).$$

<u>proof</u>: Using the series representation for $F(W(t))$ and the product rule we have that

$$(*) \quad \frac{\partial}{\partial t_j}F(W(t)) = \sum_{n=0}^{\infty}\frac{1}{n!}\left[\frac{\partial}{\partial t_j}c_n(t)\right]W^{\hat{\otimes} n}(t) + \sum_{n=0}^{\infty}\frac{1}{n!}c_n(t)nW^{\hat{\otimes} n-1}(t)\,\hat{\otimes}\,\frac{\partial W}{\partial t_j}(t)$$

By computation,

$$\frac{\partial}{\partial t_j}c_n(t) = \tfrac{1}{2}\left[\frac{\partial}{\partial t_j}\sigma(t)\right]\langle\, D_x^2 \exp(-x^2/2\sigma)/\sqrt{2\pi\sigma},\ F^{(n)}\,\rangle$$

$$= \tfrac{1}{2}\left[\frac{\partial}{\partial t_j}\sigma(t)\right]\langle\, \exp(-x^2/2\sigma)/\sqrt{2\pi\sigma},\ F^{(n+2)}\,\rangle$$

The first series in (*) then is seen to be $\frac{1}{2}\frac{\partial}{\partial t_j}(t_1t_2\cdots t_N)\hat{\otimes}F''(W(t))$. Rewriting $F^{(n)}$ as $(F')^{(n-1)}$ in the second series and adjusting the index one gets the series representation of $F'(W(t))\,\hat{\otimes}\,\frac{\partial}{\partial t_j}W(t)$.

Upon applying this theorem for the case $N = 2$, we see that

$$\frac{\partial^2}{\partial t_1\partial t_2}F(W(t)) = \tfrac{1}{2}F''(W(t)) + \tfrac{1}{4}\,t_1t_2F^{(4)}W(t)$$

$$+ \tfrac{1}{2}F'''(W(t))\,\hat{\otimes}\,[t_1\frac{\partial}{\partial t_1}W(t) + t_2\frac{\partial}{\partial t_2}W(t)]$$

$$+ F''(W(t))\,\hat{\otimes}\,\frac{\partial}{\partial t_1}W(t)\,\hat{\otimes}\,\frac{\partial}{\partial t_2}W(t) + F'(W(t))\,\hat{\otimes}\,\frac{\partial^2}{\partial t_1\partial t_2}W(t).$$

One also has the integral version of the above formula and the stochastic integrals which appear on the right are generalizations of those in the Cairoli – Walsh formula.[2] As previously mentioned, $\int_{[a,b]^2} F'(W(t))\,\hat{\otimes}\,\frac{\partial^2}{\partial t_1\partial t_2}W(t)dt$ reduces to the Ito integral. For suitable functions F, $\int_{[a,b]^2} F''(W(t))\ddot{\otimes}\,\frac{\partial}{\partial t_1}W(t)\,\dot{\otimes}\,\frac{\partial}{\partial t_2}W(t)dt$ is the integral $\int_{[a,b]^2}F''(W(t))dJ(t)$ of Cairoli and Walsh. But $\int_{[0,b]^2} F''(W(t))\hat{\otimes}\,\frac{\partial}{\partial t_1}W(t)\,\hat{\otimes}\,\frac{\partial}{\partial t_2}W(t)dt$ can also be written $\frac{1}{2}\int_{[0,b]^2\times[0,b]^2}\partial_z^*\partial_{z'}^*1_G(z,z')F''(W(z\vee z'))dzdz'$,where G is the set of unordered pairs in $[0,b]^2\times[0,b]^2$, and one recognizes this as the Wong–Zakai integral $\frac{1}{2}\int_{[0,b]^2\times[0,b]^2} F''(W(z\vee z'))dW(z)dW(z')$, see reference 6. For arbitrary N, we need the following notation.

Definition Suppose τ is a partition of $\{1,...,N\}$ of the form $\{S_1,...,S_k,T_1,...,T_m\}$ and that $\sigma = t_1 \cdots t_N$. For $p \in \mathbb{R}^N$ and $A \subseteq \{1,...,N\}$ let p_A be the projection of p onto $\mathbb{R}^A$. Also, for $S = \bigcup_{i=1}^{k} S_i$ and $T = \bigcup_{i=1}^{m} T_i$, let

$$G(T,S) = \{(p^1,...,p^{m+k}) \in (\mathbb{R}^N)^{m+k};$$
$$p^a_{T_j} < p^j_{T_j} \ \forall j \text{ in } \{1,...,m\}, \forall a \text{ in } \{1,...,m+k\}, a \neq j \text{ and}$$
$$p^a_{S_r} < p^{m+k}_{S_r} \ \forall \, r \text{ in } \{1,...,k\} \text{ and } \forall a \text{ in } \{1,...,m+k\}, a \neq m+k\}.$$

If $T = \{j_1,...,j_k\} \subseteq \{1,...,N\}$ and $\varphi(t)$ is suitably differentiable, then $\varphi_T(t)$ denotes $\dfrac{\partial^k \varphi(t)}{\partial t_{j_1} \cdots \partial t_{j_k}}$.

One obtains the N – dimensional time Ito formula by repeated applications of Theorem 1. We write it in the following form in order to compare it with the formula of Imkeller.[4]

Theorem 2

$$\frac{\partial^N}{\partial t_1 \cdots \partial t_N} F(W(t)) = \sum (\tfrac{1}{2})^k \sigma_{S_1}(t) \cdots \sigma_{S_k}(t) \, F^{(2k+m)}(W(t)) \hat{\otimes} W_{T_1}(t) \hat{\otimes} \cdots \hat{\otimes} W_{T_m}(t).$$

Here the sum is taken over all partitions τ of the indicated form. In integral form, for a rectangle A in $\mathbb{R}^N_+$, the formula becomes

$$\Delta_A F(W) = \sum \int_A (\tfrac{1}{2})^k \sigma_{S_1}(t) \cdots \sigma_{S_k}(t) \, F^{(2k+m)}(W(t)) \hat{\otimes} W_{T_1}(t) \hat{\otimes} \cdots \hat{\otimes} W_{T_m}(t) dt$$

Theorem 3

Suppose that $\varphi(t) = I_n(f(t))$ with $f \in \hat{L}^2((\mathbb{R}^N)^{n+1})$, then,

$$\int \varphi(t) \sigma_{S_1}(t) \cdots \sigma_{S_k}(t) \hat{\otimes} W_{T_1}(t) \hat{\otimes} \cdots \hat{\otimes} W_{T_m}(t) dt$$

$$= \int \varphi(p^1 \vee \cdots \vee p^m \vee q^1 \vee \cdots \vee q^k) \, 1_{G(T,S)}(p^1,...,p^m,q^1,...,q^k)$$

$$\hat{\otimes} W_{1...N}(p^1) \hat{\otimes} \cdots \hat{\otimes} W_{1...N}(p^m) dp^1 ... dp^m dq^1 ... dq^k$$

$$= \int \overset{*}{\partial}_{p^1}\overset{*}{\partial}_{p^2} \cdots \overset{*}{\partial}_{p^m}\varphi(p^1\vee\cdots\vee p^m\vee q^1\vee\cdots\vee q^k)$$

$$\times\ 1_{G(T,S)}(p^1,\ldots,p^m,q^1,\ldots,q^k)dp^1\ldots dp^m dq^1\ldots dq^k.$$

If φ satisfies the conditions for existence of the Imkeller integral, one can show that the integrals above coincide with Imkeller's integral: $\mathcal{J}_\tau(\varphi^\tau)$ where φ^τ is the τ – corner function of φ.

REFERENCES

[1] Betounes, D. and Redfern, M., 1989,"Differential forms with values in $(L^2)^-$, (part of these proceedings).

[2] Cairoli, R. and Walsh, J.B., 1975, "Stochastic integrals in the plane", Acta Math., **134**, 111–183.

[3] Hida, T., 1978, "Generalized multiple Wiener integrals", Proc. Japan Acad. Ser. A Math Sci. **54**, 55–58.

[4] Imkeller, P., 1984, "Ito's formula for continuous (N,d)–processes", Z.W., **65**, 535–562.

[5] Kubo, I., 1983, "Ito formula for generalized Brownian functionals", Lecture Notes in Control and Information Sciences, **49**, 156–166.

[6] Redfern M., 1988, "White noise approach to multiparameter stochastic integration", Journal of Multivariate Analysis, to appear.

[7] Russek, A., "Hermite Expansions of generalized Brownian Functionals", Lecture Notes in Math, **1080**, 233–244.

ON THE ASYMPTOTICS OF FIRST PASSAGE TIME DENSITIES*

LUIGI M. RICCIARDI
Dipartimento di Matematica e Applicazioni,
Università di Napoli, Napoli, Italy

VIRGINIA GIORNO and AMELIA G. NOBILE
Dipartimento di Informatica e Applicazioni,
Università di Salerno, Salerno, Italy

ABSTRACT

The integral equations given in References 1 and 3 are exploited to determine the asymptotic behaviour of the first passage time (FPT) p.d.f.'s through certain time–varying boundaries, including periodic boundaries, for a class of one–dimensional diffusion processes possessing steady–state densities. Sufficient conditions for the FPT density to exhibit an asymptotically exponential behaviour are given. Explicit formulas for the Ornstein–Uhlenbeck process are obtained and some new asymptotic results for the FPT density of the Wiener process are proved.

1. Introduction

First–passage–time (FPT) problems for diffusion processes through generally time–dependent boundaries have been often considered in the literature (see, for instance[1–4] and related references). The existing investigations have essentially proceeded along three main directions: *i*) to obtain closed form solutions under suitable assumptions on the involved diffusion processes and on the boundaries; *ii*) to device efficient algorithms to solve the integral equations satisfied by FPT densities and *iii*) to analyse the asymptotic behaviour of the FPT densities as boundary or time grow larger.

The present paper falls within the class *iii*) of problems. It represents the natural extension of some preliminary investigations earlier carried out for the Ornstein–Uhlenbeck (OU) process[8], and successively for a class of diffusion processes in the presence of time–independent boundaries[9], for which it was shown that an exponential asymptotic behaviour is exhibited by the FPT p.d.f. as the boundary is moved away from the starting point of the process. Successively, by making use of a new integral equation for the FPT p.d.f. through time–varying boundaries extensive computations were performed by us with the aim to gain some insight on the behaviour of FPT densities for the OU process through single periodic boundaries[1,3]. The results of the computations have indicated that for some particular periodic boundaries the FPT p.d.f. exhibits damped

* Work supported in part by CNR, contracts Nos. 88.276.01 and 88.333.01, and by MPI (40% funds).

oscillations having the some period T as the boundary. Indeed, to a very high degree of accuracy such FPT densities can be represented in the form

$$g(t) \simeq Z(t)\, e^{-\lambda t} \tag{1.1}$$

where λ is a constant and $Z(t)$ is a periodic function of period T. (See for instance Figs. 1–3).

On the grounds of these computational results and making use of the integral equation of Buonocore *et al.*[1], recently Sacerdote[10] has suggested that the FPT p.d.f. of an OU process $X(t)$ through a single periodic boundary of period T with $S(0) > X(0)$ for large times and large boundaries is approximately exponentially distributed, i.e.

$$g(t) \simeq -\alpha(t)\, \exp\left[\int_0^t d\tau\, \alpha(\tau)\right], \tag{1.2}$$

where $\alpha(t) < 0$ is a periodic function of period T.

It should be stressed that (1.2) can always be put in the form (1.1). Furthermore, (1.1) and (1.2) in the case of the OU process play a complementary role in the sense that (1.2) appears to be useful to approximate $g(t)$ whenever the boundary is far from the starting point of the process, in which cases the FPT p.d.f. takes much too small values to be reliably computed by numerical methods. In turn, (1.1) is an useful approximation to FPT p.d.f. whenever boundary and time fall within ranges whereby the asymptotic approximation does not hold.

In the present paper for brevity we shall confine our analysis to the case when a unique boundary is assigned, which either possesses an horizontal asymptote or is asymptotically periodic (Section 2). We shall then see that the asymptotic non–homogeneous exponential form of $g(t)$ holds for a class of diffusion processes admitting of steady–state densities. Extensions to the case of two boundaries each of which either possesses an horizontal asymptote or is asymptotically periodic will be discussed elsewhere[5]. We would like to acknowledge that our proofs have been inspired by some theorems of Mandl[7] dealing with the constant boundary case.

The basic notation that will be used throughout this paper are given hereafter. Further details can be found in the quoted references.

Let $\{X(t);\ t \geq 0\}$ be a time–homogeneous one–dimensional diffusion process defined over the interval $I = (r_1, r_2)$, with r_i $(i = 1, 2)$ natural boundaries, let $P\{X(0) = x_0\} = 1$, with x_0 in the interior of I, and let $A_1(x)$ and $A_2(x)$ be the drift and infinitesimal variance of $X(t)$, respectively. As in Reference 4, we denote by f and F, respectively, the transition p.d.f. and distribution of $X(t)$ and by g the FPT p.d.f. through a single boundary. In Reference 3 it was proved that if the unique boundary $S(t)$ takes values in I and is of class $C_2[0, t_{\max}]$, $t_{\max} < +\infty$, then g is solution of the following second–kind Volterra integral equation:

$$g(t|x_0) = \rho \left\{-2\, \psi[S(t), t|x_0] + 2 \int_0^t d\tau\, g(\tau|x_0)\, \psi[S(t), t|S(\tau), \tau]\right\} \tag{1.3}$$

with $\rho = \mathrm{sgn}[S(0) - x_0]$ and, for all $y \in I$,

$$\psi[S(t), t|y, \tau] = \left\{\frac{S'(t)}{2} - \frac{A_1[S(t)]}{2} + \frac{3}{8} A_2'[S(t)]\right\} f[S(t), t|y, \tau] + \frac{A_2[S(t)]}{2} \frac{\partial}{\partial x} f(x, t|y, \tau)\bigg|_{x=S(t)}. \tag{1.4}$$

Equation (1.3) is non–singular and

$$\lim_{\tau\uparrow t} \psi[S(t), t|S(\tau), \tau] = 0\,. \tag{1.5}$$

We finally assume that $X(t)$ admits of the steady–state p.d.f.

$$W(x) = \frac{c}{A_2(x)} \exp\left[2\int^x d\xi\, \frac{A_1(\xi)}{A_2(\xi)}\right], \tag{1.6}$$

where c is a constant such that

$$c = \left\{\int_{r_1}^{r_2} \frac{dz}{A_2(z)} \exp\left[2\int^z d\xi\, \frac{A_1(\xi)}{A_2(\xi)}\right]\right\}^{-1}. \tag{1.7}$$

2. Mathematical Results

The following two cases will be separately analysed: *i)* the boundary possesses an asymptote and *ii)* the boundary is asyptotically periodic.

Lemma 2.1. Let

$$\lim_{t\to+\infty} S(t) = S\,, \qquad S \in I\,. \tag{2.1}$$

Then, for all $y \in I$ and for all $\tau < t$ one has:

$$\lim_{t\to+\infty} \psi[S(t), t|y, \tau] = \frac{1}{2}\left[A_1(S) - \frac{A_2'(S)}{4}\right] W(S)\,. \tag{2.2}$$

Proof. Taking the limit in (1.4) as $t \to +\infty$ we obtain:

$$\lim_{t\to+\infty} \psi[S(t), t|y, \tau] = \frac{1}{2}\left[-A_1(S) + \frac{3}{4} A_2'(S)\right] W(S) + \frac{A_2(S)}{2}\frac{d}{dS} W(S)\,. \tag{2.3}$$

Since from (1.6) we have:

$$\frac{dW(x)}{dx} = \frac{2}{A_2(x)}\left[A_1(x) - \frac{A_2'(x)}{2}\right] W(x)\,, \tag{2.4}$$

equation (2.3) leads to (2.2). ◇

Let us now set:

$$R(S) = -2\rho \lim_{t\to+\infty} \psi[S(t), t|y, \tau] \equiv -\rho\left[A_1(S) - \frac{A_2'(S)}{4}\right] W(S)\,, \tag{2.5}$$

and denote by r the boundary r_1 or the boundary r_2 according to whether $\rho = -1$ or $\rho = 1$.

Lemma 2.2. $\exists S^* \in I$ such that $R(S) > 0$ if $\rho\,(S - S^*) > 0$.

Proof. For any $a, b \in I$ and $a < b$ let

$$F[a,b] = \int_a^b dz\, f(z)\,, \tag{2.6}$$

where

$$f(z) = \exp\left[-2\int^z d\xi\, \frac{A_1(\xi)}{A_2(\xi)}\right] \tag{2.7}$$

is the scale measure. Since r_1 and r_2 are natural boundaries and since the steady–state density $W(x)$ exists for all $x \in I$, one has[6]:

$$F(r_1, x] = +\infty\,, \qquad F[x, r_2) = +\infty \tag{2.8}$$

and[9]

$$\lim_{S\to r_i} A_2(S)\, W^2(S) = 0 \qquad (i = 1,2)\,. \tag{2.9}$$

From the first of (2.8) and (2.9), after use of l'Hospital rule, if $x_0 > S(0)$ we then obtain:

$$\lim_{S\downarrow r_1} F[S,b]\, A_2(S)\, W^2(S) = \lim_{S\downarrow r_1} \frac{\left[A_2(S)\, W(S)\right]^2 f(S)}{4\left[A_1(S) - A_2'(S)/4\right]} \geq 0\,.$$

Hence there exists an $S^* \in I$ such that for $S < S^*$ one has $A_1(S) - A_2'(S)/4 > 0$. This, in turn, due to (2.5) implies $R(S) > 0$. Let us now assume $x_0 < S(0)$. Then:

$$\lim_{S\uparrow r_2} F[a,S]\, A_2(S)\, W^2(S) = \lim_{S\uparrow r_2} \frac{\left[A_2(S)\, W(S)\right]^2 f(S)}{4\left[A_2'(S)/4 - A_1(S)\right]} \geq 0\,,$$

which again implies $R(S) > 0$ for $S > S^*$, $S^* \in I$. ◇

Theorem 2.1. Let $S(t)$, $t \in [0, +\infty)$, be bounded and such that (2.1) holds. If

$$\lim_{S\to r} R(S) = 0\,, \tag{2.10a}$$

$$\lim_{S\to r}\left\{-2\,\rho\,[R(S)]^{-1}\psi\left[S\left(\frac{t}{R(S)}\right), \frac{t}{R(S)}\middle| x_0\right]\right\} = 1\,, \tag{2.10b}$$

$$\lim_{S\to r}\left\{-2\,\rho\,[R(S)]^{-1}\psi\left[S\left(\frac{t}{R(S)}\right), \frac{t}{R(S)}\middle| S\left(\frac{\vartheta}{R(S)}\right), \frac{\vartheta}{R(S)}\right]\right\} = 1 \qquad (0 \leq \vartheta < t)\,, \tag{2.10c}$$

then,

$$\lim_{S\to r}\left\{[R(S)]^{-1} g\left[\frac{t}{R(S)}\middle| x_0\right]\right\} = e^{-t}. \tag{2.11}$$

Proof. Due to Lemma 2.2 we can chose S such that $R(S) > 0$. Changing t into $t/R(S)$ in (1.3) we obtain:

$$g\left[\frac{t}{R(S)}\middle| x_0\right] = -2\,\rho\,\psi\left[S\left(\frac{t}{R(S)}\right), \frac{t}{R(S)}\middle| x_0\right] + 2\,\rho\int_0^{t/R(S)} d\tau\, g(\tau|x_0)$$
$$\times\, \psi\left[S\left(\frac{t}{R(S)}\right), \frac{t}{R(S)}\middle| S(\tau), \tau\right], \tag{2.12}$$

or:

$$\frac{1}{R(S)}\, g\Big[\frac{t}{R(S)}\Big|x_0\Big] = -\frac{2\,\rho}{R(S)}\,\psi\Big[S\Big(\frac{t}{R(S)}\Big),\frac{t}{R(S)}\Big|x_0\Big] - \int_0^t d\vartheta\,\frac{1}{R(S)}\,g\Big[\frac{\vartheta}{R(S)}\Big|x_0\Big] \times\Big\{-\frac{2\rho}{R(S)}\,\psi\Big[S\Big(\frac{t}{R(S)}\Big),\frac{t}{R(S)}\Big|S\Big(\frac{\vartheta}{R(S)}\Big),\frac{\vartheta}{R(S)}\Big]\Big\}, \tag{2.13}$$

having performed the change of variable $\vartheta = \tau R(S)$. Setting

$$a(t) = \lim_{S\to r}\Big\{[\,R(S)\,]^{-1} g\Big[\frac{t}{R(S)}\Big|x_0\Big]\Big\} \tag{2.14}$$

and recalling (1.5), from (2.13) and (2.10) we obtain:

$$a(t) = 1 - \int_0^t d\vartheta\; a(\vartheta)\,, \tag{2.15}$$

which implies

$$a(t) = e^{-t}\,. \tag{2.16}$$

Equation (2.11) is thus proved. ◇

Remark 2.1. Under the assumptions of Theorem 2.1 for $S \to r$ one has asymptotically

$$g(t|x_0) \sim R(S)\exp\big[-R(S)\,t\big]\,. \quad \diamond \tag{2.17}$$

We shall now come to the case of asymptotically periodic boundaries.

Lemma 2.3. Let $h(t)$ be, for all $t > 0$, a positive periodic function of period T and let

$$\lambda = \frac{1}{T}\int_0^T d\tau\; h(\tau)\,. \tag{2.18}$$

Then, there exists a non–negative monotonically increasing function $\varphi(t)$ which is solution of

$$\int_0^{\varphi(t)} d\tau\; h(\tau) = \lambda t \tag{2.19}$$

and such that

$$\varphi(t + nT) = \varphi(t) + nT \qquad (n = 0, 1, 2, \ldots)\,. \tag{2.20}$$

Proof. Let $\mathcal{P}$ be any primitive function of $h(t)$. From (2.19) we have $\mathcal{P}[\varphi(t)] = \mathcal{P}(0) + \lambda t$. Since $h(t) > 0$, $\forall t > 0$, $\mathcal{P}(t)$ possesses an inverse so that

$$\varphi(t) = \mathcal{P}^{-1}\big[\mathcal{P}(0) + \lambda t\big]\,. \tag{2.21}$$

Since $\lambda > 0$, due to (2.19) one has $\varphi(t) \geq 0$ and

$$\varphi'(t) = \frac{\lambda}{h\big[\varphi(t)\big]} > 0\,. \tag{2.22}$$

Hence, $\varphi(t)$ is a monotonically increasing function. Furthermore, from (2.19) one has:

$$\int_0^{\varphi(t)} d\tau\, h(\tau) + \int_{\varphi(t)}^{\varphi(t+nT)} d\tau\, h(\tau) = \lambda(t+nT)$$

or, due to (2.18) and (2.19):

$$\int_{\varphi(t)}^{\varphi(t+nT)} d\tau\, h(\tau) = \lambda nT \equiv n\int_0^T d\tau\, h(\tau)\,. \tag{2.23}$$

Since $h(t)$ is periodic with period T, relation (2.23) implies (2.20). ◇

Lemma 2.4. Let

$$\lim_{n\to\infty} S(t+nT) = V(t)\,, \tag{2.24}$$

where $V(t)$ is a periodic function of period T. Then, for all $y \in I$ and for all $\tau < t+nT$ one has:

$$\lim_{n\to\infty} \psi\big[S(t+nT), t+nT|y,\tau\big] = \frac{1}{2}\Big\{V'(t) + A_1[V(t)] - \frac{A_2'[V(t)]}{4}\Big\} W\big[V(t)\big]\,. \tag{2.25}$$

Proof. From (1.4) we have:

$$\psi\big[S(t+nT), t+nT|y,\tau\big] = \frac{1}{2}\Big\{\frac{d}{dt}S(t+nT) - A_1[S(t+nT)] + \frac{3}{4}A_2'[S(t+nT)]\Big\}$$
$$\times f\big[S(t+nT), t+nT|y,\tau\big] + \frac{A_2[S(t+nT)]}{2}\frac{\partial}{\partial x} f\big[x, t+nT|y,\tau\big]\Big|_{x=S(t+nT)}\,.$$

Taking the limit of both sides as $n\to\infty$ and recalling (2.24) we obtain:

$$\lim_{n\to\infty} \psi\big[S(t+nT), t+nT|y,\tau\big] = \frac{1}{2}\Big\{V'(t) - A_1[V(t)] + \frac{3}{4}A_2'[V(t)]\Big\} W\big[V(t)\big]$$
$$+ \frac{A_2[V(t)]}{2}\frac{d}{dx}W(x)\Big|_{x=V(t)}\,. \tag{2.26}$$

From (2.4) and (2.26) the limit (2.25) follows. ◇

Let us now set

$$R[V(t)] \equiv -2\,\rho \lim_{n\to\infty} \psi\big[S(t+nT), t+nT|y,\tau\big]$$
$$= -\rho\Big\{V'(t) + A_1[V(t)] - \frac{A_2'[V(t)]}{4}\Big\} W\big[V(t)\big]\,. \tag{2.27}$$

Theorem 2.2. Let $S(t)$, $t\in[0,+\infty)$ be a bounded function such that (2.24) holds and let

$$S = \frac{1}{T}\int_0^T d\tau\, V(\tau) \tag{2.28}$$

and

$$h(t) = R[V(t)] \,. \tag{2.29}$$

If there exists $S^* \in I$ such that $R[V(t)] > 0$ for $\rho(S - S^*) > 0$ and if

$$\lim_{S\to r} R\Big\{V\Big[\varphi\Big(\frac{t}{\lambda}\Big)\Big]\Big\} = 0 \,, \tag{2.30a}$$

$$\lim_{S\to r} \frac{-2\,\rho\,\psi\Big\{S\Big[\varphi\Big(\frac{t}{\lambda}\Big)\Big], \varphi\Big(\frac{t}{\lambda}\Big)\Big| x_0\Big\}}{R\Big\{V\Big[\varphi\Big(\frac{t}{\lambda}\Big)\Big]\Big\}} = 1 \,, \tag{2.30b}$$

$$\lim_{S\to r} \frac{-2\,\rho\,\psi\Big\{S\Big[\varphi\Big(\frac{t}{\lambda}\Big)\Big], \varphi\Big(\frac{t}{\lambda}\Big)\Big| S\Big[\varphi\Big(\frac{\vartheta}{\lambda}\Big)\Big], \varphi\Big(\frac{\vartheta}{\lambda}\Big)\Big\}}{R\Big\{V\Big[\varphi\Big(\frac{t}{\lambda}\Big)\Big]\Big\}} = 1 \qquad (0 \le \vartheta < t) \,, \tag{2.30c}$$

with λ and φ defined by (2.18), (2.19), (2.27) and (2.29), then one has:

$$\lim_{S\to r} \Big\{\Big[\frac{d}{dt}\varphi\Big(\frac{t}{\lambda}\Big)\Big]\, g\Big[\varphi\Big(\frac{t}{\lambda}\Big)\Big| x_0\Big]\Big\} = e^{-t}. \tag{2.31}$$

Proof. Changing t into $\varphi(t/\lambda)$ in equation (1.3) and the integration variable by $\tau = \varphi(\vartheta/\lambda)$, after multiplying both sides by $d\varphi(t/\lambda)/dt$ we obtain:

$$\begin{aligned}\Big[\frac{d}{dt}\varphi\Big(\frac{t}{\lambda}\Big)\Big]\, g\Big[\varphi\Big(\frac{t}{\lambda}\Big)\Big| x_0\Big] &= -2\,\rho\Big[\frac{d}{dt}\varphi\Big(\frac{t}{\lambda}\Big)\Big]\psi\Big\{S\Big[\varphi\Big(\frac{t}{\lambda}\Big)\Big], \varphi\Big(\frac{t}{\lambda}\Big)\Big| x_0\Big\} - \int_0^t d\vartheta\Big[\frac{d}{d\vartheta}\varphi\Big(\frac{\vartheta}{\lambda}\Big)\Big] \\ &\times g\Big[\varphi\Big(\frac{\vartheta}{\lambda}\Big)\Big| x_0\Big]\Big\{-2\,\rho\Big[\frac{d}{dt}\varphi\Big(\frac{t}{\lambda}\Big)\Big]\,\psi\Big\{S\Big[\varphi\Big(\frac{t}{\lambda}\Big)\Big], \varphi\Big(\frac{t}{\lambda}\Big)\Big| S\Big[\varphi\Big(\frac{\vartheta}{\lambda}\Big)\Big], \varphi\Big(\frac{\vartheta}{\lambda}\Big)\Big\}\Big\}.\end{aligned} \tag{2.32}$$

Since from (2.19) and (2.29) one has:

$$\int_0^{\varphi(t/\lambda)} d\tau\, R[V(\tau)] = t \,, \tag{2.33}$$

or:

$$\frac{1}{R\Big\{V\Big[\varphi\Big(\frac{t}{\lambda}\Big)\Big]\Big\}} = \frac{d}{dt}\varphi\Big(\frac{t}{\lambda}\Big) \,, \tag{2.34}$$

from (2.32), (2.30)'s, (2.34) and (1.5) we obtain:

$$b(t) = 1 - \int_0^t d\vartheta\, b(\vartheta) \,, \tag{2.35}$$

having set:

$$b(t) = \lim_{S\to r} \Big\{\Big[\frac{d}{dt}\varphi\Big(\frac{t}{\lambda}\Big)\Big]\, g\Big[\varphi\Big(\frac{t}{\lambda}\Big)\Big| x_0\Big]\Big\}. \tag{2.36}$$

From (2.35) and (2.36) we are finally led to (2.31). ◇

Corollary 2.1. Under the assumptions of Theorem 2.2, for $S \to r$ asymptotically one has:

$$g(t|x_0) \sim R[V(t)] \exp\Big\{-\int_0^t d\vartheta\, R[V(\vartheta)]\Big\}. \tag{2.37}$$

Proof. From (2.31) and (2.34) we obtain:

$$\lim_{S\to r} \frac{g\left[\varphi\left(\frac{t}{\lambda}\right)\Big|x_0\right]}{R\left\{V\left[\varphi\left(\frac{t}{\lambda}\right)\right]\right\}} = e^{-t}.$$

Hence, for $S \to r$, using (2.33), we have asymptotically:

$$g\left[\varphi\left(\frac{t}{\lambda}\right)\Big|x_0\right] \sim R\left\{V\left[\varphi\left(\frac{t}{\lambda}\right)\right]\right\} e^{-t} = R\left\{V\left[\varphi\left(\frac{t}{\lambda}\right)\right]\right\} \exp\Big\{-\int_0^{\varphi(t/\lambda)} d\tau\, R[V(\tau)]\Big\}.$$

The asymptotic formula (2.37) is then obtained. ◇

Note that (2.37) can also be written as:

$$g(t|x_0) \sim Z(t)\, e^{-\lambda t} \tag{2.38}$$

where $Z(t)$ is a periodic function of period T given by

$$Z(t) = R[V(t)] \exp\Big\{\lambda t - \int_0^t d\tau\, R[V(\tau)]\Big\} \tag{2.39}$$

and where

$$\lambda = \frac{1}{T} \int_0^T d\tau\, R[V(\tau)]. \tag{2.40}$$

An asymptotic expression of $g(t|x_0)$ of the same form as (1.1) has thus been obtained.

Remark 2.2. The function $R\{V[\varphi(t/\lambda)]\}$ is periodic with period T.

Proof. It follows from (2.20), (2.25) and from the assumed periodicity of $V(t)$ (see Lemma 2.4). ◇

Remark 2.3. If there exists $S^* \in I$ such that $R[V(t)] > 0$ for $\rho(S - S^*) > 0$, then

$$\varphi\left(\frac{t}{\lambda}\right) = \int_0^t \frac{d\vartheta}{R\left\{V\left[\varphi\left(\frac{\vartheta}{\lambda}\right)\right]\right\}}. \tag{2.41}$$

Furthermore, if (2.30a) holds, then

$$\lim_{S\leftarrow r} \varphi\left(\frac{t}{\lambda}\right) = +\infty. \tag{2.42}$$

Proof. From (2.33) one has:

$$\int_0^{\varphi(0)} d\tau\, R[V(\tau)] = 0\ .$$

Hence, for all S such that $R[V(t)] > 0$ (i.e. for any S such that $\rho(S - S^*) > 0$) it must be $\varphi(0) = 0$. Integrating (2.34) between 0 and t, (2.41) then follows. Moreover, for any fixed t, as S approaches r we see from (2.30a) and (2.41) that $\varphi(t/\lambda)$ grows without bounds, i.e. (2.42) holds. ◇

From Remark 2.3 it follows that $\varphi(t/\lambda)$ can be viewed as a scaled time.

The function $R(S)$ defined by (2.5) plays an essential role when dealing with the asymptotic exponential behaviour of the FPT p.d.f.'s. Hereafter, we shall relate this function to the asymptotic mean first–passage time in the case of a constant boundary $S(t) = S$.

Remark 2.4. (Proof is given in Reference 5). If

$$\lim_{S \to r} R(S) = 0 \tag{2.43}$$

and

$$\lim_{S \to r} \left[A_1(S) - \frac{A_2'(S)}{4}\right]^{-2} \left\{A_2(S)\left[A_1'(S) - \frac{A_2''(S)}{4}\right] - \frac{A_2'(S)}{2}\left[A_1(S) - \frac{A_2'(S)}{4}\right]\right\} = 0 \tag{2.44}$$

then,

$$\lim_{S \to r} \left[t_1(S|x_0)\, R(S)\right] = 1\ , \tag{2.45}$$

where $t_1(S|x_0)$ denotes the mean first–passage time of $X(t)$ through S. ◇

From (2.45) we see that as $S \to r$, under the assumptions of Remark 2.4 one has

$$t_1(S|x_0) \sim \left[R(S)\right]^{-1}, \tag{2.46}$$

which is an asymptotic representation of the mean first–passage time.

3. Special Cases

The validity of conditions (2.10) and (2.30), on which Theorems 2.1 and 2.2 are based, is in general hard to prove. It is thus desirable to determine simpler conditions that insure their validity. In the remaining part of this paper we shall provide sufficient conditions for the validity of (2.10) and (2.30) for the Ornstein–Uhlenbeck process and obtain as well some new asymptotic results for the FPT density of the Wiener process.

a) Ornstein–Uhlenbeck process. Let $A_1(x) = -\alpha x + \beta$, $A_2 = \sigma^2$, $I = (-\infty, +\infty)$, with $\alpha, \sigma \in \Re^+, \beta \in \Re$. For all $x, y \in I$ and $\tau < t$ one has

$$f(x,t|y,\tau) = \left[\frac{\alpha}{\pi\sigma^2\left[1 - e^{-2\alpha(t-\tau)}\right]}\right]^{1/2} \exp\left\{-\frac{\alpha\left[x - y\,e^{-\alpha(t-\tau)} - \dfrac{\beta}{\alpha}\left(1 - e^{-\alpha(t-\tau)}\right)\right]^2}{\sigma^2\left[1 - e^{-2\alpha(t-\tau)}\right]}\right\} \tag{3.1}$$

and

$$\psi[S(t),t|y,\tau] = \frac{1}{2}\left\{S'(t) - \alpha\left[S(t) - \frac{\beta}{\alpha}\right]\frac{1+e^{-2\alpha(t-\tau)}}{1-e^{-2\alpha(t-\tau)}} + \frac{2\alpha\, e^{-\alpha(t-\tau)}}{1-e^{-2\alpha(t-\tau)}}\left(y-\frac{\beta}{\alpha}\right)\right\} f[S(t),t|y,\tau]\,, \tag{3.2}$$

where $S(t)$ denotes an arbitrary $C_2[0,+\infty)$–class function. Note that from (3.1) one immediately obtains:

$$W(x) = \left(\frac{\alpha}{\pi\sigma^2}\right)^{1/2} \exp\left\{-\frac{\alpha(x-\beta/\alpha)^2}{\sigma^2}\right\}, \tag{3.3}$$

while from (2.5) one has:

$$R(S) = \rho(\alpha S - \beta)W(S)\,, \tag{3.4}$$

where $W(S)$ is given by (3.3). Let us now take

$$S(t) = S + \gamma(t)\,, \qquad S \in \Re, \tag{3.5}$$

with $\gamma(t)$ bounded, independent of S and such that

$$\gamma(t) = O(t^{-q}) \quad \text{or} \quad \gamma(t) = o(t^{-q}) \quad \text{as} \quad t \to +\infty \tag{3.6}$$

for all $q > 0$ (asymptotically constant boundary). Note that as $S \to r$, $R(S)$ goes to zero due to (3.3) and thus condition (2.10a) of Theorem 2.1 holds. Furthermore, recalling (3.2) and the expression (3.4) of $R(S)$, it is easy to prove that conditions (2.10b) and (2.10c) of Theorem 2.1 are satisfied, so that (2.11) holds. Let us now assume

$$S(t) = S + \gamma(t) + p(t)\,, \qquad S \in \Re, \tag{3.7}$$

where $\gamma(t)$ is taken as in the previous case and where $p(t)$ is a zero–mean bounded periodic function of period T independent of S. Then,

$$\int_0^T d\tau\, p(\tau) = 0 \tag{3.8}$$

and

$$V(t) = \lim_{n\to\infty} S(t+nT) = S + p(t). \tag{3.9}$$

From (2.27) we have:

$$\frac{R[V(t)]}{W[V(t)]} = \rho\left\{\alpha[V(t) - S] - V'(t)\right\} + \rho\left(\alpha S - \beta\right). \tag{3.10}$$

The first term on the right–hand side of (3.10) is bounded as $S \to r$ while the second term goes to $+\infty$ as $S \to r$. Therefore, there exists $S^* \in I$ such that the left–hand side of

(3.10) is positive for $\rho(S - S^*) > 0$. Hence, for $\rho(S - S^*) > 0$ one has $R[V(t)] > 0$. Furthermore, from (3.3) and (3.10) we have:

$$R[V(\tilde{t})] = \rho\left(\frac{\alpha}{\pi\sigma^2}\right)^{1/2}\left\{\alpha\left[V(\tilde{t}) - S\right] - \frac{d}{d\tilde{t}}V(\tilde{t}) + \alpha S - \beta\right\}\exp\left\{-\frac{\alpha}{\sigma^2}\left[V(\tilde{t}) - \frac{\beta}{\alpha}\right]^2\right\} \tag{3.11}$$

where

$$\tilde{t} = \varphi\left(\frac{t}{\lambda}\right), \tag{3.12}$$

with λ and φ defined by (2.18), (2.19), (2.27) and (2.29). Since $V(\tilde{t}) - S$ and $dV(\tilde{t})/d\tilde{t}$ are bounded as $S \to r$, it is easy to see that (2.30a) follows from (3.11). Note that since the assumptions of Remark 2.3 are satisfied, from (2.42) and (3.12) we have:

$$\lim_{S \to r} \tilde{t} = +\infty. \tag{3.13}$$

Recalling (3.2) and the expression (3.11) of $R[V(\tilde{t})]$, it is easy to prove that conditions (2.30b) and (2.30c) of Theorem 2.2 are satisfied, so that (2.31) holds.

Note that in the special case of periodic boundaries the asymptotic behaviour of the FPT density indicated in Reference 10 immediately follows.

b) *Asymptotic behaviours for Wiener process.* Since the transformation

$$\begin{aligned} t^* &= \alpha^{-1}\left(e^{2\alpha t} - 1\right) \\ x^* &= x\, e^{\alpha t} - \frac{\beta}{\alpha} e^{\alpha t}\left(1 - e^{\alpha t}\right) \\ x_0^* &= x_0 \quad, t_0^* = t_0 = 0 \\ f^*(x^*, t^* | x_0^*) &= e^{-\alpha t} f(x, t | x_0) \end{aligned} \tag{3.14}$$

maps the Ornstein–Uhlenbeck process of a) into the unrestricted Wiener process with drift $B_1 = \beta$ and infinitesimal variance $B_2 = \sigma^2/2$, the FPT p.d.f. for the former is related to that of the latter by

$$g_{OU}\left[S(t), t | x_0\right] = 2\, e^{2\alpha t} g_W\left[S(t)\, e^{\alpha t} - \frac{\beta}{\alpha} e^{\alpha t}\left(1 - e^{\alpha t}\right), \frac{1}{\alpha}\left(e^{2\alpha t} - 1\right)\Big| x_0\right]. \tag{3.15}$$

We have already proved that for boundaries of type (3.5)–(3.6) result (2.11) holds. Then, from (3.15) and (2.11) one has:

$$\lim_{S \to r}\left\{2\omega e^{2\alpha\omega t} g_W\left[S(\omega t) e^{\alpha\omega t} - \frac{\beta}{\alpha} e^{\alpha\omega t}(1 - e^{\alpha\omega t}), \frac{1}{\alpha}(e^{2\alpha\omega t} - 1)\big| x_0\right]\right. = e^{-t}$$

with $\omega = 1/R(S)$ and $R(S)$ given by (3.4). Therefore, the following apparently new asymptotic result as S tends to r is obtained:

$$g_W\left[S\left(\frac{\ln(1+\alpha\tau)}{2\alpha}\right)\sqrt{1+\alpha\tau}-\frac{\beta}{\alpha}\sqrt{1+\alpha\tau}\left(1-\sqrt{1+\alpha\tau}\right),\tau\middle|x_0\right]$$
$$\sim\frac{R(S)}{2}\left(1+\alpha\tau\right)^{-1-R(S)/2\alpha}. \tag{3.16}$$

In the particular case $\alpha=1$, $\sigma^2=2$, and $\beta=x_0=0$ from (3.16) and (2.46) we obtain:

$$g_W[S\sqrt{1+t},t|0]\sim\frac{1}{2\,t_1(S|0)}(1+t)^{-1-1/2\,t_1(S|0)}, \tag{3.17}$$

i.e. for large t, g_W goes to zero as $t^{-1-1/2t_1(S|0)}$. This is in agreement with a result due to Sato[11] stating that for large S, g_W goes to zero as $t^{-1-p(S)}$, where $p(S)$ approaches $[2\,t_1(S|0)]^{-1}$ as S grows larger and larger. Note, however, that the asymptotic expressions (3.16) and (3.17) are normalized p.d.f.'s.

In the case of the asymptotically periodic boundaries of the type considered in *a)*, from (3.15) and (2.31) we obtain

$$\lim_{S\to r}\frac{2\,e^{2\alpha\tilde{t}}}{R[V(\tilde{t})]}\,g_W\left[S(\tilde{t})\,e^{\alpha\tilde{t}}-\frac{\beta}{\alpha}e^{\alpha\tilde{t}}\left(1-e^{\alpha\tilde{t}}\right),\frac{1}{\alpha}\left(e^{2\alpha\tilde{t}}-1\right)\middle|x_0\right]=e^{-t},$$

with $\tilde{t}$ given by (3.12). Hence, in this case as $S\to r$ the asymptotic behaviour

$$g_W\left[S\left(\frac{\ln(1+\alpha\tau)}{2\alpha}\right)\sqrt{1+\alpha\tau}-\frac{\beta}{\alpha}\sqrt{1+\alpha\tau}\left(1-\sqrt{1+\alpha\tau}\right),\tau\middle|x_0\right]$$
$$\sim\frac{1}{2}R\left[V\left(\frac{\ln(1+\alpha\tau)}{2\alpha}\right)\right](1+\alpha\tau)^{-1}\exp\left\{-\int_0^{\ln(1+\alpha\tau)/2\alpha}d\tau\,R[V(\tau)]\right\}$$

follows, with $R[V(t)]$ specified by (3.10).

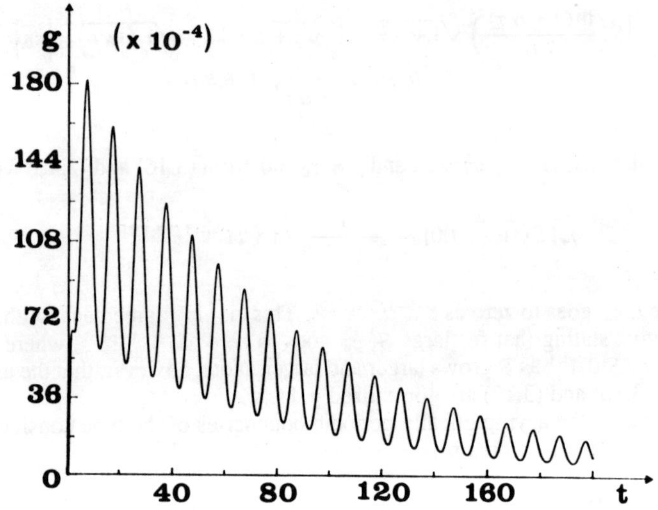

Fig. 1a OU process with drift $A_1(x) = -x$, infinitesimal variance $A_2 = 2$, originating at $x_0 = 0$ and boundary $S(t) = A + C\sin(2\pi t/T)$. The FPT density is plotted for $A = 3, C = 0.2, T = 10$.

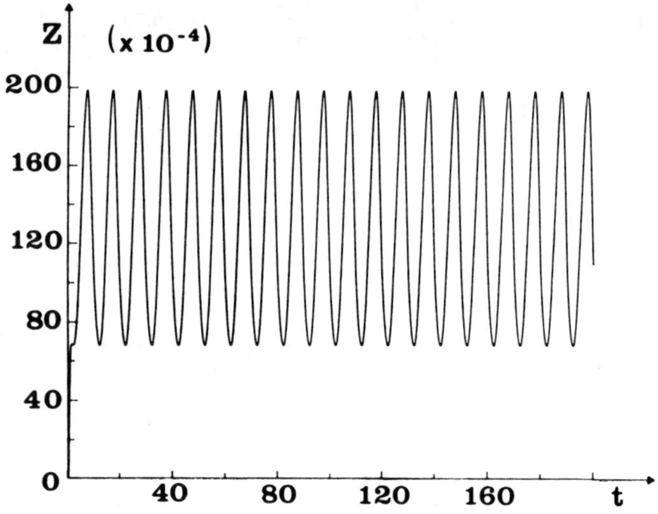

Fig. 1b The function $Z(t)$ in (1.1) for the case of Fig. 1a. Here $\lambda = 0.12343852 \times 10^{-1}$.

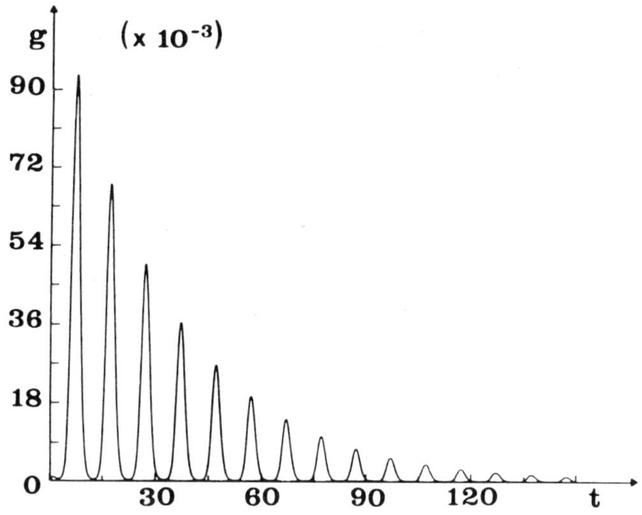

Fig. 2a Same as Fig. 1a with $A = 3, C = 1, T = 10$.

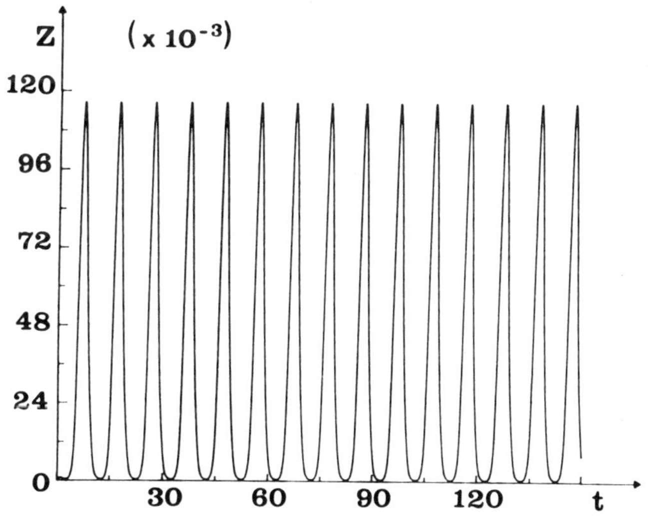

Fig. 2b Plot of $Z(t)$ for the case of Fig. 2a. Here $\lambda = 0.31411586 \times 10^{-1}$.

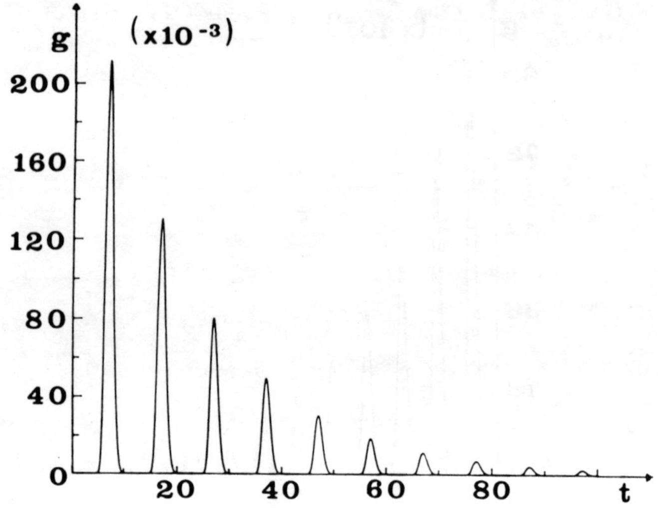

Fig. 3a Same as Fig. 1a with $A = 4, C = 2.5, T = 10$.

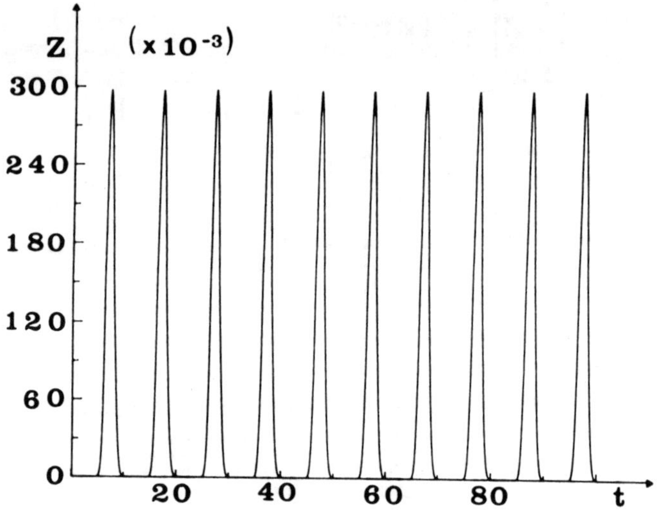

Fig. 3b Plot of $Z(t)$ for the case of Fig. 3a. Here $\lambda = 0.48428087 \times 10^{-1}$.

References

1. Buonocore, A., Nobile, A.G. and Ricciardi, L.M. (1987) A new integral equation for the evaluation of first-passage-time probability densities. Adv. Appl. Prob. **19**, 784–800.
2. Buonocore, A., Giorno, V., Nobile, A.G. and Ricciardi, L.M. (in press) On the two–boundary first–crossing–time problem for diffusion processes. J. Appl. Prob.
3. Giorno, V., Nobile, A.G., Ricciardi, L.M. and Sato, S. (1989) On the evaluation of first–passage–time probability densities via nonsingular integral equations. Adv. Appl. Prob. **21**, 20–36.
4. Giorno, V., Nobile, A.G. and Ricciardi, L.M. (in press) A symmetry–based constructive approach to probability densities for one dimensional diffusion processes. J. Appl. Prob.
5. Giorno, V., Nobile, A.G. and Ricciardi, L.M. (in preparation) On the asymptotic behaviour of first–passage–time densities for one–dimensional diffusion processes and varying boundaries.
6. Karlin, S. and Taylor, H.M. (1975) *A First Course in Stochastic Processes.* Academic Press, New York.
7. Mandl, P. (1968) *Analitycal treatment of one–dimensional Markov processes.* Springer Verlag, Berlin.
8. Nobile, A.G., Ricciardi, L.M. and Sacerdote, L. (1985) Exponential trends of Ornstein–Uhlenbeck first passage time densities. J. Appl. Prob. **22**, 360–369.
9. Nobile, A.G., Ricciardi, L.M. and Sacerdote, L. (1985) Exponential trends of first–passage–time densities for a class of diffusion processes with steady–state distribution. J. Appl. Prob. **22**, 611–618.
10. Sacerdote, L. (in press) Asymptotic behavior of Ornstein–Uhlenbeck first–passage–time density through periodic boundaries. Applied Stochastic Models and Data Analysis.
11. Sato, S. (1977) Evaluation of first–passage time probability to a square root boundary for the Wiener process. J. Appl. Prob. **14**, 850–856.

DISTRIBUTIONS OF CLASS L AND SELF-SIMILAR PROCESSES WITH INDEPENDENT INCREMENTS

KEN-ITI SATO

Department of Mathematics, College of General Education
Nagoya University, Nagoya, JAPAN

ABSTRACT.
It has recently been noticed that self-similar processes with independent increments have close connection with distributions of class L. Their distributions at any fixed time are of class L and all distributions of class L appear in this way. Results and background facts are formulated and related problems are discussed. Properties of distributions of class L are reviewed.

1. Definitions and background

The Brownian motion is a fundamental stochastic process in probability theory. It has many nice properties : it is a process with stationary independent increments, it has Gaussian distribution at any time, it is self-similar, almost all sample functions of it are continuous, and so on. Thus many directions of extension of the Brownian motion are conceivable. Among them strictly stable processes are the most direct extension : they are self-similar processes with stationary independent increments (although their sample functions have dense discontinuity in the time axis and their distributions have heavy tail). Main further extensions are in two directions : one is to processes with stationary independent increments (called Lévy processes) and the other is to self-similar processes with stationary increments. A lot of papers are devoted to the study of these processes. In this paper we consider a third extension of strictly stable processes : self-similar processes with independent increments. Namely we abandon stationarity of increments in generalization of strictly stable processes. It turns out that the distributions of such processes at each time are of class L, which is a class of infinitely divisible distributions known for fifty years after Paul Lévy's book[11]. Conversely every distribution of class L appears in this way. We will give this characterization (Section 2) and discuss its consequences. We propose to call a self-similar process with independent increments a *process of class L*. Now any distribution of class L naturally induces two processes – one is a Lévy process and the other is a process of class L. We will compare properties of these two processes (Section 3). Distributions of class L have been investigated in many papers ; some of their results will be summarized (Section 4). They are now understood as properties of processes of class L. In case that the state space is the Euclidean space of dimension ≥ 2, we will show operator generalization of the relations discussed above (Section 5). Main results of Sections 2, 3, and 5 are given in the author's paper[20] with proofs.

In the preceding paragraph we did not mention regularity assumptions that we need to impose. Let us give precise definitions. As usual ω is a general element of a probability space.

DEFINITION 1.1. A stochastic process $\{X(t,\omega) = X_t(\omega) : t \geq 0\}$ on $\mathbf{R}^d$ is called a *process with independent increments* (or ii process) if, for every $n \geq 1$ and every choice of $0 \leq t_0 < t_1 < \cdots < t_n, X(t_0), X(t_1) - X(t_0), \cdots, X(t_n) - X(t_{n-1})$ are independent.

DEFINITION 1.2. A process $\{X_t(\omega) : t \geq 0\}$ on $\mathbf{R}^d$ is called a *Lévy process* if it is an ii process satisfying the following four conditions :

(1.1) $X_{s+t} - X_s \stackrel{\mathrm{d}}{=} X_{s'+t} - X_{s'}$ for every $t, s, s' \geq 0$,

(1.2) $\{X_t\}$ is stochastically continuous (continuous in probability),

(1.3) $X_0 = 0$ a.s.,

(1.4) for almost every ω, $X_t(\omega)$ is right-continuous in $t \geq 0$ and has left limits in $t > 0$.

The notation $\stackrel{\mathrm{d}}{=}$ means that the two sides are identically distributed. It is known that, if $\{X_t\}$ is an ii process satisfying (1.2), then it has a modification satisfying (1.4).

DEFINITION 1.3. A probability measure μ on $\mathbf{R}^d$ is called *infinitely divisible* if, for each positive integer n, there is a distribution μ_n such that $\mu = \mu_n * \cdots * \mu_n$ (n times).

The following are well-known facts.

PROPOSITION 1.4. *If $\{X_t\}$ is an ii process satisfying* (1.2) *and* (1.3), *then, for each t, the distribution of X_t is infinitely divisible.*

PROPOSITION 1.5. *Let μ be infinitely divisible. Then there is a Lévy process $\{X_t\}$ such that X_1 has distribution μ. If $\{X_t\}$ and $\{X'_t\}$ are two such processes, then*

$$\{X_t\} \stackrel{\mathrm{d}}{=} \{X'_t\}. \tag{1.5}$$

The meaning of (1.5) is that the processes in both sides are identical in distribution, that is, their finite-dimensional distributions coincide. A consequence of Propositions 1.4 and 1.5 is that there is one-one correspondence between the Lévy processes and the infinitely divisible distributions. Let $\langle x, y\rangle$ denote the inner product of $x, y \in \mathbf{R}^d$, where x, y are understood as column vectors. Let $\hat{\mu}(z)$ be the characteristic function of μ :

$$\hat{\mu}(z) = \int_{\mathbf{R}^d} e^{i\langle z,x\rangle}\mu(dx), \quad z \in \mathbf{R}^d.$$

Let U be the unit disc $\{x : |x| < 1\}$.

PROPOSITION 1.6. (Lévy representation) *If μ is infinitely divisible, then $\hat{\mu}(z)$ is represented in a unique way as*

$$(1.6) \quad \hat{\mu}(z) = \exp[-2^{-1}\langle z, Az\rangle + \int_{\mathbf{R}^d}(e^{i\langle z,Az\rangle} - 1 - i\langle z,x\rangle 1_U(x))\nu(dx) + i\langle\gamma, x\rangle]$$

by (A, ν, γ), where A is a nonnegative-definite symmetric matrix, ν is a finite measure on $\mathbf{R}^d$ with $\nu(\{0\}) = 0$ and $\int(|x|^2 \wedge 1)\nu(dx) < \infty$, and γ is an element of $\mathbf{R}^d$. Conversely, for every choice of (A, ν, γ), the right-hand side of (1.6) is the characteristic function of an infinitely divisible distribution.

The measure ν is called the *Lévy measure* of μ. If $\nu = 0$, then μ is Gaussian (possibly degenerate). Every infinitely divisible distribution μ induces a convolution semigroup $\{\mu_t : t \geq 0\}$ with $\mu_1 = \mu$; thus μ_t is the distribution at time t of the Lévy process $\{X_t\}$ of Proposition 1.5. This $\{\mu_t\}$ defines a strongly continuous semigroup of linear operators with norm 1 on the Banach space of continuous functions on $\mathbf{R}^d$ vanishing at the infinity. The infinitesimal generator of the semigroup is an integro-differential operator determined by (A, ν, γ). For properties of Lévy processes see Gihman and Skorohod[2].

DEFINITION 1.7. We call a stochastic process $\{X_t : t \geq 0\}$ on $\mathbf{R}^d$ a *self-similar process* if, for any $c > 0$, there exists $a_c \in \mathbf{R}$, $a_c \neq 0$, such that

$$(1.7) \qquad \{X_{ct} : t \geq 0\} \stackrel{\mathrm{d}}{=} \{a_c X_t : t \geq 0\}$$

and if $\{X_t\}$ satisfies (1.2) and (1.3). A process $\{X_t : t \geq 0\}$ is said to have *stationary increments* if, for every $s, s' \geq 0$,

$$\{X_{s+t} - X_s : t \geq 0\} \stackrel{\mathrm{d}}{=} \{X_{s'+t} - X_{s'} : t \geq 0\}$$

Self-similarity means that change of time scale has the same effect as spatial enlargement (or contraction). Self-similar processes (mostly with stationary increments) are investigated by Lamperti[10], Dobrushin and Major[1], Vervaat[30], and many others ; see Taqqu[26] and Maejima[12].

Proposition 1.8. *If $\{X_t\}$ is self-similar, then there exists $H > 0$ such that, for any $c > 0$,*

$$\{X_{ct} : t \geq 0\} \stackrel{\mathrm{d}}{=} \{c^H X_t : t \geq 0\}. \tag{1.8}$$

If, moreover, $\{X_t\}$ is not identically zero, then H is unique.

The number H is called the *exponent* of the self-similar process $\{X_t\}$.

Definition 1.9. A Lévy process on $\mathbf{R}^d$ is called a *strictly stable process* if it is self-similar. A Lévy process $\{X_t\}$ on $\mathbf{R}^d$ is called a *stable process* if, for every $c > 0$, there are $a_c \in \mathbf{R}$, $a_c \neq 0$, and $b_c \in \mathbf{R}^d$ such that

$$\{X_{ct} : t \geq 0\} \stackrel{\mathrm{d}}{=} \{a_c X_t + t b_c : t \geq 0\}. \tag{1.9}$$

A probability measure μ on $\mathbf{R}^d$ is called *stable* [resp. *strictly stable*] if there is a stable [resp. strictly stable] process $\{X_t\}$ such that X_1 has distribution μ.

If $\{X_t\}$ is a stable process on $\mathbf{R}^d$ not of the form $X_t = tb$, $b \in \mathbf{R}^d$, then there is $H > 0$ such that (1.9) holds with $a_c = c^H$. This H is unique and not smaller than $1/2$; $\alpha = 1/H$ is called the *index* of the stable process. If $\{X_t\}$ is a strictly stable process not identically zero, then, similarly, its exponent H of self-similarity is not smaller than $1/2$ and $\alpha = 1/H$ is called the *index* of the strictly stable process.

Definition 1.10. A probability measure μ on $\mathbf{R}^d$ is called a *distribution of class L* if there are sequences of independent random variables $\{Z_n\}$ on $\mathbf{R}^d$, non-zero numbers $\{a_n\}$, and vectors $\{b_n\}$ such that the distribution of $a_n \sum_{j=1}^n Z_j + b_n$ weakly converges to μ as $n \to \infty$ and

$$\lim_{n\to\infty} \max_{1\leq j\leq n} \mathbf{P}(|a_n Z_j| > \varepsilon) = 0 \quad \text{for every } \varepsilon > 0.$$

The class L is a proper subclass of the class of infinitely divisible distributions. The class of stable distributions is a proper subclass of the class L. The following fact is given in Lévy[11] and Gnedenko and Kolmogorov[3] for $d = 1$ and in Wolfe[32] and Sato[14] for general d.

Proposition 1.11. *A probability measure μ on $\mathbf{R}^d$ is of class L if and only if it is infinitely divisible and the Lévy measure ν is expressed as*

$$\nu(E) = \int_S \lambda(d\xi) \int_0^\infty 1_E(r\xi) k_\xi(r) r^{-1} dr \text{ for any Borel set } E, \tag{1.10}$$

where λ is a finite measure on the unit sphere $S = \{\xi : |\xi| = 1\}$ and $k_\xi(r)$ is nonincreasing in $r \in (0, \infty)$ and measurable in $\xi \in S$ and

$$0 < \int_0^\infty k_\xi(r)(r \wedge r^{-1})dr = c < \infty$$

with c being independent of ξ. In this expression λ is unique and $k_\xi(r)$ is unique $\lambda(d\xi)dr$-almost everywhere, each up to a positive constant multiple.

In the class L stable and strictly stable distributions are characterized as follows, which is due to Lévy[11].

PROPOSITION 1.12. *Suppose that μ is not a delta distribution. It is stable with index 2 if and only if it is Gaussian. It is stable with index $\alpha \in (0, 2)$ if and only if it is infinitely divisible with Lévy representation (A, ν, γ) such that $A = 0$ and ν is expressed as (1.10) with $k_\xi(r) = r^{-\alpha}$.*

PROPOSITION 1.13. *Suppose that μ is not the delta distribution at 0. It is strictly stable with index 2 if and only if it is Gaussian with mean 0. It is strictly stable with index $\alpha \in (1, 2)$ if and only if*

$$\hat{\mu}(z) = \exp[\int_S \lambda(d\xi) \int_0^\infty (e^{i\langle z, r\xi\rangle} - 1 - i\langle z, r\xi\rangle)r^{-1-\alpha}dr]$$

and λ is not identically zero. It is strictly stable with index 1 if and only if

$$\hat{\mu}(z) = \exp[\int_S \lambda(d\xi) \int_0^\infty (e^{i\langle z, r\xi\rangle} - 1 - i\langle z, r\xi\rangle\, 1_U(r\xi))r^{-2}dr + i\langle \gamma, z\rangle]$$

with $\int_S \xi\lambda(d\xi) = 0$ and $\lambda(S) + |\gamma| > 0$. It is strictly stable with index $\alpha \in (0, 1)$ if and only if

$$\hat{\mu}(z) = \exp[\int_S \lambda(d\xi) \int_0^\infty (e^{i\langle z, r\xi\rangle} - 1)r^{-1-\alpha}dr]$$

and λ is not identically zero.

2. Characterization of self-similar processes with independent increments

The following two theorems are given by Sato[20].

THEOREM 2.1. *If $\{X_t : t \geq 0\}$ is a self-similar process on $\mathbf{R}^d$ with independent increments, then, for every $t \geq 0$, the distribution of X_t is of class L.*

THEOREM 2.2. *Let μ be a distribution of class L on $\mathbf{R}^d$ and let $H > 0$ and $t_0 > 0$. Then there exists a self-similar process $\{X_t : t \geq 0\}$ with exponent H*

with independent increments such that the distribution of $X(t_0)$ is μ. This $\{X_t\}$ is unique in the sense of distribution.

Keeping these theorems in mind, we use the following terminology.

DEFINITION 2.3 A stochastic process on $\mathbf{R}^d$ is called a *process of class L* if it is a self-similar process with independent increments satisfying (1.4). When it is not identically zero, the exponent H of self-similarity is called its exponent.

The condition (1.4) is not essential, since any self-similar process with independent increments has a modification satisfying (1.4). We note also that, if $\{X_t\}$ is an ii process such that there exists $H > 0$ satisfying (1.8) for any $c > 0$, then (1.2) and (1.3) are automatically fulfilled.

The characterization below of the class L due to Lévy[11] is essential to the proof of the theorems.

PROPOSITION 2.4. (Self-decomposability) *A distribution μ on $\mathbf{R}^d$ is of class L if and only if, for every $a \in (0,1)$, there is a probability measure ρ_a on $\mathbf{R}^d$ such that*

$$\hat{\mu}(z) = \hat{\mu}(az)\hat{\rho}_a(z). \tag{2.1}$$

Sketch of proof of Theorem 2.1. Let μ_t and $\mu_{s,t}$ be the distributions of X_t and $X_t - X_s$, respectively. Then $\hat{\mu}_t(z) = \hat{\mu}_s(z)\hat{\mu}_{s,t}(z)$ for $0 < s < t$. Since $\hat{\mu}_s(z) = \hat{\mu}_t((s/t)^H z)$ for some $H > 0$, this implies self-decomposability of μ.

LEMMA 2.5. *Suppose that $X_t = Y(t^H)$. Then, $\{X_t\}$ is a process of class L with exponent H if and only if $\{Y_t\}$ is a process of class L with exponent* 1.

Sketch of proof of Theorem 2.2. By the lemma above the proof reduces to the case $H = 1$. We can also assume that $t_0 = 1$. Define μ_t and $\mu_{s,t}$ for $0 < s < t$ from the given μ by $\hat{\mu}_t(z) = \hat{\mu}(tz)$ and $\hat{\mu}_{s,t}(z) = \hat{\rho}_{s/t}(z)$. Then $\mu_t = \mu_s * \mu_{s,t}$ and $\mu_{r,t} = \mu_{r,s} * \mu_{s,t}$ for $0 < r < s < t$. Use Kolmogorov's extension theorem for construction of the desired process $\{X_t\}$. The uniqueness in distribution is easy to establish.

EXAMPLE 2.6 (Kawazu[8]) Let $\{B_t : t \geq 0\}$ be the Brownian motion on $\mathbf{R}$ with $B_0 = 0$. Let M_t be the smallest u such that $B_u - \min_{s \leq u} B_s \geq t$. That is , M_t is the first passage time to t for a reflected Brownian motion. Let N_t be the smallest u such that $B_u = \min_{s \leq M(t)} B_s$, and let $V_t = -\min_{s \leq M(t)} B_s$. Then, $\{M_t\}$ and $\{N_t\}$ are both processes of class L with exponent 2. The process $\{V_t\}$ is a process of class L with exponent 1. None of them has stationary increments. Kawazu calculated also their joint distributions.

3. Lévy processes and processes of class L

Proposition 1.5 and Theorem 2.2 tell us that to any distribution μ on $\mathbf{R}^d$ of class L there correspond a Lévy process $\{X_t\}$ and a process $\{Y_t^{(H)}\}$ of class L with exponent H for each $H > 0$ in such a way that X_1 and $Y_1^{(H)}$ have distribution μ. They are determined by μ uniquely in the sense of distribution. The processes $\{Y_t^{(H)}\}$ and $\{Y_t^{(H')}\}$ for distinct H and H' transform to each other as Lemma 2.5 shows. It is an interesting problem to compare $\{Y_t^{(H)}\}$ with $\{X_t\}$. Results in this section are given by Sato[20].

PROPOSITION 3.1. *$\{Y_t^{(H)}\} \stackrel{d}{=} \{X_t\}$ if and only if μ is strictly stable with index $\alpha = 1/H$.*

This is rephrasing of Definition 1.9. If μ is a stable distribution, then the difference between $\{Y_t^{(H)}\}$ and $\{X_t\}$ is a deterministic function. In general, however, $\{Y_t^{(H)}\}$ and $\{X_t\}$ are widely different.

PROPOSITION 3.2. *If μ is stable with index $\alpha \neq 1$, then, for $H = 1/\alpha$, there is $b \in \mathbf{R}^d$ such that*

$$\{Y_t^{(H)}\} \stackrel{d}{=} \{X_t - (t - t^H)b\}.$$

If μ is stable with index 1, then, for some b,

$$\{Y_t^{(1)}\} \stackrel{d}{=} \{X_t - (t \log t)b\}.$$

Proof is obtained from Proposition 1.12.

Let us consider the case of the exponential distribution, which serves as a typical non-stable example.

THEOREM 3.3. *Let $d = 1$ and let μ be the exponential distribution with mean 1. Let $\{X_t\}$ and $\{Y_t^{(1)}\}(H = 1)$ be processes corresponding to μ. Then, both X_t and $Y_t^{(1)}$ are nondecreasing and increase only by jumps a.s. But the set $\{t > 0 : X_t \neq X_{t-}\}$ (jumping times for X_t) is dense in $(0, \infty)$ a.s., while the set $\{t > 0 : Y_t \neq Y_{t-}\}$ has accumulation only at $t = 0$ a.s. For any $c > 0$*

$$\lim_{t \downarrow 0} \frac{X_t}{t^c} = 0 \quad a.s.,$$

while

$$\limsup_{t \downarrow 0} \frac{Y_t}{t \log |\log t|} = 1 \quad a.s.$$

In general, for any given μ of class L on $\mathbf{R}^d$, let us make comparison of the corresponding processes $\{X_t\}$ and $\{Y_t^{(H)}\}$ with $H = 1$.

1° Let m and V be the mean vector and the covariance matrix of μ and let $c = \int |x|\mu(dx)$, assuming their existence. Then $\mathbf{E}X_t = \mathbf{E}Y_t^{(1)} = tm$, $\mathrm{Cov}(X_t) = tV$, $\mathrm{Cov}(Y_t^{(1)}) = t^2V$, and $\mathbf{E}|Y_t^{(1)}| = tc$. No general relation exists between $\mathbf{E}|X_t|$ and c.

2° Let $d = 1$. Then the distributions of X_t and $Y_t^{(1)}$ are unimodal by the theorem of Yamazato[35]. Let a be the mode of μ. Then the mode of $Y_t^{(1)}$ is ta, while there is no general relation between a and the mode of X_t.

3° Let (A, ν, γ) be the Lévy representation of μ. Then the distribution of X_t has representation $(tA, t\nu, t\gamma)$ on the one hand and, on the other hand, $Y_t^{(1)}$ has representation (t^2A, ν_t, γ_t) with

$$\nu_t(E) = \int 1_E(tx)\nu(dx),$$

$$\gamma_t = t\gamma + t\int (1_U(tx) - 1_U(x))x\nu(dx).$$

4° For each n and $0 \le t_1 < \cdots < t_n$, the joint distribution of $(X(t_1), \cdots, X(t_n))$ is of class L, but the joint distribution of $(Y(t_1), \cdots, Y(t_n))$ is not always of class L (although it is always infinitely divisible).

4. Properties of distributions of class L

Among infinitely divisible distributions stable distributions on $\mathbf{R}$ are most deeply investigated; see Zolotarev's monograph[39]. Many properties of distributions of class L on $\mathbf{R}$ are known as well; some are also studied on $\mathbf{R}^d$. Thus, for any process $\{Y_t\}$ of class L, the following assertions hold for the distribution of Y_t for any fixed t. A distribution μ on $\mathbf{R}^d$ is said to be *genuinely d-dimensional* if its support is not contained in any $(d-1)$-dimensional hyperplane.

1° (General d) Any genuinely d-dimensional distribution of class L is absolutely continuous. (Sato[15])

2° (General d) There is a hierarchy of subclasses

$$L = L_0 \supset L_1 \supset L_2 \supset \cdots \supset L_\infty \supset S,$$

where, for any $m < \infty$, L_m is inductively defined as the class of μ such that, in Definition 1.10, the random variables $\{Z_n\}$ can be chosen to have distributions in the class L_{m-1}. The class L_∞ is the intersection of L_m, $m < \infty$, and S is the class of stable distributions. The distributions of the increments $Y_t - Y_s$, $0 < s < t$, belong to the class L_m if and only if Y_1 has a distribution of the class L_{m+1}. A characterization of the distributions of the class L_m is given by the order of monotonicity of the function $k_\xi(r)$ of the representation of the class L in Proposition 1.11. (Urbanik[28,29], Sato[14])

3° (General d) Let ν be the Lévy measure of an infinitely divisible distribution μ. Let $\alpha, \beta > 0$. then μ has finite absolute moment of order α if and only if $\int_{|x|>1} |x|^\alpha \nu(dx)$ is finite. The integral $\int e^{\beta |x| \log |x|}\mu(dx)$ is finite [resp. infinite] if $\int_{|x|>(1/\beta)-\varepsilon} \nu(dx) = 0$ for some $\varepsilon > 0$ [resp. $\int_{|x|>1/\beta} \nu(dx) > 0$]. (Kruglov[9], Sato[13])

4° (General d) Let $\{Z_t : t \geq 0\}$ be a process of Ornstein-Uhlenbeck type on $\mathbf{R}^d$ generated by

$$G - \sum_{j=1}^{d} x_j \frac{\partial}{\partial x_j}, \tag{4.1}$$

where G is the infinitesimal generator of a Lévy process. That is, $\{Z_t\}$ is a time-homogeneous Markov process with infinitesimal generator (4.1). Let (B, ρ, b) be the Lévy representation of the Lévy process generated by G. Then Z_t has limiting distribution (as $t \to \infty$) if and only if $\int_{|x|>1} \log |x| \rho(dx) < \infty$. The limiting distribution μ is of class L with the Lévy representation (A, ν, γ) determined by

$$A = 2^{-1}B,$$

$$\nu(E) = \int_{\mathbf{R}^d} \rho(dx) \int_0^\infty 1_E(e^{-t}x)dt,$$

$$\gamma = b + \int_{|x|\geq 1} (x/|x|)\rho(dx).$$

Conversely, every distribution μ of class L is the limiting distribution of a unique process of Ornstein-Uhlenbeck type with infinitesimal generator of the form (4.1). If μ is represented as in Proposition 1.11, then ρ is determined by

$$\rho(E) = -\int_S \lambda(d\xi) \int_0^\infty 1_E(r\xi)dk_\xi(r).$$

These are results of Wolfe[33] and Sato and Yamazato[23]. Representation of class L as the distributions of stochastic integrals by Jurek and Vervaat[7] is closely connected.

In the rest of this section we assume that $d = 1$ and that μ is a distribution of class L on $\mathbf{R}$. By Proposition 1.11 we can write

$$\hat{\mu}(z) = \exp[-2^{-1}Az^2 + \int_{\mathbf{R}\setminus\{0\}} (e^{izx} - 1 - izx1_{(-1,1)}(x))k(x)|x|^{-1}dx + i\gamma z], \tag{4.2}$$

where $k(x)$ is a nonnegative function nonincreasing on $(0, \infty)$ and nondecreasing on $(-\infty, 0)$. In case $\int_{|x|<1} k(x)dx < \infty$, we rewrite (4.2) as

$$\hat{\mu}(z) = \exp[-2^{-1}Az^2 + \int_{\mathbf{R}\setminus\{0\}} (e^{izx} - 1)k(x)|x|^{-1}dx + i\gamma_0 z] \tag{4.3}$$

with some $\gamma_0 \in \mathbf{R}$. Assume that μ is not a delta measure and let $f(x)$ be the density function of μ. Denote $c = k(0+) + k(0-) \leq \infty$. If $k(0-) = 0$ and $k(x) \geq 1$ for some $x > 0$, we denote $b = \sup\{x > 0 : k(x) \geq 1\}$. When $c < \infty$, we define a constant p and two slowly varying (at 0) functions $K(x)$ and $L(x)$ of $x \neq 0$ as follows :

$$p = \exp[-c \int_0^1 (1 - e^{-x})x^{-1}dx + c \int_1^\infty e^{-x}x^{-1}dx - \int_1^\infty (k(x) + k(-x))x^{-1}dx],$$

$$K(x) = \exp \int_{|x|}^1 (c - k(u) - k(-u))u^{-1}du,$$

$$L(x) = \int_{|x|}^1 K(u)u^{-1}du,$$

5° $(d = 1)$ The distribution μ is unimodal. The mode of μ is a unique point a except the case where either

$$A = 0, k(0-) = 0, b > 0 \text{ and } k(x) = 1 \text{ on } (0, b) \tag{4.4}$$

or the distribution μ' defined by $\mu'(E) = \mu(-E)$ satisfies (4.4). If μ satisfies (4.4), then the set of modes is the interval $[\gamma_0, \gamma_0 + b]$.(Yamazato[35] and Sato and Yamazato[21])

6° $(d = 1)$ Suppose that μ has the representation (4.3) with $A = 0$ and $\gamma_0 = 0$. If $k(0-) = 0$, then the support of μ is $[0, \infty)$.

(i) If $k(0-) \leq 1$, then μ has a nonpositive mode.

(ii) If $k(0-) \leq 1$ and $k(0+) \leq 1$ then 0 is a mode of μ.

(iii) If $k(0-) = 0$ and $k(0+) = c < 1$, then $f(x)$ is strictly decreasing on $(0, \infty)$, $f(0+) = \infty$, and

$$f(x) \sim \frac{p}{\Gamma(c)} x^{c-1} K(x) \quad \text{as } x \downarrow 0. \tag{4.5}$$

(iv) Suppose that $k(0-) = 0, k(0+) = c = 1$, and $k(x) < 1$ on $(0, \infty)$. Then $f(x)$ is strictly decreasing on $(0, \infty)$. The limit $f(0+)$ is finite or infinite according as $\int_0^1 (1 - k(x))x^{-1}dx$ is finite or infinite, respectively. More precisely, (4.5) holds in this case also.

(v) If $k(0-) = 0$ and $k(0+) > 1$, then the mode a of μ satisfies $b < a < m$, where m is the mean of μ.

(vi) Suppose that $k(0+) > 0$ and $k(0-) > 0$. If $c \leq 1$, then $f(0+) = \infty$, and $f(0-) = \infty$. If $c > 1$, then $f(0+) = f(0-) < \infty$. More precisely, if $c < 1$, then

$$f(x) \sim \frac{p \sin k(0+)\pi}{\Gamma(c) \sin c\pi} x^{c-1} K(x) \quad \text{as } x \downarrow 0,$$

$$f(x) \sim \frac{p \sin k(0-)\pi}{\Gamma(c) \sin c\pi} x^{c-1} K(x) \quad \text{as } x \uparrow 0;$$

if $c = 1$, then

$$f(x) \sim \frac{p}{\pi} \left(\cos \frac{k(0+) - k(0-)}{2} \pi \right) L(x) \quad \text{as } x \to 0.$$

(Yamazato[34], Sato and Yamazato[21])

7° ($d = 1$) Let $\{X_t\}$ be the Lévy process that has the distribution μ at $t = 1$ and let a_t be the mode of X_t. If $m = \mathbf{E}X_1$ is defined (possible $+\infty$ or $-\infty$), then $t^{-1}a_t$ tends to m as $t \to \infty$. Sato[17,18] gives further asymptotic behaviors of a_t. If $\{X_t\}$ is a stable process, then, Proposition 3.2 yields an explicit formula of a_t, which shows that a_t is not necessarily monotone in t.

8° ($d = 1$) If $A > 0$ or $c = \infty$, then the density function f is of class C^∞ on $\mathbf{R}$. If $A = 0$ and $c < \infty$, then exact order of smoothness of f is as follows. Let N be the nonnegative integer such that $N < c \leq N+1$. Then, f is of class C^N on $\mathbf{R}\backslash\{\gamma_0\}$ and not of class C^N on $\mathbf{R}$. In this sense the point γ_0 is a singular point for the density function. Define $g(x)$ by $g(x) = (x - \gamma_0)f(x)$ for $x \neq \gamma_0$ and $g(\gamma_0) = 0$. Then g is of class C^N on $\mathbf{R}$. If $N \geq 1$, then f is of class C^{N-1} on $\mathbf{R}$. Asymptotic behaviors of $f^{(N)}(x)$ as $x \to \gamma_0$ are known. (Zolotarev[38], Wolfe[31], Sato and Yamazato[21])

5. Operator generalization

In the Euclidean space of dimension ≥ 2 we can give wider definitions of self-similarity, self-decomposability, and the class L property than those given in Sections 1 and 2. Namely we can consider multiplication by invertible linear operators in place of multiplication by non-zero constants. Thus we get the following three definitions. Let $M_+(\mathbf{R}^d)$ be the class of linear operators ($d \times d$ matrices) Q on $\mathbf{R}^d$ such that all eigenvalues of Q have positive real parts. Let

$$c^Q = \sum_{n=0}^{\infty} (n!)^{-1} (\log c)^n Q^n \qquad \text{for } c > 0.$$

DEFINITION 5.1. A stochastic process $\{X_t : t \geq 0\}$ on $\mathbf{R}^d$ is called *operator-self-similar* if, for every $c > 0$, there is an invertible linear operator A_c such that $\{X_{ct}\} \stackrel{\mathrm{d}}{=} \{A_c X_t\}$ and if (1.2) and (1.3) are satisfied. If $\{X_t\}$ is operator-self-similar and we can choose $A_c = c^Q$ with $Q \in M_+(\mathbf{R}^d)$, then we say that $\{X_t\}$ is operator-self-similar with *exponent* Q.

DEFINITION 5.2. A distribution μ on $\mathbf{R}^d$ is *of class OL* if all the conditions of Definition 1.10 are satisfied with the sequence $\{a_n\}$ of non-zero numbers replaced by the sequence $\{A_n\}$ of invertible $d \times d$ matrices.

DEFINITION 5.3. Let $Q \in M_+(\mathbf{R}^d)$. A distribution μ on $\mathbf{R}^d$ is *Q-self-decomposable* if, for every $a \in (0,1)$, there is a distribution ρ_a on $\mathbf{R}^d$ such that

$$\hat{\mu}(z) = \hat{\mu}(a^{Q'}z)\hat{\rho}_a(z),$$

where Q' is the adjoint operator (transposed matrix) of Q. We say that μ is *operator-self-decomposable* if μ is Q-self-decomposable for some $Q \in M_+(\mathbf{R}^d)$.

Basic results are the two propositions below.

PROPOSITION 5.4. *If $\{X_t\}$ is an operator-self-similar process on $\mathbf{R}^d$ and, for some $t > 0$, the distribution of X_t is genuinely d-dimensional, then $\{X_t\}$ has an exponent $Q \in M_+(\mathbf{R}^d)$.*

PROPOSITION 5.5. (i) *If μ is a genuinely d-dimensional distribution of class OL on $\mathbf{R}^d$, then μ is operator-self-decomposable.*

(ii) *If μ is operator-self-decomposable, then μ is of class OL.*

Proposition 5.4 is essentially a result of Hudson and Mason[4]. We note that the exponent Q of $\{X_t\}$ is not necessarily unique. Proposition 5.5 (i) is proved by Urbanik[27] with long and complicated argument. The assertion (ii) is easy to establish. Yamazato[37] remarks that distributions of class OL are not always operator-self-decomposable unless genuine d-dimensionality is assumed.

Analogues of Theorems 2.1 and 2.2 are proved by similar argument (Sato[20]).

THEOREM 5.6. *Let $Q \in M_+(\mathbf{R}^d)$ and $H > 0$. If $\{X_t\}$ is an operator-self-similar process on $\mathbf{R}^d$ with exponent HQ with independent increments, then, for every $t \geq 0$, the distribution of X_t is Q-self-decomposable.*

THEOREM 5.7. *Let $Q \in M_+(\mathbf{R}^d)$, $H > 0$, and $t_0 > 0$. If μ is a Q-self-decomposable distribution on $\mathbf{R}^d$, then there exists a unique (in the sense of distribution) operator-self-similar process $\{X_t\}$ with exponent HQ with independent increments such that $X(t_0)$ has distribution μ.*

Distribution of class OL on $\mathbf{R}^d$ are infinitely divisible and their Lévy measures have characterization similar to Proposition 1.11. They are absolutely continuous if genuinely d-dimensional. An analogue of the sequence $L_m, m = 0, 1, 2, \cdots, \infty$, is defined and analyzed. The Q-self-decomposable distributions have relation similar to 4° of Section 4 with processes of Ornstein-Uhlenbeck type generated by

$$G - \sum_{j,k=1}^{d} Q_{jk} x_k \frac{\partial}{\partial x_j}$$

in place of (4.1). Here Q_{jk} is the (j,k)-component of Q. These are results of Yamazato[36], Sato and Yamazato[23,24], Jurek[6], and others.

The first operator generalization was the work of Sharpe[25] on operator-stable distributions. We define strictly operator-stable processes as operator-self-similar Lévy processes. Operator-stable processes are defined as operator generalization of the definition of stable processes. Operator-stable [resp. strictly operator-stable] distributions are defined as the distributions at $t = 1$ of operator-stable [resp. strictly operator-stable] processes. Conditions of the Lévy measures of operator-stable distributions are found by Sharpe[25]; see also Sato[16]. Characterization of those of strictly operator-stable distributions is given by Sato[19].

ACKNOWLEDGMENT. The author thanks T. Hida for his constant encouragement and I. Kubo, T. Shimura, and S. Sato for their help in preparation of the manuscript.

REFERENCES

1. R. L. Dobrushin and P. Major, *Non-central limit theorems for non-linear functionals of Gaussian fields*, Z. Wahrsch. Verw. Geb. **50** (1979), p. 27-52.
2. I. I. Gihman and A. V. Skorohod, "The Theory of Stochastic Processes, II," Springer, Berlin, 1975.
3. B. V. Gnedenko and A. N. Kolmogorov, "Limit Distributions for Sums of Independent Random Variables," 2nd ed., Addison-Wesley, Reading, Mass., 1968.
4. W. N. Hudson and J. D. Mason, *Operator-self-similar processes in a finite-dimensional space*, Trans. Amer. Math. Soc. **273** (1982), p. 281-297.
5. Z. J. Jurek, *An integral representation of operator-selfdecomposable random variables*, Bull. Acad. Pol. Sci. Sér. Sci. Math. **30** (1982), p. 385-393.
6. Z. J. Jurek, *Limit distributions and one-parameter groups of linear operators on Banach spaces*, J. Multivar. Anal. **13** (1983), p. 578-604.
7. Z. J. Jurek and W. Vervaat, *An integral representation for self-decomposable Banach space valued random variables*, Z. Wahrsch. Verw. Geb. **62** (1983), p. 247-262.
8. K. Kawazu, *Some topics on random walk in random environment*, Seminar talk at Nagoya Univ. on March 13, 1989.
9. V. M. Kruglov, *A note on infinitely divisible distributions*, Theor. Probab. Appl. **15** (1970), p. 319-324.
10. J. W. Lamperti, *Semi-stable stochastic processes*, Trans. Amer. Math. Soc. **104** (1962), p. 62-78.
11. P. Lévy, "Théorie de l'Addition des Variables Aléatoires," 2ème ed., Gauthier-Villars, Paris, 1954; (1ère ed., 1937) .
12. M. Maejima, *Self-similar processes and limit theorems*, Sugaku Expositions **2** (1989), p. 103-123.
13. K. Sato, *A note on infinitely divisible distributions and their Lévy measures*, Sci. Rep. Tokyo Kyoiku Daigaku, Sect. A **12** (1973), p. 101-109.
14. K. Sato, *Class L of multivariate distributions and its subclasses*, J. Multivar. Anal. **10** (1980), p. 207-232.
15. K. Sato, *Absolute continuity of multivariate distributions of class L*, J. Multivar. Anal. **12** (1982), p. 89-94.
16. K. Sato, *Lectures on Multivariate Infinitely Divisible Distributions and Operator-Stable Processes*, Technical Report Series, Lab. Res. Statist. Probab. Carleton Univ. and Univ. Ottawa, No. 54, (1985).
17. K. Sato, *Bounds of modes and unimodal processes with independent increments*, Nagoya Math. J. **104** (1986), p. 29-42.
18. K.Sato, *Behavior of modes of a class of processes with independent increments*, J. Math. Soc. Japan **38** (1986), p. 679-695.
19. K. Sato, *Strictly operator-stable distributions*, J. Multivar. Anal. **22** (1987), p. 278-295.
20. K. Sato, *Self-similar processes with independent increments*, (to appear).

21. K. Sato and M. Yamazato, *On distribution functions of class L*, Z. Wahrsch. Verw. Geb. **43** (1978), p. 273-308.
22. K. Sato and M. Yamazato, *On higher derivatives of distribution functions of class L*, J. Math. Kyoto Univ. **21** (1981), p. 575-591.
23. K. Sato and M. Yamazato, *Operator-self-decomposable distributions as limit distributions of processes of Ornstein-Uhlenbeck type*, Stoch. Proc. Appl. **17** (1984), p. 73-100.
24. K. Sato and M.Yamazato, *Completely operator-self-decomposable distributions and operator-stable distributions*, Nagoya Math. J. **97** (1985), p. 71-94.
25. M. Sharpe, *Operator-stable probability distributions on vector groups*, Trans. Amer. Math. Soc. **136** (1969), p. 51-65.
26. M. S. Taqqu, *A bibliographical guide to self-similar processes and long-range dependence*, in "Dependence in Probability and Statistics," (E. Eberlein and M. S. Taqqu, eds.), Birkhäuser, Boston, 1986, p. 137-162.
27. K. Urbanik, *Lévy's probability measures on Euclidean spaces*, Studia Math. **44** (1972), p. 119-148.
28. K. Urbanik, *Slowly varying sequences of random variables*, Bull. Acad. Pol. Sci. Sér. Sci. Math. Astronom. Phys. **20** (1972), p. 679-682.
29. K. Urbanik, *Limit laws for sequences of normed sums satisfying some stability conditions*, in "Multivariate Analysis III," (P. R. Krishnaiah, ed.), Academic Press, New York, 1973, p. 225-237.
30. W. Vervaat, *Sample path properties of self-similar processes with stationary increments*, Ann. Probab. **13** (1985), p. 1-27.
31. S. J. Wolfe, *On the continuity properties of L functions*, Ann. Math. Statist. **42** (1971), p. 2064-2073.
32. S. J. Wolfe, *On the unimodality of multivariate symmetric distribution functions of class L*, J. Multivar. Anal. **8** (1978), p. 141-145.
33. S. J. Wolfe, *On a continuous analogue of the stochastic difference equation $X_n = \rho X_{n-1} + B_n$*, Stoch. Proc. Appl. **12** (1982), p. 301-312.
34. M. Yamazato, *Some theorems on infinitely divisible distributions of class L with applications to branching processes*, Sci. Rep. Tokyo Kyoiku Daigaku, Sect. A **13** (1975), p. 133-139.
35. M. Yamazato, *Unimodality of infinitely divisible distributions of class L*, Ann. Probab. **6** (1978), p. 523-531.
36. M. Yamazato, *Absolute continuity of operator-self-decomposable distributions on R^d*, J. Multivar. Anal. **13** (1983), p. 550-560.
37. M. Yamazato, *OL distributions on Euclidean spaces*, Theor. Probab. Appl. **29** (1984), p. 1-17.
38. V. M. Zolotarev, *The analytic structure of infinitely divisible laws of class L*, Litovsk. Mat. Sb. **3** (1963), p. 123-140; (Russian).
39. V. M. Zolotarev, "One-Dimensional Stable Distributions," Amer. Math. Soc., Providence, 1986.

Positivity of Donsker's Delta Function

Narn-Rueih Shieh and Yoshitaka Yokoi

Abstract

We prove that Donsker's Delta function $\delta(B(t)-x)$ is a positive generalized Brownian functional in the sense of [10].

§ 1. Introduction.

In finite dimensional distribution theory $\mathcal{S}(\mathbb{R}^k) \subset L^2(\mathbb{R}^k) \subset \mathcal{S}'(\mathbb{R}^k)$, $\delta(\cdot - x)$ is a ready and important example of generalized function. Going to the infinite dimensional case, we see in the works of Kuo [7] and Hida-Streit [3] that Donsker's δ-function plays an important role. Let B(t) be the one dimensional Brownian motion. For each fixed $t > 0$ and $x \in \mathbb{R}$, Donsker's δ-function is defined formally to be

$$\delta_{t,x} := \delta(B(t)-x) = \frac{1}{2\pi}\int_{-\infty}^{\infty} e^{iy(B(t)-x)}\,dy,$$

see Kuo [7]. Kuo proved that $\delta(B(t)-x)$ is a generalized Brownian functional in Hida's sense [1,2]. Very recently, a fundamental results for positive generalized Brownian functionals were obtained consecutively by Potthoff [8] and the second author of this note [10].

The purpose of this note is to prove that Donsker's δ-function is a positive generalized Brownian functional in the sense of [10]. In § 2, we first summarize the theory of generalized Brownian functionals briefly. Then, in § 3 and § 4 we prove our main result and derive relevant consequences.

The present work was initiated while the authors visited Nagoya University in August 1988. The authors would like to thank the hospitality of Department of Mathematics of Nagoya Univ., in particular the hospitality done by Professor Hida.

§ 2. Summary of generalized Brownian functionals.

We begin with Gel'fand triplet

$$E \subsetneq L^2_{real}(\mathbb{R}^1) \subsetneq E^*, \text{ where } E = \mathcal{S}_{real}(\mathbb{R}^1) \text{ and } E^* = \mathcal{S}'_{real}(\mathbb{R}^1).$$

Let $\mathcal{B}$ denote the Borel σ-field on E^*. By Bochner-Minlos' Theorem , a probability measure is uniquely determined by the characteristic functional

$$C(\xi) = \exp[-\frac{1}{2}\|\xi\|^2_{L^2_{real}(\mathbb{R})}] \quad \text{for } \xi \in E = \mathcal{S}_{real}(\mathbb{R}^1).$$

The measure μ is called the Gaussian white noise (with variance 1) and we denote $L^2(E^*, \mathcal{B}, \mu)$ by (L^2) for brevity. Members in (L^2) are called ordinary Brownian functionals. Each $\phi \in (L^2)$ has a unique decomposition (**Wiener-Itô decomposition**)

$$\phi = \sum_{n=0}^{\infty} I_n(f_n),$$

where $I_n(f_n)$ is the n-ple Wiener-Itô integral of $f_n \in \hat{L}^2_{complex}(\mathbb{R}^n)$ (= all symmetric square integrable functions on $\mathbb{R}^n$).

Now, we consider the differential operator

$$D\xi(u) = [1+u^2-\left(\frac{d}{du}\right)^2]\xi(u) \quad \text{for } \xi \in E.$$

For each $n \geq 0$ and $p \geq 0$, let

$$\mathcal{S}_p^{\hat{\otimes}n} := \left\{f(u_1, \cdots, u_n);\ D^p_{u_1}\cdots D^p_{u_n} f(u_1, \cdots, u_n) \in L^2_{complex}(\mathbb{R}^n)\right\},$$

$$\mathcal{S}_p^{\hat{\otimes}n} := \mathbb{C} \text{ for } n = 0,$$

and define a Hilbertian norm $\|\cdot\|_{\mathcal{S}_p^{\hat{\otimes}n}}$ in each space $\mathcal{S}_p^{\hat{\otimes}n}$ as follows:

$$\|f\|^2_{\mathcal{S}_p^{\hat{\otimes}n}} := \int\cdots\int_{\mathbb{R}^n} |D^p_{u_1}\cdots D^p_{u_n} f(u_1, \cdots, u_n)|^2 du_1 \cdots du_n.$$

Next, set for every $p \geq 0$

$$(\mathcal{S}_p) := \left\{\sum_{n=0}^{\infty} I_n(f_n);\ f_n \in \mathcal{S}_p^{\hat{\otimes}n}\ (n \geq 0) \text{ and } \sum_{n=0}^{\infty} n!\|f_n\|^2_{\mathcal{S}_p^{\hat{\otimes}n}} < \infty\right\}$$

$(\mathcal{S}_{-p})$:= the topological dual of $(\mathcal{S}_p)$.

Finally, we set

$$(\mathcal{S}) := \bigcap_{p=0}^{\infty} (\mathcal{S}_p) \quad \text{and} \quad (\mathcal{S}') := \bigcup_{n=0}^{\infty} (\mathcal{S}_{-p}),$$

and equip $(\mathcal{S})$ and $(\mathcal{S}')$ with the projective limit topology and the inductive limit one, respectively. Thus, we have a new Gel'fand triplet

$$(\mathcal{S}) \subsetneq (L^2) \subsetneq (\mathcal{S}');$$

members in $(\mathcal{S})$ and $(\mathcal{S}')$ are called test Brownian functionals and generalized Brownian ones, respectively.

It should be mentioned that our triplet is motivated by Hida's original idea of generalized Brownian functionals [1, 2]. He has established triplets such as

$$(L^2)^+ \subsetneq (L^2) \subsetneq (L^2)^-.$$

It should be noticed that $(L^2)^+$ is a Hilbert space but $(\mathcal{S})$ is a nuclear space and that $(\mathcal{S})$ is an algebra (Kubo-Takenaka).

Let $\langle \phi, \Phi \rangle$ denote the canonical bilinear form between $\phi \in (\mathcal{S})$ and $\Phi \in (\mathcal{S}')$. In particular,

$$\langle \phi_1, \phi_2 \rangle = \mathrm{E}_\mu[\phi_1(\omega)\phi_2(\omega)] := \int_{E^*} \phi_1(\omega)\phi_2(\omega)d\mu(\omega),$$

if $\phi_1, \phi_2 \in (L^2)$. The positivity of $\Phi \in (\mathcal{S}')$ is defined $\langle \phi, \Phi \rangle \geq 0$ whenever $\phi(\omega) \geq 0$ (μ-a.e. ω). It is proved in [6] that, for each $\phi \in (\mathcal{S})$, there exists a continuous version $\tilde{\phi}$ of ϕ such that

$$\phi(\omega) = \tilde{\phi}(\omega) \quad (\mu\text{-a.e. } \omega \in E^*).$$

Theorem 2. 1 ([10]). *Suppose that Φ is a positive generalized Brownian functional. Then, there exists a unique finite measure ν_Φ on $(E^*, \mathcal{B})$ which satisfies that*

$$\langle \phi, \Phi \rangle = \int_{E^*} \tilde{\phi}(\omega)d\nu_\Phi(\omega)$$

for all $\phi \in (\mathcal{S})$.

§ 3. Donsker's delta function as a positive generalized Brownian functional.

Define $B(t) = B(t,\omega) = \langle I_{[0,t]}, \omega \rangle$ for $t \geq 0$ and $\omega \in E^*$; then $[0, \infty) \ni t \longmapsto B(t) \in \mathbb{R}$ is a one-dimensional Brownian motion.

Let $g_\varepsilon(u) = (2\pi\varepsilon)^{-1/2}\exp[-\frac{u^2}{2\varepsilon}]$ for $\varepsilon > 0$. Then $g_\varepsilon(B(t)-x) \in (L^2)$. Donsker's delta function $\delta(B(t)-x)$ is defined as follows ([5]):

$$\delta(B(t)-x) = \lim_{\varepsilon\to 0} g_\varepsilon(B(t)-x) \text{ in } (\mathcal{S}').$$

We shall prove:

Theorem 3. 1 *For each $t > 0$ and $x \in \mathbb{R}$, Donsker's delta function $\delta(B(t)-x)$ is a positive generalized Brownian functional.*

Proof. We start with the following formal expansion of $\delta(B(t)-x)$, suggested by Kuo [7, (10)].

$$(3.\ 1) \qquad \delta(B(t)-x) = \frac{1}{\sqrt{2\pi t}}\, e^{-\frac{x^2}{2t}} \sum_{n=0}^{\infty} \frac{1}{n!2^n} H_n\left(\frac{x}{\sqrt{2t}}\right) H_n\left(\frac{\langle I_{[0,t]}, \omega\rangle}{\sqrt{2t}}\right),$$

where $H_n(u)$ is Hermite polynomial of n-th order;

$$H_n(u) = (-1)^n \exp[-u^2]\left(\frac{d}{du}\right)^n \exp[u^2].$$

We show that the right hand side of (3. 1) converges in $(\mathcal{S}_{-p})$ when p is large enough. Set $h_t(u) = \frac{1}{\sqrt{t}} I_{[0,\ t]}(u)$. Then by Hida [2, p. 139], $H_n\left(\frac{\langle I_{[0,\ t]}, \omega\rangle}{\sqrt{2t}}\right) = I_n(f_n)$, where $f_n = \left(\sqrt{2}\ h_t\right)^{\hat{\otimes} n}$ is the n-fold symmetric tensor product of $\left(\sqrt{2}\ h_t\right)$. So, if both $t > 0$ and $u \in \mathbb{R}$ are fixed, we have only to prove that the series

$$\|\delta(B(t)-x)\|^2_{(\mathcal{S}_{-p})} = p(t,0,x)^2 \sum_{n=0}^{\infty} n! \frac{1}{(n!)^2\ 2^{2n}} \left(H_n\left(\frac{x}{\sqrt{2t}}\right)\right)^2 \|f_n\|^2_{\mathcal{S}_{-p}^{\hat{\otimes} n}}$$

converges for sufficiently large $p \geq 0$. In order to see the convergence of this series, we prepare the following:

$$\text{(a)} \qquad \|f_n\|^2_{\mathcal{S}_{-p}^{\hat{\otimes} n}} = \left(\|\sqrt{2}\ h_t\|^2_{\mathcal{S}_{-p}}\right)^n$$

$$= \left(\sum_{k=0}^{\infty} \left|\int_{\mathbb{R}} \sqrt{2}\ h_t(u)(2k+2)^{-p} \xi_k(u)\ du\right|^2\right)^n,$$

$$= \left(2^{-2p-1}t^{-1}\sum_{k=0}^{\infty}(k+1)^{-2p}\left|\int_0^t \xi_k(u)\,du\right|^2\right)^n$$

where $\xi_k(u) = \left(k!2^k\sqrt{\pi}\right)^{-1/2} H_n(u)\exp[-\frac{u^2}{2}]$,

(b) $|\xi_k(u)| \le C$ with some constant C independent of $k \ge 0$ and $u \in \mathbb{R}$ (Hille, [4], p. 435),

(c) $|H_n(v)| \le C\,\pi^{1/4}\,2^{n/2}\sqrt{n!}\,\exp[\frac{v^2}{2}]$ for any $v \in \mathbb{R}$ ([4]),

From (a), (b) and (c) we have

$$\|\delta(B(t)-x\|^2_{(\mathcal{S}_{-p})}$$

$$\le p(t,0,x)^2\sqrt{\pi}\,C^2\exp[\frac{x^2}{4t}]\sum_{n=0}^{\infty}\left(C^2t2^{-2p}\sum_{k=0}^{\infty}(k+1)^{-p}\right)^n.$$

But this series converges if $C^2t2^{-2p}\sum_{k=0}^{\infty}(k+1)^{-p} < 1$. This condition is clearly satisfied for a sufficiently large p and a fixed $t > 0$. Consequently, we have that $\delta(B(t)-x) \in (\mathcal{S}')$.

Next we show that $\delta(B(t)-x)$ is positive. It is sufficient to prove that $\mathcal{T}$-transform $\langle e^{i\langle\xi,\,\omega\rangle},\ \delta(B(t)-x)\rangle$ of $\delta(B(t)-x)$ is positive definite in $\xi \in E$. Let $\beta_j \in \mathbb{C}$, $\xi_j \in E$ $(j = 1, 2,\cdots, N)$, and let $\xi_j^{\perp} \equiv \xi_j - (\xi_j,\ h_t)h_t$. Then, $\langle\xi_j^{\perp},\ \omega\rangle$ and $\langle h_t,\ \omega\rangle$ are independent with respect to μ and we have,

$$\sum_{j,k=1}^{N}\sum^{N}\beta_j\overline{\beta}_k\langle e^{i\langle\xi_j-\xi_k,\ \omega\rangle},\ \delta(B(t)-x)\rangle$$

$$= \sum_{j,k=1}^{N}\sum^{N}\beta_j\overline{\beta}_k\langle e^{i(\xi_j-\xi_k,\ h_t)\langle h_t,\ \omega\rangle}e^{i\langle\xi_j^{\perp}-\xi_k^{\perp},\ \omega\rangle},\ \delta(\sqrt{t}\langle h_t\omega\rangle - x)\rangle$$

$$= \sum_{j,k=1}^{N}\sum^{N}\beta_j\overline{\beta}_k\int_{E^*}e^{i(\xi_j-\xi_k,\ h_t)\langle h_t,\ \omega\rangle}\delta(\sqrt{t}\langle h_t,\ \omega\rangle - x)\,d\mu(\omega)\ C(\xi_j^{\perp}-\xi_k^{\perp})$$

$$= \sum_{j,k=1}^{N}\sum^{N}\beta_j\overline{\beta}_k\int_{\mathbb{R}}e^{i(\xi_j-\xi_k,\ h_t)q}\delta(\sqrt{t}q - x)\frac{1}{\sqrt{2\pi}}\exp[-\frac{q^2}{2}]\,dq\ C(\xi_j^{\perp}-\xi_k^{\perp})$$

$$= \sum_{j,k=1}^{N}\sum^{N}\beta_j\overline{\beta}_k\ e^{i(\xi_j-\xi_k,h_t)x/\sqrt{t}}\ \frac{1}{\sqrt{2\pi t}}\exp[-\frac{x^2}{2t}]\ C(\xi_j^{\perp}-\xi_k^{\perp})$$

$$= \frac{1}{\sqrt{2\pi t}} \exp[-\frac{x^2}{2t}] \sum_{j,k=1}^{N} \left(\beta_j e^{i(\xi_j, h_t)x/\sqrt{t}}\right) \overline{\left(\beta_k e^{i(\xi_k, h_t)x/\sqrt{t}}\right)} C(\xi_j^{\perp}-\xi_k^{\perp})$$

≥ 0. Because $C(\cdot)$ is positive definite. This rather formal computation is justified by approximating $\delta_{t,x}$ with δ-convergence sequence such as $\{\frac{n}{\sqrt{2\pi}}\exp[-\frac{n^2 (B(t)-x))^2}{2}]\}_{n=1}^{\infty}$ and considering that the space $\mathbb{A} = \{\sum_{j=1}^{N} \beta_j e^{i\langle\xi_j,\omega\rangle};\ \beta_j \in \mathbb{C},\ \xi_j \in E, \text{ and } N \in \mathbb{N}\}$ is dense in $(\mathcal{S})$.

Thus, it is proved that $\delta(B(t)-x)$ is a positive generalized Brownian functional. □

In order to supplement the above proof we note:

Theorem 3.2 *Suppose that:*

(1) $\Phi_n,\ \Phi \in (\mathcal{S}_{-p})$ *for some* $p \geq 0$,

(2) $\|\Phi_n\|_{(\mathcal{S}_{-p})} \leq K$ *(constant) and*

(3) $\lim_{n\to\infty} \mathcal{T}\Phi_n(\xi) = \mathcal{T}\Phi(\xi)$ *for* $\forall\ \xi \in E$.

Then, $\lim_{n\to\infty} \Phi_n = \Phi$ *in* $(\mathcal{S}')$ *under the inductive limit topology.*

Proof. Let $\mathbb{A} = \{\sum \beta_j e^{i\langle x,\xi_j\rangle};\ \beta_j \in \mathbb{C},\ \xi_j \in E\}$. Since $\mathbb{A}$ is dense in $(\mathcal{S}_p)$, Φ_n converges to Φ in $(\mathcal{S}_p)$. Therefore, Φ_n converges to Φ in $(\mathcal{S}')$ under the inductive limit topology. □

Let us apply this theorem to Donsker's delta function $\delta_{t,a} = \delta(B(t)-a)$ and show that $\Phi_n = \frac{n}{\sqrt{2\pi}}\exp[-\frac{n^2(B(t)-a)^2}{2}]$ converges to $\delta_{t,a}$ in $(\mathcal{S}')$. For that purpose, let us compute the norms of Φ_n's in $(\mathcal{S}_{-p})$ for some $p \geq 0$ and show that they are bounded. We use $\mathcal{T}$-transform of Φ_n to estimate the norm $\|\Phi_n\|_{(\mathcal{S}_{-p})}$. Write $h_t(u) = \frac{1}{\sqrt{t}} I_{[0,t]}(u)$, then

$$\Phi_n = \frac{n}{\sqrt{2\pi}} \exp[-\frac{n^2}{2}\left(\sqrt{t}\langle x, h_t\rangle - a\right)^2] \quad \text{and} \quad \xi = (\xi, h_t)h_t + \xi^{\perp}.$$

Therefore,

$$\mathcal{T}\Phi_n(\xi) = \langle \Phi_n, e^{i\langle x, \xi\rangle}\rangle$$

$$= \frac{n}{\sqrt{2\pi}} \langle \exp[-\frac{n^2}{2n}\left(\sqrt{t}\langle x, h_t\rangle - a\right)^2], e^{i\langle x, (\xi, h_t)h_t - \xi^{\perp}\rangle}\rangle$$

$$= C(\xi^{\perp}) \frac{n}{\sqrt{2\pi}} \int_{\mathbb{R}} \frac{1}{\sqrt{2\pi}} \exp[-\frac{1}{2}u^2 - \frac{n^2}{2}\left(\sqrt{t}u - a\right)^2 + i(\xi, h_t)u]du$$

$$= C(\xi)\frac{1}{\sqrt{2\pi}} \frac{n}{\left(1+n^2 t\right)^{1/2}} \exp[\frac{1}{2}\frac{n^2}{1+n^2 t}\left(\sqrt{t}(\xi, h_t) - ia\right)^2].$$

Let's write $\sigma_n = n^{-1}\left(1+n^2 t\right)^{1/2}$ and $p_n(t,0,a) = \frac{1}{\sqrt{2\pi}\sigma_n} \exp[-\frac{a^2}{2\sigma_n^2}]$, then we have;

$$\mathcal{T}\Phi_n(\xi) = p_n(t,0,a)\, C(\xi)\, \exp[\sigma_n^{-2}\left(ai\sqrt{t}(\xi, h_t) - \frac{1}{2}\left(i\sqrt{t}(\xi, h_t)\right)^2\right)]$$

$$= p_n(t,0,a)\, C(\xi)\, \exp[2\left(\frac{a}{\sqrt{2}\sigma_n}\right)\left(\frac{i}{\sqrt{2}\sigma_n}\int_0^t \xi(u)du\right) - \left(\frac{i}{\sqrt{2}\sigma_n}\int_0^t \xi(u)du\right)^2\Big)]$$

$$= p_n(t,0,a)\, C(\xi) \sum_{k=0}^{\infty} \frac{1}{k\,!} H_k\left(\frac{a}{\sqrt{2}\sigma_n}\right) \left(\frac{i}{\sqrt{2}\sigma_n}\int_0^t \xi(u)du\right)^k$$

$$= p_n(t,0,a)\, C(\xi) \sum_{k=0}^{\infty} i^k \frac{1}{k\,!}\left(\frac{t}{2\sigma_n}\right)^k H_k\left(\frac{a}{\sqrt{2}\sigma_n}\right) \langle \left(\sqrt{2}h_t\right)^{\hat{\otimes}k}, \xi^{\hat{\otimes}k}\rangle.$$

Therefore,

(1) $\Phi_n = p_n(t,0,a)\, C(\xi) \sum_{k=0}^{\infty} \frac{1}{k\,!}\left(\frac{t}{2\sigma_n}\right)^k H_k\left(\frac{a}{\sqrt{2}\sigma_n}\right) H_k\left(\sqrt{2}h_t\right),$

(2) $\|\Phi_n\|^2_{(\mathcal{S}_{-p})} = p_n(t,0,a)^2 \sum_{k=0}^{\infty} \frac{1}{(k!)^2} \left(\frac{t}{2\sigma_n}\right)^{2k} \left(H_k\left(\frac{a}{\sqrt{2}\sigma_n}\right)\right)^2 \| \left(\sqrt{2}h_t\right)^{\hat{\otimes}k} \|^2_{\mathcal{S}_{-p}^{\hat{\otimes}k}}$

and

(3) $\lim_{n\to\infty} \left(\mathcal{T}\Phi_n\right)(\xi) = \left(\mathcal{T}\delta_{t,a}\right)(\xi)$ for $\forall \xi \in E$.

By using the following fact about Hermite polynomials,

$$|H_n(u)| \le C\, \pi^{1/4}\, 2^{n/2} \sqrt{n!} \exp[\frac{u^2}{2}] \text{ for } \forall\, u \in \mathbb{R}, \text{ ([Hille])},$$

we can see that $\{\|\Phi_n\|_{(\mathcal{S}_{-p})};\ n \geq 1\}$ is bounded for sufficiently large $p \geq 1$. (See Theorem 3. 1 and its proof.) Thus, Φ_n converges to $\delta_{t,a}$ in $(\mathcal{S}')$.

§ 4. The measure associated with $\delta(B(t)-x)$.

We have proved that $\delta(B(t)-x)$ is a positive generalized Brownian functional, therefore by the theorem quoted in § 2 there exists a unique finite measure $\nu_{t,x}$ on $(E^*, \mathcal{B})$ such that

$$\langle \phi,\ \delta(B(t)-x)\rangle = \int_{E^*} \tilde{\phi}(\omega)d\nu_{t,x}(\omega).$$

Observe that

$$[\mathcal{T}\delta(B(t)-x)](\xi) := \langle e^{i\langle \xi,\omega\rangle},\ \delta(B(t)-x)\rangle$$

$$= p(t,0,x)C(\xi)\exp[ix\frac{1}{t}\int_0^t \xi(s)ds + \frac{1}{2}\frac{1}{t}\Big(\int_0^t \xi(s)ds\Big)^2]$$

$$= p(t,0,x)C(\xi)\sum_{n=0}^{\infty} i^n\frac{1}{n!2^n}H_n\Big(\frac{x}{\sqrt{2t}}\Big)\langle \xi^{\hat{\otimes} n},\ \big(\sqrt{2}h_t\big)^{\hat{\otimes} n}\rangle.$$

From this we see that

(a) $\displaystyle\int_{E^*} 1\, d\nu_{t,x}(\omega) = \langle 1,\ \delta_{t,x}\rangle = p(t,0,x)$,

(b) $\displaystyle\int_{E^*} I_n(f_n)(\omega)d\nu_{t,x}(\omega) = \langle I_n(f_n),\ \delta_{t,x}\rangle$

$$= p(t,0,u)H_n\Big(\frac{x}{\sqrt{2t}}\Big)\Big(2t\Big)^{-n/2}\int_0^t\cdots\int_0^t f_n(s_1,\cdots,s_n)ds_1\cdots ds_n\ ,$$

where $f_n \in \mathcal{S}_p^{\hat{\otimes} n}$ for sufficiently large $p \geq 1$. In particular, in the case of $n = 1$,

$$\int_{E^*} \langle f_1,\ \omega\rangle\, d\nu_{t,x}(\omega) = p(t,0,x)\frac{x}{t}\int_0^t f_1(s)ds,$$

and consequently,

$$\int_{E^*} \langle f_1,\ \omega\rangle^2 d\nu_{t,x}(\omega)$$

$$= p(t,0,x)\Big(H_2\Big(\frac{x}{\sqrt{2t}}\Big)\frac{1}{2}\frac{1}{t}\Big(\int_0^t f_1(s)ds\Big)^2 + \int_{\mathbb{R}}|f_1(s)|^2 ds\Big).$$

here, we used the relation $H_n\left(\frac{\langle\eta,\omega\rangle}{\sqrt{2}}\right) = 2^{n/2} I_n(\eta^{\hat{\otimes} n})(\omega)$ if $\|\eta\|_{\mathcal{S}_0} = 1$.

Corollary 4. 1 $\frac{\nu_{t,x}}{p(t,0,x)}$ *is a probability measure on* $(E^*, \mathcal{B})$ *and* $\langle\xi, \omega\rangle$ *is a Gaussian random variable on* $(E^*, \mathcal{B}, \frac{\nu_{t,x}}{p(t,0,x)})$ *with mean* $\frac{x}{t}\int_0^t \xi(s)ds$ *and variance* $\int_{\mathbb{R}} |\xi(u)|^2 du - \frac{1}{t}\left(\int_0^t \xi(u)du\right)^2$.

References

1. Hida, T. : Analysis of Brownian Functionals, Carleton Math. Lecture Notes, No. 13, Carleton University, Ottawa, 1975.
2. Hida, T. : Brownian Motion, Application of Mathematics, Vol. 11, Springer Verlag, 1980.
3. Hida, T. and Streit, L. : Generalized Brownian Functionals and Feynman Integral, *Stochastic Processes and Their Applications*, Vol. 16 (1983), 55-69.
4. Hille, E. : A Class of Reciprocal Functions, *Annals of Math.*, (2) 27(1926), 427-464.
5. Kubo, I. : Ito Formula for Generalized Brownian Functionals, *Lecture Notes in Control and Information Science*, 49, Springer Verlag, New York/Heidelberg/Berlin, 1983, 156-166.
6. Kubo, I. and Yokoi, Y. : A Remark on the Space of Testing Functionals in the White Noise Calculus, *To appear (Nagoya Math. J.)*.
7. Kuo, H-H. : Donsker's Delta Function as a Generalized Brownian Functional and Its Application, Lecture Notes in Control and Information, Vol. 49 (1983), 167-178.
8. Potthoff, J. : On Positive Generalized Functionals, *J. Funct. Annal.*, Vol. 74 (1987), 81-95.
9. Szego, G. ; Orthogonal Polynomials, *Amer. Math. Soc. Colloq. Publ*. Vol. XXIII (1939).
10. Yokoi, Y. : Positive Generalized White Noise Functionals, *To appear (Hiroshima Math. J.)*.

LINEAR KERNELS ASSOCIATED WITH DISTRIBUTIONS OF STOCHASTIC PROCESSES.

W. SMOLEŃSKI
Department of Mathematics
Warsaw University
PKiN IX p.
00-901 WARSAW, POLAND

ABSTRACT

With the distribution of a stochastic process one can associate a certain linear set of trajectories, in general P-negligible but with interesting properties and applications.

1. Introduction

The notion of a linear kernel of a probability measure on a locally convex space was introduced by Hoffmann-Jørgensen [12] in the case of a product measure on R^∞. However it had been implicitly used earlier (Dudley [8], Borell [4], Słowikowski [23]).

For a Gaussian measure this linear kernel equals to the reproducing kernel Hilbert space (RKHS). So in this very important case it has been studied for decades and its importance was illuminated in Gross [10], [11], Kallianpur [15] and Kuo [16].

In a more general setting the kernel was studied by Chevet, Tortrat, Okazaki, Takahashi, Kwapień, Smoleński, Suchanecki (cf. References).

An excelent survey of results about the kernel was given by Chevet [7]. It includes many examples of kernels explicitely calculated.We do not repeat them here trying to avoid too big overlap with [7].

Let us recall first some different equivalent ways of defining RKHS $\mathcal{H}_\mu$ of a tight symmetric Gaussian measure μ on a locally convex space E (cf. [2], [5]).

Let E' be the topological dual of E and let s_μ denote the topology of the convergence in μ.

Theorem 1.1.

a) $\mathcal{H}_\mu = (E', s_\mu)'$

b) $\mathcal{H}_\mu = \bigcap\{L \subset E : L -$ linear, $\mu -$ measurable, $\mu(L) = 1\}$

c) $\mathcal{H}_\mu = \bigcap\{L \subset E : L -$ linear, $L = \bigcup_{n=1}^{\infty} K_n, K_n -$ convex, compact, $\mu(L) = 1\}$.

d) $\mathcal{H}_\mu$ *is the minimal element in the sense of inclusion ordering in the net* $\{U\}$ *of linear subsets of* E *with the following property:*

(*) $\forall \varepsilon > 0 \exists K_\varepsilon \subset U$, *convex, compact such that* $\forall f \in E'$ *if*

$$K_\varepsilon \subset \{x \in E : |f(x)| \leq 1\} \text{ then}$$
$$\mu\{x \in E : |f(x)| > 1\} < \varepsilon.$$

e) $\mathcal{H}_\mu = AS(\mu) = \{x \in E : \mu_x = \mu * \delta_x \ll \mu\}$, *i.e. RKHS is the Cameron-Martin space.*

For a general tight probability measure μ on a quasi-complete locally convex space E (a) can be taken as a definition of a linear kernel J_μ of μ. (Quasi-complete means that bounded closed subsets are complete).
R. Dudley [8] was probably the first who make an implicit use of the kernel by proving the following theorem:

Theorem 1.2.

a) $AS(\mu) \subset (E', s_\mu)'$

b) *if* $AS(\mu) = (E', s_\mu)'$ *then*

$$\mu_x \approx \mu \text{ iff } x \in (E', s_\mu)'$$
$$\mu_x \perp \mu \text{ iff } x \notin (E', s_\mu)'.$$

2. Pre-supports and kernels of cylinder probability measures on locally convex spaces.

Let E be a locally convex linear topological space. A subset Z of E is called a cylinder set if it is of the form $Z = T^{-1}(B)$, where T is a continuous linear mapping from E into R^n and B is a Borell subset of R^n. A positive normed set function μ on the algebra of cylinder sets is a cylinder measure if for every T as above $\mu \circ T^{-1}$ is a σ-additive Borell measure on R^n. With every cylinder measure μ we can associate a linear mapping L_μ from E' into some $L_0(\Omega, \mathcal{M}, \mathbf{P})$ such that for every $f \in E'$ and every $t \in R$ $\mu(f < t) = \mathbf{P}(L_\mu f < t)$. L_μ is so-called "adjoint linear stochastic process" cf.[1].
A cylinder measure μ is said to be continuous if L_μ is continuous from E' endowed with the topology τ_E of the uniform convergence on convex compact subsets of E into L_0 endowed with the usual topology of the convergence in probability.

Definition 2.1. *A linear subspace U of E is a pre-support of μ iff the condition $(*)$ from Theorem 1.1.d holds.*

It is easy to observe that μ is continuous if and only if it has a pre-support i.e. if E is a pre-support of μ.

Definition 2.2. *The kernel J_μ of μ is defined as the dual of E' equipped with the topology of the convergence in μ;*

$$J_\mu = (E', s_\mu)'.$$

Theorem 2.3. [7,19,20,21,24,25,27]. *Let μ be a continuous cylinder measure on a locally convex space E. Then*
 a) J_μ is the intersection of all pre-supports;
 b) intersection of any countable family of pre-supports is a pre-support;
 c) J_μ is a pre-support itself iff (E', s_μ) is locally convex;
 d) μ is σ-additive and $\mu(J_\mu) = 1$ iff (E', s_μ) is nuclear.

It is interesting to compare theorems 1.1. and 2.3. For instance if μ is Gaussian then (E', s_μ) is a pre-Hilbert space. This and the zero-one law for values of Gaussian measure on linear sets (cf. [9] or [26] for very short and simple proofs) gives a well known fact that $\mu(H_\mu) = 0$ unless H_μ is finite dimensional [15].

Theorem 2.4. [4] *If μ is a tight probability measure on a quasi-complete locally convex space E then*

$$\begin{aligned} J_\mu = \bigcap \{ & L \subset E : L - \text{linear} \\ & \text{and } \forall \varepsilon > 0 \ \exists K_\varepsilon \subset L,\ \mu(K_\varepsilon) > 1 - \varepsilon,\ K_\varepsilon - \text{ convex, compact}\}. \end{aligned}$$

(In a quasi-complete locally convex space a closed convex envelope of a compact set is compact and this is what is really necessary as far as the completness is concerned).

Remark 2.5. *From the above theorem 2.4. one can easily derive 1.2. (cf.[17]).*

3. Pre-supports and kernels of stochastic processes

A stochastic process $X = \{X_t, t \in T\}$ defines in a well known way a probability measure μ on a σ-algebra $\mathcal{R}$ generated by cylinder subsets of R^T. R^T is a complete nuclear locally convex space with the topological dual

$$R_0^T = \{f : T \longrightarrow R : \ \text{card}\{t : f(t) \neq 0\} < \infty\}$$

and duality $f(x) = \sum f(t)x(t)$, $f \in R_0^T$, $x \in R^T$.

Unfortunately R^T is non-separable and non-metrizable unless T is countable. Moreover, interesting sets of functions are not μ-measurable. For instance if $T = [0,1]$ and X_t is such that $P(X_t = 0) < 1$ for uncountably many t then $C[0,1]$ is not μ-measurable i.e. does not belong to the extended σ-field $\mathcal{R}_\mu$ generated by $\mathcal{R}$ and subsets of sets of measure zero.

Let F be a linear subset of R^T such that it has a "natural" structure of, say, a separable Banach space. It may happen that a Borell σ-field $B(F)$ is uncomparable with the "trace" of $\mathcal{R}$ on F but the outer measure μ^* is a tight probability $B(F)$.

Let $J_\mu(R^T)$ be the kernel of μ on R^T i.e. $J_\mu(R^T) = (R_0^T, s_\mu)'$ and $J_\mu(F)$ be the kernel of μ on F. The next theorem gives a sufficient condition for the following useful equalities

$$B(F)_\mu = \mathcal{R}_\mu \cap F \; ; \; J_\mu(R^T) = J_\mu(F).$$

Theorem 3.1. *Let F be such a linear subset of R^T which is a separable Banach space under some norm which we assume to be fixed. Let μ be a probability on R^T such that $\mu^*(F) = 1$. Suppose there exists a Banach space E continuously imbeded in R^T and such that $E \cap F$ is dense in F. More over we assume that the intersection of F and the unit ball of E is a weakly compact subset of F.*
If E is a pre-support of μ then
a) $R_\mu \cap F = B(F)_\mu$
b) $J_\mu(R^T) = J_\mu(F)$
c) *value at t is an almost sure limit of elements of F'.*

Sketch of a proof:

Let $f \in F'$. Then $f_{|E} \in E'$. By the Grothendieck Completness Theorem ([14],p.248) there exists a sequence of elements of R_0^T converging to f uniformly on every ball in E.

Since E is a pre-support of μ if follows that this sequence converges to f in the measure μ (cf. [27] Remark 3.2.) this and the separability of F entail the inclusion $B(F)_\mu \subset \mathcal{R}_\mu \cap F$.

Conversely, let θ_t denote the element of R_0^T given by $\theta_t(x) = x(t)$.

Now we can repeat the above argument to get the sequence of elements of F' converging in μ to θ_t. This gives (a) and (c).

From (a) and the fact that F is a Polish space we get (Ulam's Theorem) that μ is tight on $B(F)$. From [27] (Corollary 3.4) and the assumption that balls from E intersected with F are weakly compact in F follows that $E \cap F$ is a pre-support of μ on F. Now (b) is an immediate consequence of Theorem 2.3.a,b above.

Comment. *So if the assumption of 3.1 are fullfiled we can consider $(F, B(F), \mu)$ as a basic probability space for the proces X_t which was represented by $(R^T, \mathcal{R}, \mu)$.*

Example 3.2. *Let $T = [0,1]$ and let $\{X_t\}$ be such that $\sup E|X_t| < \infty$.*
Then $E = B[0,1] = \{x \in R^T : \sup |x(t)| < \infty\}$ is a pre-support of μ.

Proof:

Let K_r be the ball in E with center at zero and radius r. Let $\sup E|X_t| = M$. Then for $f \in R_0^T$, $f(x) = \sum_{n=1}^{N} a_n x(t_n)$,

$$\mu(|f| > 1) = P(|\sum a_n X_{t_n}| \geq 1) \leq M \sum |a_n| = \frac{M}{r} \sup\{f(x) : x \in K_r\}.$$

Since K_r is a compact subset of R^T it fulfills condition (*) of the definition of pre-support for $\varepsilon = \frac{M}{r}$.

Example 3.3. *Let $T = [0,1]$ and let X_t be such that $X_0 = 0$, $EX_t = 0$, $E(X_t - X_s)^2 \leq C|t-s|$.*
Then $E = W^{1,2}[0,1] = \{x \in R^T :$ x is abs. cont. and $\int_0^1 |\dot{x}(t)|^2 dt < \infty\}$ is a pre-support of $\mu = \mathcal{L}(X_t)$.

Proof:

Let f and K_r be as above. Assume that for $x \in K_r$ $f(x) \leq 1$. Let us put $c_N = a_N$, $c_{N-1} = a_N + a_{N-1}, \dots$ $c_1 = a_N + \dots + a_1$ and $t_0 = 0$.

Then

$$\mu(|f(x)| > 1) = P\Big(\Big|\sum_{n=1}^{N} a_n X_{t_n}\Big| > 1\Big) =$$

$$P\Big(\Big|\sum_{n=1}^{N} c_n(X_{t_n} - X_{t_{n-1}})\Big| > 1\Big) \leq E\Big(\sum c_n(X_{t_n} - X_{t_{n-1}})\Big)^2 \leq$$

$$2 \cdot \sum {c_n}^2 E(X_{t_n} - X_{t_{n-1}})^2 \leq 2 \cdot C \cdot \sum {c_n}^2 \Big|t_n - t_{n-1}\Big| \leq \frac{2C}{r^2}$$

so K_r works in (*) for $\varepsilon = \frac{2C}{r^2}$.

(The last inequality holds because $W^{1,2}$ is continuously imbeded in R^T and

$$\|f\|_{(W^{1,2})'} = \sqrt{\sum c_n^2 |t_n - t_{n-1}|}\ ).$$

Remark. *Having in mind Theorem 3.1. we can take for instance $F = L^p[0,1]$, $p \geq 1$ to work with E from 3.2. and $F = C[0,1]$ to work with E from 3.3. (There is an obstacle arising from the fact that L^p is really a quotient of a subset of R^T. It can be overcome but to make 3.1. general enough to cover the case would be to make its formulation awkwardly complicated).*

4. Examples of applications

Example 4.1.

Let $X_1, X_2, \ldots$ be a sequence of independent random variables. Let Y be a random variable of the form $Y = \sum_{n=1}^{\infty} a_n X_n$, where a_n's are real coefficients and the series is assumed to converge. Then Y is mesurable with respect to (X_n) and $P \times P \times P$ a.e. the following implication holds

$$\Big\{\forall n \quad X_n(w_3) = \alpha X_n(w_1) + \beta X_n(w_2)\Big\} \Longrightarrow Y(w_3) = \alpha Y(w_1) + \beta Y(w_2).$$

In other words Y is a measurable linear functional of the sequence $X_1, X_2, \ldots$

One can pose the following question: Is every measurable linear functional of (X_n) representable by a series as above? The answer, in general, is "NO".

For instance it is not so for $X_n = Z_n + \frac{1}{n}$, where $P(Z_n = 1) = P(Z_n = -1) = \frac{1}{2}$.

However, for $X_n = Z_n$ and for $X_n = Z_n + n$ the answer is "Yes". The explanation is the following. Consider the law μ of the sequence $Z_1, Z_2, \ldots$ as a probability on R^T, $T = \{1, 2, 3, \ldots\}$. Then, in our terminology, the kernel J_μ is

$$J_\mu = \Big\{x \in R^T : \ \sum {x_n}^2 < \infty\Big\} = l_2.$$

However

$$S_\mu = \bigcap \{L \subset R^T, \quad L - \text{linear}, \ L - \text{measurable}, \ \mu(L) = 1\} = R_0^T.$$

The criterion here is the following [28]:

Let $X_n = Z_n + b_n$. Then every linear measurable functional is representable by a series iff $(b_n) \notin l_2 \setminus R_0^T = J_\mu \setminus S_\mu$.

One can prove the following general result ([29]).

Theorem. *Let μ be a symmetric tight probability on a locally convex quasi-complete space E. Assume that every μ-measurable linear functional on E is approximable μ. - a.e. by a sequence of elements of E'. Then the same property holds for $\mu_x = \mu * \delta_x$ iff $x \notin J_\mu \setminus S_\mu$. Since for μ - Gaussian $J_\mu = S_\mu$, so if Z_n were Gaussian in the example 4.1. then the answer would be always "Yes".*

Example 4.2.

Let $\mathcal{S} \subset L^2(R) \subset \mathcal{S}^*$ be the Hida triple ([13]). Let μ be a probability on $\mathcal{S}^*$ with

$$\widehat{\mu}(\xi) = \exp\Big\{-\frac{1}{2}\int_R \xi(t)^2 dt + \int_R \int_R \Big(e^{i\xi(t)u} - 1 - \frac{i\xi(t)u}{1+u^2}\Big) d\eta(u) dt\Big\},$$

where

$$\int \frac{u^2}{1+u^2} d\eta(u) < \infty \quad \text{and} \quad \int u^2 d\eta(u) < \infty.$$

Then for any $w \in S^*$ either $\mu_w \approx \mu$ or $\mu_w \perp \mu$. Moreover $\mu_w \approx \mu$ iff $w \in L^2(R)$.

The above result is taken from [17] (Th.1).

The proof is based on the following lemma.

Lemma. *Let $\mu = \rho * \lambda$ be a tight probability on a locally convex Frechet space E. Suppose that*

(i) ρ is symmetric;

(ii) $AS(\mu) = AS(\mu * \widetilde{\lambda}) = J_{\mu * \widetilde{\lambda}}$, where

$$\widetilde{\lambda}(A) = \lambda(-A), \ \forall A \in B(E).$$

Then $\forall x \in E \quad \mu_x \approx \mu \quad$ or $\mu_x \perp \mu$.

Example 4.3. ([6]).

Theorem. *Let $X = \{X(t), \ t \in T\}$ be a pth order process with $0 < p \leq 2$. Then all admissible shifts of X belong to the following space $F(X)$:*

$$F(X) = \left\{ s \in R^T : \ \sup \frac{\left| \sum_{n=1}^{N} a_n s(t_n) \right|}{\left(E \left| \sum_{n=1}^{N} X(t_n) \right|^p \right)^{\frac{1}{p}}} < \infty \right\}$$

Proof:

It is easy to see that $F(X) = (R_0^T, s_p)'$, where s_p is the topology of the convergence in p^{th} moments.

Hence denoting $\mu = \mathcal{L}(\mathcal{X})$, $AS(\mu) \subset J_\mu \subset F(X)$.

Example 4.4.

There is a deep connection between notions of kernel and pre-support and those of Radonifying and absolutely summing operators.

For instance the celebrated Duality Principle of Schwartz [18], [22] says, roughly, that if a linear operator maps a kernel of a full cylinder measure μ to a support of a continuous measure $\mu \circ T^{-1}$ then the transpose of T is Radonifying. Let us quote the following recent result [30]:

Theorem. *Let E and F be Banach spaces and T be a continuous linear mapping from E into F.*

Suppose that E is separable and of type 2. Then the following statements are equivalent:

(1) There exists a Borell probability measure on E which is quasi-invariant with respect to $T(F)$;

(2) There exists a Borell probability measure μ on E such that $T(F) \subset J_\mu$;

(3) $T^* : E^* \longrightarrow F^*$ is 2-summing

(4) $T^* : E^* \longrightarrow F^*$ is Radonifying.

5. Concluding remarks

The kernel of a stochastic process with the law μ defined as $(R_0^T, S_\mu)'$ is a natural lower bound for "nice" supports of μ and upper bound for admissible shifts of μ. The theory of this kernel for probability measures (or even cylinder measures) on locally convex spaces is well developed but its useful specification for the use of the theory of stochastic processes it is still an open task.

6. References

1. A. Badrikian, *Séminaire sur les fonctions aléatoires linéaires at les mesures cylindriques,* Springer's LNM **139** (1970).
2. A. Badrikian, S. Chevet, *Mesures Cylindriques, Espaces de Wiener et Fonctions Aléatoires Gaussiennes,* Springer's LNM **379** (1974).
3. P. Billingsley, *Probability and Measure,* John Wiley & Sons, (1979).
4. C. Borell, *Random linear functionals and subspaces of probability one,* Arkiv for matematik **14** (1976), 79-92.
5. C. Borell, *Gaussian Radon measures on locally convex spaces,* Math.Scand. **38** (1976), 265-284.
6. S. Cambanis, M. Marques, *Admisible and singular translates of stable processes,* Center for Stochastic Processes, (Chapel Hill, North Carolina, Technical Report 201,August 1987).
7. S. Chevet, *Kernel associated with a cylindrical measure,* Springer's LNM **860** (1981), 51-84.
8. R. M. Dudley, *Singularity of measures on linear spaces,* Z.Wahrsch.Verw.Gebiete **6** (1966), 129-132.
9. X. Fernique, *Regularité des trajectoires des fonctions aléatoires Gaussieanes,* Springer's LNM **480** (1974).
10. L. Gross, *Harmonic Analysis on Hilbert Space,* Memoirs AMS **46** (1963).
11. L. Gross, *Abstract Wiener Spaces,* (Prac. 5th Berkeley Sym. Math. Stat. Prob.) **2** (1965), 31-42.
12. J. Hoffmann-Jørgensen, *Integrability of semi-norms the 0-1 law and the affine kernel for product measures,* Studia Math. **61** (1977), 137-159.
13. T. Hida, *Brownian Motion,* Application of Math. 11, Springer, (1980).

14. J. Horvath, *Topological vector spaces and distributions,* Addisson-Whesley, (1966).
15. G. Kallianpur, *The role of reproducing kernel Hilbert space in the study of Gaussian processes,* Advances in Prob. and related topics **2** (1970), 49-83.
16. H. H. Kuo, *Gaussian measures on Banach spaces,* Springer's LNM **463** (1975).
17. H. H. Kuo, W. Smoleński, *On admissible shifts of generalized white noises,* J.Multiv.Analysis **12** (1982), 80- 88.
18. S. Kwapień, *On a theorem of L.Schwartz and its applications to absolutely summing operators,* Studia Math. **38** (1970), 193-201.
19. S. Kwapień, W. Smoleński, *On the nuclearity of dual space with convergence in probability topology,* Z.Wahrsch. Verw.Gebiete **59** (1982), 197-201.
20. D. H. Mouchtari, *On almost sure convergence in linear spaces of random variables,* Theory Prob. Appl. **15** (1970), 337- 342.
21. Y. Okazaki, Y. Takahashi, *Nuclear subspaces of L^0 and the kernel of a linear measure,* J. Mult. Analysis **22** (1987), 65-73.
22. L. Schwartz, *Applications p-radonifiantes et théorème de dualité,* Studia Math. **38** (1970), 203-213.
23. W. Słowikowski, *Pre-supports of linear probability measures and linear Lusin-measurable functionals,* Dissertationes Math. (Rozprawy Matematyczne) **93** (1972), 1- 43.
24. W. Słowikowski, *The second quantization, the stochastic integratrion and measures in linear spaces,* Aarhus Universitet, Matematisk Institut, Preprint Series (1976/77) No.5.
25. W. Smoleński, *Pre-supports and kernels of probability measures in Frechet spaces,* Demonstratio Math. **10** (1977), 751-762.
26. W. Smoleński, *A new proof of the zero-one law for stable measures,* Prac. AMS **83** (1981), 398-399.
27. W. Smoleński, *Linear Lusin-measurable functionals in case of a continuous cylinder measure,* Ann.Inst.Henri Poincare **19** (1983), 311-321.
28. W. Smoleński, *On the approximation of measurable linear functionals,* Stat.Prob. Letters **3** (1985), 205-207.
29. W. Smoleński, *On the Riesz property of a shifted measure,* Bull. Polish Acad. Sc. **33** (1985), 659- 661
30. Y. Takahashi, *On the relation between Radonifying mappings and kernels of probability measures on Banach Spaces,* Hokaido Math. Journal **14** (1985), 97-106.
31. Z. Suchanecki, *Linear kernels of p−stable measures,* Prob.Math.Stat. **8** (1987), 77-80.
32. A. Tortrat, *Sur la comparaison d'une mesure μ dans un espace vectorial X avec ses translatées,* Ann.Inst.H.P.**14** (1978), 61-83.

CONSTRUCTION OF GAUGE FIELDS IN FOUR DIMENSIONS FROM GROUND STATE PROCESSES

R. Vilela Mendes
CFMC - Instituto Nacional de Investigação Científica
Av. Gama Pinto, 2 - 1699 Lisboa Codex

ABSTRACT

Stochastic processes are constructed from ground state functionals, for abelian and non-abelian gauge fields without fermions.

Using lattice versions, the processes are shown to lead to well defined quantum theories, in the Schrödinger representation.

Two types of functionals are considered. For the non-abelian process that generalizes in a gauge invariant manner the abelian ground state, a qualitative analysis implies quasi-free behaviour at short distances and maximally disordered field strength configurations for the low momentum components.

Scaling relations for the mass gap are inferred from the theory of small random perturbations of dynamical systems.

1. INTRODUCTION

The construction of quantum theory through ground state measures and the corresponding Dirichlet forms has its historical origin in the work of Coester and Haag[1] for quantum fields. However it is in the quantum theory of systems with finitely many degrees of freedom[2,3], not in quantum fields, that it has been most sucessful. Not only is one able to deal with singular interactions that cannot be consistently described by potentials, but also the stochastic process provides a natural framework for the evaluation of non-perturbative effects[4,5]. The singular nature of quantum fields and the fact that the formal definition of an infinite-dimensional theory through Hamiltonian or Lagrangian densities is an ambiguous specification[6], makes quantum fields natural candidates for a description through ground state measures.

Euclidean fields are defined by a probability measure $d\mu$ on the space of vector-valued distributions $\prod_{\otimes\alpha} \mathcal{D}'(\mathbf{R}^d)$, d being the dimension of space-time and α denoting the tensor and internal degrees of freedom of the fields. The densities S_n of the n-th order moments of $d\mu$ are the Schwinger functions. From the path space $L_2(\prod_{\otimes\alpha} \mathcal{D}'(\mathbf{R}^d), d\mu)$ for quantum operators, reflection positivity yields[7] the positivity of the inner product in the Hilbert space where Minkowski fields act, and thus the existence of the Schrödinger representation

$$\mathcal{H} = L_2 \left(\prod_{\otimes\alpha} \mathcal{D}'(\mathbf{R}^{d-1}), d\nu\right)$$

where $d\nu$, the measure defined by the ground state of the Hamiltonian operator, is the restriction of $d\mu$ to the time t=0 subspace.

Conversely, given a measure $d\nu$ on the space of fixed time field configurations one defines the form

$$\mathcal{E}(u,v) = \int Du.Dv \, d\nu$$

where Du denotes the generalized gradient in the space E of t=0 field configurations. If $\mathcal{E}$ is a closable form, its closure $\overline{\mathcal{E}}$ is a Dirichlet form[8]. Then, there is a positive definite self-adjoint operator H in $L_2(E,d\nu)$ associated to $\overline{\mathcal{E}}$ by $\overline{\mathcal{E}}(u,v)=(\sqrt{H}\, u, \sqrt{H}\, v)$. $T_t=\exp(-tH)$ is a symmetric contraction semigroup and a Markov process $((X_t)_{t\geq o}, (P_z)_{z\in E})$ with state space E will be associated to $\overline{\mathcal{E}}$ if for any bounded measurable function u on E

$$\int u(X_t) \, dP_z = T_t \, u(z) \qquad \text{for} \quad \mu\text{-a.e.} \quad z\in E$$

Then $d\nu$ is the invariant measure for $(X_t)_{t\geq o}$. Therefore closability of the form defined by the measure $d\nu$ allows the reconstruction of an Hilbert space and a Hamiltonian operator. Schwinger functions are associated to the stochastic correlations of the Markov process X_t.

For quantum fields one faces, in this construction, the difficult mathematical problems of closability of infinite dimensional Dirichlet forms and the associated existence problem for the infinite dimensional Markov process. It is only very recently that reasonable progress has been made towards a satisfactory formulation of the necessary tools[9-12].

The existence problem of the Markov process associated to $\bar{\mathcal{E}}$ may be turned around by defining the theory directly in terms of a stochastic differential equation. The Hamiltonian is the generator of the process and the problem is then one of proving the existence of solutions and existence of an ergodic invariant measure, i. e. a measure for a unique ground state. This point of view is perhaps the most appropriate one for quantum fields[13].

An additional difficulty in the application of the ground state process method to quantum fields is the non-existence of non-trivial physically reasonable examples of ground state functionals, which might serve as a guide to the type of concrete mathematical problems one has to face in the construction of physical theories. In this paper some attempts to deal with this problem, for gauge fields, are described. In Sect.2 the abelian (free field) case is reviewed and in Sect.3 after describing a general framework for the construction of gauge invariant non-abelian functionals, two concrete examples are studied. More details may be found in Refs.[13] and [14].

2. ABELIAN GAUGE FIELDS

Consider continuum QED (without fermions) in the Schrödinger picture and in the temporal gauge, $A_o(x)=0$[15]. The Hamiltonian operator is

$$H = \frac{1}{2} \int d^3x \quad \{- \frac{\delta^2}{\delta A_i(x)^2} + B_i(x)^2 \} \tag{1}$$

with the magnetic field $B_i(x)=\epsilon_{ijn}\ \partial_j\ A_n(x)$. Gauss' law $\partial_i F^{oi}=0$ is imposed as a constraint, $\partial_i \frac{\delta}{i\delta A_i(x)} \psi[A]=0$, and is equivalent to require gauge invariance of the wave functionals.

It is convenient to use both the configuration space fields and their Fourier transforms

$$A_i(k) = (\tfrac{1}{2\pi})^{\frac{3}{2}} \int d^3x\ e^{-ik.x}\ A_i(x)$$

In these coordinates the Hamiltonian becomes

$$H = \tfrac{1}{2} \int d^3k \ \{- \frac{\delta}{\delta A_i(k)} \frac{\delta}{\delta A_i(-k)} + B_i(k)\ B_i(-k)\ \} \tag{2}$$

with $B_j(k)=-i\ \epsilon_{jmn}\ k_m\ A_n(k)$.

Up to a constant the Hamiltonian of Eq.(2) admits two decompositions in the form

$$H + \text{const}=\tfrac{1}{2}\int d^3k\ \{- \frac{\delta}{\delta A_i(k)} + L_i(-k)\}\{\frac{\delta}{\delta A_i(-k)} + L_i(k)\} \tag{3}$$

where the $L_i(k)$ are gradients of a functional of the fields, $L_i^{(\alpha)}(k)=-\frac{\delta}{\delta A_i(-k)}\sigma^{(\alpha)}[A]$. They are[15-18]

$$L_i^{(1)}(k) = -i\ \epsilon_{imn}\ k_m\ A_n(k) \tag{4a}$$

$$L_i^{(2)}(k) = |k|\ A_i(k) - \frac{k_i k_j}{|k|}\ A_j(k) \tag{4b}$$

with

$$\sigma^{(1)}[A] = \tfrac{i}{2} \int d^3k \quad \epsilon_{imn}\ A_i(-k)\ \ k_m\ \ A_n(k) \tag{5a}$$

$$\sigma^{(2)}[A]=-\tfrac{1}{2}\int d^3k\ \{\ A_i(-k)\ |k|\ A_i(k)-A_i(-k)\ \frac{k_i k_j}{|k|}\ A_j(k)\ \} \tag{5b}$$

A decomposition of the type of Eq.(3) means that $\phi_{(\alpha)}=e^{\sigma^{(\alpha)}[A]}$ are eigenstates of H. Making the transformation $H\to H'_{(\alpha)}= \phi^{-1}_{(\alpha)}\ H\ \phi_{(\alpha)}$ one obtains

$$H'_{(\alpha)} = \int d^3k\ \ \{-\tfrac{1}{2} \frac{\delta}{\delta A_i(k)} \frac{\delta}{\delta A_i(-k)} + L_i^{(\alpha)}(k) \frac{\delta}{\delta A_i(k)}\ \} \tag{6}$$

implying that $-H'_{(\alpha)}$ is the generator of a diffusion process

$$dA_i^{(\alpha)}(k) = -L_i^{(\alpha)}(k)\ \ dt + dW_i(t) \tag{7}$$

with drift $-L_i^{(\alpha)}(k)$ and invariant density $e^{2\sigma^{(\alpha)}[A]}$ (in the space of fixed time field configurations).

Of course, the correcteness of the above statements presumes that one can carry to the infinite dimensional setting of fields the corresponding manipulations of finite dimensional analysis. The justification of these constructions may be carried out using a lattice and approaching the continuum limit through scaling relations[13], or directly by infinite dimensional techniques[12,19] which, at least in the (free field) abelian gauge field case, are expected to suceed.

The wave functional $\phi_{(2)}$ is the usual perturbative ground state corresponding to an infinite number of harmonic modes. The corresponding stochastic process describes fluctuations around the $\{A_i(k)=0\}$ configuration, which is an attractive fixed point for the classical (noiseless) motion. By contrast for the process associated to $\phi_{(1)}$ the $\{A_i(k)=0\}$ configuration is an hyperbolic (unstable) fixed point.

These two functionals have an interesting geometrical relation. The drift $-L_i=\frac{\delta\sigma^{(\alpha)}}{\delta A_i}$ associated to any wave functional $e^{\sigma^{(\alpha)}}$, satisfying Gauss' law constraint, must be orthogonal to an arbitrary gauge transformation

$$\delta A_i(k) = i\, k_i\, \delta\alpha(k) \tag{8a}$$

One finds that $L_i^{(2)}(k)$ is the exterior product

$$\vec{L}^{(2)}(k) = \delta\vec{A}(k) \times \vec{L}^{(1)}(k)$$

for $\delta\alpha(k) = -\frac{1}{|k|}$.

3. NON-ABELIAN GAUGE FIELDS

For non-abelian gauge fields the Gauss' law constraint is also equivalent to impose gauge invariance on the wave functionals, i. e. invariance for

$$\delta A_i^\alpha(x) = (D_i)_{\alpha\beta}\ \delta\alpha^\beta = (\partial_i\ \delta_{\alpha\beta} - g\ f_{\alpha\gamma\beta}\ A_i^\gamma(x))\ \ \delta\alpha^\beta(x) \tag{8b}$$

A drift vector $-L_i^\beta(x)=\frac{\delta\sigma}{\delta A_i^\beta(x)}$ will be orthogonal to an arbitrary $\delta A_i^c(x)$ if

$$\int d^3x\ L_i^\beta\ (D_i\ \delta\alpha)^\beta = 0\ ,\ \text{i. e.}$$

$$(D_i\ L_i(x))^\alpha = (\partial_i\ \delta_{\alpha\beta} - g\ f_{\alpha\gamma\beta}\ A_i^\gamma(x))\ \ L_i^\beta(x) = 0 \tag{9}$$

The general solution to Eq.(9) is

$$L_i^\alpha(x) = \left(\frac{1}{1-M}\right)_{i\alpha,k\beta}\ \Gamma_k^\beta \tag{10}$$

where $M_{i\alpha,k\beta} = g\ f_{\alpha\gamma\beta}\ \frac{\partial_i}{\Delta}\ A_k^\gamma(x)$ and Γ_k^β is any solution of $\partial_k\ \Gamma_k^\beta=0$.

Eq.(10) may be used to construct general gauge invariant diffusion processes for non-abelian gauge fields. Here we will limit ourselves to the non-abelian generalizations of $L_i^{(1)}$ and $L_i^{(2)}$ of Eqs.(4). For this purpose we rewrite the exponents $\sigma^{(\alpha)}[A]$ of the corresponding wave functionals $(\exp(\sigma^{(\alpha)}))$ in terms of the configuration space fields and in a form that exhibits their gauge transformation properties.

$$\sigma^{(1)}[A] = \tfrac{1}{2}\int d^3x\ \ \epsilon_{oijk}\ A_i(x)\ \partial_j\ A_k(x) \tag{5c}$$

$$\sigma^{(2)}[A] = -\frac{1}{4\pi^2}\int d^3x\ d^3y\ \ B_i(x)\ \frac{1}{|x-y|^2}\ \ B_i(y)$$

$$= -\tfrac{1}{2}\int d^3x\ B_i(x)\ \frac{1}{\sqrt{-\Delta}}\ B_i(x) \tag{5d}$$

$\sigma^{(1)}[A]$ generalizes easily to the non-abelian case as

$$\sigma^{(1)}[A]=\int d^3x\ \epsilon_{oijk}\mathrm{Tr}\Big\{\underset{\sim}{A}_i(x)\Big(\partial_j\ \underset{\sim}{A}_k(x)+i\tfrac{2}{3}\ \underset{\sim}{A}_j(x)\ \underset{\sim}{A}_k(x)\Big)\Big\} \tag{11}$$

$\sigma^{(1)}[A]$ is proportional to the Pontryagin index of the field configuration and leads to a gauge invariant drift even for topologically non-trivial transformations. This "winding number functional" first mentioned in Loos[18], has been discussed by several authors[15,20]. As in the abelian case the $\{A_i^\alpha=0\}$ configuration is an unstable fixed point, and in the continuum formulation with unbounded variables it is not possible to use the corresponding drift

$$-L_{i,\alpha}^{(1)}=\frac{\delta\sigma^{(1)}}{\delta A_i^\alpha}$$

to define a recorrent process with an ergodic ground state measure. However if the variables are compactified in a lattice version[13], existence (and uniqueness) of the solutions to the stochastic differential equation may be proved, as well as existence of a unique invariant measure and a well defined scaling limit at weak coupling. Although this process is unrelated to the perturbative ground state it nevertheless defines a well behaved quantum theory.

Physically more interesting is probably the non-abelian generalization[21] of $\sigma^{(2)}[A]$

$$\sigma^{(2)}[A] = -\tfrac{1}{2}\int d^3x\ B_k^\alpha(x)\left(\frac{1}{\sqrt{-D_iD_i}}\right)_{\alpha\beta} B_k^\beta(x) \tag{12}$$

For Eq.(12) to make sense a meaning has to be assigned to $(-D_iD_i)^{-\frac{1}{2}}$. The abelian version $(-\Delta)^{-\frac{1}{2}}$ becomes well defined through the Fourier transform. A similar transform does not exist for the non-abelian case because the operator set $\{D_i, i=1,2,3\}$ has a common system of eigenvalues only if A_i^α is a gauge transform of $A_i^\alpha(x)=0$.

Due to the antisymmetry of the structure constants, in the space of vector-valued functions of compact support in $\mathbf{R}^3$ with norm

$$(u,v) = \int d^3x\ u^\alpha(x)\ v^\alpha(x)$$

one has $(u,D_iv) = -(D_iu,v)$. Therefore one may use the integral representation for fractional powers of positive operators[23] and define

$$\frac{1}{\sqrt{-D_iD_i}} = \frac{1}{\pi}\int_0^\infty \lambda^{-\frac{1}{2}}\ \frac{1}{\lambda-D_iD_i}\ d\lambda \tag{13}$$

Using (13) in Eq.(12) and a formal series expansion of $\frac{1}{\lambda-D_iD_i}$ to compute the functional derivative, one obtains

$$\frac{\delta\sigma^{(2)}}{\delta A_i^\alpha(x)} = -\epsilon_{ijk}(D_j)_{\alpha\beta}\left(\frac{1}{\sqrt{-D_sD_s}}B_k\right)^\beta + \frac{g}{\pi}\int_0^\infty \lambda^{-\frac{1}{2}}\left(\frac{1}{\lambda-D_mD_m}B_k\right)^\beta\left(F_\alpha D_i\frac{1}{\lambda-D_sD_s}B_k\right)^\beta d\lambda \tag{14}$$

for the drift of the process. The quantities involved in Eq.(14) are

$$(D_i\ )_{\alpha\beta} = \partial_i\ \delta_{\alpha\beta} - g\ f_{\alpha\gamma\beta}\ A_i^\gamma(x)$$

$$B_k^\alpha = \epsilon_{kmn} \left(\partial_m A_n^\alpha(x) - \frac{g}{2} f_{\alpha\beta\gamma} A_m^\beta(x) A_n^\gamma(x) \right)$$

$$(F_\alpha)_{\beta\gamma} = f_{\beta\alpha\gamma}$$

The task now is to obtain a precise meaning for the stochastic process with drift given by (14). We use a lattice formulation, define the variables $\theta^\alpha(n,n+\hat{i})$ corresponding to the scaled fields $A_i^\alpha(n)$ and make the replacements

$$g\, a\, A_i^\alpha(n) \to \theta^\alpha(n,n+\hat{i}) \tag{15a}$$

$$ga^2 B_i^\alpha(n) \to \beta_i^\alpha(n) = \tfrac{1}{4}\gamma_{ijk}\Big(\theta^\alpha(n+\hat{j},n+\hat{j}+\hat{k}) - \tfrac{1}{2} f_{\alpha\beta\gamma}\theta^\beta(n,n+\hat{j})\theta^\gamma(n,n+\hat{k})\Big) \tag{15b}$$

where $\gamma_{ijk} = (\text{sign } i)(\text{sign } j)(\text{sign } k)\ \epsilon_{|i||j||k|}$; $i,j,k \in \{1,2,3\}$, $\hat{i}$ denotes the unit lattice vector along the i-direction and a is the lattice spacing.

On the other hand the covariant derivative of a quantity $v^\alpha(n)$ transforming under the adjoint representation of the internal symmetry group, becomes

$$(D_i)_{\alpha\beta} v^\beta(n) \to \tfrac{1}{a}\{\tfrac{1}{2}(v^\alpha(n+\hat{i}) - v^\alpha(n-\hat{i})) - f_{\alpha\gamma\beta}\ \theta^\gamma(n,n+\hat{i})\ v^\beta(n)\}$$

$$= \tfrac{1}{a}\ (\mathfrak{D}_i)_{\alpha\beta}\ v^\beta(n) \tag{16}$$

With these definitions and using Eq.(13) the lattice version of $\sigma^{(2)}[A]$ is

$$\sigma^{(2)}[\theta] = -\frac{1}{2\pi g^2} \sum_n \int_0^\infty d\lambda\ \lambda^{-\frac{1}{2}}\ \beta_k^\alpha(n) \left(\frac{1}{\lambda - \mathfrak{D}_i\mathfrak{D}_i}\right)_{\alpha\gamma} \beta_k^\gamma(n) \tag{17}$$

and for the drift of the lattice process one obtains

$$-L(\theta^\alpha(n,n+\hat{i})) = \frac{1}{\pi a} \int_0^\infty d\lambda\ \lambda^{-\frac{1}{2}} \Bigg\{ -\epsilon_{ijk}\ (\mathfrak{D}_j)_{\alpha\gamma} \left(\frac{1}{\lambda - \mathfrak{D}_s\mathfrak{D}_s}\ \beta_k(n)\right)^\gamma +$$

$$\left(\frac{1}{\lambda - \mathfrak{D}_s\mathfrak{D}_s}\ \beta_k(n)\right)^\gamma \left(F_\alpha\ \mathfrak{D}_i\ \frac{1}{\lambda - \mathfrak{D}_m\mathfrak{D}_m}\ \beta_k(n)\right)^\gamma \Bigg\} \tag{18}$$

Notice that, for simplicity, we have written for $-L(\theta^\alpha(n,n+\hat{i}))$ the lattice version of Eq.(14) rather than recomputing

$$\frac{g^2}{a}\ \frac{\delta\sigma^{(2)}[\theta]}{\delta\theta^\alpha(n,n+\hat{i})}$$

The two quantities should however coincide in the continuum limit.

We have now two ways to study the lattice model. The first takes the energy form associated to the density $\rho(\theta)=\exp(2\sigma^{(2)}[\theta])$

$$E(u,v) = \int \prod_{n,i,\alpha} d\theta^{\alpha}(n,n+\hat{i}) \quad \rho(\theta) \quad u(\theta) \quad v(\theta) \tag{19}$$

and proceeds, through the proof of closability, to the existence of a positive self-adjoint generator of time translations and a Markov process.

The second studies directly the stochastic differential equation

$$d\theta^{\alpha}(n,n+\hat{i}) = - L(\theta^{\alpha}(n,n+\hat{i}))\, dt + \frac{g}{\sqrt{a}}\, dW^{\alpha}_{t}(n,n+\hat{i}) \tag{20}$$

trying to insure existence of solutions and an invariant measure.

In both cases to insure the existence of a well defined Markov process is an essential step to be able to use the (non-perturbative) Wentzell and Freidlin technique[24,6] to infer the continuum behavior through scaling relations for the mass gap.

Before showing that $\exp(2\sigma^{(2)}[\theta])$ as the density of a Dirichlet form (or (18) as the drift of a stochastic differential equation) lead to a well defined quantum theory, we examine the physical significance of the ground state wave functional $\exp(\sigma^{(2)}[A])$ (with $\sigma^{(2)}[A]$ as in Eq.(12)).

For field configurations for which $A^{\alpha}_{i}(x)$ varies slowly in space, the effect of the derivative terms in the denominator of $\frac{1}{\lambda - D_i D_i}$ is small and the operator becomes a local one. I. e. in the neighbourhood of these slowly varying field configurations the ground state wave functional factors out into a product $\prod_{x} e^{\Gamma(x)}$ where each $\Gamma(x)$ depends only on the fields at the point x.

$$\Gamma(x) \sim -\tfrac{1}{2} \int_{0}^{\infty} d\lambda \quad \lambda^{-\frac{1}{2}}\, B^{\alpha}_{k}(x) \left(\frac{1}{\lambda - g^2\, F_{\gamma}\, F_{\xi}\, A^{\gamma}_{i}(x)\, A^{\xi}_{i}(x)} \right)_{\alpha\beta} B^{\beta}_{k}(x)$$

Considering the Fourier transform of $A^{\alpha}_{i}(x)$ it is clear that the slow varying configurations are the low momentum components that control the long distance behaviour. Hence the main correlations between field configurations at different points in space, imposed by the low momentum component of the ground state wave functional, are only those that follow from the Bianchi identity. I. e. the long distance behaviour is associated to a maximally disordered field strength configuration.

Conversely for high-momentum (short distance) the derivative terms in $\frac{1}{\lambda - D_i D_i}$ are dominant and the kernel will approach the $\frac{1}{|x-y|^2}$ of the U(1) free theory . From the qualitative limiting behaviour of the operators one already expects the $\exp(2\sigma^{(2)}[A])$ ground state measure to describe both asympotic freedom at short distances and a maximally disordered situation at low momentum.

The exponent $\sigma^{(2)}[A]$ is negative semi-definite. Therefore the maximum of the (real) ground state functional is reached for the $B_k^\alpha(x)=0$ configurations, the functional being peaked at all homotopically non-equivalent classical vacua.

For the quantum theory defined by the ground state measure $\exp\{2\sigma[\theta]\}$ (i. e. the lattice version) we now make a few exact statements and sketch their proofs.

#1."For a finite lattice with periodic boundary conditions the ground state measure $\rho[\theta]=\exp\{2\sigma[\theta]\}$ with $\sigma[\theta]$ of eq.(17) characterizes a well-defined quantum theory"

In a finite lattice with periodic boundary conditions

$$(u,\mathfrak{D}_i v) = \sum_{n,\alpha,\beta} u_k^\alpha(n)\ (\mathfrak{D}_i)_{\alpha\beta}\ v_k^\beta(n) = -\sum_{n,\alpha,\beta}\ (\mathfrak{D}_i)_{\alpha\beta}\ u_k^\beta(n)\ \ v_k^\alpha(n)$$

and $-\mathfrak{D}_i\mathfrak{D}_i$ is a positive operator.

$\sigma^{(2)}[\theta]$ being ≤ 0 , $\rho(\theta)$ is bounded.

If μ_0 is the lowest positive eigenvalue of $-\mathfrak{D}_i\mathfrak{D}_i$, the norm of the operator

$\frac{1}{\pi}\int_0^\infty \lambda^{-\frac{1}{2}} \frac{1}{\lambda-\mathfrak{D}_i\mathfrak{D}_i}\, d\lambda$ is bounded by $(\mu_0)^{-\frac{1}{2}}$. However zero eigenvalues of $-\mathfrak{D}_i\mathfrak{D}_i$ may exist and one worries about possible zeros of $\rho(\theta)$, more precisely about the measure of the singular set $S(\rho)$

$$S(\rho)=\left\{\{\theta\}:\int\rho^{-1}\prod_{n,\alpha,\hat{i}} d\theta^{\alpha}(n,n+\hat{i})=\infty \text{ for any neighbourhood of}\{\theta\}\right\}$$

i. e. the set of field configurations $\{\theta\}$ for which ρ^{-1} is not locally integrable.

Let $v_k^{\beta}(n)$ be a zero eigenvalue of $-\mathfrak{D}_i\mathfrak{D}_i$. Then from $(\mathfrak{D}_i v,\mathfrak{D}_i v)=0$ it follows :

$$\tfrac{1}{2}(v_k^{\alpha}(n+\hat{i})-v_k^{\alpha}(n-\hat{i}))-f_{\alpha\beta\gamma}\theta^{\gamma}(n,n+\hat{i})v_k^{\beta}(n)=0 \quad \forall n,\alpha,k,i \quad (21)$$

Let the lattice have N^3 sites and the Lie algebra dimensionality be d_L. Eq.(21) is a system of $9N^3d_L$ equations for $3N^3d_L$ variables. Consider the Nd_L-dimensional subsystems associated to N sites along one coordinate direction and fixed k. For generic θ the equations in these subsystems are linearly independent. If (21) has a non-trivial solution then, at least one of the subsystems should have a non-zero solution. This imposes a determinant relation on the θ coordinates that reduces the dimensionality by one unit, at least. I. e. the subset in θ-space associated to a zero eigenvalue of $-\mathfrak{D}_i\mathfrak{D}_i$ has dimensionality less than $3N^3d_L$.

Therefore, the singular set $S(\rho)$ has zero measure and the density $\rho(\theta)$ satisfies the conditions of theorem 1 in Ref.[25]. The form E(u,v) of Eq. (19) is closable, there is associated to it a diffusion process and, in the Hilbert space of square-integrable field configurations, its generator defines a positive self-adjoint Hamiltonian operator. I. e. a well defined quantum theory is associated to $\rho(\theta)$.

#2. "The constant field configurations contribution to the Wilson loop of the theory defined in #1 has area law decay".

By constant field contribution one means precisely the contribution of the amplitudes

$$\psi_o(\theta)=\exp\left\{-\frac{1}{128\pi g^2}\sum_n\int_0^\infty d\lambda\lambda^{-\frac{1}{2}}\gamma_{ijk}\ \gamma_{ij'k'}\ f_{\alpha\beta\gamma}\ \theta^\beta(n,n+\hat{j})\ \theta^\gamma(n,n+\hat{k})\right.$$

$$\left.\left(\frac{1}{\lambda-F_\xi F_{\xi'}\theta^\xi(n,n+\hat{r})\theta^{\xi'}(n,n+\hat{r})}\right)_{\alpha\alpha'} f_{\alpha'\beta'\gamma'}\ \theta^{\beta'}(n,n+\hat{j}')\ \theta^{\gamma'}(n,n+\hat{k}')\right\} \tag{22}$$

which are obtained from Eq.(17) in the constant field limit.

One can now either consider spatial Wilson loops and use ψ_o^2 as a dimensionally reduced partition function, or compute the time correlation of two strings, which is the form taken by the space-time Wilson loop in the temporal gauge.

$$W(N,T)=\langle Tr\left\{U^\dagger(N,T)\ldots U^\dagger(1,T)\ U(1,0)\ldots U(N,0)\right\}\rangle =$$
$$=\langle Tr\left\{U^\dagger(N,0)\ldots U^\dagger(1,0)\ e^{-HT}\ U(1,0)\ldots U(N,0)\right\}\rangle \tag{23}$$

Transforming the Hamiltonian

$$H=\frac{g^2}{2a}\sum_{n,i,\alpha}\left\{-\frac{\partial}{\partial\theta^\alpha(n,n+\hat{i})}-\frac{\partial\sigma^{(2)}}{\partial\theta^\alpha(n,n+\hat{i})}\right\}\left\{\frac{\partial}{\partial\theta^\alpha(n,n+\hat{i})}-\frac{\partial\sigma^{(2)}}{\partial\theta^\alpha(n,n+\hat{i})}\right\} \tag{24}$$

by the string operator one obtains

$$W(n,T)\ =\ \langle Tr\left\{\ e^{-\ \hat{H}T}\ \right\}\rangle \tag{25}$$

where $\hat{H}$ differs from H by the replacement

$$\frac{\partial}{\partial\theta^\alpha(n,n+\hat{i})}\ \rightarrow\ U^\dagger(s,0)\ \frac{\partial}{\partial\theta^\alpha(n,n+\hat{i})}\ U(s,0)$$

whenever the link $s=(n,n+\hat{i})$ belongs to the string , and remains unchanged otherwise. To compute Eq.(25) one uses (22) with

$$\theta^\alpha(n,n-\hat{i})\ =\ -\ \theta^\alpha(n,n+i)\ +\ a\ \frac{\partial\theta^\alpha}{\partial\hat{i}}$$

the contribution of the second term vanishing in the constant field limit. One then finds that the operator in Eq.(25) factors out in a product of independent site operators leading to a $\exp\{-cNT\}$ area law.

Remark : The reason why this is not a confinement result follows from the fact that constant field configurations have zero measure in the space of all configurations. The crucial step missing is, for example, proof that a positive measure neighbourhood of the constant field configurations has the same area law behaviour.

The last statement pertains to the approach to the continuum limit through scaling relations :

#3. "The mass gap of the theory defined in #1 scales as $am \sim \exp\{-c/g^2\}$ when $a \to 0$, $g(a) \to 0$ ".

The drift $-L$ (Eq.(18)) of the lattice process is independent of g, therefore the model satisfies the necessary condition for the mass gap to have the stated behaviour[7] when $g \to 0$. The proof of statement #3 is based on the asymptotics of the lowest eigenvalue of elliptic operators, as obtained from the theory of small random perturbations of dynamical systems[24,6]. The application of the small random perturbations theory is relevant to the continuum limit only if $a \to 0$ implies also $g(a) \to 0$. We refer to Ref.[6] for details on the application of this technique to lattice problems.

A sufficient condition for $am \sim \exp\{-c/g^2\}$ when $g \to 0$ is that the ω-limit set of the deterministic problem

$$\frac{d\theta^{\alpha}(n,n+\hat{i})}{dt} = - L(\theta^{\alpha}(n,n+\hat{i}))$$

be attractive in the domain of the Dirichlet problem. This is fulfilled because $\sigma^{(2)}$ is negative definite with maxima in the $\beta=0$ field configurations.

From $am \sim \exp\{-c/g^2\}$ one sees that for the physical mass gap to remain fixed when $a \to 0$, one should require $g^2(a) \sim \left|\frac{c}{\log a}\right|$. Therefore when $a \to 0$ $g(a) \to 0$, consistent with the use of the theory of small random perturbations.

REFERENCES AND FOOTNOTES

[1]F. Coester and R. Haag; Phys. Rev.117 (1960) 1137.

[2]S. Albeverio, R. Høegh-Krohn, L. Streit; J. Math. Phys. 18(1977) 907.

[3]S. Albeverio, M. Fukushima, W. Karwowski, L. Streit; Commun. Math. Phys. 81 (1981) 501.

[4]G. Jona-Lasinio, F. Martinelli, E. Scoppola; Phys. Rep. 77 (1981) 313 ; Commun. Math. Phys. 80 (1981) 233.

[5]S. M. Eleutério, R. Vilela Mendes; J. Phys. A20 (1987) 6411.

[6]See for example the discussion in R. V. Mendes; Phys. Rev. D37 (1988) 1608.

[7]See for example J. Glimm, A. Jaffe; "Quantum physics. A functional integral point of view",ch.6,Springer,N. Y.,1987.

[8]M. Fukushima; "Dirichlet forms and Markov processes", Kodansha and North-Holland, 1980.

[9]S. Kusuoka; J. Fac. Science, Univ. Tokyo, Se. 1A, 29 (1982) 79.

[10]S. Albeverio, M. Röckner; "Classical Dirichlet forms on topological vector spaces -closability and a Cameron-Martin formula" Bielefeld BiBoS preprint 312/88 and references therein.

[11]S. Albeverio, M. Röckner; "Dirichlet forms, quantum fields and stochastic quantisation" Bochum SFB237 preprint 15/88.

[12]T. Hida, J. Potthoff, L. Streit; Commun. Math. Phys. 116 (1988) 235.

[13]R. Vilela Mendes; Phys. Lett. B202 (1988) 369.

[14]R. Vilela Mendes; Phys. Lett. B223 (1989) 83.

[15]J. P. Greensite; Nucl. Phys. B158 (1979) 469.

[16]J. A. Wheeler; "Geometrodynamics",Academic Press,N.Y.,1962.

[17]A. Katz; Il Nuovo Cimento 37 (1965) 342.

[18]H. G. Loos; Phys. Rev. 188 (1969) 2342.

[19]S. Albeverio, T. Hida, J. Potthoff, L. Streit; "The vacuum of the Høegh-Krohn model as a generalized white noise functional", Bielefeld BiBoS preprint 322/88.

[20] M. Claudson and M. B. Halpern; Phys. Lett. B151 (1985) 281.

[21] Previous attempts[15,22] to generalize the abelian ground state functional $e^{\sigma^{(2)}[A]}$ start from

$$\sigma^{(2)}[A] = -\frac{1}{4\pi^2}\int d^3x\, d^3y \quad B_k(x) \quad \frac{1}{|x-y|^2} \quad B_k(y)$$

and attempt to generalize the kernel $|x-y|^{-2}$. The problems encountered by these authors are probably related to the difficulty in finding a well defined explicite form for the (symbolic) kernel $\left(\frac{1}{\sqrt{-D_iD_i}}\right)\delta^3(x-y)$.

[22] R. P. Feynman; Nucl. Phys. B188 (1981) 479.

[23] K. Yosida; "Functional analysis", Ch. IX, Springer, Berlin, 1974.

[24] M. I. Freidlin, A. D. Wentzell ; "Random perturbations of dynamical systems", Springer, N. Y., 1984.

[25] M. Fukushima ; "Energy forms and diffusion processes" in "Mathematics + Physics" vol. 1, page 65, ed. L. Streit, World Scientific, Singapore, 1985.

POSITIVE GENERALIZED BROWNIAN FUNCTIONALS

YOSHITAKA YOKOI

Department of Mathematics, Faculty of General Education,
Kumamoto University, Kumamoto, JAPAN

ABSTRACT

The linear form corresponding to every positive generalized Brownian functional is expressed in terms of the integral form by a positive finite measure defined on the space of all real tempered distributions. The renormalizations like : $\exp[c \int \dot{B}(u)^2 du]$: $(c < 1/2)$ or : $\exp[\lambda \dot{B}(t)]$: $(\lambda, t \in \mathbf{R})$ correspond to Gaussian measures which are singular with respect to the original one.

§1. INTRODUCTION.

Let E be the Schwartz nuclear space which consists of all real valued, C^∞-, and rapidly decreasing functions. The functional on E

$$C(\xi) = \exp[-\frac{1}{2}\int_{\mathbf{R}} |\xi(u)|^2 du] \quad (\xi \in E)$$

induces Gaussian measure μ on the topologically dual space E^* of E ([3],[21]). That is, the characteristic functional $C(\xi)$ of μ is given by

$$\int_{E^*} e^{i<x,\xi>} d\mu(x) = \exp[-\frac{1}{2}\int_{\mathbf{R}} |\xi(u)|^2 du] \text{ for } \xi \in E.$$

The elements of $(L^2) \equiv L^2(E^*, \mu)$ are called *Brownian functionals* or *white noise functionals* ([7]). In Hida's paper [6], one can see the idea of generalized Brownian functionals constructed on the theories of Sobolev spaces and Gel'fand triplets. Not a few papers have been published treating the analysis of ordinary or generalized Brownian functionals along the idea of [6], e.g., [7],[8], [15]~ [18], [20], [21], [23] and [26].

In this paper, we will treat the problem of *positive* generalized Brownian functionals and will show that every positive generalized Brownian functional, as a continuous linear form on a suitable space of test Brownian functionals, is expressed in terms of the integral form by some positive finite measure on E^*. This is an improvement of Potthoff's result (THEOREM 3. 2 and its *Remark* in [23]).

Relating to this problem, we should mention Sugita's important work ([27]). He has already obtained, in the framework of Malliavin calculus, that every positive generalized *Wiener* functional in the abstract Wiener space is associated with a measure. The result in this paper resembles his very much. However, Malliavin calculus and white noise analysis (Hida calculus) have their own purposes. For instance, Hida calculus can naturally explain the functionals of $\dot{B}(t)$, e.g., a kind of renormalizations like

$$:\exp[\lambda\dot{B}(t)]: \quad \text{for } \lambda, t \in \mathbf{R} \quad \text{or} \quad :\exp[c\int_{\mathbf{R}}\dot{B}(u)^2du]: \quad \text{for } c < \frac{1}{2}.$$

Let $\mathcal{S}(\mathbf{R}^n)$ be the Schwartz space of complex valued C^∞ functions of rapid decrease. As to the triplet $\mathcal{S}(\mathbf{R}^n) \subset L^2(\mathbf{R}^n) \subset \mathcal{S}'(\mathbf{R}^n)$, there is a beautiful result due to L. Schwartz: Every positive generalized function $F \in \mathcal{S}'(\mathbf{R}^n)$ on the space $\mathcal{S}(\mathbf{R}^n)$ is given by a tempered measure λ_F on $\mathbf{R}^n$;

$$< F, f >= \int_{\mathbf{R}^n} f(x)d\lambda_F(x)$$

for all test functions $f \in \mathcal{S}(\mathbf{R}^n)$. In white noise analysis, after choosing a suitable Gel'fand triplet $(\mathcal{S}) \subset (L^2) \subset (\mathcal{S}')$, we can prove an analogous theorem: *Every positive generalized Brownian functional $\Psi \in (\mathcal{S}')$ is given by a finite measure ν_Ψ on E^*(= the space of all real tempered distributions);*

$$< \Psi, \psi >= \int_{E^*} \tilde{\psi}(x)d\nu_\Psi(x),$$

for all test Brownian functionals $\psi \in (\mathcal{S})$, where $\tilde{\psi}$ is a continuous version of $\psi \in (\mathcal{S})$.

We will explain the definition of *positive* generalized Brownian functionals. We will use two properties of $(\mathcal{S})$ to prove the above theorem; (1) $(\mathcal{S})$ is an algebra and (2) every functional in $(\mathcal{S})$ has a continuous version as a function defined on E^* (see [19] and [28]).

Although our argument is developed concretely for simplicity in this paper, it can be done in a little more general situation. (See [28], H.M.J.)

§2. MEASURE OF GAUSSIAN WHITE NOISE AND BROWNIAN FUNCTIONALS.

Let E be the real Schwartz space $\mathcal{S}_{real}(\mathbf{R})$ and E^* be its dual space $\mathcal{S}'_{real}(\mathbf{R})$. Let $H_k(u)$ be Hermite polynomials;

$$H_k(u) = (-1)^k \exp[u^2]\left(\frac{d}{du}\right)^k \exp(-u^2), \ k = 0, 1, 2, \cdots. \tag{2.1}$$

Write

$$\zeta_k(u) = \left(2^k k!\sqrt{\pi}\right)^{-1/2} H_k(u)\exp[-\frac{1}{2}u^2], \ k = 0, 1, 2, \cdots. \tag{2.2}$$

We call $\zeta_k(u)$ Hermite functions. Let us introduce a sequence of inner products and a sequence of norms in E :

$$(\xi, \zeta)_p = \int_{\mathbf{R}}\{D^p\xi(u)\}D^p\eta(u)du, \ \ \|\xi\|_p^2 = \int_{\mathbf{R}}|D^p\xi(u)|^2du \tag{2.3}$$

for $\xi, \eta \in E$ and $p = 0, 1, 2, \cdots$, with $D\xi(u) = \left(1 + u^2 - \left(\frac{d}{du}\right)^2\right) \xi(u)$.

If we complete the space E with respect to the norm $\|\cdot\|_p$, we obtain a real Hilbert space E_p for every $p \geq 0$. We denote the dual space of E_p by E_{-p}. Then E_{-p} is a Hilbert space with norm

$$\|x\|_{-p} = \sup\{|x(\xi)|; \|\xi\|_p \leq 1, \xi \in E\} \quad \text{for} \quad x \in E^*.$$

We note that

$$D\zeta_k = 2(k+1)\zeta_k \tag{2.4}$$

and that $\{\zeta_k\}_{k=0}^{\infty}$ is a complete orthonormal system of $L^2_{real}(\mathbf{R})$. It is readily shown that

$$\|\xi\|_p^2 = \sum_{k=0}^{\infty} \{2(k+1)\}^{2p} |(\xi, \zeta_k)_0|^2 \tag{2.5}$$

and that there exists a constant $\rho \in (0, 1)$ for which

$$\|\xi\|_p \leq \rho \|\xi\|_{p+1} \tag{2.6}$$

holds for any $\xi \in E$ and any $p \geq 0$. In fact, we can choose $1/2$ as the constant ρ. The increasing sequence

$$E \subset \cdots \subset E_p \subset \cdots \subset E_0 = L^2(\mathbf{R}; \mathbf{R}) \subset \cdots \subset E_{-p} \subset \cdots \subset E^* \tag{2.7}$$

constitutes a real rigged Hilbert space $E \subset L^2_{real}(\mathbf{R}) \subset E^*$ ([1],[2]). The inclusion mapping

$$\iota_{p,p+1} : E_{p+1} \longmapsto E_p \quad (p \in \mathbf{Z}) \tag{2.8}$$

is of Hilbert-Schmidt type and its Hilbert-Schmidt norm is

$$\|\iota_{p,p+1}\|_{H.S.} = \left(\sum_{k=0}^{\infty} \{2(k+1)\}^{-2}\right)^{1/2} = \frac{\pi}{\sqrt{24}}. \tag{2.9}$$

Since this is independent of p, we denote it simply by $\|\delta\|$. It comes from the following fact ([16]). Dirac's δ-function $\delta_u(\cdot)$ at $u \in \mathbf{R}$ belongs to E_{-1} and

$$\int_{\mathbf{R}} (\|\delta_u\|_{-1})^2 du = \|\delta\|^2.$$

By (2. 9), the rigged Hilbert space $E \subset L^2_{real}(\mathbf{R}) \subset E^*$ is a Gel'fand triplet. Namely, E is a nuclear space and E^* is topologized by the inductive limit of E_{-p} as $p \to \infty$. The canonical bilinear form for $(x, \xi) \in E^* \times E$ is denoted by $< x, \xi >$. Bilinear forms $< \cdot, \cdot >$ define cylinder sets. For any $n \geq 0, \xi_1, \cdots, \xi_n \in E$ and an n-dimensional Borel set B, the set of the form $\{x \in E^*; (< x, \xi_1 >, \cdots, < x, \xi_n >) \in \mathrm{B}\}$ is called a cylinder set. We denote the smallest σ -algebra containing all cylinder sets by $\mathcal{B}$.

By Bochner-Minlos' theorem a probability measure μ on $(E^*, \mathcal{B})$ is uniquely determined by the equation

$$\int_{E^*} \exp(i < x, \xi >) d\mu(x) = \exp(-\frac{1}{2}\|\xi\|_0^2) \equiv C(\xi), \quad \text{for} \quad \xi \in E. \tag{2.10}$$

This measure μ is called the Gaussian white noise (with variance 1) and we will denote $L^2(E^*, \mathcal{B}, \mu)$ briefly by (L^2) according to the notation in [7]. The functionals in (L^2) are called *Brownian functionals.* Using Fourier-Hermite polynomials, (L^2) can be decomposed as follows:

$$(L^2) = \sum_{n=0}^{\infty} \oplus \mathcal{H}_n \quad \text{(Wiener-Itô decomposition)}, \tag{2.11}$$

where $\mathcal{H}_n$ is the closed subspace of (L^2) spanned by

$$\{\prod_j H_{n_j}\left(\frac{< x, \zeta_j >}{\sqrt{2}}\right); \sum_j n_j = n\} \quad (n \geq 0).$$

For $\varphi \in (L^2)$, $\mathcal{T}$*-transform* of φ is defined by

$$(\mathcal{T}\varphi)(\xi) \equiv \int_{E^*} \varphi(x) \exp(i < x, \xi >) d\mu(x) \quad \text{for} \quad \xi \in E. \tag{2.12}$$

Suppose that $\varphi = \sum_{n=0}^{\infty} \varphi_n$ $(\varphi_n \in \mathcal{H}_n)$. Then there exists a sequence of functions $\{f_n\}_{n=0}^{\infty}$ satisfying the following (2. 13) and (2. 14).

$$(\mathcal{T}\varphi)(\xi) = \sum_{n=0}^{\infty} i^n C(\xi)(f_n, \xi^{\hat{\otimes} n})_{L^2(\mathbf{R}^n)}, \tag{2.13}$$

where $\xi^{\hat{\otimes} n}$ is the n-fold symmetric tensor product of $\xi \in E$: f_n is a symmetric L^2-function of n-variables, i.e., $f_n \in \hat{L}^2(\mathbf{R}^n)$ $(n \geq 0)$.

$$\|\varphi\|^2_{(L^2)} = \sum_{n=0}^{\infty} n! \|f_n\|^2_{L^2(\mathbf{R}^n)} \tag{2.14}$$

Furthermore, we have

$$\varphi(x) = \sum_{n=0}^{\infty} I_n(f_n)(x), \quad \mu\text{-a.e.} \quad x \in E^*. \tag{2.15}$$

Here, $I_n(f_n)(x)$ is the n-ple Wiener-Itô integral of f_n with respect to the following Gaussian random measure $W(\cdot)$. (See [11].)

$$W(\Delta) \equiv < x, I_\Delta > = (L^2)\text{-} \lim_{n \to \infty} < x, \xi_n > \quad \text{for} \quad \Delta \subset \mathbf{R}, \tag{2.16}$$

where $\xi_n \in E$ satisfying that $I_\Delta = \text{l.i.m.}_{n \to \infty} \xi_n$.

§3. BROWNIAN FUNCTIONALS AND GENERALIZED ONES.

In this section we consider the generalized Brownian functionals by choosing a suitable Gel'fand triplet $(\mathcal{S}) \subset (L^2) \subset (\mathcal{S}')$ and discuss the properties of $(\mathcal{S})$ and $(\mathcal{S}')$. The contents of this section can be found in more general situation in [19] or [28].

Let $\mathcal{S} = \mathcal{S}_{complex}(\mathbf{R})$ be the Schwartz space and $\mathcal{S}'$ be its dual space. The canonical bilinear form for $(F, f) \in \mathcal{S}' \times \mathcal{S}$ is also denoted by $< F, f >$. Let $\mathcal{S}_p$ be the completion of $\mathcal{S}$ by the scalar product

$$(f, g)_p = \int_{\mathbf{R}} \{D^p f(u)\} \overline{D^p g(u)} du \tag{3.1}$$

and $\mathcal{S}_{-p}$ be the dual space of $\mathcal{S}_p$ $(p \geq 0)$. Then, as in §2, we can regard $\mathcal{S} \subset L^2(\mathbf{R}; \mathbf{C}) \subset \mathcal{S}'$ as a complex Gel'fand triplet. Let $\mathcal{S}_p^{\hat{\otimes} n}$ be the symmetric n-fold tensor product of the Hilbert space $\mathcal{S}_p$. Since $\mathcal{S}_p$ is a subspace of $L^2(\mathbf{R}; \mathbf{C}), \mathcal{S}_p^{\hat{\otimes} n}$ is identified with a subspace of symmetric functions of n-variables, i.e.,

$$\mathcal{S}_p^{\hat{\otimes} n} = \{f(u_1, \cdots, u_n); D_{u_1}^p \cdots D_{u_n}^p f(u_1, \cdots, u_n) \in \hat{L}^2(\mathbf{R}^n; \mathbf{C})\}. \tag{3.2}$$

We see that the dual space of $\mathcal{S}_p^{\hat{\otimes} n}$ is the n-fold symmetric tensor product $\mathcal{S}_{-p}^{\hat{\otimes} n}$ of $\mathcal{S}_{-p}$. Further if we set $\mathcal{S}^{\hat{\otimes} n} = \cap_{p=0}^{\infty} \mathcal{S}_p^{\hat{\otimes} n}$, then $\mathcal{S}^{\hat{\otimes} n}$ is a nuclear countably Hilbert space and its dual space, denoted by $\mathcal{S}'^{\hat{\otimes} n}$, is equal to $\cup_{p=0}^{\infty} \mathcal{S}_{-p}^{\hat{\otimes} n}$. The triplet $\mathcal{S}^{\hat{\otimes} n} \subset \hat{L}^2(\mathbf{R}^n; \mathbf{C})$ $\subset \mathcal{S}'^{\hat{\otimes} n}$ is also a Gel'fand triplet. We notice that $\mathcal{S}^{\hat{\otimes} n} = \hat{\mathcal{S}}(\mathbf{R}^n; \mathbf{C})$(see [13] or [24, Ex.7, §7, Chap.7]).

Now, let us construct a new complex Gel'fand triplet $(\mathcal{S}) \subset (L^2) \subset (\mathcal{S}')$. For $p \geq 0$, define

$$(\mathcal{S}_p) \equiv \{\sum_{n=0}^{\infty} I_n(f_n)(x); f_n \in \mathcal{S}_p^{\hat{\otimes} n} \ (n \geq 0), \sum_{n=0}^{\infty} n! \|f_n\|^2_{\mathcal{S}_p^{\hat{\otimes} n}} < \infty\} \tag{3.3}$$

and for $\varphi(x) = \sum_{n=0}^{\infty} I_n(f_n)(x) \in (\mathcal{S}_p)$ write

$$\|\varphi\|^2_{(\mathcal{S}_p)} = \sum_{n=0}^{\infty} n! \|f_n\|^2_{\mathcal{S}_p^{\hat{\otimes} n}}. \tag{3.4}$$

Clearly $(\mathcal{S}_p)$ is a complex Hilbert space with a Hilbertian norm $\| \cdot \|_{(\mathcal{S}_p)}$. Therefore its dual space can be defined. We denote it by $(\mathcal{S}_{-p})$. It is also clear that $(\mathcal{S}_p)$ is isomorphic to a Fock space

$$\Phi_p = \sum_{n=0}^{\infty} \oplus \sqrt{n!} \mathcal{S}_p^{\hat{\otimes} n}$$

for any $p \in \mathbf{Z}$.

Proposition 3.1 *If $p > q$, then* (1) $(\mathcal{S}_p) \subset (\mathcal{S}_q)$, (2) *the inclusion mapping*

$$\iota_{(\mathcal{S}_q),(\mathcal{S}_p)} : (\mathcal{S}_p) \longmapsto (\mathcal{S}_q)$$

is of Hilbert-Schmidt type and (3) $(\mathcal{S}) \equiv \cap_{p=0}^{\infty}(\mathcal{S}_p)$ *is a nuclear space.*

Proof. These properties are clear from the fact that each Fock space Φ_p has a complete orthonormal system

$$\cup_{n=0}^{\infty}\{(\underset{\substack{\uparrow \\ 0\text{-}th}}{0},\cdots,0,\ \underset{\substack{\uparrow \\ n\text{-}th}}{(\textstyle\prod_k n_k!)^{-1/2}(\{2(k+1)\}^{-p}\zeta_k)^{\hat{\otimes} n_k}}\ ,0,0,\cdots);\textstyle\sum_k n_k = n\} \tag{3.5}$$

where ζ_k are Hermite functions (see §2). □

It is easily shown that $(\mathcal{S}_0) = (L^2)$ and that if the dual space of $(\mathcal{S})$ is denoted by $(\mathcal{S}')$, then $(\mathcal{S}') = \cup_{p=0}^{\infty}(\mathcal{S}_{-p})$. Thus we have obtained a new complex Gel'fand triplet

$$(\mathcal{S}) \subset (L^2) \subset (\mathcal{S}').$$

The spaces $(\mathcal{S})$ and $(\mathcal{S}')$ are called the space of *test Brownian functionals* and the space of *generalized Brownian functionals* respectively. The space (L^2) is the space of *ordinary Brownian functionals.*

Proposition 3.2 *For any $\zeta \in E, \exp(i < x, \zeta >)$ is in $(\mathcal{S})$ and the mapping $\zeta \longmapsto \exp(i < x, \zeta >)$ from E into $(\mathcal{S})$ is continuous.*

Proof. Suppose $\zeta \in E$ and consider $\mathcal{T}$-transform of $\exp(i < x, \zeta >)$. We see that

$$\{\mathcal{T}\exp[i < x, \zeta >]\}(\xi) = C(\xi + \zeta)$$

$$= \sum_{n=0}^{\infty} i^n C(\xi) < \frac{i^n}{n!} C(\zeta)\zeta^{\hat{\otimes} n}, \xi^{\hat{\otimes} n} > \quad (\xi \in E) \tag{3.6}$$

and that

$$\|\exp[i < x, \zeta >]\|_{(\mathcal{S}_p)} = C(\zeta)\exp[\frac{1}{2}\|\zeta\|_p^2] < \infty \quad \text{for any} \quad p \geq 0. \tag{3.7}$$

Thus $\exp[i < x, \zeta >] \in (\mathcal{S})$. The continuity of $e^{i<x,\zeta>}$ in $\zeta \in E$ follows from the inequalities

$$\|\exp[i < x, \zeta_0 >] - \exp[i < x, \zeta >]\|_{(\mathcal{S}_p)}^2$$

$$\leq 2\sum_{n=0}^{\infty} \frac{1}{n!}|C(\zeta_0) - C(\zeta)|^2\|\zeta_0\|_p^{2n} + 2C(\zeta)^2 \sum_{n=0}^{\infty} \frac{1}{n!}\|\zeta_0^{\hat{\otimes} n} - \zeta^{\hat{\otimes} n}\|_{\mathcal{S}_p^{\hat{\otimes} n}}^2 \tag{3.8}$$

and

$$\sum_{n=0}^{\infty} \frac{1}{n!}\|\zeta_0^{\hat{\otimes} n} - \zeta^{\hat{\otimes} n}\|_{\mathcal{S}_p^{\hat{\otimes} n}}^2 \leq \|\zeta_0 - \zeta\|_p^2 \exp[1 + \|\zeta_0\|_p^2]. \quad \square \tag{3.9}$$

Proposition 3.3 *Let* $\mathbf{A} = \{\sum_j \beta_j \exp[i < x, \xi_j >];$ *finite sum,* $\beta_j \in \mathbf{C}, \xi_j \in E\}$. *Then* $\mathbf{A}$ *is dense in* $(\mathcal{S})$.

Proof. Since $(\mathcal{S})$, as a nuclear space, is topologized by the countable Hilbertian norms $\{\|\cdot\|_{(\mathcal{S}_p)}\}_{p=0}^{\infty}$, we have only to prove that $\mathbf{A}$ is dense in $(\mathcal{S}_p)$ for any $p \geq 0$. Let us denote the isomorphism of $(\mathcal{S}_p)$ onto Φ_p by Γ. Then, it follows

$$\Gamma\{\exp[i < x, \zeta >]\} = \left(C(\zeta)\frac{i^n}{n!}\zeta^{\hat{\otimes} n}\right)_{n=0}^{\infty}.$$

We see that $\Gamma^{-1}(\{g(t+h)-g(t)\}/h)$ is in $\mathbf{A}$ for $g(t) \equiv \left(\frac{i^n}{n!}(t\zeta)^{\hat{\otimes} n}\right)_{n=0}^{\infty}$, that $\{g(t+h)-g(t)\}/h$ converges (as $h \to 0$) to $g'(t)$ in Φ_p for $t \in \mathbf{R}$ and that $g'(0) = (0, i\zeta, 0, 0, \cdots)$ is in Φ_p. Inductively $g^{(n)}(0) = (0, \cdots, 0, i^n\zeta^{\hat{\otimes} n}, 0, \cdots)$ is in Φ_p. It is well-known that the Fock space Φ_p is generated by elements of the form $(0, \cdots, 0, \zeta^{\hat{\otimes} n}, 0, \cdots)$ with $\zeta \in E$ and $n \geq 0$. These facts show that $\mathbf{A}$ is dense in $(\mathcal{S}_p)$. (See [4]). □

The following proposition shows that $(\mathcal{S})$ is an algebra.

Proposition 3.4 (See Th. 7.4 and Th. 7.5 of [16].) *Suppose* φ *and* $\psi \in (\mathcal{S})$. *Then* $\bar{\varphi}$ *and* $\varphi \cdot \psi \in (\mathcal{S})$. *Moreover,*

(1) $\|\varphi\|_{(\mathcal{S}_p)} = \|\bar{\varphi}\|_{(\mathcal{S}_p)}$ *for any* $p \geq 0$.

(2) *There exist a positive number* $K > 0$ *and an integer* $q > 0$, *both independent of* φ, ψ *and* p, *such that*

$$\|\varphi \cdot \psi\|_{(\mathcal{S}_p)} \leq K\|\varphi\|_{(\mathcal{S}_{p+q})}\|\psi\|_{(\mathcal{S}_{p+q})}.$$

Proof. We give a brief outline of the proof only about $\varphi \cdot \psi \in (\mathcal{S})$. Suppose

$$\varphi = \sum_{m=0}^{\infty} I_m(g_m) \text{ and } \psi = \sum_{n=0}^{\infty} I_n(f_n) \text{ with } g_m \in \mathcal{S}^{\hat{\otimes} m} \text{ and } f_n \in \mathcal{S}^{\hat{\otimes} n}.$$

Let

$$\varphi_m = I_m(g_m) \text{ and } \psi_n = I_n(f_n).$$

Define

$$\begin{aligned} & g_m \otimes_k f_n(u_1, \cdots, u_{m+n-2k}) \\ = & \frac{1}{(m+n-2k)!} \sum_{\sigma \in s(m+n-2k)} \int \cdots \int_{\mathbf{R}^k} g_m(u_{\sigma(1)}, \cdots, u_{\sigma(m-k)}, v_1, \cdots, v_k) \\ & \times f_n(u_{\sigma(m-k+1)}, \cdots, u_{\sigma(m+n-2k)}, v_1, \cdots, v_k) dv_1 \cdots dv_k, \end{aligned} \tag{3.10}$$

where $s(n)$ denotes the permutation group of $\{1, \cdots, n\}$. Then,

$$I_m(g_m)I_n(f_n) = \sum_{k=0}^{m \wedge n} k! \binom{m}{k} \binom{n}{k} I_{m+n-2k}(g_m \otimes_k f_n). \tag{3.11}$$

Thus,

$$\|I_m(g_m)I_n(f_n)\|_{(\mathcal{S}_p)}$$

$$\le \sum_{k=0}^{m\wedge n} \left(\binom{m}{k} \binom{n}{k} \binom{m+n-2k}{n-k} \right)^{1/2} \left(\|\delta\|^2 \rho^{2(p-1)} \right)^k \|\varphi_m\|_{(\mathcal{S}_p)} \|\psi_n\|_{(\mathcal{S}_p)}$$

$$\le \left(2 + \|\delta\|^2 \rho^{2p-2}\right)^{(m+n)/2} \rho^{q(m+n)} \|\varphi_m\|_{(\mathcal{S}_{p+q})} \|\psi_n\|_{(\mathcal{S}_{p+q})}. \tag{3.12}$$

Consequently,

$$\|\varphi \cdot \psi\|_{(\mathcal{S}_p)} \le \sum_{m,n=0}^{\infty} \|I_m(g_m) I_n(f_n)\|_{(\mathcal{S}_p)}$$

$$\le \left(1 - \left(2 + \|\delta\|^2 \rho^{2p-2}\right) \rho^{2q}\right)^{-1} \|\varphi\|_{(\mathcal{S}_{p+q})} \|\psi\|_{(\mathcal{S}_{p+q})} \tag{3.13}$$

if $(2 + \|\delta\|^2 \rho^{2p-2}) \rho^{2q} < 1$. □

We cite another theorem from [19] which guarantees the existence of continuous versions for functionals in $(\mathcal{S})$.

Proposition 3.5 *Suppose that $\psi \in (\mathcal{S})$ and that*

$$\psi(x) = \sum_{n=0}^{\infty} I_n(f_n)(x) \quad (\mu\text{-a.e. } x \in E^*) \quad \textit{with} \quad f_n \in \mathcal{S}^{\hat{\otimes} n} \quad (n \ge 0).$$

Define

$$f_{n|n-2k}(u_1, \cdots, u_{n-2k})$$
$$= \int \cdots \int_{\mathbf{R}^k} f_n(u_1, \cdots, u_{n-2k}, v_1, v_1, \cdots, v_k, v_k) dv_1 \cdots dv_k \tag{3.14}$$

and define

$$\mathcal{I}_n(f_n)(x) = \sum_{k=0}^{[n/2]} \frac{n!(-2)^{-k}}{(n-2k)!k!} < x^{\hat{\otimes} n}, f_{n|n-2k} > \quad \textit{for every} \quad x \in E^*. \tag{3.15}$$

Then,

(1) $$I_n(f_n)(x) = \mathcal{I}_n(f_n)(x) \quad \mu\text{-a.e. } x \in E^*$$

(2) *$\sum_{n=0}^{\infty} \mathcal{I}_n(f_n)(x)$ converges uniformly and absolutely on any bounded set of E_{-p} for p large enough such as $\|\delta\|\rho^{p-1} < 2^{-1/2}$. Moreover, let $\tilde{\psi}(x) = \sum_{n=0}^{\infty} \mathcal{I}_n(f_n)(x)$ for every $x \in E^*$, then $\tilde{\psi}(x)$ is continuous in $x \in E^*$ and $\tilde{\psi}(x) = \psi(x)$ μ-a.e. (Let $\tilde{\psi}$ denote this continuous version of $\psi \in (\mathcal{S})$ from now on.)*

Proof. We prove only the convergence of $\tilde{\psi}(x) = \sum_{n=0}^{\infty} \mathcal{I}_n(f_n)(x)$. The rest of the proof is referred to [19] or [28]. Using the estimation of $f_{n|n-2k}$ in [19] and Mehler's formula [10],

$$\sum_{n=0}^{\infty} \frac{1}{n!} \left(\frac{t}{2}\right)^n H_n(x) H_n(y) = (1-t^2)^{-1/2} \exp[\{2xyt - (x^2+y^2)t^2\}/(1-t^2)]$$

for $|t| < 1$, it follows that if $x \in E_{-p}$ and $\|\delta\|\rho^{p-1} < 2^{-1/2}$

$$|\sum_{n=0}^{\infty} \mathcal{I}_n(f_n)(x)|$$

$$= |\sum_{n=0}^{\infty}\sum_{k=0}^{[n/2]}(-1)^k \frac{n!2^{-k}}{(n-2k)!k!} < x^{\hat{\otimes}(n-2k)}, f_{n|n-2k} > |$$

$$\leq \sum_{n=0}^{\infty}\sum_{k=0}^{[n/2]} \frac{n!2^{-k}}{(n-2k)!k!}||x||_{-p}^{n-2k}||f_n||_{\mathcal{S}_p^{\hat{\otimes}n}}||\delta||^{2k}\rho^{2k(p-1)}$$

$$= \sum_{n=0}^{\infty} ||f_n||_{\mathcal{S}_p^{\hat{\otimes}n}} \left(\frac{||\delta||\rho^{p-1}}{\sqrt{2}i}\right)^n \sum_{k=0}^{[n/2]}(-1)^k \frac{n!}{(n-2k)!k!}\left(2\frac{i||x||_{-p}}{\sqrt{2}||\delta||\rho^{p-1}}\right)^{n-2k}$$

$$= \sum_{n=0}^{\infty} ||f_n||_{\mathcal{S}_p^{\hat{\otimes}n}} \left(\frac{||\delta||\rho^{p-1}}{\sqrt{2}i}\right)^n H_n\left(\frac{i||x||_{-p}}{\sqrt{2}||\delta||\rho^{p-1}}\right)$$

$$\leq ||\psi||_{(\mathcal{S}_p)} \left(\sum_{n=0}^{\infty}\frac{1}{n!}\left(\frac{||\delta||\rho^{p-1}}{\sqrt{2}}\right)^{2n} |H_n\left(\frac{i||x||_{-p}}{\sqrt{2}||\delta||\rho^{p-1}}\right)|^2\right)^{1/2}$$

$$= ||\psi||_{(\mathcal{S}_p)} \left(\sum_{n=0}^{\infty}\frac{1}{n!}\left(\frac{||\delta||^2\rho^{2(p-1)}}{2}\right)^{n} H_n\left(\frac{i||x||_{-p}}{\sqrt{2}||\delta||\rho^{p-1}}\right) H_n\left(\frac{-i||x||_{-p}}{\sqrt{2}||\delta||\rho^{p-1}}\right)\right)^{1/2}$$

$$= ||\psi||_{(\mathcal{S}_p)} \left(1-||\delta||^4\rho^{4(p-1)}\right)^{-1/4} \exp[\frac{1}{2}\left(1-||\delta||^2\rho^{2(p-1)}\right)^{-1}||x||_{-p}^2]$$

$$\leq (4/3)^{1/4}||\psi||_{(\mathcal{S}_p)}\exp[||x||_{-p}^2]. \tag{3.16}$$

Therefore, $\tilde{\psi}(x)$ converges absolutely and uniformly in any bounded set of E_{-p} if p satisfies $||\delta||\rho^{p-1} < 2^{-1/2}$. □

This inequality (3.16) leads us to the next two corollaries.

Corollary 1 *For any $x \in E^*$ (fixed), $\tilde{\psi}(x)$ defines a functional δ_x of $\psi \in (\mathcal{S})$, i.e., $< \delta_x(\cdot), \psi(\cdot) >= \tilde{\psi}(x)$ and if $x \in E_{-p}$ for p such as $||\delta||\rho^{p-1} < 2^{-1/2}$, then*

$$\delta_x \in (\mathcal{S}_{-p}) \quad \text{and} \quad ||\delta_x||_{(\mathcal{S}_{-p})} \leq (4/3)^{1/4}\exp[||x||_{-p}^2].$$

Corollary 2 *Suppose that $\psi, \psi_n \in (\mathcal{S})$ $(n = 1, 2, \cdots)$ and that ψ_n converges to ψ in $(\mathcal{S})$. Then $\tilde{\psi}_n(x)$ converges to $\tilde{\psi}(x)$ for any $x \in E^*$.*

Remark 1 *In our setting the condition $||\delta||\rho^{p-1} < 2^{-1/2}$ is automatically satisfied for any $p \geq 1$ by (2.9).*

§4. POSITIVE GENERALIZED BROWNIAN FUNCTIONALS. MAIN THEOREM.

We begin by defining the *positive* generalized Brownian functionals.

Definition 4.1 A generalized Brownian functional $\Psi \in (\mathcal{S}')$ is said to be *positive*, denoted by $\Psi \geq 0$, if $< \Psi, \psi > \geq 0$ for all $\psi \in (\mathcal{S})$ such that $\tilde{\psi}(x) \geq 0$ at every point $x \in E^*$.

If we recall the well-known fact that $\mu(U)$ is positive (> 0) for every nonempty open subset U of E^*, the condition for $\psi \in (\mathcal{S})$ in Def. 4. 1 can be replaced by the condition that $\psi(x) \geq 0$ for μ-a.e. $x \in E^*$.

Let $(\mathcal{S}')_{>}$ denote the set $\{\Psi; \Psi \in (\mathcal{S}'), \Psi \geq 0\}$ ([25]).

Definition 4.2 A generalized Brownian functional $\Psi \in (\mathcal{S}')$ is said to be *multiplicatively positive* on $\mathbf{A}$ if $< \Psi, \alpha \cdot \bar{\alpha} > \geq 0$ for all $\alpha \in \mathbf{A}$.

Notice that $\alpha \in \mathbf{A}$ is continuous; i.e., $\tilde{\alpha} = \alpha$. Further, we easily see that $\Psi \in (\mathcal{S}')$ is multiplicatively positive on $\mathbf{A}$ if and only if $< \Psi, e^{i<\cdot,\xi>} >$ is a positive-definite function of $\xi \in E$.

Theroem 4.1 *Suppose $\Psi \in (\mathcal{S}')_{>}$. Then there exists a unique finite measure ν_Ψ on $(E^*, \mathcal{B})$ which satisfies that*

$$< \Psi, \psi >= \int_{E^*} \tilde{\psi}(x) d\nu_\Psi(x) \tag{4.1}$$

for all $\psi \in (\mathcal{S})$.

Proof. Let $\Psi \in (\mathcal{S}')_{>}$. Since the relations $\tilde{\alpha}(x) = \alpha(x)$ and $\tilde{\alpha}(x)\overline{\tilde{\alpha}(x)} \geq 0$ hold for any $x \in E^*$ and $\alpha \in \mathbf{A}$, we see that $< \Psi, \alpha \cdot \bar{\alpha} > \geq 0$ for $\alpha \in \mathbf{A}$, i.e., $< \Psi, e^{i<\cdot,\xi>} >$ is positive-definite in $\xi \in E$. Further by Proposition 3. 2, $< \Psi, e^{i<\cdot,\xi>} >$ is continuous in $\xi \in E$. Therefore, Bochner-Minlos' theorem defines a finite measure ν_Ψ on $(E^*, \mathcal{B})$ uniquely by the equation;

$$< \Psi, e^{i<\cdot,\xi>} >= \int_{E^*} e^{i<x,\xi>} d\nu_\Psi(x). \tag{4.2}$$

From the linearity in φ of $< \Psi, \varphi >$ and $\int_{E^*} \varphi d\nu_\Psi$ the following equality holds;

$$< \Psi, \alpha >= \int_{E^*} \alpha(x) d\nu_\Psi(x) \quad \text{for any} \quad \alpha \in \mathbf{A}. \tag{4.3}$$

We want to extend (4. 3) to the form (4. 1). Now let $\psi \in (\mathcal{S})$. Using Proposition 3. 3, we can approximate ψ in $(\mathcal{S})$ by a sequence $\{\alpha_n\}_{n=1}^{\infty} \subset \mathbf{A}$. Then $|\alpha_n - \psi|^2$ converges to 0 in $(\mathcal{S})$, because the inequality

$$\|(\alpha_n - \psi)(\overline{\alpha_n - \psi})\|_{(\mathcal{S}_p)} \leq K \|\alpha_n - \psi\|^2_{(\mathcal{S}_{p+q})} \text{ for any p } \geq 0 \tag{4.4}$$

follows from Proposition 3. 4. Next, to show that $\{\alpha_n\}_{n=0}^{\infty}$ is a Cauchy sequence in $L^2(E^*, \mathcal{B}, \nu_\Psi)$, we note that $|\alpha_n - \alpha_m|^2$ is in **A** and that

$$\|(\alpha_n - \psi)(\overline{\alpha_m - \psi})\|_{(\mathcal{S}_p)} \leq K\|\alpha_n - \psi\|_{(\mathcal{S}_{p+q})}\|\alpha_m - \psi\|_{(\mathcal{S}_{p+q})}. \quad (4.5)$$

We have

$$\begin{aligned}
&\int_{E^*} |\alpha_n - \alpha_m|^2 d\nu_\Psi \\
&=< \Psi, |\alpha_n - \alpha_m|^2 > \\
&=< \Psi, |\alpha_n - \psi|^2 - (\alpha_n - \psi)(\overline{\alpha_m - \psi}) - (\alpha_m - \psi)(\overline{\alpha_n - \psi}) + |\alpha_m - \psi|^2 > \\
&=< \Psi, |\alpha_n - \psi|^2 > - < \Psi, (\alpha_n - \psi)(\overline{\alpha_m - \psi}) > \\
&- < \Psi, (\alpha_m - \psi)(\overline{\alpha_n - \psi}) > + < \Psi, |\alpha_m - \psi|^2 > . \quad (4.6)
\end{aligned}$$

The last four terms tend to 0 as $m, n \to \infty$ because of (4. 4) and (4. 5). Since $L^2(E^*, \mathcal{B}, \nu_\Psi)$ is a Banach space, there exists an element $\chi \in L^2(E^*, \mathcal{B}, \nu_\Psi)$ such that $\alpha_n \to \chi \quad (n \to \infty)$ in $L^2(E^*, \mathcal{B}, \nu_\Psi)$. We can choose a subsequence $\{\alpha_{n_j}\}_{j=1}^{\infty}$ of $\{\alpha_n\}_{n=1}^{\infty}$ satisfying that

$$\alpha_{n_j}(x) \to \chi(x) \quad (j \to \infty) \quad \nu_\Psi\text{-a.e.} \quad x \in E^*.$$

On the other hand, by Corollary 2 the continuous versions satisfy that

$$\tilde{\alpha}_{n_j}(x) \to \tilde{\psi}(x) \quad (j \to \infty) \quad \text{for any} \quad x \in E^*.$$

It follows that $\chi(x) = \tilde{\psi}(x) \quad \nu_\Psi$-a.e. $x \in E^*$. Clearly,

$$\int_{E^*} \alpha_{n_j} d\nu_\Psi \to \int_{E^*} \chi d\nu_\Psi \quad (j \to \infty).$$

Therefore,

$$< \Psi, \psi >= \int_{E^*} \chi d\nu_\Psi = \int_{E^*} \tilde{\psi} d\nu_\Psi. \ \square$$

Theroem 4.2 *Let $\Psi \in (\mathcal{S}')$. Then, $\Psi \geq 0$ if and only if $< \Psi, e^{i<\cdot,\xi>} >$ is a positive-definite functional of $\xi \in E$.*

Proof. Suppose that $\Psi \in (\mathcal{S}')_{>}$. For $\alpha \in \mathbf{A}, \tilde{\alpha} = \alpha$ and $\tilde{\alpha}(x) \cdot \overline{\tilde{\alpha}(x)} \geq 0$ at every $x \in E^*$. Thus $< \Psi, \alpha \cdot \bar{\alpha} > \geq 0$. Conversely, suppose that $\Psi \in (\mathcal{S}')$ and that $< \Psi, \exp[i < \cdot, \xi >] >$ is positive-definite in $\xi \in E$. Then, the same argument as in the proof of Theorem 4. 1 holds and there exists a unique measure ν_Ψ such that

$$< \Psi, \psi >= \int_{E^*} \tilde{\psi} d\nu_\Psi \quad \text{for all} \quad \psi \in (\mathcal{S}).$$

In particular, for $\psi \in (\mathcal{S})$ satisfying $\tilde{\psi}(x) \geq 0$ for any $x \in E^*$

$$< \Psi, \psi >= \int_{E^*} \tilde{\psi}(x) d\nu_\Psi(x) \geq 0. \ \square$$

§5. EXAMPLES OF POSITIVE FUNCTIONALS

We will conclude by giving some examples of positive Brownian functionals.

(A) The Ordinary Case.

Let $\Psi \in (L^2)$. The following $(a), (b), (c)$ and (d) are equivalent.

(a) $\Psi \geq 0$.

(b) $\Psi(x) \geq 0$ μ-a.e. $x \in E^*$.

(c) $(\mathcal{T}\Psi)(\xi)$ is a positive-definite function of $\xi \in E$.

(d) $< \Psi, \varphi \cdot \bar{\varphi} > \geq 0$ for all $\varphi \in (L^2)$. Namely, Ψ is multiplicatively positive on (L^2).

The equivalence of $(a), (b), (c)$ and (d) is easily proved by using Bochner-Minlos' theorem. The measure ν_Ψ corresponding to Ψ satisfies $d\nu_\Psi(x) = \Psi(x)d\mu(x)$, i.e., ν_Ψ is absolutely continuous with respect to μ.

(B) The Generalized Case.

(B.1) The indicator function $I_{[a,b]}$ of the interval $[a, b]$ for $a < b$ is approximated in $L^2(\mathbf{R})$ by a sequence of C^∞ and rapidly decreasing functions $\{\xi_n\}_{n=1}^\infty \subset E$. Therefore, $< x, I_{[a,b]} >$ can be defined in (L^2) as the limit

$$< x, I_{[a,b]} > = (L^2)\text{-} \lim_{n\to\infty} < x, \xi_n >$$

(see [7]). Let us define the "*Brownian motion*" $B(t)$:

$$B(t) = B(t, x) = \begin{cases} < x, I_{[0,t]} > & \text{if } t \geq 0 \\ - < x, I_{[t,0]} > & \text{if } t \leq 0. \end{cases}$$

Adopting this definition of Brownian motion $B(t)$, the Gaussian random measure $W(\cdot)$ at the end of §2 satisfies $W([s, t]) = B(t) - B(s)$ for $s < t$.

Let us show that some of the multiplicative renormalizations stated in [21] or [26] are positive.

(i) $\quad : \exp[\lambda \dot{B}(t)] :$ (λ and t are fixed real numbers.)

This (generalized) functional is defined as a limit functional

$$\lim_{h\to 0} \left(E[\exp(\lambda \frac{B(t+h) - B(t)}{h})] \right)^{-1} \exp\left(\lambda \frac{B(t+h) - B(t)}{h}\right) \quad \text{in } (\mathcal{S}')$$

and its $\mathcal{T}$-transform is $C(\xi)\exp[i\lambda\xi(t)]$. As is well-known, $C(\xi)\exp[i\lambda\xi(t)]$ is positive-definite in $\xi \in E$. Accordingly, $\Psi = : \exp[\lambda\dot{B}(t)] :$ is a positive generalized Brownian functional. The measure ν_Ψ corresponding to Ψ is the one such as the distribution of $< x, \xi >$ with respect to ν_Ψ is the normal distribution $N(\lambda\xi(t), \|\xi\|_0^2)$. ν_Ψ and μ are mutually singular.

$$(ii) \qquad \Psi =: \exp[c \int_{\mathbf{R}} \dot{B}(u)du] : \text{ for } c < \frac{1}{2}$$

Let $\Delta = \{\Delta_k\}$ be a partition of a finite interval T into subintervals Δ_k with the same length $|\Delta|$ and write

$$\Psi_\Delta = \exp\left(c \sum_k |\frac{\Delta_k B}{|\Delta|}|^2 |\Delta| \right), \text{where } \Delta_k B = B(t_k, \cdot) - B(t_{k-1}, \cdot).$$

The $\mathcal{T}$-transform of $\Psi_\Delta / E[\Psi_\Delta]$ is $C(\xi)\exp\left(\frac{1}{2c-1}\sum_k |\Delta|(\xi, I_{\Delta_k})_0^2\right)$, which is positive definite in $\xi \in E$ and converges also to a positive definite function $C_T(\xi)$ of $\xi \in E$ as $|\Delta| \to 0$:

$$C_T(\xi) = C(\xi)\exp\left(\frac{1}{2c-1}\int_T \xi(u)^2 du\right).$$

Moreover,

$$C_T(\xi) \to \exp\left(-\frac{1}{2}\frac{1}{1-2c}||\xi||_0^2\right) \text{ as } T \to (-\infty, \infty).$$

This is the characteristic functional of the measure of Gaussian white noise with variance $\frac{1}{1-2c}$. We can prove that there exists a positive generalized Brownian functional $\Psi \in (\mathcal{S}')$, denoted by $: \exp[c\int_{\mathbf{R}} \dot{B}(u)^2 du] :$, for which $< \Psi, e^{i<\cdot,\xi>} >= \exp\left(-\frac{1}{2}\frac{1}{1-2c}||\xi||_0^2\right)$.

To see how $\Psi =: \exp[c\int_{\mathbf{R}} \dot{B}(u)^2 du] :$ acts on $\psi = \sum_{n=0}^{\infty} I_n(f_n) \in (\mathcal{S})$, let us expand $< \Psi, e^{i<\cdot,\xi>} >$ in the following way. (We can obtain the Fock representation of Ψ by doing this.)

$$\begin{aligned} &< \Psi, e^{i<\cdot,\xi>} > \\ &= \exp\left(-\frac{1}{2}\frac{1}{1-2c}||\xi||_0^2\right) \\ &= C(\xi)\exp\left(\frac{c}{2c-1}||\xi||_0^2\right) \\ &= C(\xi)\sum_{n=0}^{\infty}\frac{1}{n!}\left(\frac{c}{2c-1} < \delta(u-v), \xi\hat{\otimes}\xi(u,v) >\right)^n \\ &= \sum_{n=0}^{\infty} i^{2n} C(\xi)\frac{1}{n!}\left(\frac{c}{1-2c}\right)^n < \{\delta(\cdot - *)\}^{\hat{\otimes}n}, \xi^{\hat{\otimes}2n} > \end{aligned} \tag{5.1}$$

Thus,

$$< \Psi, \psi >= \sum_{n=0}^{\infty}\frac{(2n)!}{n!}\left(\frac{c}{1-2c}\right)^n \int\cdots\int_{\mathbf{R}^n} f_{2n}(u_1, u_1, \cdots, u_n, u_n)du_1\cdots du_n.$$

As expected, this is certainly the integral of $\tilde{\psi}(x)$ with respect to the Gaussian white noise with variance $\frac{1}{1-2c}$.

(B.2) We give another example: δ_x for an arbitrarily fixed $x \in E^*$.

As mentioned in Corollary 2, this functional belongs to $(\mathcal{S}')$. The $\mathcal{T}$-transform of δ_x is given by $< \delta_x, e^{i<\cdot,\xi>} >= e^{i<x,\xi>}$ $(\xi \in E^*)$. Obviously this is a positive-definite function of $\xi \in E$. Thus δ_x is a positive generalized functional. The measure ν corresponding to δ_x satisfies $\nu(\{x\}) = \nu(E^*) = 1$. Further, if $x \in E^*$ is fixed as above, we have

$$\begin{aligned}
&e^{i<x,\xi>} \\
&= C(\xi)\exp(i < x, \xi >)\exp\left(\frac{1}{2} < \delta(u-v), \xi \otimes \xi(u,v) >\right) \\
&= C(\xi)\left(\sum_{m=0}^{\infty} \frac{i^m}{m!} < x, \xi >^m\right)\left(\sum_{k=0}^{\infty} \frac{i^k}{k!} < \frac{1}{2}\delta(u-v), \xi \otimes \xi(u,v) >^k\right) \\
&= \sum_{n=0}^{\infty} i^n C(\xi)\left(\sum_{m+2k=n} \frac{(-2)^{-k}}{m!k!} < x, \xi >^m < \delta(u-v), \xi \otimes \xi(u,v) >^k\right) \\
&= \sum_{n=0}^{\infty} i^n C(\xi) < \sum_{k=0}^{[n/2]} \frac{(-2)^{-k}}{(n-2k)!k!} x^{\hat{\otimes}(n-2k)} \hat{\otimes} \{\delta(\cdot - *)\}^{\hat{\otimes}k}, \xi^{\hat{\otimes}n} >,
\end{aligned}$$

where

$$\begin{aligned}
&x^{\hat{\otimes}(n-2k)} \hat{\otimes} \{\delta(\cdot - *)\}^{\hat{\otimes}k}(u_1, \cdots, u_{n-2k}, \cdots, u_n) \\
&= \frac{1}{n!} \sum_{\sigma \in s(n)} x(u_{\sigma(1)}) \cdots x(u_{\sigma(n-2k)}) \delta\left(u_{\sigma(n-2k+1)} - u_{\sigma(n-2k+2)}\right) \times \cdots \\
&\qquad \cdots \times \delta\left(u_{\sigma(n-1)} - u_{\sigma(n)}\right).
\end{aligned}$$

From the above we can see how δ_x acts on $\psi \in (\mathcal{S})$. In fact, let

$$\psi(\cdot) = \sum_{n=0}^{\infty} I_n(f_n)(\cdot) \text{ with } f_n \in \mathcal{S}^{\hat{\otimes}n},$$

then

$$< \delta_x, \psi >= \sum_{n=0}^{\infty} n! < \sum_{k=0}^{[n/2]} \frac{(-2)^{-k}}{(n-2k)!k!} x^{\hat{\otimes}(n-2k)} \hat{\otimes} \{\delta(\cdot - *)\}^{\hat{\otimes}k}, f_n > .$$

On the other hand the definition of δ_x gives $< \delta_x, \psi >= \tilde{\psi}(x)$. Hence,

$$\tilde{\psi}(x) = \sum_{n=0}^{\infty} < \sum_{k=0}^{[n/2]} \frac{n!(-2)^{-k}}{(n-2k)!k!} x^{\hat{\otimes}(n-2k)} \hat{\otimes} \{\delta(\cdot - *)\}^{\hat{\otimes}k}, f_n > .$$

However, we can easily see that the n-th term of the right hand side of this equality is equal to $\mathcal{I}_n(f_n)(x)$.

Acknowledgement

I wish to thank Professor T. Hida for his encouragement and kind advice. I am also grateful to Professor I. Kubo for his invaluable suggestions, in particular, in proving Theorem 4. 1. I would also like to thank Professors L. Streit, S. Takenaka and N. Obata for their stimulating and helpful discussions.

References

[1] I. M. Gel'fand and G. E. Shilov, "Generalized Functions", Vol.2, Academic Press, New York/London, 1964.

[2] I. M. Gel'fand and N. Ya. Vilenkin, "Generalized Functions", Vol. 4, Academic Press, New York/London, 1964.

[3] L. Gross, Abstract Wiener Spaces, *Proc. 5th Berkeley Symp. on Math. Stat. and Prob.* II, Part 1, Univ. of Calif. (1967), 31-42.

[4] A. Guichardet, "Symmetric Hilbert Spaces and Related Topics", Lecture Notes in Math., 261, Springer-Verlag, Berlin/Heidelberg/New York, 1975.

[5] T. Hida, "Stationary Stochastic Processes", Princeton Univ. Press, Princeton NJ, 1970.

[6] T. Hida, "Analysis of Generalized Brownian Functionals", Carleton Math. Lecture Notes, No.13, July, 1975.

[7] T. Hida, "Brownian Motion", Appl. Math., 11, Springer-Verlag, New York/ Heidelberg/ Berlin, 1980.

[8] T. Hida, Generalized Brownian Functionals, *Lecture Notes in Control and Information Science*, 49, Springer-Verlag, Berlin/Heidelberg/New York, (1983), 89-95.

[9] T. Hida, J. Potthoff and L. Streit, Dirichlet Forms and White Noise Analysis, *Commun. Math. Phys.* 116, (1988), 235-245.

[10] E. Hille, A Class of Reciprocal Functions, Annals of Math., (2), 27 (1926), 427-464.

[11] K. Itô, Multiple Wiener Integral, *J. Math. Soc. Japan,* Vol. 3, No. 1, May, (1951), 157-169.

[12] K. Itô, Stochastic Analysis in Infinite Dimensions, *Proc. Internat. Conf., Northwest. Univ., Evanston,* Ill, Academic Press, New York, (1978),187-197.

[13] K. Itô, "Foundations of Stochastic Differential Equations in Infinite Dimensional Spaces", Society for Industrial and Applied Mathematics, Philadelphia, Pensylvania, 1984.

[14] I. Kubo, Itô Formula for Generalized Brownian Functionals, *Lecture Notes in Control and Information Science*, 49, Springer-Verlag, New York/ Heidelberg/ Berlin, (1983), 156-166.

[15] I. Kubo and S. Takenaka, Calculus on Gaussian White Noise I, *Proc. Jap. Acad.* 56A (1980), 376-380.

[16] I. Kubo and S. Takenaka, Calculus on Gaussian White Noise II, *ibid.*, 56A (1980), 411-416.

[17] I. Kubo and S. Takenaka, Calculus on Gaussian White Noise III, *ibid.*, 57A (1981), 433-437.

[18] I. Kubo and S. Takenaka, Calculus on Gaussian White Noise IV, *ibid.*,58A (1982), 186-189.

[19] I. Kubo and Y. Yokoi, A Remark on the Space of Testing Random Variables in the White Noise Calculus, *Nagoya Math. J.*, Vol 115, (1989), 139-149.

[20] H-H. Kuo, Donsker's Delta Function as a Generalized Brownian Functional and its Application, *Lecture Notes in Control and Information Science*, 49, Springer-Verlag, Berlin/Heidelberg/New York, (1983), 167-178.

[21] H-H. Kuo, Brownian Functionals and Applications, *Acta Appl. Math.* 1 (1983), 175-188.

[22] R. A. Minlos, Generalized Random Processes and Their Extension to a Measure, *Selected Transl. in Math. Stat. & Prob.*, Vol. 3 (1962), 291-313.

[23] J. Potthoff, On Positive Generalized Functionals, *J. Funct. Anal.*, 74, (1987), 81-95.

[24] L. Schwartz, "Théorie des Distributions", 2ème éd., Herman Cie, 1966.

[25] L. Streit, Lecture at Nagoya Univ. in March 1987.

[26] L. Streit and T. Hida, Generalized Brownian Functionals and Feynman Integral, *Stochastic Process. Appl.*, 16 (1983), 55-69.

[27] H. Sugita, Positive Generalized Wiener Functionals and Potential Theory over Abstract Wiener Spaces, *Osaka J. Math.*, (1987) to appear.

[28] Y. Yokoi, Positive Generalized White Noise Functionals, Hiroshima Math. J., Vol. 20, (1990) to appear.

List of Participants

S. Albeverio	Ruhr–Universität Bochum, Bochum
K. Baumann	Universität Göttingen, Göttingen
D. Betounes	University of Southern Mississippi, Hattiesburg
Ph. Blanchard	Universität Bielefeld, Bielefeld
E. Carlen	Princeton University, Princeton
D. Castrigiano	Technische Universität München, München
P.-L. Chow	Wayne State University, Detroit
O. Cohendet	BiBoS, Bielefeld
N.J. Cutland	University of Hull, Hull
D. Dawson	Carleton University, Ottawa
D. de Falco	Università di Milano, Milano
J.C.R. Ferreira	Universidade de Lusiada, Porto
R. Gielerak	University of Wrocław, Wrocław
L. Gross	Cornell University, Ithaca
Z. Haba	University of Wrocław, Wrocław
T. Hashimoto	Universität Heidelberg, Heidelberg
M. Hazewinkel	Centrum v. Wiskunde en Informatica, Amsterdam
K. Helmes	University of Kentucky, Lexington
T. Hida	Nagoya University, Nagoya
K. Itô	Kyoto University, Kyoto
G. Kallianpur	University of North Carolina, Chapel Hill
D.C. Khandekar	Bhaba Atomic Research Center, Bombay
H. Korezlioglu	E.N.S.T., Paris
P. Krée	Université de Paris VI, Paris
P. Kröger	Universität Erlangen–Nürnberg, Erlangen
I. Kubo	Hiroshima University, Hiroshima
H.-H. Kuo	Louisiana State University, Baton Rouge

R. Léandre	Université Louis Pasteur, Strasbourg
Y.-L. Lee	National Cheng Kung University, Tainan
V. Manko	P.N. Lebedev Institute of Physics, Moscow
X. Mao	University of Warwick, Coventry
J. Naudts	Universiteit Antwerpen, Antwerpen
D. Ocone	Rutgers University, New Brunswick
Y. Okabe	Hokkaido University, Sapporo
F. Oosawa	Nagoya University, Nagoya
C. Pallenberg	Universität Göttingen, Göttingen
S. Paycha	Ruhr-Universität Bochum, Bochum
J. Potthoff	BiBoS, Bielefeld
N. Privault	E.N.S.T., Paris
P. Protter	Purdue University, West Lafayette
M. Redfern	University of Southern Mississippi, Hattiesburg
F. Redig	Universiteit Antwerpen, Antwerpen
L. Ricciardi	Università di Napoli, Napoli
M. Röckner	University of Edinburgh, Edinburgh
F. Russo	E.N.S.T., Paris
K. Sato	Nagoya University, Nagoya
N.-R. Shieh	National Taiwan University, Taipei
W.H. Smoleński	Warsaw Technical University, Warsaw
F. Stärk	Technische Universität München, München
L. Streit	Universität Bielefeld, Bielefeld
R. Vilela Mendes	CFMC - INIC, Lisboa
H. Watanabe	Kyushu University, Fukuoka
J.A. Westerdijk	Twente University of Technology, Enschede
Y. Yokoi	Kumamoto University, Kumamoto
B. Zegarlinski	Ruhr-Universität Bochum, Bochum
G. Zimmermann	Loma Linda University, Loma Linda